Landmark Papers in Clinical Chemistry

Landmark Papers
in Clinical Chemistry

Edited by

RICHARD M. ROCCO

San Francisco State University
San Francisco, USA

2005

ELSEVIER

Amsterdam - Boston - Heidelberg - London - New York - Oxford
Paris - San Diego - San Francisco - Singapore - Sydney - Tokyo

ELSEVIER B.V.
Radarweg 29
P.O. Box 211, 1000 AE Amsterdam
The Netherlands

ELSEVIER Inc.
525 B Street, Suite 1900
San Diego, CA 92101-4495
USA

ELSEVIER Ltd
The Boulevard, Langford Lane
Kidlington, Oxford OX5 1GB
UK

ELSEVIER Ltd
84 Theobalds Road
London WC1X 8RR
UK

First edition 2006

Library of Congress Cataloging in Publication Data
A catalog record is available from the Library of Congress.

British Library Cataloguing in Publication Data
A catalogue record is available from the British Library.

ISBN: 0-444-51950-5

Transferred to digital print on demand, 2008
Printed and bound by CPI Antony Rowe, Eastbourne

Contents

Section I
Immunoassay Technology

RIA (Radioimmunoassay)

FPIA (Fluorescence Polarization Immunoassay)

CPB (Competitive Protein Binding)

ELISA (Enzyme-Linked Immunosorbent Assay)

EMIT™ (Enzyme Immunoassay Technique)

Section II

Therapeutic Drug Monitoring (TDM)

Section III

Enzymology

Section IV

Specific Analytes

Section V

Instrumentation and Techniques

Section VI
Chemometrics

Test Kits

Section VII

Molecular Diagnostics

Biochemical Genetics

Nucleic Acid Probes

Microarrays

Foreword

There are members of the clinical chemistry profession who still remember the modest facilities they once occupied. These laboratories provided a few chemical tests on blood and urine that appeared to have some value in clinical diagnosis. These tests were developed mainly by Ottto Folin at Harvard Medical School and Donald Van Slyke at the Rockefeller Institute. Methods were usually presented in the format of cookbook recipes and were carried out by persons of limited skill in chemistry. Today a vast amount of literature on clinical chemistry provides services to the thousands of personnel in hospital and industrial laboratories. Looking at today's clinical chemistry laboratory with its assortment of complex instruments, equipment and methodology, I can't help recall my first hospital laboratory. Located in a single room of moderate size, there were test tubes and pipettes, and no AutoAnalyzer™. The protein-free filtrate was the starting point of nearly all the tests. The Duboscq visual comparison colorimeter shouldered the bulk of the work. A flame photometer provided sodium and potassium analyses three times a week; carbon dioxide combining power was a manual procedure, with hands-on from start to finish; and some procedures were reported in varying degrees of opacity, turbidity, or flocculation. The technologies were often cumbersome and labor-intensive. But there was an excitement of learning new methods, developing new skills, and using different apparatus.

The routine work—less than a dozen analytical procedures—required only simple manual operations. The rapid advance of clinical chemistry has been made possible by innovative technology and its application to clinical settings. These have helped narrow the gap between medical science and medical practice. The papers identified in this volume are significant turning points and signposts in the growth and development of clinical chemistry during the twentieth century. This is an interesting concept with an impressive list of citations. Most of them are from the second half of the century, which indicates how recent has been the phenomenal expansion and widening versatility of clinical chemistry in its preeminent role in "laboratory medicine."

To avoid being swept into the future by the impact of continuous change, before we completely understand the present, this collection identifies ancestral landmarks which dot the landscape and illuminate the road as we prepare for the future.

The history of an analytical procedure is often as fascinating and instructive as the biography of a scientist, and many teachers as well as their students would profit greatly by such studies. They would then learn the great truth that analytical procedures are not found inscribed on stone but are products of evolution. Their inherent faults are corrected largely by those who work, not for their own fame, but for the making of a better science and a better world.

Louis Rosenfeld, PhD
Brooklyn, NY

Introduction

Many excellent articles on the history of clinical chemistry in the 20th century have been published in the last few years (1–6). In addition, a collection of essays by Büttner (7) and a book of biographies by Rosenfeld (8) review the development of the discipline during this period. The purpose of this book is to present the history of clinical chemistry in the 20th century in the words of the men and woman who helped create this field. Thirty-nine original articles have been chosen and reprinted in this collection. They range from the introduction of the Duboscq colorimeter into clinical chemistry by Folin in the early 1900s to the description of oligonucleotide hybridization on glass supports by Southern in 1992. Each paper is reproduced from its original and is accompanied by a short commentary intended to help set the article within the historical context of the period.

The criteria used to select the articles for compilations like this will often reflect the editor's interests and opinions. This collection is no exception. Hundreds of articles published in the last century were reviewed for possible inclusion. Yalow and Berson's first complete description of a radioimmunoassay (RIA) in 1960 and which led to a Nobel Prize is an obvious choice. Papers that presented a new technology still in widespread use today were also included. Two examples are Engvall and Perlmann's development of the enzyme-linked immunosorbent assay (ELISA) in 1971 and Jorgenson and Lukacs' first description of a capillary electrophoresis system in 1981. Citation index services and editors often publish lists of the most highly cited papers in a particular field or journal. This also helped in the selection process. Huggett and Nixon's 1957 enzymatic glucose assay, for example, was the second most cited paper published in *The Lancet* between 1961 and 1983. Friedewald, Levy and Fredrickson's calculation of low-density lipoprotein (LDL) cholesterol was the highest cited paper published in the journal *Clinical Chemistry* between 1955 and 1998. A paper that is ahead of its time is also a strong contender for inclusion. An example is Wuth's paper on serum bromide levels in the *Journal of the American Medical Association* in 1927. In this paper, he demonstrated, 50 years before it became standard practice, the importance of therapeutic drug monitoring (TDM) for improved chemotherapeutic control of seizures.

Finally, the articles chosen, along with any major omissions, are the sole responsibility of the editor. The goal was to present a selection of original articles that covered many of the major developments in clinical chemistry in the 20th century. A secondary objective was to select articles that opened up a new way of thinking about an old problem or presented a creative new technology that improved the practice of clinical chemistry. The papers chosen, for the most part, stimulated clinical chemists to action and adoption. Their impact became apparent when the original articles, not yet available in electronic form, were read in library holdings. Many of these journal articles contained notations and markings that brought to mind the poem *Marginalia* by the former Poet Laureate of the United States, Billy Collins. In this poem, Collins writes about the notations penned by readers along the margins in the library copies of famous books (9). Written, he says, by, "fans who cheer from the empty bleachers."

Many of the landmark articles from 1919 to the 1960s chosen for this collection contained penciled marginalia along with thumb prints, torn pages repaired with archival tape, folded-over corners and a general worn and well-used appearance. Faded graphite pencil notations ran along the margins. In a Folin article published in 1919, next to the description of the Dubsocq colorimeter are the words, "we need to get one of these." In many of the papers from the 1930s to 1940s, the Results sections have multiple penciled check marks next to a line of data. The Conclusion sections in some are enclosed in long vertical double wavy lines. Multiple "yes" notations with three exclamation marks are common marginalia in papers of this period. In articles from the 1960s through the 1980s, the marginalia are penned in ink and iridescent markers, garish and destructive but none-the-less telling. These papers were read by large numbers of interested scientists. Of all the marginalia found in these articles however, the words that resonated the most, the words that have come to mean for competitive scientists the highest form of compliment, were in a paper from the 1960s that simply said, "I wish I had thought of that."

Richard M. Rocco, PhD
San Francisco State University
August 2005

References

(1)   Moss, M.L., Horton, C.A., and White, J.C. (1971) Clinical biochemistry. Annual Review of Biochemistry. 40:573–604.
(2)   Caraway, W.T. (1973) The scientific development of clinical chemistry to 1948. Clinical Chemistry. 19(4):373–383.
(3)   Martinek, R.G. (1973) Developments and trends in clinical chemistry, in *Progress in Clinical Pathology*, Stefanini, M. (ed), Grune & Stratton, New York, pgs 49–58.
(4)   Varley, H. (1974) Methodology-past and present. Annals of Clinical Biochemistry. 11(5):161–165.
(5)   Savory, J., Bertholf, R.L., Boyd, J.C., Bruns, D.E., Felder, R.A., Lovell, M., Shipe, J.R., and Wills, M.R. (1986) Advances in clinical chemistry over the past 25 years. Analytica Chimica Acta. 180:99–135.
(6)   Sunderman, F.W. (1994) The foundation of clinical chemistry in the Untied States. Clinical Chemistry. 40(5):835–842.
(7)   Büttner, J. (ed) (1983) *History of Clinical Chemistry*. Walter de Gruyter, Berlin.
(8)   Rosenfeld, L. (ed) (1999) *Biographies and Other Essays on the History of Clinical Chemistry*. American Association for Clinical Chemistry Inc., Washington, DC.
(9)   Collins, B. (2001) Marginalia, in *Sailing Alone Around the Room*, Random Hous, New York, pgs 94–96.

Acknowledgments

The inspiration for this project came from the graduate students at San Francisco State University who over the years have attended the *Landmark Papers* seminar. They deserve many thanks. The book itself was made a reality through the commitment and support of two editors at Elsevier, K. Noelle Gracy and Anne Russum. A special thank you is expressed to them. I appreciate the help I received from the staff at four important libraries in the San Francisco Bay area; the J. Paul Leonard Library at San Francisco State University especially from Gina Castro, Nina Hagiwara, David Hellman, Kathleen Messer and Alex Perez; the Kalmanovitz Library at the University of California San Francisco; the Marian Koshland Bioscience & Natural Resources Library at the University of California Berkeley and the San Jose Public Library System in San Jose. I also wish to thank the authors and publishers for granting rights to reproduce these thirty-nine papers. Finally, this book is dedicated with love to my two children, Denise and John, and to Susan for her constant support.

Thus the genius of inventiveness, so precious in the sciences, may be diminished or even smothered by a poor method, while a good method may increase and develop it. In short, a good method promotes scientific development and forewarns [persons] of science against those numberless sources of error which they meet in the search for truth; this is the only possible object of the experimental method.

Claude Bernard (1813–1878)

An Introduction to the Study of Experimental Medicine, 1865
H.C. Green translation, Dover Publications, 1957, pg 35.

Section I

Immunoassay Technology

RIA (Radioimmunoassay)

    1. Yalow, R. S. and Berson, S. A. **(1960)**
        Immunoassay of endogeneous plasma insulin in man.

FPIA (Fluorescence Polarization Immunoassay)

    2. Dandliker, W. B. and Feigen, G. A. **(1961)**
        Quantification of the antigen-antibody reaction by the polarization of fluorescence.

CPB (Competitive Protein Binding)

    3. Murphy, B. E. P. and Pattee, C. J. **(1964)**
        Determination of thyroxine utilizing the property of protein-binding.

ELISA (Enzyme-Linked Immunosorbent Assay)

    4. Engvall, E., Jonsson, K. and Perlmann, P. **(1971)**
        Enzyme-linked immunosorbent assay II.

EMIT™ (Enzyme Immunoassay Technique)

    5. Rubenstein, K. E., Schneider, R. S. and Ullman, E. F. **(1972)**
        "Homogeneous" enzyme immunoassay. A new immunochemical technique.

COMMENTARY TO

1. Yalow, R. S. and Berson, S. A. (1960)
Immunoassay of Endogenous Plasma Insulin in Man. Journal of Clinical Investigation 39(7): 1157–1175.

Albert Szent-Gyorgyi the biochemist and Nobel Laureate wrote that, "most of the new observations I made were based on wrong theories. My theories collapsed, but something was left afterwards (1)." This describes how Rosalyn S. Yalow and Solomon A. Berson discovered radioimmunoassay (RIA). They set out to test a hypothesis, not a theory, but the results were the same. Their data destroyed the hypothesis and instead they discovered the most important new technology in laboratory science in the twentieth century.

In 1952, I.A. Mirsky the chairperson of the Department of Clinical Sciences at the University of Pittsburgh School of Medicine proposed that Type II diabetics cleared insulin from their systems faster than non-diabetics due to an abnormal increase in the activity of their hepatic enzyme, insulinase (2). Yalow and Berson who worked in the Radioisotope Service at the Veterans Administration (VA) Hospital in Bronx, New York, decided to test this hypothesis by measuring the clearance of radio labeled insulin in diabetics. Isotopic labeling of proteins was a technique they were familiar with. In 1952, they had used ^{131}I labeled human albumin to accurately measure blood volume (3). To test the Mirsky hypothesis they prepared beef ^{131}I-insulin. They injected this insulin tracer into three subject groups; non-diabetic normal controls, diabetics who had never received insulin and diabetics on insulin therapy. The clearance rates in the normal non-diabetics and the diabetics who had never received insulin were the same. Clearance times in insulin treated diabetics were markedly increased compared to the other two groups (4). Data from ultracentrifugation, salt and ethanol fractionation and paper electrophoresis demonstrated that an insulin antibody was present in insulin treated patients. It was the insulin–antibody complex in insulin treated diabetics that took longer to clear (5). The results of these experiments overturned the Mirsky hypothesis, opened up a whole new line of diabetes research based on immune response factors and disproved the belief, held at the time, that small proteins like insulin with a molecular weight of 6 000 could not elicit an immune response. Their paper on these results was accepted in the prestigious *Journal of Clinical Investigation* in 1956 (4) only after they agreed to replace the term "insulin antibody" with the term "insulin binding globulin" in the title (6).

Yalow and Berson then focused their research on the use of insulin antibody as a laboratory tool. This was the critical insight that led to the discovery of RIA. They studied the binding kinetics between radio iodinated crystalline beef insulin and the insulin antibody found in the serum of diabetic patients previously treated with insulin (7). They made their own insulin antibody by immunizing guinea pigs with beef insulin and demonstrated that this antibody cross-reacted with human insulin (8). They showed that the binding of radio labeled insulin was inversely proportional to the concentration of native insulin in the unknown sample being tested. Bound from free insulin tracer was separated in these first competitive immunoassays with paper electrophoresis. In 1959 a short two-page article appeared in *Nature* that described, for the first time, the quantitative measurement of insulin levels in serum by an immunoassay technique (9). The paper presented here is their first full report describing an RIA method, a twenty-page report that reads like a methods manual. It went on to become the most highly cited paper in the history of the *Journal of Clinical Investigation* through 2004 (6). Yalow and Berson chose not to patent their new technology. Instead, they opened up their laboratory to more than 100 researchers whom they trained over the next 5 years in the technique of RIA. By 1978, the sales of RIA test kits in the US had reached $125 million dollars per year (10).

Berson died of a heart attack while attending a scientific conference in 1972. At Yalow's request, the laboratory at the VA Hospital was named in his honor (11). In 1976 she was the first woman to win the Albert Lasker Award for Basic Medical Research for the development of RIA. In 1977 she became the first American woman to receive the Nobel Prize in Physiology or Medicine, awarded for her discovery of RIA.

References

(1) Szent-Gyorgyi, A. (1963) Lost in the twentieth century. Annual Review of Biochemistry. 32:1–14.

(2) Mirsky, I.A. (1952) The etiology of diabetes in man. Recent Progress in Hormone Research. 7:437–467.

(3) Berson, S.A. and Yalow, R.S. (1952) The use of K^{42} labeled erythrocytes and I^{131} tagged human serum albumin in simultaneous blood volume determinations. Journal of Clinical Investigation. 31(6):572–580.

(4) Berson, S.A., Yalow, R.S., Bauman, A., Rothschild, M.A., and Newerly, K. (1956) Insulin-I^{131} metabolism in human subjects: demonstration of insulin binding globulin in the circulation of insulin treated subjects. Journal of Clinical Investigation. 35(2):170–190.

4

(5) Berson, S.A. and Yalow, R.S. (1957) Ethanol fractionation of plasma and electrophoretic identification of insulin-binding antibody. Journal of Clinical Investigation. 36(5):642–647.

(6) Kahn, C.R. and Roth, J. (2004) Berson, Yalow, and the JCI: the agony and the ecstasy. Journal of Clinical Investigation. 114(8):1051–1054.

(7) Berson, S.A. and Yalow, R.S. (1959) Quantitative aspects of the reaction between insulin and insulin-binding antibody. Journal of Clinical Investigation. 38(11):1996–2016.

(8) Berson, S.A. and Yalow, R.S. (1959) Recent studies on insulin-binding antibodies. Annals of the New York Academy of Sciences. 82(Article 2):338–344.

(9) Yalow, R.S. and Berson, S.A. (1959) Assay of plasma insulin in human subjects by immunological methods. Nature. 184(4699):1648–1649.

(10) [Anonymous]. (1980) Frost and Sullivan report: the clinical diagnostic reagents and test kit markets. Surgical Business. 43(3):44–47.

(11) Straus, E. (1998) *Rosalyn Yalow Nobel Laureate, Her Life and Work in Medicine*. Plenum Trade, New York, pg 234.

IMMUNOASSAY OF ENDOGENOUS PLASMA INSULIN IN MAN

By ROSALYN S. YALOW AND SOLOMON A. BERSON

(From the Radioisotope Service, Veterans Administration Hospital, New York, N. Y.)

(Submitted for publication March 7, 1960; accepted March 22, 1960)

For years investigators have sought an assay for insulin which would combine virtually absolute specificity with a high degree of sensitivity, sufficiently exquisite for measurement of the minute insulin concentrations usually present in the circulation. Methods in use recently depend on the ability of insulin to exert an effect on the metabolism of glucose *in vivo* or in excised muscle or adipose tissue. Thus, the insulin concentration in plasma has been estimated: *a*) from the degree of hypoglycemia produced in hypophysectomized, adrenalectomized, alloxan-diabetic rats (1); *b*) from the augmentation of glucose uptake by isolated rat hemidiaphragm (2); or *c*) from the increased oxidation of glucose-1-C^{14} by the rat epididymal fat pad (3). Since there have been reports indicating the presence, in plasma, of inhibitors of insulin action (4) and of non-insulin substances capable of inducing an insulin-like effect (5, 6), these procedures, while yielding interesting information regarding the effects of various plasmas on glucose metabolism in tissues, are of doubtful specificity for the measurement of insulin per se (5).

Recently it has been shown (7, 8) that insulins from various species (pork, beef, horse and sheep) show quantitative differences in reaction and cross reaction with antisera obtained from human subjects treated with commercial insulin preparations (beef, pork insulin mixtures). An immunoassay method for beef insulin has been reported in which the insulin content is determined from the degree of competitive inhibition which the insulin offers to the binding of beef insulin-I^{131} by human antisera (9–12). Although human insulin reacts with human antibeef, pork insulin antiserum and displaces beef insulin-I^{131} by competitive inhibition (7, 8, 10), the reaction is too weak to permit measurement of the low insulin concentrations present in human plasma (7, 8, 11–13). In preliminary communications we have reported that the competitive inhibition by human insulin of binding of crystalline beef insulin-I^{131} to guinea

pig antibeef insulin antibodies is sufficiently marked to permit measurement of plasma insulin in man (11, 12, 14), and to be capable of detecting as little as a fraction of a microunit of human insulin (12, 14). Preliminary data on insulin concentrations in man before and after glucose loading have been reported (12, 14, 15). The present communication describes in detail the methods employed in the immunoassay of endogenous insulin in the plasma of man, and reports plasma insulin concentrations during glucose tolerance tests in nondiabetic and in early diabetic subjects and plasma insulin concentrations in subjects with functioning islet cell tumors or leucine-sensitive hypoglycemia.

METHODS

Immunization of guinea pigs. Guinea pigs were injected subcutaneously at 1 to 4 week intervals with 5 to 10 units of either protamine zinc beef insulin (Squibb) or commercial regular beef insulin (Squibb) emulsified with mannide mono-oleate. Insulin-binding antibodies were detected in all animals after 2 to 3 injections. The antiserum employed in the present study (GP 49, serum 6-25-59) was obtained from a guinea pig immunized with protamine zinc beef insulin without adjuvant and was selected for its relatively high antibody concentration and other suitable characteristics described below.

Preparation of insulin-I^{131}. Because of the desirability of keeping the concentration of added insulin-I^{131} as low as possible and yet assuring an adequate counting rate, it is necessary to prepare the insulin-I^{131} with a high specific activity. The lots of insulin-I^{131} employed in this study had specific activities of 75 to 300 mc per mg at the time of use. The preparation of such highly labeled preparations entails difficulties not encountered when the specific activity is very much lower. The Newerly modification (16) of the Pressman-Eisen method (17) was used for labeling with several further modifications designed to increase specific activity and to minimize damage to the insulin from irradiation and other causes. To approximately 0.3 ml chloroform in a 50 ml separatory funnel are added in turn, 0.2 ml of 2.5 N HCl, 20 μl of 10^{-8} M KI, 30 to 80 mc I^{131} (as iodide) and 1 drop of 1 M $NaNO_2$. Immediately after addition of the last reagent, the funnel is stoppered to prevent loss of I^{131} into the atmosphere and is shaken vigorously for 2 to 3 minutes. The chloroform layer (bottom) is then drawn

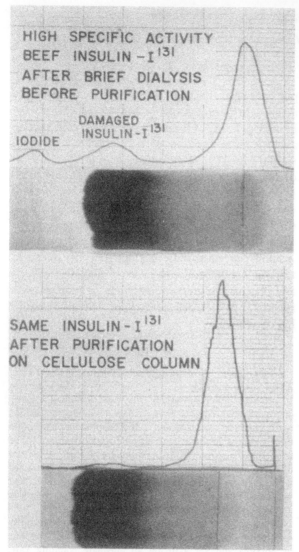

FIG. 1. PURIFICATION OF DAMAGED INSULIN-I[131] BY CELLULOSE COLUMN ADSORPTION. *Top:* Chromato-electrophoretograms of beef insulin-I[131], with specific activity about 300 mc per mg, after 30 minutes' dialysis. *Bottom:* Same preparation after elution from cellulose column with control (nonimmune) plasma. At significantly lower specific activities the preparations appear as in the bottom figure *without* purification.

into a test tube beneath a layer of a few drops of water (to prevent loss of I[131] into the air) and assayed for radioactivity in any low-sensitivity counting device. If much less than one-half of the starting radioactivity has been extracted, a second extraction with 0.2 to 0.3 ml chloroform is performed. The total amount of chloroform should be kept as small as possible to facilitate the subsequent extraction of iodine into the aqueous protein solution. The volume of the aqueous phase in the

separatory funnel also should be kept small to favor the initial extraction of iodine into the chloroform. The chloroform-iodine mixture is added to 0.5 ml of 0.2 M borate buffer, pH 8, containing 20 μg of crystalline beef insulin in a 50 ml centrifuge tube, which provides for a broad interface between the two phases. The tube is shaken briskly but not violently for not more than 2 to 3 minutes following which an additional 1.0 ml borate buffer is mixed into the contents. A barely visible flocculate appears occasionally and should be allowed to settle, whereupon the top 0.5 ml (one-third of total) of the water phase is quickly removed and dialyzed against 2 L of distilled water.[1] Owing to the high concentration of radioactivity and low concentration of protein, the insulin is very susceptible to radiation damage (18, 19); therefore, exposure to I[131] at this stage should be as brief as possible, not more than 5 to 10 minutes elapsing between addition of I[131] to the insulin and the start of dialysis. Of the total radioactivity in the dialysis bag, approximately 65 to 80 per cent represents unbound I[131] which is reduced to less than 1 per cent of the I[131] bound to insulin after 2 hours of dialysis. Between 20 and 60 per cent of the insulin-I[131] is adsorbed to the dialysis membrane during this time so that the procedure yields approximately 3 to 5 μg insulin labeled with about 1.0 to 2.0 mc I[131]. Considerable sacrifices in total yield are made to expedite the procuring of a highly labeled preparation which usually contains no more than 4 to 6 per cent damaged components. We have the impression that the addition of 10^{-3} M. KI or phenol (as radical scavengers) to the dialyzing solution may help to minimize radiation damage, but this has been difficult to establish since other factors are also responsible for damage to the protein during the procedure. Distilled water is used in the last dialysis following which 1 drop of human serum albumin (250 mg per ml) is added to the insulin-I[131] solution to prevent losses of labeled insulin by adsorption to glassware (20, 21) and to minimize any further irradiation damage (18, 19). Solutions are kept frozen when not in use.

If the insulin-I[131] solution is surveyed for radioactivity at completion of dialysis, the specific activity of the insulin-I[131] may be estimated approximately. If the yield of labeled insulin has been sufficient to produce a specific activity in excess of 150 mc per mg, it can be anticipated that damage will be significantly in excess of 4 to 6 per cent, and at 300 mc per mg may be as great as 15 to 18 per cent. It is then necessary to effect partial purification of the insulin-I[131]. Since the damaged components do not adsorb to paper but are observed to migrate with serum proteins on paper strip chromatography or electrophoresis (22), it is possible to use a cellulose column for the purification procedure as follows: The dialyzed insulin-I[131] solution is added to 0.1 ml control (nonimmune) serum and the mixture is then passed through a column packed

---

[1] Removal of unbound iodide[131] by anionic exchange resins is usually unsatisfactory because much of the insulin-I[131] at this low concentration is lost by adsorption to the resin.

with a cellulose powder [2] about 1 ml in volume following which the column is washed 3 or 4 times with 1 ml of veronal buffer, 0.1 ionic strength. Most of the damaged components pass through the column with the serum while the undamaged insulin remains adsorbed to the cellulose in the column and can now be eluted slowly with undiluted control serum or plasma. Usually 3 to 4 eluates (each 0.5 ml of plasma) are collected and diluted immediately 1:20 to 1:100 with veronal buffer containing 0.025 per cent serum albumin to prevent further damage to the insulin by the concentrated plasma. Although the elution of insulin-I[131] from the column is far from complete, adequate amounts are obtained for almost any number of insulin assays. Most of the damaged fraction is removed by this procedure (Figure 1).

Principles of immunoassay. The basis of the technique resides in the ability of human insulin to react strongly with the insulin-binding antibodies present in guinea pig antibeef insulin serum (11, 12, 14), and by so doing, to inhibit competitively the binding of crystalline beef insulin-I[131] to antibody. The assay of human insulin in unknown solutions is accomplished by comparison with known concentrations of human insulin. The use of I[131]-labeled animal insulin as a tracer is necessitated by the lack of a crystalline preparation of human insulin.

The determination of antibody-bound insulin-I[131] and free insulin-I[131] by paper chromato-electrophoresis has been described previously (22). Briefly, the separation of antibody-bound insulin from unbound insulin in plasma results from the adsorption of all free insulin (when present in amounts less than 1 to 5 μg) to the paper at the site of application ("origin"), while the antibody-bound insulin migrates toward the anode with the inter-β-γ-globulins. Thus, in the presence of insulin-I[131] there appear two separate peaks of radioactivity; measurement of the areas beneath the two peaks (by planimetry) yields the relative proportion of bound insulin-I[131] (migrating with serum globulins) and free insulin-I[131] (remaining at origin). The ratio of bound insulin-I[131] to free insulin-I[131] (B/F) is a function of the concentration of insulin-binding antibodies, of both insulin concentrations, and of the characteristic kinetic and thermodynamic constants for the reactions between the insulins and the particular antiserum (23). Selection of an antiserum for purposes of this assay is determined primarily by the desirability of obtaining a relatively marked decrease in B/F ratio with small increments in the concentration of human insulin. Although the antibody concentration is of only secondary importance, it should be high enough to permit at least 1:100 dilution of the antiserum (preferably 1:1,000 dilution or greater). On the basis of preliminary tests the antiserum is diluted appropriately to yield an initial B/F ratio between 2 and 4 for tracer beef insulin-I[131] alone, in the absence of added human insulin. Provided that the amount of the beef insulin-I[131] used is truly a tracer quantity, the initial B/F ratio is inversely proportional to the dilution factor

(23). In the presence of human insulin, the B/F ratio decreases progressively with increase in insulin concentration; with sensitive antisera the B/F ratio is reduced by about 50 per cent in the presence of 15 μU per ml human insulin.

Standard curves. Two preparations of human insulin were employed as standards. The first ("Tietze human insulin")[3] is reported (24) to have a potency of 1.8 U per mg crude preparation; the second ("Fisher human insulin"),[3] was assayed at 6.8 U per mg in 1956 (25), but it was believed that the activity of the latter preparation might have decreased slightly since its initial preparation (25). A tentative value of 6 U per mg for the Fisher insulin was assigned. However, since a value as low as 22 U per mg could be placed on a crystalline sample of the latter preparation (25), whereas the Tietze insulin was assayed relative to a standard of 27 to 29 U per mg, we have regarded the Tietze crude insulin preparation as $1.8/28 \times 100 = 6.45$ per cent pure insulin by weight, and the Fisher insulin powder preparation as $6/22 \times 100 = 28.2$ per cent pure insulin by weight. When compared on this basis, no consistent differences in potencies of the two preparations were observed in the immunoassay procedure and the value of 6 U per mg for the Fisher preparation was accepted as the correct value. Since the Fisher preparation is the more highly purified, it was employed as standard in most of the studies.

All dilutions of insulin and antiserum are prepared in 0.1 ionic strength veronal buffer containing 0.25 per cent human serum albumin to prevent adsorption of reactants to glassware. (There is no detectable insulin in commercial supplies of human serum albumin.) Standard solutions each contain identical concentrations of tracer beef insulin-I[131] (about 0.05 to 0.15 mμg per ml but differing in different runs) and antiserum, but varying concentrations of human insulin ranging from 0.05 to 5.0 mμg per ml (calculated as "pure" human insulin). The antiserum is added last in all cases. Mixtures are refrigerated at 4° C for 4 days. These conditions provide sufficient time to reach equilibrium between bound and free insulin. The mixtures are then subjected to chromato-electrophoresis (22) in a cold room at 4° C (Whatman 3 MM paper, veronal buffer, 0.1 ionic strength, pH 8.6, constant voltage 20 to 25 v per cm, cover of apparatus open), which produces a satisfactory separation of the peaks of bound and free insulin-I[131] in about 1 to 1.5 hours. Earlier immunoassays (10) were performed after prolonged incubation at 37° C. However, it has since been shown (23) that the standard free energy change of the reaction in the direction of antigen-antibody complex formation is increased considerably at 4° C, which results in an approximately twofold greater slope in the B/F versus insulin concentration curves at low insulin concentrations. Just prior to chromato-

---

[2] Genuine Whatman Cellulose Powder, W & R Balston Ltd., England.

[3] We are greatly indebted to Dr. F. Tietze of the National Institutes of Health and Dr. A. M. Fisher of the Connaught Laboratories, Toronto, Canada, for these preparations.

electrophoresis, control (nonimmune) guinea pig plasma is added to the mixtures to prevent trailing of antibody-bound insulin on the paper strips, since the very low concentrations of serum proteins in these mixtures are insufficient in themselves to prevent adsorption of the serum proteins (including antibody) to the paper.

The chromato-electrophoretograms are developed until the albumin band has moved about 2.5 to 3 inches from the origin, which, under the conditions employed here, usually takes about an hour. The peak of antibody-bound insulin-I[131] moves about 2.25 inches under these conditions. The use of several large boxes, each with a capacity for 16 strips, makes it possible to run 250 to 300 strips a day. After drying, the strips are assayed for radioactivity in an automatic strip counter (Figure 2A). A "standard curve" is obtained by plotting the B/F ratio as a function of the concentration of added human insulin (Figure 2B) after correction for damaged components of insulin-I[131]. From 3 to 6 per cent of the insulin-I[131] was damaged after final preparation of the lots employed in this study. These damaged components

migrate nonspecifically with the serum proteins, primarily with the α-globulins (22), and are demonstrably not available for binding by antibody. The short run chromato-electrophoresis does not resolve the serum proteins well enough to distinguish between antibody-bound insulin-I[131] and damaged insulin-I[131] so that the damaged fraction is determined by using either control (nonimmune) plasma, or antiserum whose binding capacity for undamaged insulin-I[131] is completely saturated with beef insulin. Since the antiserum used here has a maximal beef insulin-binding capacity of about 1 mμg per ml at the dilutions employed, it has been general practice to include one or more samples made up with 1 to 4 μg per ml beef insulin for the purpose of determining the damaged fraction. Damaged insulin-I[131] migrating with serum proteins is corrected for by subtracting the fraction damaged from the total area in the chromato-electrophoretogram. The area under the free insulin peak is then divided by the corrected total area to yield the fraction of free insulin. The fraction "bound insulin" is then 1.00 minus the fraction "free insulin."

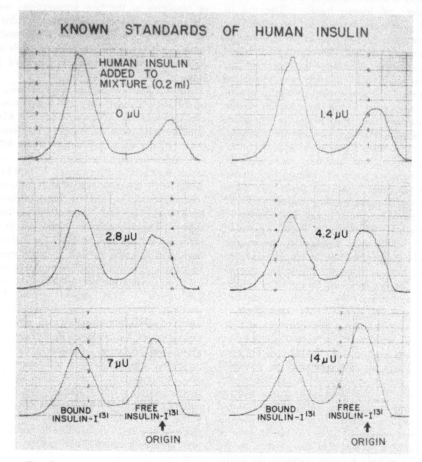

Fig. 2. A: Radiochromato-electrophoretograms of antiserum, insulin mixtures. Mixtures contained the same concentrations of guinea pig antibeef insulin serum and beef insulin-I[131] but varying concentrations of human insulin as indicated.

It is evident that variation in the volumes of solution applied to the paper strips is of no consequence. Generally 100 to 200 μl is applied, the larger volume permitting use of a smaller quantity of tracer beef insulin-I[131] for the same counting rate.

Assay of insulin in plasma. Mixtures containing unknown samples are prepared at the same time and in the same way as are standard solutions except that the unknown sample is substituted for the human insulin. Plasma insulin is best determined in a 1:10 final dilution unless the insulin concentration is unusually high; then a 1:20 or 1:40 dilution may be used. Mixtures may be made up to any desired volume. However, since only 100 to 200 μl is applied to the paper strips, it is convenient to prepare all mixtures in 0.2 or 0.5 ml volumes containing 20 or 50 μl of plasma, respectively.

Since insulin may be damaged by plasma during incubation (22), an effect which is more marked in concentrated plasma than in diluted plasma, and at 37° rather than at 4° C,[4] it is advisable to run a control mixture with unknown plasma but without antiserum to correct for "incubation damage." However, at 1:10 dilution of plasma after 4 days at 4° C, incubation damage amounts only to 0 to 3 per cent, an observation which contributed to the selection of these conditions. Therefore, only a negligible additional correction for damage is required in the plasma samples.

The insulin concentration in each plasma sample is determined from the standard curve by referring to the insulin concentration which corresponds to the corrected B/F ratio observed in the plasma sample (10–12, 14, 15).

Subjects for glucose tolerance tests. Subjects were chosen at random from patients sent to the general laboratory for glucose tolerance tests and from known diabetic and apparently nondiabetic patients on the wards of the Veterans Administration Hospital, Bronx, N. Y. Patients who had *ever* been treated with insulin were excluded from this study in order to obviate effects of antibodies in their own serum (22). Other than the exception noted below, subjects were classified as diabetic or nondiabetic on the basis of the following criteria applied to the 2-hour blood sugar curve following oral ingestion of 100 g of glucose: *diabetic*—a peak blood sugar concentration of 180 mg per 100 ml or greater, and a 2-hour blood sugar concentration of 120 mg per 100 ml or greater; *nondiabetic*—a peak blood sugar concentration not exceeding 160 mg per 100 ml, and 2-hour level no more than 120 mg per 100 ml. One subject with marginal ulcer and a dumping syndrome, with a blood sugar concentration of 286 mg per 100 ml at 0.5 hour falling to 134 mg per 100 ml at 1 hour and 44 mg per 100 ml at 2 hours, is included in this group. Because of the exclusion of insulin-treated patients, only mild or early maturity-onset diabetes is represented in the diabetic group. Subjects who did not qualify by these criteria for

---

[4] For this reason plasma is separated in a refrigerated centrifuge immediately after withdrawal of blood and is used immediately or kept frozen until used in order to minimize loss of the endogenous insulin present.

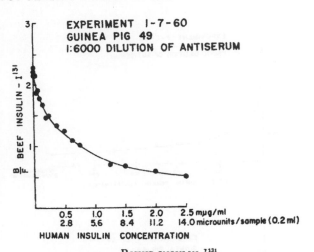

FIG. 2. B: RATIO, $\dfrac{\text{BOUND INSULIN-I}^{131}}{\text{FREE INSULIN-I}^{131}}$, AS A FUNCTION OF THE CONCENTRATION OF ADDED HUMAN INSULIN. The ratios were obtained from the complete series of radiochromato-electrophoretograms, a few of which are shown in Figure 2A.

either group are considered in an *"undetermined status."* The criteria employed are modified from those suggested by Fajans and Conn (26) and are designed to eliminate questionable cases from diabetic and nondiabetic categories.

All subjects were to have fasted for 14 hours prior to the glucose tolerance test, but from the fasting blood sugar concentration in one subject (Ri) it is suspected that this restriction was not observed in his case. All subjects were to have consumed a diet containing at least 300 g carbohydrate per day for 3 days preceding the glucose tolerance test, but there is no assurance that this regimen was followed in all cases. Blood samples were obtained in the fasting state immediately before, and 0.5 hour, 1 hour and 2 hours following glucose feeding. In a small group of cases an additional 50 g glucose was administered at 1.5, 2 and 2.5 hours, and blood collections were continued to 3 hours.

Blood sugar determinations were determined according to the method of Somogyi (27).

RESULTS

Standard curves. Several representative standard curves are shown in Figure 3. The amount of insulin-I[131] employed as tracer varied somewhat from experiment to experiment. In the experiments shown in Figure 4, the effects of Tietze and Fisher insulins are compared with each other and with the effect of crystalline beef insulin. As in other experiments no significant differences between the two human insulin preparations were observed. Since 100 to 200 μl of solution was as-

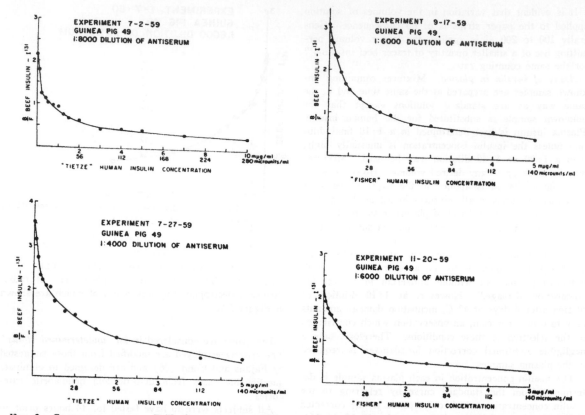

FIG. 3. STANDARD CURVES: B/F (BEEF INSULIN-I^{131}) RATIO AS A FUNCTION OF THE CONCENTRATION OF TIETZE OR FISHER HUMAN INSULIN.

sayed, less than 1 μU of human insulin was readily detectable with this antiserum. At low insulin concentrations, random variations in B/F produce only small errors in the absolute quantity of insulin but the percentage error is high; conversely, at high insulin concentrations the absolute

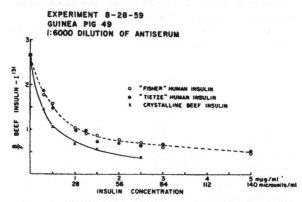

FIG. 4. COMPARISON OF THE EFFECTIVENESS OF VARIOUS CONCENTRATIONS OF TIETZE AND FISHER HUMAN INSULINS AND CRYSTALLINE BEEF INSULIN IN REDUCING THE B/F RATIO FOR BEEF INSULIN-I^{131}.

error is likely to be higher but the percentage error lower. By increasing the dilution of the antiserum, the entire concentration range is easily scaled down by a factor of 2 or 3 and the limit of sensitivity increased to about 0.1 to 0.2 μU of insulin. However, the conditions employed are suitable for determination over the 50- to 100-fold range of insulin concentrations ordinarily encountered in man.

It is evident that beef insulin reacts about two to four times more strongly (depending on the insulin concentration employed) with the guinea pig antibeef insulin serum than does human insulin (Figure 4). Other guinea pig antisera to beef insulin have shown even greater differences in reaction of beef insulin and human insulin. On this account *beef insulin cannot be used as a standard for the assay of human insulin in the guinea pig antibeef insulin system.* Because of the differences in reactivity of human and beef insulin, differences in the specific activity of the beef insulin-I^{131} preparations result in different

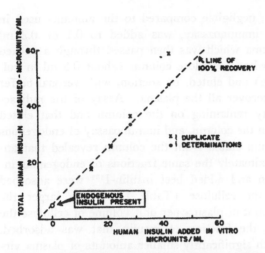

Fig. 5. A: Recovery of human insulin added *in vitro* to a fasting plasma sample. Endogenous insulin concentration in undiluted plasma was 48 μU per ml. All assays were performed in 1:10 dilution of plasma.

TABLE I

Effect of cysteine on endogenous plasma insulin*

		Insulin concentration		
Subj.	Plasma sample	Original plasma before incubation and dialysis	Control sample incubated and dialyzed without cysteine	Sample incubated with cysteine and dialyzed
		μU/ml	μU/ml	μU/ml
Yo.	1 hr	324	238	14
Un.	1 hr	337	216	0

* See text for conditions of experiments.

initial B/F values and somewhat differently shaped curves even at the same dilution of antiserum if approximately the same radioactivity (and therefore different amounts of beef insulin) is used. These differences could be abolished if each lot of beef insulin-I¹³¹ were assayed for its beef insulin concentration and if the same amount of beef insulin were employed, independent of its content of radioactivity. However, it is more expedient to include a standard curve with human insulin for each run of unknowns. When 250 or more unknown samples have been run in a single

experiment, an added set of 15 to 16 standard solutions is a negligible addition.

Recovery of added human insulin and effect of plasma dilution. The virtually quantitative recovery of human insulin added to plasma *in vitro* (Figure 5A) indicates that the plasma has neither an inhibitory nor an augmentative effect and this conclusion is confirmed by the proportionate decrease in measured insulin concentration when the plasma is diluted over a large range (Figure 5B).

Effect of cysteine and cellulose on endogenous plasma insulin. Since insulin is destroyed by incubation with cysteine at alkaline pH and is adsorbed by powdered cellulose, the effects of these agents on endogenous insulin were tested. Plasmas of relatively high insulin concentration were incubated at 37° C with 0.02 M cysteine at pH 8 for 1.5 hours and then dialyzed against normal saline for 3 hours to remove the cysteine. Aliquots of the same serum samples were treated

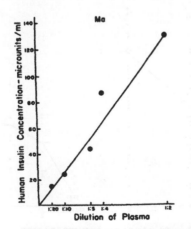

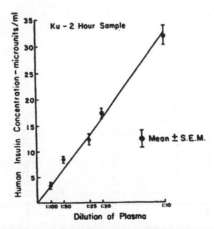

Fig. 5. B: Effect of dilution of plasma on measured concentration of endogenous plasma insulin. Four replicate determinations were made for each point in the experiment on the right.

1164　　　　　ROSALYN S. YALOW AND SOLOMON A. BERSON

TABLE II

Cellulose adsorption of beef insulin-I[131] and endogenous human plasma insulin

| | | Per cent adsorbed by cellulose column | |
| | | Beef insulin-I[131] | Endogenous human insulin |
Subj.	Plasma		
	ml		
Y.	0.1	69	64
	0.2	49	44
U.	0.1	71	88
E.	0.1	81	84

similarly except that cysteine was omitted. Although incubation and dialysis alone led to a 26 to 36 per cent loss in endogenous insulin concentration in the control samples, cysteine was almost completely effective in destroying the endogenous insulin (Table I). In simultaneous experiments insulin-I[131] was found to be virtually completely destroyed under these conditions as determined by paper chromato-electrophoresis.

To evaluate cellulose adsorption of endogenous insulin, a minute amount of tracer beef insulin-

I[131], negligible compared to the amounts used in the immunoassay, was added to 0.1 or 0.2 ml plasma which was then passed through a packed powdered cellulose column (about 0.5 ml in volume) and eluted, by suction, with veronal buffer to recover all the plasma. Assay of the radioactivity remaining on the column and that eluted from the column, and immunoassay of endogenous insulin eluted from the column revealed that approximately the same fractions of endogenous insulin and added beef insulin-I[131] were adsorbed by the cellulose (Table II). The larger the amount of plasma per unit volume of cellulose the less the fraction of insulin that was adsorbed. With significantly smaller amounts of plasma virtually all insulin-I[131] and endogenous insulin are adsorbed, but the insulin concentrations then become unmeasurable. Only negligible fractions of albumin-I[131] and γ-globulin-I[131] are adsorbed by cellulose under these conditions.

Insulin concentrations in early maturity-onset diabetic and control subjects. The average fasting insulin concentrations tended to be only slightly

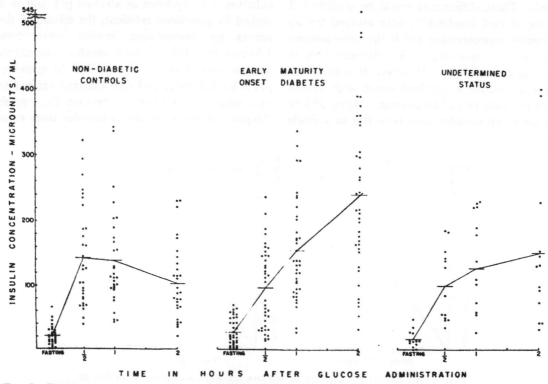

FIG. 6. PLASMA INSULIN CONCENTRATIONS DURING STANDARD 100 G (P.O.) GLUCOSE TOLERANCE TEST IN VARIOUS GROUPS OF SUBJECTS.

TABLE III

Blood sugar and plasma insulin concentrations during a standard 100 g oral glucose tolerance test

Subj.	Blood sugars						Plasma insulin concentrations					
	F	0.5 hr	1 hr	2 hrs	3 hrs	4 hrs	F	0.5 hr	1 hr	2 hrs	3 hrs	4 hrs
	mg/100 ml						μU/ml					
A. Nondiabetic controls												
La.	87	122	78	91			31	162	112	89		
Pa.	73	109	68	77			25	270	154	155		
Hu. J.	85	120	113	113			7	103	95	116		
Wh.	78	110	85	88			66	294	168	174		
Don.	88	108	135	115			45	67	98	39		
St.	89	120	104	93			13	95	46	43		
Kan.	83	113	103	83			28	62	229	117		
Ra.	91	131	135	90			20	176	128	112		
Jo.	92	113	113	80			25	124	118	21		
Wa.	98	128	147	118			20	68	93	75		
Dor.	90	140	148	118			31	81	126	90		
Ei.	83	100	70	95			50	235	101	79		
Cor.	110	133	133	115			21	130	155	145		
Cal.	91	117	147	119			7	39	76	47		
Te.	83	143	155	73			17	78	190	126		
Ru.	83	103	113	98			3	145	204	222		
Dam.	90	118	120	110			18	187	205	180		
Kas.	98	150	135	108			11	48	112	65		
Kr.	98	115	125	120			2	71	95	117		
Him.	93	140	120	108			22	322	252	232		
Hu. J. J.	95	128	110	105			9	126	56			
Dan.	83	115	73	100			0	224	18	98		
Sc.	83	158	140	113			22	91	114	135		
Un.	88	133	153	115			0	247	337	233		
Ke.	103	138	128	112			34	163	174	67		
Hig.	100	148	88	120			14	241	42	70		
Con.	90	133	135	115			11	104	45	36		
Pop.	93	148	143	95			39	67	101	79		
Ry.	90	143	150	110			11	84	107	81		
Al.	96	296	134	44			14	188	342	42		
Mean							21	143	139	106		
B. Early maturity-onset diabetes												
Ri.	245	346	436	472			51	123	173	179		
Wa. D.	96	178	218	150			6	32	97	166		
Mor.	93	110	180	218			3	77	158	300		
Moh.	95	145	193	135			59	162	339	355		
Ko.	93	240	360	120			3	14	294	364		
Fl.	138	173	248	266			56	70	190	270		
Fel.	100	152	190	141			19	107	121	102		
Fr.	113	238	310	195			35	113	316	378		
Sh.	118	232	300	190			13	38	175	76		
Go.	100	146	218	173			22	54	156	216		
Bl.	114	168	236	223			19	59	108	221		
Ma.	163	240	256	320			51	46	35	54		
We.	96	178	218	150			6	32	97	166		
Qu.	93	155	180	178			56	120	121	283		
Ok.	143	202	244	204			11	24	79	160		
Ha.	110	155	193	200			50	154	185	482		
Cr.	92	200	245	225			0	31	75	70		
Mi.	105	180	211	170			25	117	112	187		
Ro.	113	177	233	240			0	182	224	266		
Wo.	100	163	170	180			42	160	168	207		
Poi.	103	188	243	190			5	17	83	173		
Fo. F.	152	244	266	380			11	22	28	23		
No.	105	195	215	185	91	83	59	190	238	392	159	59
Ny.	93	158	193	138	60		11	53	90	154	17	
Moo.	113	153	205	268			63	148	91	131		
Fo. J.	90	169	193	185	83		34	238	249	490	173	
Br.	90	167	198	163	86	67	20	62	133	201	79	29
Pl.	100	183	193	160	63		42	168	241	308	191	
Mu.	130	233	326	374			3	17	68	126		
Fla.	93	140	170	178			22	140	126	220		

TABLE III—*Continued*

Subj.	Blood sugars						Plasma insulin concentrations					
	F	0.5 hr	1 hr	2 hrs	3 hrs	4 hrs	F	0.5 hr	1 hr	2 hrs	3 hrs	4 hrs
			mg/100 ml						*μU/ml*			
B. *Early maturity-onset diabetes*—Continued												
Car.	108	165	238	183	90		42	34	106	247	106	
How.	125	198	223	250			14	132	140	245		
Coo.	95	145	163	178	128		28	213	205	350	233	
Fele.	100	193	266	235			0	31	84	81		
Le.	90	173	250	193			65	65	294	392		
Ga.	105	148	178	135			70	148	210	302		
Hor.	95	165	147	195	211		8	148	140	364	386	
Eh.	91	156	211	309	246		17	129	176	545	531	
Mean							27	97	156	243		
C. *Undetermined status*												
Wa. D.	85	100	150	148			0	35	56	42		
Wr.	98	123	158	138			27	86	75	75		
Ric.	100	138	158	143			13	140	226	100		
Ba.	88	140	150	129			14	144	230	405		
Har.	88	148	178	140			36	187	232	134		
Hew.	83	145	163	105			4	56	64	94		
Jos.	95	147	171	136			49	130	192	157		
Leh.	100	152	162	145			14	186	224	395		
Wis.	103	150	158	128			0	155	182	148		
Doh.	70	110	155	128			8	48	28	33		
Lut.	100	118	160	130			28	51	112	132		
Sa.	100	170	143	110			11	48	140	42		
Wi.	90	123	164	118			14	56	104	233		
Mean							17	101	128	153		
D. *Decompensated cirrhosis*												
Man.	96	164	146	136			54	240	356	486		
Mar.	92	168	150	100			8	140	226	143		
Cara.	95	143	140	93			14	40	57	30		
Cro.	85	143	118	85			2	16	14	8		
Di.	78	100	82	65			25	33	19	5		
Fo.	80	135	100	80			5	36	22	8		
E. *Pituitary tumors*												
Sil.*	92	122	147	134			32	350	570	175		
Sin.†	90	135	140	117			5	177	192	180		
Led. (Acromegaly)	93	163	174	103			8	156	203	109		
F. *Thyrotoxicosis*												
El.	103	215	240	210			38	275	230	240		
Yo.	93	170	184	103			65	247	324	81		
G. *Others*												
Ku. (Hemochromatosis)	73	135	175	160			48	121	321	330		
Cra. (Acute pancreatitis)	100	120	138	125			5	20	100	135		
Bl.‡	48	108	148	135			28	90	56	67		
Coh.§	36	52	78	76			118	190	199	98		

* Chromophobe adenoma.
† Eosinophilic and chromophobe adenoma with acromegaly.
‡ Hypoglycemia, cause undetermined, after partial pancreatectomy.
§ Proven islet cell adenoma (courtesy of Dr. H. Epstein).

higher in the diabetic (mean, 27 μU per ml) than in the nondiabetic (mean, 21 μU per ml) subjects, although 34 per cent of the diabetics exceeded 40 μU per ml in contrast to only 10 per cent of the nondiabetics. In none of the 68 patients in both of these groups did the fasting level exceed 70 μU per ml (Table III, A and B, Figure 6). These values are in good agreement with those reported earlier in a smaller series of subjects (15). The responses to orally administered glucose in diabetic and nondiabetic patients differed more markedly than did the fasting insulin concentrations. Nondiabetic subjects were about equally divided in showing the peak insulin concentration at 0.5 hour or 1 hour (Table IIIA), whereas with few definite exceptions diabetic patients showed the maximal insulin concentration at 2 hours (Table IIIB). The average insulin concentration at 0.5 hour was lower in diabetic (mean, 97 μU per ml) than in nondiabetic (mean, 143 μU per ml) subjects, but the diabetics appeared to form two groups at this point (Figure 6). A delayed insulin response is suggestive in the lower of these two groups.

Although there is a large scatter of individual values, the mean curves for the two groups illustrate these differences clearly (Figure 6). The average integrated insulin concentration during the 2 hour glucose tolerance test was 26 per cent higher for the diabetic (147 μU per ml) than for the nondiabetic (117 μU per ml) group. The "undetermined" group (Table IIIC) probably represents a mixture of early diabetic and nondiabetic subjects and nothing can be concluded definitely about the variable insulin response to glucose loading.

Insulin concentrations in four diabetic and five nondiabetic subjects given an additional 50 g of glucose at half hour intervals from 1.5 to 2.5 hours are shown in Figure 7 and Table IV. Insulin concentrations rose to higher levels in both

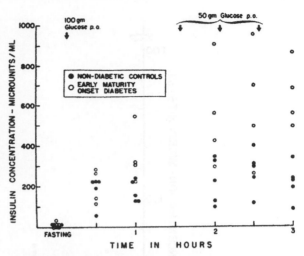

FIG. 7. PLASMA INSULIN CONCENTRATIONS DURING HEAVY GLUCOSE LOADING EXPERIMENTS IN DIABETIC AND NONDIABETIC SUBJECTS.

groups but more marked increases were observed in the diabetic subjects.

It should be emphasized that insulin-I[131] when administered intravenously exhibits a rapid fall in concentration due to a marked and continuous increase in its apparent volume of distribution for a period of about 30 to 60 minutes and to a metabolic turnover rate with a half-time of about 35 minutes (22). It may be reasonably expected that endogenously secreted insulin behaves similarly,[5] and therefore that any particular peak concentra-

---

[5] Endogenously secreted insulin is, in addition, subject to removal by the liver before it reaches the peripheral circulation (28, 29).

TABLE IV

Effect of heavy glucose loading* on blood sugars and plasma insulin levels

Subj.	Blood sugars						Plasma insulin concentrations					
	F	0.5 hr	1 hr	2 hrs	2.5 hrs	3 hrs	F	0.5 hr	1 hr	2 hrs	2.5 hrs	3 hrs
	mg/100 ml						*μU/ml*					
Nondiabetic controls												
Hea.	84	136	106	130	108	110	3	190	129	345	308	190
Gas.	78	118	86	108	104	98	14	224	129	322	300	341
Rei.	90	127	129	105	103	101	15	224	224	125	400	225
Keh.	94	158	142	98	90	86	0	224	238	224	241	235
McC.	100	144	150	104	102	90	17	56	151	98	118	84
Maturity-onset diabetes												
Cri.	100	188	214	208	188	162	17	265	548	910	960	685
All.	86	140	152	166	126	122	12	284	223	560	496	496
Ab.	90	143	170	167	177	155	31	112	309	420	700	870
Ti.	86	170	181	155	149	145	3	140	313	294	255	578

* Glucose 100 g p.o., immediately after fasting specimen; glucose, 50 g p. o., at 1.5, 2 and 2.5 hours.

1168 ROSALYN S. YALOW AND SOLOMON A. BERSON

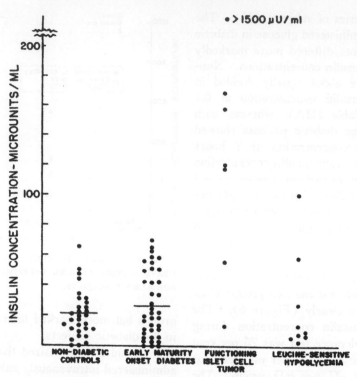

FIG. 8. FASTING PLASMA INSULIN CONCENTRATIONS IN VARIOUS
GROUPS OF SUBJECTS. The subject with plasma insulin concentration
greater than 1,500 μU per ml had an islet cell adenocarcinoma with
widespread metastases (patient of Dr. J. Field).

tion depends on the precise moment of sampling. A very rapid and pronounced fall from the peak concentration would be anticipated in the case of a single secretory spurt. Conversely, a sustained elevation or continued rise in insulin concentration implies a continued secretion during the time interval under observation.

Insulin concentrations in patients with islet cell tumors or leucine-induced hypoglycemia. Insulin concentrations in fasting plasmas from five of seven patients [6] with proven islet cell tumors were elevated above normal levels (Figure 8), but the response to glucose was normal in the one patient studied during a glucose tolerance test (Coh., Table IIIG).

Four of six subjects [7] with leucine-induced hypoglycemia showed increased insulin concentra-

tions following administration of L-leucine (75 to 150 mg per kg) in six of nine experiments (Figure 9), although fasting insulin concentrations were elevated in only a single patient (Figure 8), the only adult in the series and the one patient suspected on clinical grounds to have an islet cell tumor.[8] The peaks of insulin concentration, when observed, were in good time correspondence with the induced hypoglycemia.

Plasma insulin in cirrhosis, acromegaly and hyperthyroidism. Six patients with decompensated cirrhosis were studied (Table IIID). In two cases the glucose curves were high, but not within the diabetic range, and were associated with relatively high insulin concentrations. In three cases insulin concentrations were very low throughout the 2 hour glucose tolerance test, and in two of these the glucose concentration curves were quite flat. Intravenous glucose tolerance tests are necessary before it can be decided whether the observed association in the latter cases is to be at-

[6] We are indebted to Doctors H. Epstein, J. Field, E. D. Furth, E. Gordon, A. Renold and J. Steinke for these sera.

[7] We are indebted to Doctors A. DiGeorge, M. Goldner, M. Grumbach, I. Rosenthal and S. Weisenfeld for these sera.

[8] Courtesy of Doctors S. Weisenfeld and M. Goldner.

tributed to poor glucose absorption or to heightened insulin sensitivity.

In three patients with pituitary tumors, two of whom had clinical acromegaly, and in two thyrotoxic subjects, insulin concentrations during the glucose tolerance test were in the high normal range (Table III, E and F).

Results in a few individual cases that do not fall into the other categories are also included in Table IIIG.

In the absence of glucose loading, plasma insulin concentrations did not change significantly in two control subjects (Fra. and Gre., Table V).

In seven cases, sera were refrozen and repeat determinations were performed one or more months later with a different lot of insulin-I[131]. The reproducibility of determinations performed under these conditions is shown in Figure 10.

DISCUSSION

The demonstration that unlabeled insulin could displace insulin-I[131] from complexes with insulin-binding antibody (22) and that the fraction of insulin-I[131] bound to antibody decreases progres-

TABLE V

*Blood sugar and plasma insulin concentrations in the absence of glucose loading**

Nondia-betic controls	Time	Blood sugar	Plasma insulin concentrations
	min	*mg/100 ml*	*μU/ml*
Gre.	0	93	17
	20	90	19
	40	90	18
	60	88	18
	120	78	25
Fra.	0	85	3
	20	85	5
	40	80	5
	60	85	3
	120	88	3

* Subjects were fasted overnight and throughout the period of blood sampling.

sively with increase in insulin concentration (22) laid the foundation for the immunoassay of insulin employing isotopically labeled insulin. In initial reports describing results with the present method for immunoassay of beef insulin (9, 10) it was emphasized that species differences in the reaction of insulin with insulin antisera exist and that

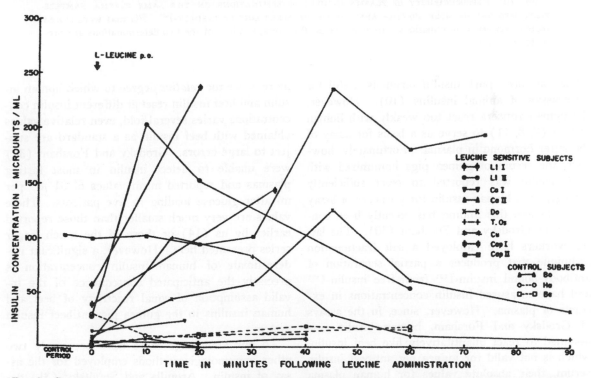

FIG. 9. PLASMA INSULIN CONCENTRATIONS FOLLOWING ADMINISTRATION OF L-LEUCINE TO CONTROL AND LEUCINE-SENSITIVE HYPOGLYCEMIC SUBJECTS.

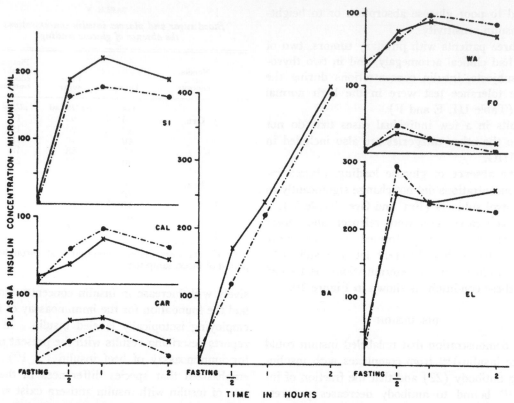

Fig. 10. Reproducibility of plasma insulin determinations on the same plasma samples performed one or more months apart with different lots of insulin-I^{131}. Plasmas were stored frozen between determinations. In these cases, the average values of the two determinations are presented in Table III.

human antibeef, pork insulin serum is useful for microassay of animal insulins (10). However, the human antisera react too weakly with human insulin (7, 8, 11) to serve as a basis for assay of the latter hormone in plasma. Fortunately, however, the serum of guinea pigs immunized with beef insulin was reported to react sufficiently strongly with human insulin for purposes of assay (11, 14) and this finding has recently been confirmed by Grodsky and Forsham (30). The latter workers have employed a salt fractionation technique that produces a partial separation of antibody-bound insulin-I^{131} from free insulin-I^{131} and have measured insulin concentrations in extracts of plasma. However, since, in the assays of Grodsky and Forsham, human insulin was assumed to react quantitatively like beef insulin, which is not valid for guinea pig antibeef insulin serum, their absolute values for human plasma insulin concentrations are questionable. Further-

more, since the *relative* degree to which human insulin and beef insulin react at different insulin concentrations varies several-fold, even relative values obtained with beef insulin as a standard are subject to large errors. Grodsky and Forsham (30) were unable to detect insulin in most fasting plasmas and reported mean values of 31 μU per ml after glucose loading in five patients. These values are very much smaller than those reported earlier by us (14) or those of the much larger series presented here. However, a significant underestimate of human insulin concentration is precisely the anticipated consequence of the invalid assumption of equal reactivity of beef and human insulins in the guinea pig antibeef insulin system.

To our knowledge there have been only two other immunologic methods employed for the assay of insulin. Arquilla and Stavitsky (31) developed an assay for insulin based on the inhibition of hemolysis of insulin-sensitized red blood cells; however, the lower limit of detectability by this technique was approximately 0.1 μg (2.8 mU) making it unsuitable for determination of plasma insulin. Loveless (32) has used certain normal human subjects, in whom the skin can be locally sensitized to insulin (by the intracutaneous injection of human anti-insulin serum) to assay insulin by the whealing response obtained. Aside from the inconvenience associated with this method, the lower limit of detectability was 200 μU beef insulin per ml and human plasma insulin was not detectable, a result attributed in part to the lesser reactivity of human insulin (32).

Reported estimates of plasma insulin concentrations, derived from the various biological assay procedures, have varied widely. Thus, the *in vitro* diaphragm assay has yielded values ranging from 40 to 80 μU per ml (33) to as high as 4,600 μU per ml (34) in fasting plasmas and from about 130 to 800 μU per ml (33) to 9,000 to 22,000 μU per ml (35) after glucose in normal subjects. Measuring the increase in oxidation of glucose-1-C^{14} to $C^{14}O_2$ by rat epididymal adipose tissue *in vitro,* Martin, Renold and Dagenais (3) found that the insulin-like activity of fasting normal plasma in this preparation corresponded to 50 to 350 μU of insulin per ml. Pfeiffer, Pfeiffer, Ditschuneit and Ahn (36), using the same assay, found that plasma diluted 1:2 gave higher and more consistent insulin concentrations and reported normal fasting levels of 135 to 680 μU per ml in 15 normal human subjects, with concentrations frequently exceeding 2,000 to 4,000 μU per ml after tolbutamide and metahexamide. Employing the immunoassay method we have observed generally much smaller increases in peripheral insulin concentration after large doses of sodium tolbutamide, administered intravenously or by mouth, than after glucose given by the same routes to normal or diabetic men (37).

It is generally agreed (5, 34) that dilution of plasma or serum increases markedly the estimated insulin concentration in the diaphragm assay and similar observations have been made in the rat epididymal fat pad assay (36). This phenomenon has been either attributed to the presence of inhibitory substances in the plasma (5, 34) or interpreted as indicating that insulin-like activity of serum as measured by the isolated rat diaphragm

is not specific for insulin per se. Randle has found that albumin and other proteins may exert a nonspecific stimulation of glucose uptake by rat diaphragm (5) and that 1 ml of plasma exhibiting an insulin concentration of 13,000 μU per ml in the diaphragm assay had no effect on blood sugar when injected into alloxan-diabetic hypophysectomized rats, whereas 2,000 μU insulin produced a marked hypoglycemia (35).

The recent report by Leonards (6), that insulin-like activity in plasma, when tested on the rat epididymal fat pad, persists after total pancreatectomy and that insulin-neutralizing antiserum from guinea pigs has no inhibitory effect on the insulin-like activity of human serum in this system, has raised a serious question as to what part of the insulin-like effect on fat tissue is due to insulin itself.

In vivo insulin assays have also yielded variable estimates of plasma insulin concentration. Measuring the fall in blood sugar induced in adreno-demedullated alloxan-diabetic hypophysectomized rats, Anderson, Lindner and Sutton (38) were unable to detect circulating insulin in fasting plasma although the method was sensitive to 125 μU insulin. Bornstein and Lawrence (1) using adrenalectomized hypophysectomized diabetic rats reported plasma insulin concentrations 2 hours after glucose to average about 340 μU per ml in normal subjects and 100 to 320 μU per ml in diabetic patients not subject to ketosis, but Randle (5) was unable to confirm the suitability of these animals for insulin assay. More recently, Baird and Bornstein (39), employing adrenalectomized alloxan-diabetic mice, have found that normal fasting plasma extracted with acid ethanol-*n*-butanol-toluene (which is thought to separate insulin from insulin antagonists) contains about 1,000 μU per ml. Values about three times as high were observed after glucose feeding. Values as high or higher were found in four of six diabetic subjects.

At the present time it does not appear possible to resolve all the apparently divergent findings summarized here. It is necessary, however, to point out that plasma insulin concentrations determined by the immunoassay technique are in agreement with the lowest estimates derived from other methods of assay, notably the *in vivo* bioassays of Bornstein and Lawrence (1), and of Anderson and co-workers (38), and the diaphragm assay of Vallance-Owen and Hurlock (33). By comparison with the biological effects of exogenous insulin the lower concentrations appear most reasonable.

A rough estimate of the amount of insulin secreted can be derived from the insulin concentrations reported here according to the following considerations. From the area under the mean insulin curve in nondiabetic subjects it is found that the average insulin concentration during the 2 hour period following glucose administration was 117 μU per ml. It has been shown previously (22) that I^{131}-labeled insulin in man is metabolized at a rate of about 2 per cent per minute and is distributed into an apparent volume of distribution of about 37 per cent of body weight in about 45 minutes, distribution being about half completed at 15 minutes. If we now assume that the distribution and metabolism of endogenous insulin that reaches the peripheral circulation is similar quantitatively to that of exogenous insulin [9] and make the conservative estimate that, on the average, the endogenous insulin was distributed in a volume corresponding to 30 per cent of body weight (21 L) over the 2 hour period, we can calculate that 0.117 U per L $\times$ 21 L was being degraded at the rate of 2 per cent per minute during these 120 minutes. This computation leads to the estimate that approximately 6 U of insulin reached the peripheral circulation during the 2 hour glucose tolerance test. Madison, Combes, Unger and Kaplan (28) have found that approximately 50 per cent of insulin given into the portal vein is removed from the circulation during its first passage through the liver, and this value is in good agreement with the figure of 40 per cent given by Mortimore and Tietze in the rat (29). If we accept the 50 per cent value for the liver of man, it can be concluded that an average of about 12 U of insulin was secreted during the glucose tolerance test in nondiabetic subjects. This is what might

[9] It has been established that exogenous unlabeled crystalline beef insulin and I^{131}-labeled crystalline beef insulin show virtually identical plasma disappearance curves in the rabbit (10). Furthermore, the precipitous fall in insulin concentration from peak levels, observed in many patients of the present study (Table IIIA), even when insulin secretion may be presumed to be continuing, is evidence that endogenous insulin also is rapidly removed from the circulation.

be expected in each of three feedings per day. If, also, there is added into the calculation (on the same basis) the amount of insulin necessary to maintain a fasting level of 0.021 U per L for the other 18 hours, we arrive at the estimate of 36 U (postprandial secretion) plus 19 U (fasting secretion) = 55 U for the average total insulin secretion per day in nondiabetic subjects.[10] Since, even at the end of the 2 hour glucose tolerance test the insulin concentration was still above fasting levels, calculations over a more extended time period would yield even slightly higher estimates. These figures are certainly consistent with the insulin requirement of 30 to 40 U daily in totally depancreatized human subjects (40), since exogenously administered insulin does not experience the initial hepatic removal to which endogenously secreted insulin is subjected.

In order to resolve the much higher estimates of plasma insulin concentration given by Willebrands, van der Geld and Groen (34), Randle (35), and Pfeiffer and associates (36) with these considerations, we must assume either that the turnover rate of endogenous insulin is very much slower than that of exogenously administered insulin (in which event it is difficult to understand why human subjects do not remain in prolonged jeopardy of hypoglycemia from the high insulin concentrations that follow glucose administration) or that endogenous insulin is confined almost exclusively to the plasma. Even if the latter alternative (which also is in strong conflict with the results on distribution of exogenous insulin) were true, a fasting level as high as 4,600 μU per ml (34) would mean that there is almost 14 U of insulin in the circulation of fasting human subjects, a conclusion which is still difficult to accept. Randle's (35) values of 9,000 to 22,000 μU per ml in normal plasma 2.5 hours following glucose would mean a total of 27 to 66 U in plasma alone, neglecting insulin in extravascular space. at a time when the blood sugar is usually at a normal level. However, as already noted, Randle has indicated his conviction that this "insulin-like" activity is not due entirely to insulin alone (5).

To return now to the results of the present study, it will be noted that the high insulin con-

[10] These calculations ignore any increase in insulin secretion that would result from small feedings between meals.

centrations observed in diabetic subjects during the glucose tolerance test are not inconsistent with the less extensive data of Bornstein and Lawrence (1) and Baird and Bornstein (39). Very recently Seltzer and Smith (41), employing the rat diaphragm assay of Vallance-Owen and Hurlock (33), have reported insulin concentrations one hour after glucose, in tolbutamide-sensitive adult diabetics, almost in the normal range, but significantly lower values were observed in juvenile diabetics and adult tolbutamide-insensitive diabetics. To resolve the present finding of a higher than normal integrated insulin output in diabetics during the glucose tolerance test with sustained hyperglycemia in these patients, it must be concluded that the tissues of the maturity-onset diabetic do not respond to his insulin as well as the tissues of the nondiabetic subject respond to his insulin. However, from these observations it cannot be concluded that the early diabetic has the same maximal potential insulin output as the nondiabetic, since in the latter the return of blood sugar to normal levels does not allow for the continued stimulus of prolonged hyperglycemia as in the diabetic. The attempt to produce a sustained stimulus to insulin secretion by repeated administration of glucose to a total of 250 g did result in a more marked insulin secretion in nondiabetic subjects. However, the response of diabetics was still greater indicating that their insulin reserve is not depleted during the 100 g glucose tolerance test. The experiments failed, however, to test maximal insulin secretory capacity of the nondiabetic subjects since a sustained hyperglycemia was not achieved in these patients.

Appreciation of the lack of responsiveness of blood sugar, in the face of apparently adequate amounts of insulin secreted by early maturity-onset diabetic subjects, is obviously of importance in the interpretation of the pathogenesis of this type of diabetes. However, the data at hand can only indicate that absolute insulin deficiency per se is not the cause of the hyperglycemia and suggest other possibilities that merit investigation, namely, 1) abnormal tissues with a high threshold for the action of insulin; 2) an abnormal insulin that acts poorly with respect to hormonal activity *in vivo* but reacts well immunologically *in vitro*; 3) an abnormally rapid inactivation of hormonally active sites [a suggestion in accord with the ideas expressed by Mirsky (42)] but not of immunologically active sites on the insulin molecules; and 4) the presence of insulin antagonists. The last suggestion has been made many times by previous workers. A joint attack on the problem, utilizing both the specific immunoassay for plasma insulin and an assay method that measures the net biological effect of insulin and its inhibitors would seem to be indicated.

The high fasting insulin concentrations observed in hypoglycemia associated with functioning islet cell tumors are not unexpected. However, the normal response to glucose in the one patient studied suggests that the insulin-producing tumor may be secreting insulin continuously or sporadically but that it is not stimulated specifically by hyperglycemia. The failure to detect high plasma concentrations of insulin in two cases can possibly be explained by the normally rapid turnover of insulin and the sampling at a time when insulin production by the tumor had been quiescent for an hour or two previously.

Leucine-induced hypoglycemia in children with idiopathic hypoglycemia was first reported by Cochrane, Payne, Simpkiss and Woolf (43) but a satisfactory interpretation of the disturbance has not been given. From the results of the present study it appears that leucine serves as an abnormal stimulus to insulin secretion in these subjects but may also have other effects. Most of the patients whose sera were assayed here have been studied in detail in other respects as well by the various investigators who supplied the sera and are to be reported by them individually.

SUMMARY AND CONCLUSION

1. An immunoassay for plasma insulin in man is presented, based on the reaction of human insulin, competing with beef insulin-I^{131}, with insulin-binding antibodies in the sera of guinea pigs immunized with beef insulin. The method is sensitive to less than 1 μU of insulin, permitting measurement of insulin concentrations in 10 to 20 μl of plasma.

2. Human insulin added *in vitro* to plasma is recovered quantitatively, and measured endogenous insulin concentrations decrease proportionately on dilution of plasma over the range 1:2 to 1:100.

3. Endogenous plasma insulin is destroyed by incubation with cysteine and endogenous insulin adsorption by cellulose columns is quantitatively similar to the adsorption of added beef insulin-I^{131}.

4. Repeat determinations of insulin concentrations on the same plasma samples (stored frozen in the interim) one or more months apart, with different lots of insulin-I^{131}, were generally in good agreement.

5. Fasting plasma insulin concentrations in early maturity-onset diabetic patients who had never been treated with insulin, (mean, 27 μU per ml) and in nondiabetic subjects (mean, 21 μU per ml) did not differ markedly. Following 100 g of glucose by mouth, nondiabetic subjects usually showed peak insulin concentrations at 0.5 hour (mean, 143 μU per ml) or 1 hour (mean, 139 μU per ml) and a decline by 2 hours (mean, 106 μU per ml). In contrast, insulin concentrations in diabetic subjects showed a lesser increase at 0.5 hour (mean, 97 μU per ml) but continued to rise to a peak at 2 hours (mean, 243 μU per ml). The integrated average insulin concentration during the 2 hour glucose tolerance test was 26 per cent higher in diabetics (mean, 147 μU per ml) than in nondiabetics (mean, 117 μU per ml).

6. In a small series of patients subjected to additional glucose loading at 1.5, 2 and 2.5 hours, very high insulin concentrations were observed in both groups, but levels in diabetic patients far exceeded those in nondiabetic subjects.

7. Fasting insulin concentrations were elevated in five of seven subjects with functioning islet cell adenomas but insulin secretory response to glucose was normal in the one patient studied.

8. Four of six subjects with leucine-sensitive hypoglycemia showed increased insulin concentrations following administration of leucine in six of nine experiments.

9. Insulin responses were generally in the high normal range in three patients with pituitary tumors (two associated with acromegaly) and in two patients with thyrotoxicosis.

10. Plasma insulin concentrations measured by imunoassay are compared with values obtained by other assay methods and found to yield the lowest estimates.

11. Calculation of the average normal daily insulin secretion rate, on the basis of the data pre-

sented, yields an estimate of about 55 U of insulin per day.

ACKNOWLEDGMENTS

We are indebted to the investigators mentioned, who sent us sera on the unusual cases reported here. We also wish to thank Mr. Manuel Villazon for technical assistance, Mr. Paul Newman for the charts, Mr. David Lubin, Mr. Glenn Harahan and Mr. Lawrence Steur for the figures, and Miss Eve Spelke and Mrs. Frieda Steiner for secretarial assistance. Finally, the cooperation of the Medical and Laboratory Services of the Bronx Veterans Administration Hospital is gratefully acknowledged.

REFERENCES

1. Bornstein, J., and Lawrence, R. D. Plasma insulin in human diabetes mellitus. Brit. med. J. 1951, 2, 1541.
2. Groen, J., Kamminga, C. E., Willebrands, A. F., and Blickman, J. R. Evidence for the presence of insulin in blood serum. A method for an approximate determination of the insulin content of blood. J. clin. Invest. 1952, 31, 97.
3. Martin, D. B., Renold, A. E., and Dagenais, Y. M. An assay for insulin-like activity using rat adipose tissue. Lancet 1958, 2, 76.
4. Baird, C. W., and Bornstein, J. Plasma-insulin and insulin resistance. Lancet 1957, 1, 1111.
5. Randle, P. J. Insulin in blood. Ciba Found. Coll. Endocr. 1957, vol. XI, p. 115.
6. Leonards, J. R. Insulin-like activity of blood. What it is. Fed. Proc. 1959, 18, 272.
7. Berson, S. A., and Yalow, R. S. Cross reactions of human anti-beef pork insulin with beef, pork, sheep, horse and human insulins. Fed. Proc. 1959, 18, 11.
8. Berson, S. A., and Yalow, R. S. Species-specificity of human anti-beef, pork insulin serum. J. clin. Invest. 1959, 38, 2017.
9. Berson, S. A. In Résumé of Conference on Insulin Activity in Blood and Tissue Fluids, R. Levine and E. Anderson, Eds. Bethesda, Md. 1957, p. 7.
10. Berson, S. A., and Yalow, R. S. Isotopic tracers in the study of diabetes in Advances in Biological and Medical Physics, J. H. Lawrence and C. A. Tobias, Eds. New York, Academic Press Inc., 1958, vol. VI, p. 349.
11. Berson, S. A., and Yalow, R. S. Recent studies on insulin-binding antibodies. Ann. N. Y. Acad. Sci. 1959, 82, 338.
12. Berson, S. A., and Yalow, R. S. Immunoassay of insulin in Hormones in Human Plasma, H. N. Antoniades, Ed. Boston, Little, Brown & Co. In press.
13. Berson, S. A., and Yallow, R. S. Immunologic reactions to insulin in Diabetes, R. H. Williams, Ed. New York, Paul B. Hoeber, Inc. 1960, p. 272.

14. Yalow, R. S., and Berson, S. A. Assay of plasma insulin in human subjects by immunological methods. Nature (Lond.) 1959, 184, 1648.
15. Yalow, R. S., and Berson, S. A. Plasma insulin concentrations in non-diabetic and early diabetic subjects determined by a new sensitive immunoassay technique. Diabetes. In press.
16. Bauman, A., Rothschild, M. A., Yalow, R. S., and Berson, S. A. Distribution and metabolism of I^{131} labeled human serum albumin in congestive heart failure with and without proteinuria. J. clin. Invest. 1955, 34, 1359.
17. Pressman, D., and Eisen, H. N. The zone of localization of antibodies. V. An attempt to saturate antibody-binding sites in mouse kidney. J. Immunol. 1950, 64, 273.
18. Yalow, R. S., and Berson, S. A. Effect of x-rays on trace-labeled I^{131}-insulin and its relevance to biologic studies with I^{131}-labeled proteins. Radiology 1956, 66, 106.
19. Berson, S. A., and Yalow, R. S. Radiochemical and radiobiological alterations of I^{131}-labeled proteins in solution. Ann. N. Y. Acad. Sci. 1957, 70, 56.
20. Ferrebee, J. W., Johnson, B. B., Mithoefer, J. C., and Gardella, J. W. Insulin and adrenocorticotropin labeled with radio-iodine. Endocrinology 1951, 48, 277.
21. Newerly, K., and Berson, S. A. Lack of specificity of insulin-I^{131} binding by isolated rat diaphragm. Proc. Soc. exp. Biol. (N. Y.) 1957, 94, 751.
22. Berson, S. A., Yalow, R. S., Bauman, A., Rothschild, M. A., and Newerly, K. Insulin-I^{131} metabolism in human subjects: Demonstration of insulin binding globulin in the circulation of insulin-treated subjects. J. clin. Invest. 1956, 35, 170.
23. Berson, S. A., and Yalow, R. S. Quantitative aspects of the reaction between insulin and insulin-binding antibody: Relation to problem of insulin resistance. J. clin. Invest. 1959, 38, 1996.
24. Field, J. B., Tietze, F., and Stetten, D., Jr. Further characterization of an insulin antagonist in the serum of patients in diabetic acidosis. J. clin. Invest. 1957, 36, 1588.
25. Fisher, A. M. Personal communication.
26. Fajans, S. S., and Conn, J. W. The early recognition of diabetes mellitus. Ann. N. Y. Acad. Sci. 1959, 82, 208.
27. Somogyi, M. Determination of blood sugar. J. biol. Chem. 1945, 160, 69.
28. Madison, L. L., Combes, B., Unger, R. H., and Kaplan, N. The relationship between the mechanism of action of the sulfonylureas and the secretion of insulin into the portal circulation. Ann. N. Y. Acad. Sci. 1959, 74, 548.
29. Mortimore, G. E., and Tietze, F. Studies on the mechanism of capture and degradation of insulin-I^{131} by the cyclically perfused rat liver. Ann. N. Y. Acad. Sci. 1959, 82, 329.
30. Grodsky, G., and Forsham, P. An immunochemical assay of total extractable insulin in man. J. clin. Invest. 1960, 39, 000.
31. Arquilla, E. R., and Stavitsky, A. B. The production and identification of antibodies to insulin and their use in assaying insulin. J. clin. Invest. 1956, 35, 458.
32. Loveless, M. H. A means of estimating circulating insulin in man. Quart. Rev. Allergy 1956, 10, 374.
33. Vallance-Owen, J., and Hurlock, B. Estimation of plasma-insulin by the rat diaphragm method. Lancet 1954, 1, 68.
34. Willebrands, A. F., v. d. Geld, H., and Groen, J. Determination of serum insulin using the isolated rat diaphragm. The effect of serum dilution. Diabetes 1958, 7, 119.
35. Randle, P. J. Assay of plasma insulin activity by the rat-diaphragm method. Brit. med. J. 1954, 1, 1237.
36. Pfeiffer, E. F., Pfeiffer, M., Ditschuneit, H., and Ahn, C. Clinical and experimental studies of insulin secretion following tolbutamide and metahexamide administration. Ann. N. Y. Acad. Sci. 1959, 82, 479.
37. Yalow, R. S., Black, H., Villazon, M., and Berson, S. A. Comparison of plasma insulin levels following administration of tolbutamide and glucose. Diabetes. In press.
38. Anderson, E., Lindner, E., and Sutton, V. A sensitive method for the assay of insulin in blood. Amer. J. Physiol. 1947, 149, 350.
39. Baird, C. W., and Bornstein, J. Assay of insulin-like activity in the plasma of normal and diabetic human subjects. J. Endocr. 1959, 19, 74.
40. Goldner, M. G., and Clark, D. E. The insulin requirement of man after total pancreatectomy. J. clin. Endocr. 1944, 4, 194.
41. Seltzer, H. S., and Smith, W. L. Plasma insulin activity after glucose. An index of insulogenic reserve in normal and diabetic man. Diabetes 1959, 8, 417.
42. Mirsky, I. A. The etiology of diabetes mellitus in man. Recent Progr. Hormone Res. 1952, 7, 437.
43. Cochrane, W. A., Payne, W. W., Simpkiss, M. J., and Woolf, L. I. Familial hypoglycemia precipitated by amino acids. J. clin. Invest. 1956, 35, 411.

COMMENTARY TO

2. Dandliker, W. B. and Feigen, G. A. (1961)
Quantification of the Antigen–Antibody Reaction by the Polarization of Fluorescence. Biochemical and Biophysical Research Communications 5(4): 299–304.

Fluorescence polarization (FP) was discovered in the 1920's; was used to monitor an antigen antibody reaction for the first time in 1961; commercialized in an automated immunoassay analyzer in 1981 and within 20 years was installed in over 70 000 clinical analyzers worldwide.

FP is based on the observation that low molecular weight fluorescent molecules like rhodamine and fluorescein freely tumble and rotate in solution. When excited by plane-polarized light they emit fluorescent light with decreased polarization due to the free movement of the fluorophore during excitation. If a small molecule like fluorescein binds to a large molecule like a protein its free rotational movement is slowed down. When excited in this bound state its emitted fluorescent light remains polarized in the same plane as the excitation light. The degree of polarization of the emitted light is directly related to the amount of fluorophore that has bound. Laurence in 1952 used FP to study the binding of various fluorescent dyes to bovine serum albumin (1). Steiner measured the binding of a soybean inhibitor-fluorescein conjugate to the enzyme trypsin in 1954 (2).

Dandliker and Feigen were the first to use FP to measure antigen antibody binding. Their paper was presented at the 45th Annual Meeting of the Federation of American Societies for Experimental Biology and published as an abstract in 1961 (3). A full paper was published that same year and is presented here. The binding of fluorescein conjugated ovalbumin to rabbit anti-ovalbumin antibody was monitored. This represents the first description of a true homogeneous assay in which the antigen–antibody event is measured directly in real time. Haber and Bennett extended these observations to include the binding of insulin, ribonuclease and bovine serum albumin to their respective antibodies (4). Over the next twelve years in numerous papers Dandliker expanded on the theoretical basis of FP, coined the term fluorescence polarization immunoassay (FPIA) and extended the applications of the technology (5–7). In 1973 Spencer *et al.* (8) described the construction of an FPIA analyzer and applied it to assays for antitrypsin enzyme–inhibitor and insulin–insulin antibody. Researchers at Abbott Laboratories under Michael Jolley and David Kelso described the development of a fully automated FPIA analyzer with applications for the therapeutic drug monitoring of aminoglycoside antibiotics and the anticonvulstants, phenytoin and phenobarbital (9–11). The Abbott TDx™ FPIA analyzer was introduced in 1981. It went on to become one of the most successful immunoassay analyzers in clinical chemistry.

References

(1) Laurence, D.J.R. (1952) A study of the adsorption of dyes on bovine serum albumin by the method of polarization of fluorescence. Biochemical Journal. 51(2):168–177.

(2) Steiner, R.F. (1954) Reversible association processes of globular proteins VI. The combination of trypsin with soybean inhibitor. Archives of Biochemistry and Biophysics. 49(1):71–92.

(3) Dandliker, W.B. and Felgen, G.A. (1961) Detection of the antigen–antibody reaction by fluorescence polarization. Federation Proceedings. 20(1 Part 1):11, Abstracts.

(4) Haber, E. and Bennett, J.C. (1962) Polarization of fluorescence as a measure of antigen–antibody interaction. Proceedings of the National Academy of Sciences. 48(11):1935–1942.

(5) Dandliker, W.B., Schapiro, H.C., Meduski, J.W., Alonso, R., Feigen, G.A., and Hamrick, J.R. Jr. (1964) Application of fluorescence polarization to the antigen–antibody reaction. Theory and experimental method. Immunochemistry. 1(3): 165–191.

(6) Dandliker, W.B. and de Saussure, V.A. (1970) Review article: fluorescence polarization in immunochemistry. Immunochemistry. 7(9):799–828.

(7) Dandliker, W.B., Kelly, R.J., Dandliker, J., Farquhar, J., and Levin, J. (1973) Fluorescence polarization immunoassay. Theory and experimental method. Immunochemistry. 10(4):219–227.

(8) Spencer, R.D., Toledo, F.B., Williams, B.T., and Yoss, N.L. (1973) Design, construction, and two applications for an automated flow-cell polarization fluorometer with digital read out: enzyme–inhibitor (antitrypsin) assay and antigen–antibody (insulin–insulin antiserum) assay. Clinical Chemistry. 19(8):838–844.

(9) Jolley, M.E., Stroup, S.D., Wang, C-H.J., Panas, H.N., Keegan, C.L., Schmidt, R.L., and Schwenzer, K.S. (1981) Fluorescence polarization immunoassay I. Monitoring aminoglycoside antibiotics in serum and plasma. Clinical Chemistry. 27(7):1190–1197.

(10) Popelka, S.R., Miller, D.M., Holen, J.T., and Kelso, D.M. (1981) Fluorescence polarization immunoassay II. Analyzer for rapid, precise measurement of fluorescence polarization with use of disposable cuvettes. Clinical Chemistry. 27(7): 1198–1201.

(11) Jolley, M.E., Stroupe, S.D., Schwenzer, K.S., Wang, C.J., Lu-Steffes, M., Hill, H.D., Popelka, S.R., Holen, J.T., and Kelso, D.M. (1981) Fluorescence polarization immunoassay. III. An automated system for therapeutic drug determination. Clinical Chemistry. 27(9):1575–1579.

Biochemical and Biophysical Research Communications. 1961, 5(4): 299–304,
Copyright 1961 Reprinted with permission from Elsevier.

QUANTIFICATION OF THE ANTIGEN-ANTIBODY REACTION BY THE POLARIZATION OF FLUORESCENCE*

W. B. Dandliker ** and G. A. Feigen

Department of Biochemistry, University of Miami, Coral Gables, Florida
and Department of Physiology, Stanford University, Stanford, California.

Received June 13, 1961

The basic theory concerning the polarization of fluorescence was developed in a series of important papers by Perrin (1926). Perrin's results have been used experimentally by Singleterry and co-workers (1951) and greatly extended, both theoretically and experimentally, by Weber (1952). Subsequent applications by Laurence (1952) and Steiner (1957) follow implicitly from the work of Weber.

The concept underlying previous work by Dandliker and Feigen (1961) and the present results is to utilize the change in rotary diffusion constant which occurs when an antigen and antibody combine in solution. An essential feature of the method is that either the antigen or antibody is made fluorescent depending upon which component is to be detected. This feature makes it possible to follow the reaction in spectral regions where adventitious fluorescence is of minor importance and also permits a choice of fluorescence lifetime, appropriate to the range of molecular size involved; it is thus distinct from other fluorescence techniques in use, cf. Coons, et al. (1941), Boroff and Fitzgerald (1958) and Velick, et al. (1960).

EXPERIMENTAL. Crystalline ovalbumin was labeled with fluorescein using fluorescein isothiocyanate (Riggs, et al. 1958); the product contained between

---

*Supported by Grant A-2984 and Grant H-3693(C2), U.S. Public Health Service.
**Howard Hughes Medical Institute.

Vol. 5, No. 4, 1961   BIOCHEMICAL AND BIOPHYSICAL RESEARCH COMMUNICATIONS

one and two fluorescein molecules per molecule of protein assuming the molar extinction coefficient of free and bound fluorescein to be the same. Eight albino rabbits, weighing about six pounds each, were immunized to fluorescein-labeled ovalbumin (F-ovalbumin) by a series of twenty intravenous injections of 10 mg. each administered on alternate days. One week after the last injection, blood was drawn by cardiac puncture and, after clotting, the serum was separated by centrifugation. A γ-globulin fraction from pooled serum was prepared by two successive precipitations in one-third saturated ammonium sulfate. For control purposes, γ -globulin was also prepared from normal animals. The immune globulin preparation contained 19% specifically precipitable anti-F-ovalbumin as estimated at optimal proportions by means of the quantitative precipitin method; this antibody also reacted strongly with native ovalbumin.

Measured volumes of antigen solution were added from a microburette to a constant quantity of antibody contained in a cuvette. The intensity and polarization of fluorescence were measured in a modified Brice-Phoenix apparatus using the unpolarized 4358 Å mercury line for excitation.

RESULTS AND DISCUSSION. The reaction between F-ovalbumin and its antibody produces two fluorescence effects. First, there is a pronounced diminution of the fluorescence due, no doubt, to the close juxtaposition with many atomic groupings of the antibody, thus establishing favorable conditions for loss of the electronic excitation energy before fluorescence takes place; possibly a transition to the triplet state is involved. The second effect is the change in polarization of fluorescence caused by the increase in relaxation time.

For quantitative pupodes, it is convenient to define the polarization (p) and a parameter (Q) which is proportional to fluorescence intensity divided by incident intensity. If the vertical and horizontal components in the fluorescent light are denoted by V and H respectively, and if M_F denotes the molar concentration of fluorescent antigen, then $p = \dfrac{V - H}{V + H}$

and $Q = \dfrac{V + H}{M_F}$.

In a typical experiment (Figure 1), 4 ml. of antibody solution was mixed with portions of F-ovalbumin to give final concentrations ranging upward from about 5×10^{-9}M.

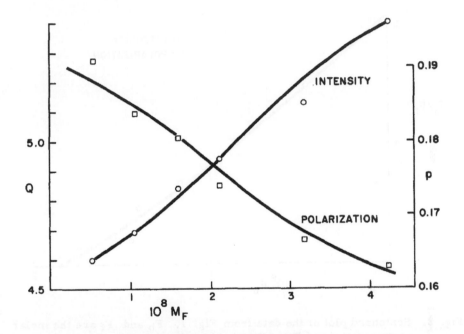

Fig. 1. Q and p as a function of M_F(see text) for a solution containing 7 μg. of precipitable anti-F-ovalbumin / ml. The curves were calculated from the constants derived from Figure 2.

If it is assumed that the bound form of the antigen may be characterized by a limiting value for Q and p equal to k_b and p_b respectively, and if the free form of the antigen is similarly characterized by k_f and p_f, then a simple mass law analysis of the data may be made according to the reaction

$F + Ab \rightleftharpoons FAb$. The association constant, $K = \dfrac{(FAb)}{(F) \, (Ab)}$

For this analysis, all concentrations are measured in terms of the molarity of antigen. With these assumptions, it is possible to express results

30

(Figure 2) as suggested by Scatchard (1949). The data for both intensity and polarization are concordant and indicate some spread in the association constants. In principle, the curvature of the Scatchard plot should be capable of yielding information concerning the range of association constants, but the analysis requires very accurate data.

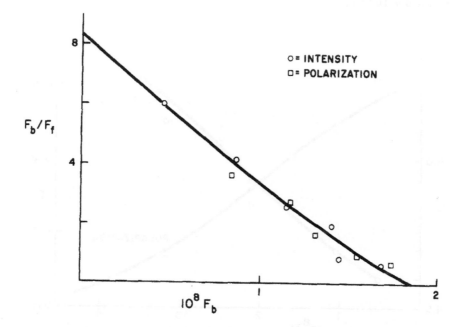

Fig. 2. Scatchard plot of the data from Fig. 1. F_b and F_f are the molar concentrations of F-ovalbumin, bound and free respectively. The association constant, $K = 4.3 \times 10^8$ and the maximum value of F_b is 1.9×10^{-8}.

For the method described here to have general applicability, the fluorescent antigen or antibody must be capable of reacting in solution with antibody to the native antigen or with the native antigen itself, respectively. Figure 3 gives the quantitative results for the system F-ovalbumin and antibody to native ovalbumin (anti-ovalbumin). There are distinct and important differences between the behavior of this system and that shown in Figure1. First of all, there is little or no variation of Q, i.e. no quenching

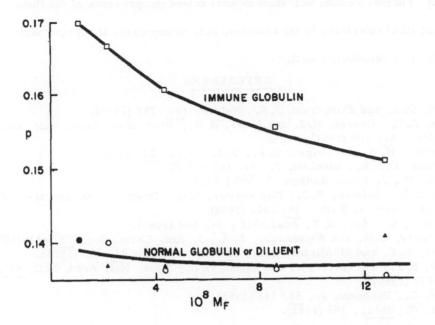

Fig. 3. Polarization as a function of M_F. F-ovalbumin was added to anti-
ovalbumin (squares) or to normal Y -globulin (circles) of
the same protein concentration or to diluent, i. e. 0.15M sodium
chloride, 0.01M disodium hydrogen phosphate and 0.005M sodium dihydrogen phosphate (triangles).

or enhancement of fluorescence. Physically, this behavior has a simple

and straightforward interpretation because the antibody in this case has no

site for the specific binding of fluorescein. The polarization effects are simi-

lar to those in Figure 1. No detectable interaction at these concentrations

between normal globulin and antigen was found, the results being the same

within experimental error as those obtained by addition to buffered saline.

SUMMARY. We have concluded that it is possible to determine by measure-

ments of fluorescence polarization two important parameters, namely the

equilibrium constant and the combining capacity characterizing the anti-

gen-antibody reaction. The combining capacity is proportional to the number

of antibody sites in a preparation and the equilibrium constant, together per-

haps with certain kinetic quantities, constitutes a quantitative measure of

Vol. 5, No. 4, 1961   BIOCHEMICAL AND BIOPHYSICAL RESEARCH COMMUNICATIONS

avidity. Further results will show to what extent the presence of the fluo-

rescent label interferes in the reaction, but, in any case, it appears that

the effect is relatively small.

REFERENCES

Boroff, D. A. and Fitzgerald, J. E., Nature, 181, 751 (1958).

Coons, A. H., Creech, H. J. and Jones, R. N., Proc. Soc. Exptl. Biol. and
    Med., 47, 200 (1941).

Dandliker, W. B. and Feigen, G. A., Fed. Proc., 20, 11 (1961).

Laurence, D. J. R., Biochem J., 51, 168 (1952).

Perrin, F., J. Phys. Radium, 7, 390 (1926).

Riggs, J. L., Seiwald, R. J., Burckhalter, J. H., Downs, C. M. and Metcalf,
    T. G., Am. J. Path., 34, 1081 (1958).

Scatchard, G., Ann. N. Y. Acad. Sci., 51, 660 (1949).

Singleterry, C. R. and Weinberger, L. A., J. Am. Chem. Soc., 73, 4574 (1951).

Steiner, R. F. and McAlister, A. J., J. Polymer Sci., 24, 105 (1957).

Velick, S. F., Parker, C. W. and Eisen, H. N., Proc. Nat. Acad. Sci., 46,
    1470 (1960).

Weber, G., Biochem. J., 51, 145 (1952).

Weber, G., ibid., 155 (1952).

COMMENTARY TO

3. Murphy, B. E. P. and Pattee, C. J. (1964)
Determination of Thyroxine Utilizing the Property of Protein-Binding.
Journal of Clinical Endocrinology and Metabolism 24:187–196.

R oger Ekins in 1960 was the first to describe an assay for thyroxine (T4) based on the binding between a radio labeled T4 ligand and thyroid binding globulin (TBG). Assays that used natural endogenous binding proteins instead of antibodies as the binding agent were termed competitive protein-binding (CPB) assays to distinguish them from immunoassays. Ekins incubated ^{131}I thyroxine with human serum and measured its binding to endogenous TBG in the serum. Free and bound radio ligand were separated using paper electrophoresis. The patient's thyroxine levels were inferred from the bound and free data (1). Barakat and Ekins employed a similar method using ^{57}Co-vitamin B12 and intrinsic factor binding protein to measure endogenous vitamin B12. Free and bound ligand were separated using dialysis (2). In 1962, Rothenberg also described an assay for vitamin B12 but he used protein precipitation to separate free from bound tracer (3).

In 1963, Beverley E. P. Murphy at McGill University measured plasma corticoids using corticosteroid-binding globulin as the binder and ^{14}C-cortisol as the ligand. Dialysis was used for separation of free from bound (4). Later, Sephadex™ G25 gel filtration columns were used for the separation step (5). The paper presented here is the application of a CPB assay for total T4 in human serum. Murphy was an MD and the work on cortisol and T4 by CPB were part of her PhD thesis. The simplicity of the assay as presented in this T4 paper resulted in its being quickly accepted in the clinical laboratory. She further improved the assay with the use of anion exchange resin beads for the separation of free from bound (6). Within a few years of Murphy's original paper, commercial assays for T4 were widely available in test kit form (7).

RIA eventually replaced the CPB method for the measurement of T4 and other hormones in serum (8). Murphy's T4 paper, however, is a watershed in the development of clinical laboratory methods. Her procedure was a direct measure of circulating T4 that was free of interference from radio opaque iodine X-ray dyes and other sources of endogenous free iodine. Because of this, the protein bound iodine (PBI) assay was soon replaced. In addition, the demand for this T4 assay stimulated the introduction of commercial low cost gamma counters and this helped facilitate the rapid adoption of RIA assays for a wide variety of analytes other than T4.

References

(1) Ekins, R.P. (1960) The estimation of thyroxine in human plasma by an electrophoretic technique. Clinica Chimica Acta. 5(4):453–459.

(2) Barakat, R.M. and Ekins, R.P. (1961) Assay of vitamin B12 in blood. A simple method. The Lancet. 278(7192):25–26.

(3) Rothenberg, S.P. (1961) Assay of serum vitamin B12 concentration using Co^{57}-B$_{12}$ and intrinsic factor. Proceedings of the Society for Experimental Biology and Medicine. 108(1):45–48.

(4) Murphy, B.P., Engelberg, W., and Pattee, C.J. (1963) Simple method for the determination of plasma corticoids. Journal of Clinical Endocrinology and Metabolism. 23(3):293–300.

(5) Murphy, B.E. and Pattee, C.J. (1964) Determination of plasma corticoids by competitive protein-binding analysis using gel filtration. Journal of Clinical Endocrinology and Metabolism. 24(9):919–923.

(6) Murphy, B.P. and Jachan, C. (1965) The determination of thyroxine by competitive protein-binding analysis employing an anion-exchange resin and radiothyroxine. Journal of Laboratory and Clinical Medicine. 66(1):161–167.

(7) Godwin, I.D. and Swoope, H.B. (1968) Comparison of a T$_4$ resin sponge uptake method with a protein-bound iodine procedure. American Journal of Clinical Pathology. 50(2):194–197.

(8) Chopra, I.J. (1972) A radioimmunoassay for measurement of thyroxine in unextracted serum. Journal of Clinical Endocrinology and Metabolism. 34(6):938–947.

J. Clin. Endo. Metab. 1964; 24: 187–196

Determination of Thyroxine Utilizing the Property of Protein-Binding

BEVERLEY E. PEARSON MURPHY, M.D.,[1]
AND CHAUNCEY J. PATTEE, M.D.,
with the technical assistance of Sorel Cohen

Clinical Investigation Unit, Queen Mary Veterans' Hospital, and Department of Investigative Medicine McGill University, Montreal, Canada

ABSTRACT. A simple, rapid method for the determination of serum thyroxine has been described, which is based on the specific binding properties of thyroxine-binding globulin (TBG). One-half ml of the test sample is deproteinized and the thyroxine thus freed is measured according to its competition with a fixed amount of thyroxine-I^{131} for a fixed amount of TBG. The method is highly specific for thyroxine, and is unaffected by iodine or mercury contamination. The mean value expressed as thyroxine iodine for euthyroid individuals was 6.6 ± 1.3 $\mu g/100$ ml, with a range of 4.0–9.2, corresponding to a mean thyroxine level of 10.1 $\mu g/100$ ml (range 6.1–13.8). A good correlation was obtained with the clinical status of patients studied. (*J Clin Endocr* **24**: 187, 1964)

THE RELATIONSHIP between thyroxine and thyroxine-binding globulin (TBG) is in many respects similar to that between cortisol and corticosteroid-binding globulin (CBG). When methods based on the principle of protein-bound isotope competition were developed by the authors for cortisol and other steroids in plasma [(1) and observations in this laboratory], this similarity prompted the investigation of the application of the same principle to the determination of plasma thyroxine. The general aspects of this principle for the determination of minute quantities of substances in blood are discussed at length elsewhere (observations in this laboratory).

Although thyroxine constitutes the principal product of thyroid biosynthesis and the major circulating thyroid hormone, no simple method of assay has been hitherto available for it. The only direct estimates of thyroxine have been made using the double isotope derivative technique (2), and all the clinically feasible methods are indirect, being measures of iodine—thus, the familiar terms, PBI (protein-bound iodine), SPI (serum precipitable iodine) and BEI (butanol-extractable iodine). Although column chromatography and butanol extraction are useful in reducing contamination, all of the indirect methods are adversely influenced by the presence of iodine from sources other than thyroxine (3). Since this type of contamination is extremely common, their usefulness is seriously limited.

Briefly, the principle of the method is this. Since there is only a small amount of TBG in plasma, the binding sites can be readily saturated by adding small amounts of thyroxine. If a small amount of thyroxine-I^{131} is added, the fraction which is protein-bound can be determined. As more unlabeled thyroxine is added, the amount of isotope bound decreases, since both labeled and unlabeled forms compete for the same binding sites. If, instead of pure thyroxine, a sample of deproteinized plasma is added, the thy-

Received May 6, 1963; accepted October 10, 1963.
[1] Medical Research Council Fellow, Medical Research Council, Canada.

roxine which it contains may be measured according to the fall in bound isotope which it causes.

Materials and Methods

Subjects. Plasma and serum samples were obtained from patients and healthy laboratory workers in the Queen Mary Veterans' Hospital and, through the kindness of Dr. J. M. McKenzie, Dr. R. Schucher and Dr. A. Gold, from patients in the Royal Victoria, the Jewish General and the Montreal General Hospitals, Montreal, respectively.

Materials. The dextran polymer gel Sephadex G25, medium grade, was obtained from Pharmacia, Sweden (also available from Pharmacia Fine Chemicals, Ltd., Box 1010, Rochester, Minnesota). About 10 g of the gel powder was put into 500 ml of distilled water in a beaker, mixed, and left to stand overnight. After removing the fines (*i.e.*, the fine particles which do not settle readily on shaking), the gel was poured into columns 1.4 cm in diameter to a depth of 2.5 cm. The gel was never allowed to dry out and was used hundreds of times with no change in its binding characteristics.

Thyroxine-I^{131} was obtained from Abbott Laboratories, North Chicago, Illinois, with an initial specific activity of 24–60 mc/mg. After adding 0.5 ml 0.1N NaOH and 1 ml of propylene glycol, the quantity required was diluted to a volume of 100 ml with distilled water to give a concentration of 0.01 μg/ml. Propylene glycol was added to increase the solubility and stability of thyroxine.

Powdered thyroxine (nonradioactive), in vials containing 1 mg, was obtained through the kindness of Dr. A. Gold from Glaxo-Allenburys (Canada) Limited, Weston, Ontario. The contents of 1 vial were dissolved in a few drops of 0.1N NaOH and 1 ml propylene glycol and then diluted to a concentration of 100 μg/ml with distilled water. After adding 0.5 ml 0.1N NaOH and 1 ml propylene glycol, 0.1 ml of this solution was diluted to give 100 ml of a working solution containing 0.1 μg/ml.

Both thyroxine solutions were made up at 2-week intervals on receipt of a new supply of radioactive material.

Barbital buffer, pH 8.6, ionic strength 0.075, was made up in 1 liter quantities daily, or as required.

"Isotope solution," containing thyroxine-I^{131} and "standard" serum, was made up at 2-week intervals, or oftener, as follows: 15 ml of pooled serum, 20 ml of thyroxine-I^{131} solution (0.01 μg/ml), 4 ml propylene glycol and 441 ml of barbital buffer were mixed and allowed to stand for 10 min before using.

All solutions containing thyroxine were kept in the dark and refrigerated as much of the time as possible. Before use, they were allowed to warm to room temperature (23 C). Shaking was avoided since it increases adsorption to glass. All glassware was cleaned ultrasonically.

Monoiodotyrosine, diiodotyrosine, diiodothyronine and triiodothyronine were obtained as dry powders through the courtesy of Dr. J. M. McKenzie, Royal Victoria Hospital, and were prepared for use in a manner similar to that used for thyroxine.

Procedure for the determination of serum thyroxine. The final procedure was as follows: 1 ml of plasma (or serum) was added to 2 ml of 95% ethanol in a small tube, approximately ½"×3", mixed with a fine steel wire, and centrifuged for 5 min at 1500 rpm. Two ½-ml aliquots of the supernatant were evaporated to dryness in similar test tubes with a stream of air in order to remove all the ethanol. One ml of isotope solution was added to each, and each tube was incubated at 45 C for 10 min to ensure solution of the thyroxine and equilibration with the isotope.

After cooling to room temperature, the tube contents were transferred with a Pasteur pipette to a column of Sephadex and allowed to enter the gel. When the level of liquid exactly reached the top of the gel, a counting cuvette was placed underneath the column and the next 5 ml of eluate was collected. During this collection 1 ml of buffer was used to rinse the test tube and was then transferred to the column and allowed to pass into the gel. This was followed with 4 ml of barbital buffer. When the level had again reached the top of the gel, the column stopped automatically. After removing the counting cuvette, 2.0 ml of pooled serum was put into the column, followed by 10–15 ml of barbital buffer.

Thus, the first ml of eluate, the void volume, was discarded. The next 5 ml, constituting the protein-bound fraction, was collected and the remaining eluate was discarded. The cuvette containing the protein-bound fraction was covered and inverted twice to ensure mixing. (Surprisingly large errors occurred if this simple step was omitted.) The sample was then counted to 3000 in a well-type scintillation counter. Three cuvettes, each containing 1 ml of isotope solution plus 4 ml of buffer (similarly inverted to ensure mixing), were counted to 10,000, giving a measure of the total counts added. The percentage isotope bound was then given by the net cpm (*i.e.*, observed cpm minus background) in the protein-bound fraction, divided by the total net cpm added.

A standard curve was determined by evaporating known amounts of nonradioactive thyroxine, *i.e.*, 0, 5, 10, 20 and 30 mμg, to dryness and carrying through the whole procedure as for the ethanol supernatant samples, calculating the percentage isotope bound as before, and plotting it against the amount of thyroxine added as shown in Fig. 1 (dilution 1/32). These were done in quadruplicate and were counted to 10,000.

The amount of thyroxine in the serum samples was determined from the curve according to the percentage bound and multiplied by 10/12.8 to express it in μg/100 ml. (The factor 10/12.8 was determined as described in the section *Recovery of thyroxine-I[131] added to serum from ethanol precipitation*.) The thyroxine iodine was calculated as 65.3% of this value.

Determination of protein-bound iodine. PBI determinations were carried out by the Royal Victoria Hospital, the Hormone Assay Laboratory Ltd., or the Montreal Medical Laboratory, using the standard chloric acid wet digestion procedures (4) which give normal ranges of 4–8 μg/100 ml—uncorrected for recovery.

Results

Preliminary Experiments

Elution of thyroxine by gel filtration. The elution curves for thyroxine-I[131] in barbital buffer at pH 8.6 and in serum were obtained by counting successive 1-ml fractions of eluate. These are shown in Fig. 2, A and B. The void volume (*i.e.*, the volume before any radioactivity was recovered) was 1 ml, and the peak of the protein-bound fraction occurred at 2 ml. Very little unbound thyroxine could be

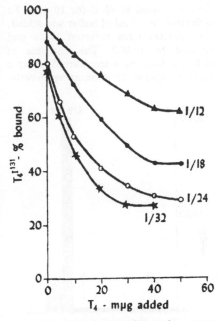

FIG. 1. Standard curves, using various serum dilutions. The percentage bound T_4I^{131} is plotted *vs.* mμg unlabeled T_4 added.

eluted with buffer alone. Therefore, after all the protein-bound fraction was obtained, 1–2 ml of pooled serum was added with a prompt peak at 2 ml after it was added to the column. Deproteinized serum samples to which thyroxine-I[131] had been added showed curves similar to those of thyroxine-I[131] in buffer, indicating that the deproteinization (with respect to thyroxine-binding proteins) was complete.

Redissolving of thyroxine after evaporation to dryness. To each of 5 counting cuvettes and 5 test tubes was added 0.5 ml of water containing an amount of thyroxine-I[131] which gave 4200 cpm. The samples in the test tubes were evaporated to dryness and 1 ml pooled serum diluted 1/32 with barbital buffer was added to each test tube and cuvette. All were incubated at 45 C for 10 min. To each cuvette 3.5 ml of buffer was added, the cuvettes were inverted twice and counted to 10,000. The contents of each test tube were transferred, using a Pasteur pipette, to a counting cuvette.

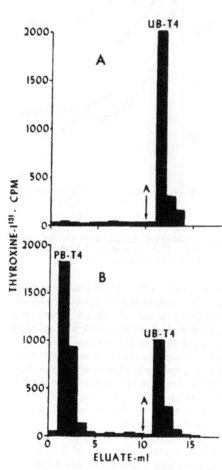

FIG. 2. Elution of thyroxine through Sephadex columns. Thyroxine-I^{131} as cpm plotted *vs.* eluate in ml. Elution was carried out using barbital buffer at pH 8.6 until the point indicated by A, at which time 2.0 ml pooled serum was added. The elution was then continued with barbital buffer. (A) Elution of thyroxine-I^{131} in buffer; (B) elution of thyroxine-I^{131} in serum.

The test tube was rinsed with 1 ml of buffer and this was added with the Pasteur pipette to the cuvette, after which 3.5 ml buffer was added, the cuvettes were inverted twice and counted to 10,000. The recovery of radioactive material transferred from the test tubes was $97.5 \pm$ SD 2.2%. Since the efficiency of transferring the contents of test tubes with Pasteur pipettes was previously determined to be $98.6 \pm$ SD 0.7%, the degree of re-solution of the dried thyroxine-I^{131} was 99.0%.

Effect of temperature on thyroxine-binding. Standard curves were determined after equilibrating the samples at various temperatures and passing them through columns maintained at these temperatures by means of water jackets connected to a thermostatically controlled bath equipped with a recirculation pump. The results, shown in Fig. 3, indicate that temperature is an important factor in binding, the binding being increased at lower temperatures. Thus, it is necessary to carry out the determinations either in a room which has a reasonably constant temperature or using temperature-controlled columns.

Establishment of a suitable standard curve. The character of the standard curve depends primarily on the amount of serum used in making up the isotope solution. Various dilutions of serum are shown in Fig. 1. That generally used was 1/32, *i.e.*, 15 ml serum in 480 ml isotope solution. If low or high values were of special interest, the amount of supernatant was altered to give these values on the steeper part of the curve. Using 0.3 ml of supernatant (*i.e.*, corresponding to approximately 0.1 ml serum), the standard curve for the dilution 1/32 covers a useful range of 0 to 30 μg/100 ml (in this case the value read from the curve is multiplied by 10/7.69 instead of 10/12.8 to give the value in μg); if 0.5 ml supernatant is used, the useful range is 0 to 18 μg/100 ml. A difference in the values obtained for the standard was found if the thyroxine was added in liquid form, isotope solution was added, and buffer was added to make a volume of 1 ml instead of evaporating the thyroxine, then redissolving it in 0.5 ml isotope solution and 0.5 ml buffer. The unevaporated thyroxine standards usually gave higher values for percentage bound than the evaporated thyroxine standards, the reverse effect of that which might be expected on the basis of incomplete solution of stable thyroxine following evaporation.

Recovery of thyroxine under various conditions

Recovery of thyroxine-I^{131} passed through the columns. The sums of the counts of the protein-bound and un-

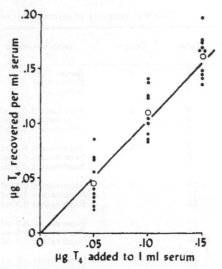

FIG. 4. Recovery of thyroxine added to serum. The mean recoveries are represented by open circles.

bound fractions for duplicate determinations on six different supernatants processed as described above were compared with the counts added. The mean recovery of radioactive material added to the columns was 98.3 ± sd 1.7% with a range of 95.4 to 100.3%.

Recovery of thyroxine-I^{131} added to serum from ethanol precipitation. Thyroxine-I^{131} solution, 0.5 ml, was added to each of 18 test tubes and evaporated to dryness. One ml of various sera was added to each and these were incubated for 10 min at 45 C to ensure solution of the thyroxine. To the first three of these 4 ml of water was added, and they were counted to 10,000. To each of the remaining tubes, 2 ml 95% ethanol was added. After centrifuging for 5 min, 0.5 ml of each supernatant was transferred to another tube, 4.5 ml of water was added, and these were counted to 10,000. The results are shown in Table 1. The mean recovery of thyroxine-I^{131} was

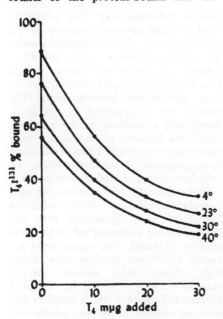

FIG. 3. Effect of temperature on thyroxine-binding.

TABLE 1. Recovery of thyroxine-I^{131} added to serum using ethanol precipitation

Tube	Serum	Sample counted	Net cpm	Net cpm as % added cpm	Mean net cpm as % added cpm
1	1	Serum 1 ml	6120		
2		Serum 1 ml	5910		
3		Serum 1 ml	6010		
4	1	Supernatant 0.5 ml	775	12.9	
5		Supernatant 0.5 ml	745	12.4	12.8
6		Supernatant 0.5 ml	782	13.0	
7	2	Supernatant 0.5 ml	748	12.4	
8		Supernatant 0.5 ml	785	13.0	12.7
9		Supernatant 0.5 ml	770	12.8	
10	3	Supernatant 0.5 ml	745	12.4	
11		Supernatant 0.5 ml	755	12.6	12.8
12		Supernatant 0.5 ml	810	13.5	
13	4	Supernatant 0.5 ml	765	12.7	
14		Supernatant 0.5 ml	800	13.3	12.8
15		Supernatant 0.5 ml	750	12.5	
16	5	Supernatant 0.5 ml	765	12.7	
17		Supernatant 0.5 ml	800	13.3	12.7
18		Supernatant 0.5 ml	735	12.2	

12.8% for 0.5 ml supernatant obtained from the initial volume of 3 ml (*i.e.*, 77%).

Recovery of thyroxine added to serum. Various amounts of unlabeled thyroxine solution were added to test tubes and evaporated to dryness. Pooled serum (1 ml) was added to each and the tubes were incubated for 10 min at 45 C to ensure solution of the thyroxine. The thyroxine content of each sample and of the pooled serum alone was determined by the usual procedure. The amounts of added thyroxine recovered are compared in Fig. 4 with the amounts originally added. The mean recovery for 30 determinations (10 at each of 3 levels) was 103%.

Specificity of the method for thyroxine in serum

To 0.5-ml samples of serum of which the thyroxine content had previously been measured, small amounts of various substances were added and the thyroxine content of the samples was meas-

ured in the usual way. The substances used included monoiodotyrosine (MIT), diiodotyrosine (DIT), diiodothyronine (T_2), triiodothyronine (T_3), potassium iodide (KI), cholesterol, glucose, bilirubin, acetylsalicylic acid, diphenylhydantoin, and several anticoagulants. The results are summarized in Table 2. Significant alterations in the apparent thyroxine content of the serum were found only with the addition of relatively large amounts of triiodothyronine, diiodothyronine and diphenylhydantoin.

Some difficulty was encountered with hemolyzed samples. With moderate hemolysis, good agreement with the corresponding unhemolyzed values was usually obtained, but occasionally severely hemolyzed samples gave falsely high values. Such samples usually showed a red precipitate in the dried ethanol supernatant which failed to dissolve on addition of the isotope solution. Mildly hemolyzed samples showed no such precipitate and gave values which did not differ significantly from those in unhemolyzed samples obtained at the same time. Moderately hemolyzed samples gave correct values if the amount of ethanol used to precipitate the protein in the sample was increased from 1 to 2 ml.

A study of the effects of radiopaque dyes *in vivo* was carried out in seven apparently euthyroid subjects. The type of dye used, and the thyroxine iodine values determined before and after its administration, are indicated in Table 3. All values fell within the normal range and showed no significant trend to higher or lower values.

Precision of the method for serum thyroxine. Duplicate determinations over three ranges, *viz.*, 0–5, 5–10 and 10–15 µg thyroxine/100 ml, were compared, as shown in Table 4, and the standard deviations about the mean for each range were calculated according to the formula $SD = \sqrt{Ed^2/2n}$ (5). For the three ranges these were ±0.73, ±0.87 and ±1.75, respectively.

Correlation of thyroxine iodine with protein-bound iodine (PBI). Samples in which PBI and total iodine determinations had recently been carried out, in which these two determinations did not differ by more than 2 µg/100 ml, were

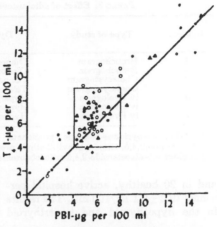

FIG. 5. Correlation of thyroxine iodine (T$_4$I) with protein-bound iodine (PBI). The different laboratories are indicated by different symbols. The square encloses normal values.

contributed by three different laboratories and these values are compared in Fig. 5 with those of thyroxine iodine obtained by the method described above. A good correlation was obtained in many instances, but the values obtained by the present method are generally higher than those obtained for the PBI. Values of the PBI for samples from clinically euthyroid subjects who had recently received iodine or mercury in some form are compared with thyroxine iodine values in Table 5. Normal values were obtained for thyroxine iodine in all these cases. In the two cases of mercury contamination, 2 ml of Mercuhydrin had been administered on the day before the blood was drawn.

Correlation of serum thyroxine values with the clinical status of the patient. Thyroxine values in clinically typical cases of hypothyroidism and hyperthyroidism were compared with values obtained in 20 clinically euthyroid patients with various diagnoses exclusive of hepatic disease (11.0 ± 1.8 µg/100 ml)

TABLE 2. Specificity of TBG for thyroxine

Substance added	Amount added, mg/100 ml	Thyroxine content, µg/100 ml
0	0	7.5 ± 0.5 SD
T$_4$	0.040	11.6
T$_3$	0.040	9.0
DIT	0.040	7.2
MIT	0.040	7.1
KI	100.	7.0
Cholesterol	100.	7.5
Glucose	100.	7.4
Bilirubin	100.	8.0
Acetylsalicylic acid	600.	7.0
Diphenylhydantoin	600.	12.2
Heparin .	excess	7.6
{Ammonium oxalate	{120.	7.5
{potassium oxalate	{100.	
Potassium oxalate	500.	7.2

TABLE 3. Effect of administered radiopaque dyes on T_4I values

Subject	Type of study	Dye	T_4I—$\mu g/100$ ml		
			Control	1–24 hr	48–72 hr
1	Bronchogram	Dionosil*	4.7	6.2	—
2	Bronchogram	Dionosil	4.2	6.2	6.6
3	Cholecystogram	Telepaque†	4.0	7.0	5.6
4	Cholecystogram	Telepaque	4.0	4.9	7.7
5	iv Pyelogram	Hypaque‡	7.7	7.3	8.2
6	iv Pyelogram	Hypaque	6.0	4.8	—
7	iv Pyelogram	Hypaque	7.7	5.2	—

* 3:5-Diido-4-pyridone-N-acetic propionate.
† 3-(3-Amino,2,4,6,-triiodophenyl)-2-ethyl propanoic acid.
‡ Sodium 3,5-diacetamido-2,4,6-triiodobenzoate.

and in 20 healthy, active hospital personnel (9.2 ± 2.1 $\mu g/100$ ml). The results in the hypothyroid and hyperthyroid

TABLE 4. Duplicate determinations of serum thyroxine

Range	Determination No. 1, $\mu g/100$ ml	Determination No. 2, $\mu g/100$ ml	Difference, $\mu g/100$ ml
0–5	3.7	4.9	1.2
	4.8	4.2	0.6
	3.4	2.2	1.2
	4.6	2.9	1.7
	0.6	1.2	0.6
	0.0	1.0	1.0
	0.7	0.7	0.0
	0.5	0.7	0.2
Mean	2.26		0.81
5–10	7.0	6.1	0.9
	9.3	10.0	0.7
	7.9	9.8	1.9
	8.8	7.1	1.7
	8.6	9.0	0.4
	8.0	8.1	0.1
	6.9	5.1	1.8
	6.3	6.6	0.3
	7.6	7.8	0.2
	9.3	7.2	2.1
	7.4	5.2	2.2
	7.3	8.5	1.2
	9.4	8.5	0.9
Mean	7.80		1.1
10–20	15.8	14.6	1.2
	16.2	12.7	3.5
	12.0	11.1	0.9
	10.2	11.1	0.9
	14.5	18.3	3.8
	18.3	19.1	0.8
Mean	14.5		1.8

groups of patients are shown in Table 6. The mean value for thyroxine iodine in the 20 euthyroid chronically ill patients was 7.2 ± 1.2 $\mu g/100$ ml, with a range of 5.2–9.0, and was slightly but significantly higher ($P < 0.01$) than that for the 20 healthy subjects, 6.0 ± 1.4, with a range of 4.0–9.2. Values in hypothyroid subjects ranged from 0.3 to 4.0, while those in hyperthyroid subjects ranged from 12.1 to 16.3 $\mu g/100$ ml. There was good general agreement as to clinical status between these values and those for the PBI, when the latter were available.

Discussion

Thyroxine is bound by at least three different proteins in serum. These include TBG, to which binding is strongest, prealbumin, to which binding is of intermediate affinity, and albumin, to which binding is relatively weak. Binding to prealbumin is known to be inhibited by barbital buffer (6). In the method described, barbital buffer was used to minimize the binding of prealbumin, since it was found that prealbumin interfered with the determination. The weak binding to albumin was decreased by dilution, so that the binding obtained may be attributed almost entirely to TBG.

The alteration of binding with temperature was an unexpected finding, and

TABLE 5. Thyroxine content of samples contaminated with iodine (*i.e.*, difference between PBI and TI > 2.0 μg/100 ml) and with mercury

	T_4	T_4I	PBI	Total I
	μg/100 ml			
A. Iodine contamination:				
1	6.0	4.0	>40	>40
2	10.7	7.2	>40	>40
3	6.2	4.2	21.4	>40
4	7.9	5.3	21.3	27.5
5	10.9	7.3	11.6	>40
6	9.9	6.6	9.3	13.0
B. Mercury contamination:				
1	11.2	7.5	0.2	0.2
2	11.9	7.8	0.2	3.6

no previous report of it was found in the literature. It is similar to that of cortisol binding by CBG (observations in this laboratory).

The specificity of the binding of thyroxine by TBG has been reviewed by Robbins and Rall (7). Danowski *et al.* studied the effect of thyroxine and other iodinated compounds on the binding of triiodothyronine-I^{131} using electrophoresis in barbital buffer at pH 8.6 (8). They observed that thyroxine had a much higher affinity for TBG than any of its analogues, including triiodothyronine, diiodothyronine and diiodotyrosine. Several drugs have been observed to lower PBI values by competing with thyroxine for binding sites on TBG, notably diphenylhydantoin (9). These results are in keeping with our own observations as to the specificity of TBG binding.

At its present stage of development, the precision of the method described is somewhat less than that obtainable for protein-bound iodine, although it is felt that this disadvantage is more than offset by its freedom from the effects of iodine and other forms of contamination. It should be pointed out, however, that the prior administration of large amounts of radioactive substances may invalidate the test, but, if suspected, this can be readily ruled out simply by count-

ing the subject's serum directly. The standard I^{131}-uptake test using 300 μc I^{131} had only a slight effect, and this effect was negligible 48 hours after administration.

The values for thyroxine iodine obtained by the present method are slightly higher than the corresponding PBI values, although Bodansky has estimated that only 87% of the protein-bound iodine in serum is recovered as the PBI. In 100 euthyroid subjects he obtained a mean PBI of 6.2 ± 1.3 μg/100 ml (10).

Using the double isotopic derivative technique for the determination of serum

TABLE 6. Serum thyroxine values in thyroid disease

Clinical diagnosis	T_4	T_4I	PBI	TI
	μg/100 ml			
Hypothyroid	0.4	0.3	1.8	3.2
	1.9	1.3	0.4	1.0
	2.1	1.3	—	—
	2.3	1.6	0.9	2.3
	2.6	1.7	1.7	2.7
	2.9	1.8	3.2	28.4
	3.0	2.0	1.6	2.8
	3.8	2.5	3.3	5.9
	5.8	3.8	3.3	5.2
Hyperthyroid	18.5	12.1	9.6	11.7
	20.0	12.3	15.6	16.0
	20.0	13.0	9.4	12.8
	21.2	13.8	8.5	15.0
	21.3	13.9	9.5	10.3
	24.3	15.9	14.4	15.7
	23.6	23.6	17.0	18.0
	24.9	16.3	14.6	15.6

TABLE 7. Comparison of thyroxine-iodine estimations in euthyroid
individuals using different methods

Method	No. of subjects	Thyroxine-iodine mean ±sd	Range
		μg per 100 ml	
Whitehead & Beale Double isotope —euthyroid patients	14	3.1 ±0.6	2.0–4.1
Present method —euthyroid patients and healthy subjects	40	6.6 ±1.3	4.0–9.2
Bodansky PBI (uncorrected for recovery) —apparently normal persons	100	6.2 ±1.3	3.6–8.8

thyroxine (2), Whitehead and Beale obtained low values in 14 euthyroid subjects, *viz.*, a mean of 4.6 μg thyroxine, 100 ml, corresponding to a thyroxine iodine of 3.1. Comparing this value with the mean protein-bound iodine value (corrected for recovery) in 100 euthyroid subjects obtained by Bodansky *et al.*, *viz.*, a mean of 7.1 ±1.5, our own combined values for thyroxine iodine in 20 healthy subjects and in 20 euthyroid patients, *viz.*, 6.6 ±1.3 μg/100 ml, are intermediate between the two but much closer to those for the PBI. These values are summarized in Table 7.

The simplicity and rapidity of the gel filtration method make it a promising tool for routine clinical use. A single well-trained technician has been able daily to carry out 30 determinations in duplicate. A series of five tests in duplicate may be carried out in approximately one hour. Further investigations of the clinical applicability of the test are in progress.

Acknowledgments

The authors are indebted to Drs. A. Gold, R. Schucher, and J. M. McKenzie for their very helpful cooperation in providing much of the material for these studies, to Mr. Wan-Ching Sun for technical assistance, and to Professor J. S. L. Browne for reviewing the manuscript.

References

1. Murphy, B. E. P., W. Engelberg, and C. J. Pattee, *J Clin Endocr* **23**: 293, 1963.
2. Whitehead, J. K., and D. Beale, *Clin Chim Acta* **4**: 710, 1959.
3. Williams, R. H., Textbook of Endocrinology, W. B. Saunders Co., Philadelphia, 1962, p. 259.
4. Chaney, A. L., *Advance Clin Chem* **1**: 82, 1958.
5. Braunsberg, H., and V. H. T. James, *J Clin Endocr* **21**: 1146, 1961.
6. Antoniades, H. D. (ed.), Hormones in Human Plasma, Little, Brown and Co., Boston, Mass., 1960, p. 531.
7. Robbins, J., and J. E. Rall, *Physiol Rev* **40**: 415, 1960.
8. Danowski, T. S., R. D. D'Ambrosia, J. Dobcak, and M. Nakano, *Metabolism* **11**: 443, 1962.
9. Wolff, J., M. E. Standaert, and J. E. Rall, *J Clin Invest* **40**: 1373, 1961.
10. Bodansky, O., R. S. Benun, and G. Pennacchia, *Amer J Clin Path* **30**: 375, 1958.

COMMENTARY TO

4. Engvall, E., Jonsson, K. and Perlmann, P. (1971)
Enzyme-Linked Immunosorbent Assay II. Quantitative assay of protein antigen, immunoglobulin G, by means of enzyme-labelled antigen and antibody-coated tubes. Biochimica et Biophysica Acta Protein Structure 251(3): 427–434.

I n 1970, Eva Engvall was a graduate student in Peter Perlmann's laboratory at the University of Stockholm in Sweden. Her assignment was to develop a quantitative non-isotopic immunoassay as an alternative to radioimmunoassay (RIA), a technology that had been in worldwide use for almost a decade. In successive stages of refinement and creativity she invented today's non-isotopic immunoassay format and named it the enzyme-linked immunosorbent assay (ELISA).

Engvall created ELISA by weaving together technical advances in the three different fields of serology, immuno-histochemistry and RIA. First, in the early 1950', Gyola Takatsy in Hungary (1) and John L. Sever along with others at the National Institutes of Health in the US (2) developed the plastic micro titer plate. These multi-well plates were designed to replace test tubes and to speed up the running of large numbers of serology assays. Second, Engvall took advantage of the development of enzyme labelled antibodies in immunohistochemistry. Avrameas (3) along with Nakane and Pierce (4) were the first to describe the use of enzyme labelled antibodies in histochemistry. Finally, in 1967, Catt in Australia introduced a solid phase RIA in which the antibody was coated onto the inside of 12×75 mm plastic test tubes. He used his test tube format for human placental lactogen and human growth hormone assays with ^{125}I labelled hormone as tracer (5).

The first ELISA report by Engvall was published as a Communication to the Editors in the journal *Immunochemistry* in 1971 (6). Rabbit immunoglobulin G (IgG) was conjugated to calf intestine alkaline phosphatase (ALP, EC 3.1.3.1) as the tracer. Sheep–anti-rabbit IgG antibody was coupled to microcrystalline cellulose in order to provide a solid phase support for the antibody. Following incubation with the unknown serum the antibody-cellulose particles were centrifuged and washed. The amount of IgG–ALP bound to the cellulose was then measured with a colorimetric substrate. Van Weeman and Schuurs independently of Engvall published a similar ELISA method for human chorionic gonadotropin (HCG). They coated antibody onto cellulose particles and utilized an HCG-peroxidase conjugate (7). The publication of the paper by Van Weeman and Schuurs preceded the Engvall and Perlmann paper by a few months, however, Engvall's first description of an ELISA format was submitted for publication 5 months before that of Van Weeman and Schuurs.

Engvall's second paper improved on her IgG assay by coating the antibody onto the walls of 11×80 mm polystyrene plastic test tubes. That paper is presented here. Bound IgG–ALP conjugate was measured with colorimetric substrate following three washings of the tubes with a saline-Tween-20™ buffer. The third paper in the series by Engvall and Perlmann changed the format of the assay in order to measure unknown antibody concentrations. Antigens were coated into plastic test tubes, incubated with the unknown sera followed by washing of the tubes. Bound antibodies if present were measured using specific anti-IgG antibody conjugated with ALP (8). In 1974, Ljungstrom, Engvall and Ruitenberg used this test tube ELISA to screen for antibodies to the trichinosis parasite in human serums (9). Engvall also described an ELISA for detection of antibodies to the malarial parasite, *Plasmodium falciparum*, in human serums (10). The parasite antigens were coated onto the walls of 96 well micro titer plates. Antibodies in the unknown serums were detected using anti-human IgG ALP conjugates. Plates were read visually or with a colorimeter. Engvall took the ELISA assay on a trip to East Africa and used it to screen for a number of parasitic infections.

Less than 4 years after the publication of the Engvall and Perlmann paper presented here, *The Lancet* in an editorial predicted that ELISA would soon replace RIA (11). Louis Pasteur is credited with having said that the best proof that a researcher has struck the right path is the constant fruitfulness of their work. Between 1971 and 1974 a total of six papers were published using the ELISA technique. Four of them were by Engvall. The fifth paper was for human IgG by Hoffman who referenced Engvall (12) and the sixth paper was the Van Weeman and Schurs paper (7). In the next four-year-period, 1975 through 1978, a total of 277 ELISA papers were published. A similar literature search for the word ELISA in the title or abstract for the years 2000 through 2003 produced a total of 24 516 citations (13).

References

(1) Takatsy, G. (1955) The use of spiral loops in serological and virological micro-methods. Acta Microbiologica Academiae Scientiarum Hungaricae. 3(1–2):191–202.

(2) Sever, J.L. (1962) Application of a microtechnique to viral serological investigations. Journal of Immunology. 88(3): 320–329.

(3) Avrameas, S. and Uriel, J. (1966) Methode de Marquage D'Antigenes et D'Anticorps avec des enzymes et son application en immunodiffusion. Comptes Rendus Hebdomadaires des Seances de l'Academie des Sciences D, Sciences Naturelles. 262(24):2543–2545.

(4) Nakane, P.K. and Pierce, G.B. Jr. (1966) Enzyme-labeled antibodies: preparation and application for the localization of antigens. Journal of Histochemistry and Cytochemistry. 14(12):929–931.

(5) Catt, K. and Tregear, G.W. (1967) Solid-phase radioimmunoassay in antibody-coated tubes. Science. 158(808):1570–1572.

(6) Engvall, E. and Perlmann, P. (1971) Enzyme-linked immunosorbent assay (ELISA). Quantitative Assay of Immungloblin G. Immunochemistry. 8(9):871–874.

(7) VanWeeman, B.K. and Schuurs, A.H.W.M. (1971) Immunoassay using antigen-enzyme conjugates. Federation of American Societies for Experimental Biology (FEBS) Letters. 15(3):232–236.

(8) Engvall, E. and Perlmann, P. (1972) Enzyme-linked immunosorbent assay, ELISA. III. Quantitation of specific antibodies by enzyme-labeled anti-immunoglobulin in antigen-coated tubes. Journal of Immunology. 109(1):129–135.

(9) Ljungstrom, I., Engvall, E., and Ruitenberg, E.J. (1974) ELISA, enzyme linked immunosorbent assay — a new technique for sero-diagnosis of trichinosis. Parasitology. 69(Part 2):xxiv.

(10) Voller, A., Bidwell, D., Huldt, G., and Engvall, E. (1974) A microplate method of enzyme-linked immunosorbent assay and its application to malaria. Bulletin of the World Health Organization. 51(2):209–211.

(11) [Editorial] (1976) ELISA: a replacement for radioimmunoassay? The Lancet. 308(7982):406–407.

(12) Hoffman, D.R. (1973) Estimation of serum IgE by an enzyme-linked immunosorbent assay (ELISA). Journal of Allergy and Clinical Immunology. 51(5):303–307.

(13) PubMed Bibliographic Database. National Center for Biotechnology Information (NCBI) at the National Library of Medicine (NLM), Bethesda, MD.

47

BBA 35998

ENZYME-LINKED IMMUNOSORBENT ASSAY

II. QUANTITATIVE ASSAY OF PROTEIN ANTIGEN, IMMUNOGLOBULIN G, BY MEANS OF ENZYME-LABELLED ANTIGEN AND ANTIBODY-COATED TUBES

EVA ENGVALL, KARIN JONSSON AND PETER PERLMANN

From the Department of Immunology, The Wenner-Gren Institute, University of Stockholm, Stockholm (Sweden)

(Received July 21st, 1971)

SUMMARY

Alkaline phosphatase from calf intestinal mucosa has been conjugated to a protein antigen, rabbit IgG. Such conjugates, prepared by glutardialdehyde, have been used in a competitive solid phase immunoassay. In this test native antigen inhibits the binding of the conjugate to homologous antibodies adsorbed to plastic tubes. Using this assay 1–100 ng/ml of the antigen could be determined.

INTRODUCTION

Radioimmunoassays have found wide application, especially for determination of proteins such as hormones and immunoglobulins. The main techniques used are based on competitive inhibition of binding of radioactively labelled antigen to its homologous antibody by unlabelled antigen in standard preparations or unknown samples, respectively (for references see ref. 1).

We have used the same principle in an assay for determination of rabbit IgG. However, instead of labelling the antigen with a radioactive isotope we have conjugated it to an enzyme, alkaline phosphatase from calf intestinal mucosa. Enzyme-coupled antigens and antibodies have been used for localization in tissue sections of homologous antibodies or antigens[2], and recently also for quantitative determination of antigen or antibody in gel diffusion[3].

In competitive immunoassays it is necessary to separate free and antibody-bound antigen in order to quantitate the latter. This can easily be done by fixing the antibody to a solid phase. In a previous report[4], in which antigen conjugated with alkaline phosphatase was used in a competitive immunoassay, we used antibodies coupled to BrCN-activated cellulose[5] as in RIST radioimmunosorbent technique[6]. In the present work we used the technique of CATT AND TREGEAR[7] in which the antibodies are physically adsorbed to the test tubes in which the entire assay takes place. In addition to thus simplifying the test, we also increased its sensitivity by using a purer enzyme preparation for conjugation to antigen.

MATERIALS AND METHODS

Antigen, immunogloublin G (IgG) from rabbit

Rabbit IgG was purified from normal rabbit serum by precipitation with $^1/_3$ saturated $(NH_4)_2SO_4$. After dialysis against phosphate-buffered saline (0.15 M NaCl–0.015 M sodium phosphate buffer (pH 7.2)) the material was used for immunization of sheep.

Antigen used as standard and for conjugation to enzyme was further purified by batch adsorbtion of impurities on DEAE-cellulose (SERVA, Heidelberg, W. Germany) at pH 6.8, followed by gel-permeation chromatography in phosphate-buffered saline on Sephadex G-200 (Pharmacia Fine Chemicals, Uppsala, Sweden). Fractions containing IgG were concentrated to 50 mg/ml by ultrafiltration. In immunoelectrophoresis this preparation formed one single precipitin arc when assayed with sheep antiserum against whole rabbit serum.

Antiserum, immunoglobulin fraction, and specific antibodies

Antiserum to rabbit IgG was prepared in sheep by repeated injections of $(NH_4)_2SO_4$ precipitated rabbit γ-globulin in Freund's complete adjuvant (Difco Laboratories, Detroit, Mich.). The immunoglobulin fraction of the antiserum was prepared by precipitation with Na_2SO_4 according to KEKWICK[8].

Specific anti-IgG antibodies were obtained by use of an immunosorbent, consisting of rabbit IgG, insolubilized by glutardialdehyde (Merck, Darmstadt, W. Germany) according to AVRAMEAS AND TERNYNCK[9]. We used a batchwise procedure and eluted the antibodies from the immunosorbent with 0.1 M glycine–HCl buffer (pH 2.8).

Protein determinations of purified immunoglobulin, rabbit IgG as well as sheep antibodies, were performed by measuring the absorbance at 280 nm in a 1-cm cuvette, using the factor 13.8 as the absorbance of a 1% solution of the protein.

Iodinated proteins

Iodination was done with ^{125}I (Radiochemical Centre, Amersham, England) by the Chloramin-T method[10]. The immunoglobulin fraction of sheep anti-rabbit IgG was labelled to give a radioactivity of approx. 10^8 counts/min per mg.

Enzyme

Alkaline phosphatase from calf intestinal mucosa (Sigma Type VII, Sigma Co., St. Louis, Mo.) was used for conjugation to antigen. The specific activity of this preparation was 410 units/mg.

Other reagents

Human serum albumin was kindly supplied by AB Kabi, Stockholm, and *p*-nitrophenylphosphate was obtained from Sigma.

Biochim. Biophys. Acta, 251 (1971) 427–434

Conjugation of enzyme to antigen

The conditions for conjugation, by means of glutardialdehyde[11], of alkaline phosphatase (Sigma Type II, 16 units/mg) have been described[4]. In the present study we used the Sigma Type VII enzyme, which is a suspension of 5 mg/ml protein in 2.6 M $(NH_4)_2SO_4$. The procedure of conjugation was as follows: 0.3 ml of the suspension was centrifuged in the cold at about 1000 rev./min for 10 min. 0.2 ml of the clear supernatant was discarded. To the remaining 0.1 ml, containing the enzyme as a pellet, was added 0.1 ml of a solution containing 0.5 mg pure rabbit IgG. The mixture thus obtained contained 10 mg protein per ml at a IgG/alkaline phosphatase ratio of 1:3. After dialysis overnight against phosphate-buffered saline, 10 μl of 4.2% glutardialdehyde in phosphate-buffered saline were added, yielding a final glutardialdehyde concentration of 0.2%. The reaction was allowed to proceed for 2 h at room temperature. The mixture was diluted to 1 ml with phosphate-buffered saline, dialyzed overnight against phosphate-buffered saline, and finally chromatographed on a 1.5 cm $\times$ 90 cm column of Sepharose 6B (Pharmacia Fine Chemicals) in 0.05 M Tris–HCl buffer (pH 8.0). Almost all the protein was present in the form of high molecular weight complexes which were eluted from the column with the void volume. To this fraction (4 ml), which contained about 40% of the total protein, was added human serum albumin to 5% for stabilization. It was stored at 4° with 0.02% NaN_3 as preservative.

Immunoassay

11 mm $\times$ 80 mm disposable polystyrene tubes (LIC, Solna, Sweden) were each coated with 1 ml of antibody solution, containing either whole γ-fraction, or specific antibodies prepared from the antiserum as described. The antibody solutions were diluted with 0.1 M sodium carbonate buffer (pH 9.8) or phosphate-buffered saline. The tubes with the coating solution were standing upright at 37° for 3 h. They were then stored with the coating solution in the cold. Before assay, a suitable number of tubes were washed with 0.9% NaCl containing 0.05% Tween-20. To each of them was added 0.5 ml of either (1) standard IgG solution, (2) unknown sample or (3) buffer only, followed by 0.1 ml of a dilution of the enzyme conjugated IgG. All dilutions were made in phosphate-buffered saline containing 1% human serum albumin and 0.02% NaN_3. The tubes were incubated over night (16 h) at room temperature in a roller drum so that the 0.6 ml present in the tubes covered the surface precoated with antibody. The tubes were washed 3 times with NaCl–Tween. The amount of enzyme-linked IgG bound to the antibody-coated tube was determined by adding 1 ml of 0.05 M sodium carbonate buffer (pH 9.8), containing 1 mg/ml p-nitrophenylphosphate and 1 mM $MgCl_2$. When the colour intensity was considered suitable, or after about 30 min, the reaction was stopped by adding 0.1 ml 1 M NaOH. The absorbance at 400 nm was measured in a 1-cm microcuvette. A standard curve was constructed by plotting the enzyme activities (increase in absorbance per unit of time) of individual samples against their content of standard IgG.

RESULTS AND DISCUSSION

Adsorption of protein to plastic tubes

The adsorbtion of proteins to polystyrene tubes was investigated by means of

430   E. ENGVALL *et al*

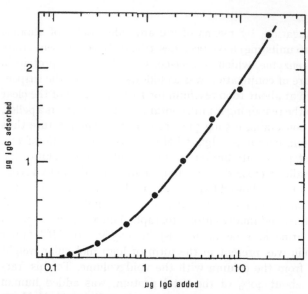

Fig. 1. Adsorbtion of sheep immunoglobulin to plastic tubes from 1 ml aliquots of different concentrations.

the radioactively labelled anti-rabbit immunoglobulin. 1 ml of a solution of this in 0.1 M sodium carbonate buffer (pH 9.8) was added to 11 mm × 80 mm tubes. These were incubated at 37° for 3 h and were then washed 3 times with 0.9% NaCl containing 0.05% Tween-20. The radioactivity remaining was measured. The results are shown in Fig. 1. At an immunoglobulin concentration of 2 μg/ml or lower, about 60% was adsorbed to the tubes. When increasing the concentration of the coating solution the percentage of protein adsorbed decreased considerably. We have found that the adsorbtion of protein to the tubes is completed within 3 h at 37° and this time was chosen for routine coating in all experiments.

The adsorbtion of protein to the polystyrene surface is not irreversible. Tubes were coated with radioactively labelled sheep immunoglobulin, washed and incubated for 16 h at room temperature in the roller drum with 0.6-ml portions of different solutions. The radioactivity in the tubes after three washings was measured and compared with that before incubation. The results are shown in Table I. One can see that at pH around neutrality 20–30% of the coat was lost. At high or low pH, these losses increased considerably. The percentage protein lost during incubation seemed to be independent of the total amount adsorbed. This is of practical importance for the immunoassay. If relatively large amounts of antibody are used for coating the tubes, the amount of antibody desorbed during the assay will also be large. Labelled and unlabelled antigen added for the assay will preferentially combine with antibody in solution and this will reduce the uptake of antigen to the coat.

During the 3-h coating time an equilibrium between the protein adsorbed to the tube and that remaining in solution will be established. By storing the tubes in the cold with the coating solution present, this equilibrium is the least disturbed and the amount of protein adsorbed to the tubes will remain the same for at least 3 weeks.

Biochim. Biophys. Acta, 251 (1971) 427–434

TABLE I

DESORBTION OF PROTEIN FROM TUBES COATED WITH A SOLUTION CONTAINING 2 μg/ml OF RADIO-ACTIVELY LABELLED SHEEP IMMUNOGLOBULIN IN DIFFERENT SOLUTIONS DURING 16 h AT ROOM-TEMPERATURE

0.6 ml solution	Percent of coat remaining
Phosphate-buffered saline,	77
Phosphate-buffered saline,	71
Phosphate-buffered saline, 0.05% Tween 20	58
Normal human serum	83
NaCl, 2 M	78
Urea, 2 M	70
Glycin–HCl buffer 0.1 M (pH 2.7)	48
NaOH, 0.1 M	27

As the coating procedure is so simple, we have felt no need of storing tubes for any longer time.

Conjugate

When rabbit IgG was conjugated to alkaline phosphatase, very high molecular weight complexes were formed. The yield in terms of antigenic activity of the IgG was low (approx. 1%), presumably due to steric hindrance. In contrast the enzyme activity was well preserved[4]. It was also noticed that approximately half of the total enzyme activity in such a conjugate was bound to immunoreactive antigen (unpublished observation).

Other bifunctional reagents have been used for conjugation, *e.g.* tolylene-2,4-diisocyanate (Fluka, Buchs, Switzerland) applied according to SINGER AND SCHICK[12], and 1-(3-dimethylaminopropyl)-3-ethylcarbodiimide (Ott Chemical Co., Muskegon, Mich.) applied according to LIKHITE AND SEHON[13]. However, satisfactory results have thus far only been obtained with glutardialdehyde.

When stored as described, the conjugates prepared with glutardialdehyde are very stable. We have used one conjugate for a year and have noticed only a moderate decrease in its activity during this time.

Immunoassay

For a sensitive system, the amount of antibody present during assay should be limited; it should bind only part of the conjugate added. The enzyme activity of 0.1 ml of the stock solution of the conjugate used corresponded to an increase in absorbance at 400 nm of 48 per min at room temperature. For the assay this stock solution was diluted 250 times and 0.1 ml of this was added to tubes which were coated with a solution containing 2 μg/ml of the immunoglobulin fraction from the sheep anti-rabbit IgG serum. Under these conditions the uptake of conjugate to the tubes corresponded to an increase in absorbance at 400 nm of 0.7 in 30 min (= 12% of the enzyme activity added). In the presence during incubation of various dilutions of the standard preparation of rabbit IgG, the typical inhibition curve shown in Fig. 2 was obtained. A similar standard curve was obtained when a solution con-

432   E. ENGVALL *et al.*

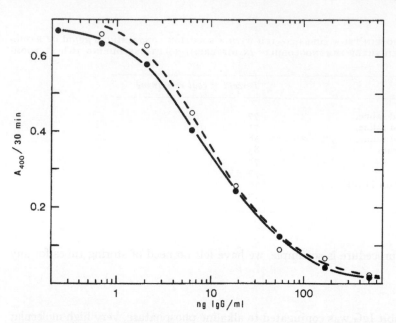

Fig. 2. Inhibition of binding of enzyme-conjugated rabbit IgG to antibody-coated tubes by rabbit IgG from a standard preparation. ●, tubes coated with a solution containing 2 μg/ml of an immunoglobulin fraction from sheep anti-rabbit IgG; ○, tubes coated with a solution containing 0.1 μg/ml of specifically purified antibodies.

taining 0.1 μg/ml of specifically purified antibodies was used for coating (Fig. 2 hatched line).

When conjugate and only buffer were incubated in antibody coated tubes for different periods of time the uptake of enzyme to the tubes followed the curve shown in Fig. 3. Equilibrium was obtained within 16 h and over night incubations were used routinely for the assay.

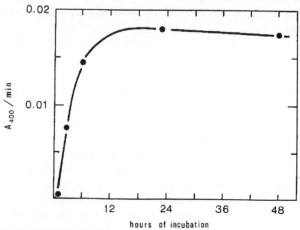

Fig. 3. The kinetics of the binding of enzyme-conjugated rabbit IgG to tubes coated with antibodies.

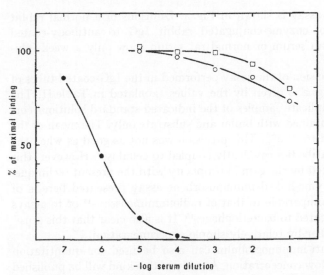

Fig. 4. Binding of enzyme-conjugated rabbit IgG to antibody-coated tubes in the presence of dilutions of various normal sera. ●, rabbit serum; ○, rat serum; □, human serum.

TABLE II

IgG (ng/ml) Standard	$A_{400\,nm}/30\,min$*	Mean $A_{400\,nm}/30\,min$	$\mid d \mid$**	S.E.***	$\dfrac{S.E.}{mean} \times 100$
500	0.020	0.018	0.002	0.002	12.0
	0.015		0.003		
	0.019		0.001		
167	0.045	0.047	0.002	0.007	14.8
	0.057		0.010		
	0.040		0.007		
55.6	0.138	0.124	0.014	0.011	9.2
	0.124		0		
	0.110		0.014		
18.5	0.288	0.247	0.041	0.029	11.7
	0.225		0.022		
	0.227		0.020		
6.17	0.393	0.405	0.012	0.009	2.3
	0.415		0.010		
	0.407		0.002		
2.06	0.520	0.579	0.059	0.047	8.1
	0.636		0.057		
	0.582		0.003		
0.686	0.581	0.636	0.055	0.047	7.4
	0.696		0.060		
	0.630		0.006		
0.229	0.623	0.672	0.049	0.036	5.4
	0.708		0.036		
	0.684		0.012		

* Values corrected for background (0.200).
** $\mid d \mid$, numerical value of difference between measured value and mean.
*** Standard error of mean, $(\sqrt{\Sigma d^2/3})$.

Biochim. Biophys. Acta, 251 (1971) 427–434

434   E. ENGVALL *et al.*

The specificity of the assay is shown in Fig. 4. Dilutions of a normal rabbit serum inhibit the binding of enzyme-conjugated rabbit IgG to antibody-coated tubes, whereas normal human serum or normal rat serum show only a week inhibition at low dilutions.

An estimate of the precision of the assay performed in the IgG-coated tubes of the experiment shown in Fig. 2 is given by the values tabulated in Table II. The mean values are based on triplicate samples of the indicated standard solutions corrected for the background obtained with buffer and substrate only. The mean of the standard errors corresponded to $\pm 9\%$. This precision was not as good as when the assay was performed with antibodies covalently coupled to cellulose[4]. However, this loss of precision is compensated by the gain in simplicity with the present technique.

In conclusion, the enzyme-linked immunosorbent assay presented here is of very high sensitivity, fully comparable to that of radioimmunoassays[14] or to assays making use of antigen conjugated to bacteriophages[15]. It is also clear that this sensitivity may be increased further by relatively simple experimentation.

Enzyme-conjugated anti-immunoglobulin can also be used for quantitation of specific antibodies at very low concentration. A study of this kind will be published elsewhere[16].

ACKNOWLEDGEMENT

This work was supported by grant No. 2032-035 from the Swedish Natural Science Research Council and by a grant from Pharmacia AB, Uppsala, Sweden.

REFERENCES

1 K. E. KIRKHAM AND W. M. HUNTER, *Radioimmunoassay Methods, European Workshop, Sept. 15–17, 1970*, Churchill Livingstone, London, 1971.
2 S. AVRAMEAS, *Int. Rev. Cytol.*, 27 (1970) 349.
3 M. STANISLAWSKI, *C.R. Acad. Sci. Paris, Ser. D*, 271 (1970) 1452.
4 E. ENGVALL AND P. PERLMANN, *Immunochemistry*, 8 (1971) 871.
5 R. AXEN, J. PORATH AND S. ERNBACH, *Nature*, 214 (1967) 1302.
6 L. WIDE AND J. PORATH, *Biochim. Biophys. Acta*, 130 (1966) 257.
7 K. CATT AND G. W. TREGEAR, *Science*, 158 (1967) 1570.
8 R. A. KEKWICK, *Biochem. J.*, 34 (1940) 1248.
9 S. AVRAMEAS AND T. TERNYNCK, *Immunochemistry*, 6 (1969) 53.
10 W. M. HUNTER AND F. C. GREENWOOD, *Nature*, 194 (1962) 495.
11 S. AVRAMEAS, *Immunochemistry*, 6 (1969) 43.
12 S. J. SINGER AND A. F. SCHICK, *J. Biophys. Biochem. Cytol.*, 9 (1961) 519.
13 V. LIKHITE AND A. SEHON, in C. A. WILLIAMS AND M. W. CHASE, *Methods in Immunology and Immunochemistry*, Vol. I, Academic Press, New York, 1967, p. 155.
14 S. G. O. JOHANSSON, H. BENNICH AND L. WIDE, *Immunology*, 14 (1968) 265.
15 J. HAIMOVICH, E. HURWITZ, N. NOVIK AND M. SELA, *Biochim. Biophys. Acta*, 207 (1970) 125.
16 E. ENGVALL AND P. PERLMANN, *J. Immunol.*, submitted.

COMMENTARY TO

5. Rubenstein, K. E., Schneider, R. S. and Ullman, E. F. (1972).
"Homogeneous" Enzyme Immunoassay. A New Immunochemical Technique. Biochemical and Biophysical Research Communications 1972, 47(4): 846–851.

The enzyme multiplied immunoassay technique (EMIT®) was developed following the introduction of an earlier homogeneous immunoassay technology called the free-radical assay technique (FRAT®). Both of these homogeneous assay technologies were invented at the Syva Corporation under the research direction of E.F. Ullman. They were initially designed for drugs of abuse screening. In the FRAT assay, a nitroxide stabilized free radical or spin-label is conjugated to morphine and mixed with anti-morphine antibody along with patient's urine. The binding of the spin-label to the antibody slows down its tumbling rate as monitored with an electron paramagnetic resonance (EPR) spectrometer (1–2). Free morphine in the patient's urine competes with the spin-label conjugate and caused a positive response in the EPR spectrum indicating increased free label in solution. Each EPR instrument weighed 225 Kg (450 lbs) however the assay was simple and could easily run 120 drug screens per hour with limited technician training. In 1970 at the request of the United States Government three EPR instruments and three technicians with reagents were sent to Vietnam from Syva to begin screening military personnel for abused morphine. This was the largest field use of an immunoassay ever attempted (3).

The Syva group then conceived of the idea of linking the antibody-binding event to the catalytic activity of an enzyme instead of a spin-labeled conjugate. In the first EMIT description, presented here, the enzyme lysozyme (EC 3.2.1.17) was conjugated to morphine as the tracer. When the lysozyme-morphine conjugate was bound to its antibody the activity of the enzyme was inhibited. When released from the antibody, activity was restored. Release of the conjugate could be monitored in a homogeneous real-time assay by mixing all the reagents along with patient urine in a suspension of killed bacteria. The decrease in turbidity caused by the liberated lysozyme was quantitatively related to how much lysozyme-morphine tracer was released by the free morphine in the urine as it bound to the antibody (4). New drug assays were added to the EMIT platform and it soon replaced thin layer chromatography (TLC) as the method of choice for urine drug screening.

Lysozyme requires the measurement of the decrease in turbidity. It was soon replaced with enzymes that catalyzed an optical assay based on reduction of NAD at 340 nm. This change improved the detection limit of the assay. Hapten drug conjugates were made with malate dehydrogenase (MDH, EC 1.1.1.37) (5) or microbial glucose-6-phosphate dehydrogenase (G6PD, EC 1.1.1.49) (6). The release of the G6PD enzyme drug conjugate could be easily monitored in a standard spectrophotometer by following the reduction of the coenzyme NAD at 340 nm in a standard G6PD assay. This is still the format used in the EMIT assay today.

References

(1)   Leute, R., Ullman, E.F., and Goldstein, A. (1972) Spin immunoassay technique for determination of morphine. Nature New Biology. 236(64):93–94.

(2)   Leute, R., Ullman, E.F., and Goldstein, A. (1972) Spin immunoassay of opiate narcotics in urine and saliva. Journal of the American Medical Association. 221(11):1231–1234.

(3)   Ullman, E.F. (1999) Homogeneous immunoassays: historical perspective and future promise. Journal of Chemical Education. 76(6):781–788.

(4)   Schneider, R.S., Lindquist, P., Wong, E.T., Rubenstein, K.E., and Ullman, E.F. (1973) Homogeneous enzyme immunoassay for opiates in urine. Clinical Chemistry. 19(8):821–825.

(5)   Rowley, G.L., Rubenstein, K.E., Huisjen, J., and Ullman, E.F. (1975) Mechanism by which antibodies inhibit hapten-malate dehydrogenase conjugates. Journal of Biological Chemistry. 250(10):3759–3766.

(6)   Walberg, C. and Gupta, R. (1979) Quantitative estimation of phencyclidine in urine by homogeneous enzyme immunoassay (EMIT). Clinical Chemistry. 25(6):1144, (Abstract No. 407).

Biochemical and Biophysical Research Communications. 1972, 47(4): 846–851,
Copyright 1972, reprinted with permission from Elsevier.

Reprinted from BIOCHEMICAL AND BIOPHYSICAL RESEARCH COMMUNICATIONS, Vol. 47,
No. 4, May 1972
Copyright © 1972 Academic Press, Inc. *Printed in U.S.A.*

"HOMOGENEOUS" ENZYME IMMUNOASSAY.

A NEW IMMUNOCHEMICAL TECHNIQUE

Kenneth E. Rubenstein, Richard S. Schneider and Edwin F. Ullman

Syva Research Institute*, 3221 Porter Drive,
Palo Alto, California 94304

Received April 17, 1972

Summary. Addition of morphine antibodies to a conjugate of
morphine and lysozyme resulted in inhibition of lysozyme activity.
Addition of free morphine to a mixture of the conjugate and
morphine antibodies reduced the inhibition of enzyme activity in
proportion to the quantity of free morphine added. As little as
1×10^{-9} M morphine could be detected in this manner. The method
constitutes a powerful new immunochemical technique for the
quantitative determination of haptens.

Certain naturally occurring molecules serve to regulate the
activity of enzymes. For example, the binding of antibodies to
cellular antigens in the presence of the family of proteins
known as complement sets off a complex sequence of events
culminating in the activation of esterase activity. Other
examples are the numerous reversible inhibitors of enzymes (1),
and the control of enzyme activity by antibodies to enzymes (2).
We wish to report an example of a related artificial regulatory
phenomenon in which enzyme activity can in principal be controlled
by practically any desired substance. This phenomenon provides
the basis for a powerful new immunoassay technique.

The extent of inhibition of enzyme activity by antibodies
to an enzyme has been related to the size of the enzyme substrate.
This has led to the proposal that antibodies to enzymes sterically
hinder the access of the substrate to the active site of the

*Contribution No. 44

enzyme (2). Steric inhibition of an enzyme by an antibody might also be envisioned if a hapten were attached to the enzyme near its active site. In a properly selected system binding of a hapten-directed antibody to the hapten might have the same effect as binding of an enzyme-directed antibody directly to the enzyme. In such a system the enzyme activity could be regulated by the amount of antibody available for binding to the enzyme-bound hapten. The amount of free antibody could in turn be controlled by the addition of free hapten which would compete with the enzyme-bound hapten for antibody binding sites. Thus the enzymic activity of a mixture of enzyme and antibody would be directly related to the amount of free hapten introduced.

The above system possesses all the elements needed for an immunoassay. The relevant reactions are illustrated in Equations 1-2, where H is the unknown, Ab is the antibody, and Enz is the enzyme.

$$H + Ab-(H-Enz) \rightleftarrows Ab-H + H-Enz^* \qquad (1)$$

$$Substrate \xrightarrow{H-Enz^*} product \qquad (2)$$

The system provides an intrinsic amplification since one molecule of free hapten frees one molecule of enzyme which in turn can catalyze the conversion of many molecules of substrate to product. We propose the term "homogeneous" enzyme immunoassay to distinguish the method from most other immunochemical methods that are "heterogeneous" and depend at some stage on the physical separation of antigen bound to antibody from unbound antigen. Such a separation is necessary when the label on the antigen is detected equally in both bound and free states; e.g., a radio-active atom (3). One other "homogeneous" immunoassay technique known as spin immunoassay has recently been described (4).

Hen egg-white lysozyme appeared to be an ideal enzyme for

this study. The natural substrate, bacterial peptidoglycan, is
a high polymer which should be highly susceptible to steric
effects, and also, one of the six lysine residues of lysozyme
(lysine 97) is very near the active site (5) and can be modified
without seriously inhibiting the enzyme (6). Carboxymethyl-
morphine (CMM) (4,7) was selected as a hapten. Conjugation to
lysozyme was achieved by combining lysozyme with CMM-isobutyl-
chloroformate mixed anhydride (4) (2 equivalents mixed anhydride
per lysine residue) in aqueous solution at pH 9.5-10.0 followed
by dialysis against water. A magnetic circular dichroism spectrum
of the conjugate indicated an average of four haptens per enzyme
molecule.* This is consistent with the observation that four
of the six lysine residues of lysozyme react readily with
iodoacetate whereas the remaining two are relatively inert (8).
Rabbit anti-morphine γ-globulin was prepared as described
previously (4). Lysozyme activity was determined by changes in
light transmission of a suspension of the substrate, Micrococcus
luteus, by a method similar to that of Shugar (9).

Anti-morphine γ-globulin did not affect the activity of
native lysozyme. However addition of morphine antibodies to
CMM-lysozyme resulted in up to 98% inhibition of enzyme activity.
Thus the antibody specifically inhibits hapten-labelled enzyme.
The result of titration of the antibody with CMM-lysozyme is
given in Figure 1. The CMM-lysozyme concentrations were estimated
from the absorption of stock solutions at 280 nm (10). As
expected, with no antibody present enzyme activity was
proportional to enzyme concentration. With a fixed amount of
antibody present the activity increased proportionately only at

---

*We wish to thank Dr. G. Barth, Dept. of Chemistry, Stanford
University for this measurement.

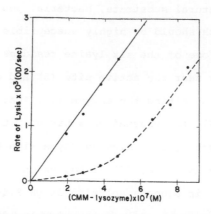

Figure 1. Effect of CMM-lysozyme on the rate of lysis of
          M. luteus (Miles) at pH 6.0 in the presence.....
          and absence   of anti-morphine γ-globulin
          (4.2×10^{-7}M binding sites).

high CMM-lysozyme concentrations where nearly all the antibody
binding sites were saturated. The horizontal distance between
the resulting two parallel lines yields the concentration of
inhibited enzyme. Based on the antibody titer as determined by
spin immunoassay (4) the data require that about two antibody
binding sites per enzyme molecule must be occupied to completely
inhibit activity.

Free morphine competes with the enzyme for binding sites
causing an increase in enzyme activity. Figure 2 shows the effect
of morphine on the enzyme activity of a solution that contained
an antibody binding site/enzyme ratio of 1.7 and was initially
96.5% inhibited. By following the enzyme kinetics for 10 sec at
30° the minimum morphine concentration detectable in the assay
mixture was approximately 3×10^{-8}M. Fifty fold lower concentrations of the reagents and a 90 minute measurement permitted
detection of as little as 1×10^{-9}M morphine.

The "homogeneous" enzyme immunoassay technique offers the
virtues of simplicity, reagent stability and quantifiability.

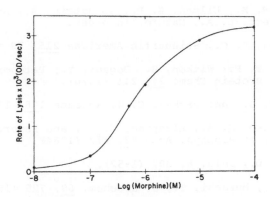

Figure 2. Effect of morphine on the rate of lysis of
M. luteus by 1.11 x 10^{-7}M CMM-lysozyme containing
1.9 x 10^{-7}M antibody binding sites at pH 6.0.

Although the present assay employs lysozyme which can be detected
above about 10^{-10}M, the use of other more readily detected
enzymes should permit the technique to rival even radioimmunoassay (3) in sensitivity. Moreover the new method requires no
physical separation of antibodies and antigens which are drawbacks both to radioimmunoassay and a recently reported "heterogeneous" enzyme immunoassay utilizing an uninhibitable enzyme-
labelled antigen (11). Preliminary studies have demonstrated
that other enzymes such as amylase and horseradish peroxidase
can also be employed and may provide increased sensitivity. In
addition the technique has been found applicable to other haptens.
Details of these studies will be reported elsewhere.

REFERENCES

1.   Hammes, G. G., and Wu, C., Science 172, 1205 (1971).

2.   Cinader, B., in Proceedings of the Second Meeting of the
      Foundation of European Biochemical Societies, Vienna, 1965,
      vol. 1, "Antibodies to Biologically Active Molecules",
      p. 85 (Pergamon, Oxford, 1967)

3.   Yalow, R. S., and Berson, S. A., J. Clin. Invest. 39, 1157
      (1960).

4.   Leute, R. K., Ullman, E. F., Goldstein, A., and
     Herzenberg, L. A., Nature, in press.

5.   Phillips, D. C., Scientific American 215, 78 (1966).

6.   Spande, T. F.; Witkop, B.; Degani, Y.; Patchornick, A.
     · Adv. in Protein Chem. 24, 241 (1970).

7.   Spector, S., and Parker, C. W. Science 168, 1347 (1970).

8.   Kravchenko, N. A., Kléopina, G. V., and Kaverzneva, E. D.
     Biochim. et Biophys. Acta 92, 412 (1964).

9.   Shugar, D., ibid. 8, 302 (1952).

10. Kato, K., Murachi, T., J. Biochem. 69, 725 (1971).

11. Van Weeman, B. K., and Schuurs, A.H.W.M., FEBS Letters 15,
     232 (1971).

Therapeutic Drug Monitoring (TDM)

Bromide

5. Wuth, O. (1927).
   Rational bromide treatment. New method for its control.

Sulfonamide

6. Marshall, A. C. and Marshall, E. K. Jr. (1939).
   A new coupling component for sulfonilamide determination.

Quinacrine

8. Brodie, B. B. and Udenfriend, S. (1943).
   The estimation of Atabrine in biological fluids and tissues.

Digoxin

9. Smith, T. W., Butler, V. P. and Haber, E. (1969).
   Determination of therapeutic and toxic serum digoxin concentrations by radioimmunoassay.

Theophylline

10. Thompson, R. D., Nagasawa, H. R. and Jenne, J. W. (1974).
    Determination of theophylline and its metabolites in human urine and serum by high-performance liquid chromatography.

COMMENTARY TO

6. Wuth, O. (1927)
Rational Bromide Treatment. New Methods for its Control. Journal of the American Medical Association 88(26): 2013–2017.

S tarting in the 1850s and for the next 150 years, bromide salts were the main ingredient in most over-the-counter and patent medicine sedatives. The half-life of bromide in humans is 12 to 15 days. Toxicity from abuse progresses from delirium, delusions and hallucinations to deep sedation followed by coma. Evelyn Waugh vividly described bromide psychosis in 1957 in his autobiographical novella *The Ordeal of Gilbert Pinfold* (1). It has been estimated that 2% of all admissions to mental hospitals were once due to bromide psychosis (2). McDanal *et al.* (3) reported on six cases of bromide psychosis from over-the-counter sedatives in San Diego County, California between 1970 and 1972. By the mid-1970s bromide was removed from all over-the-counter drugs. Miles Pharmaceutical stopped selling Nervine® and Emerson Drug took bromide out of Bromo Seltzer®. In 1990 the incidence of bromide toxicity was very low (4). Occasional reports of toxicity from imported herbal medicines that contain bromide continue to appear in the current literature (5).

In addition to over-the-counter use, prescription bromides have a long history that also dates back to the 1850s. Bromide was the only effective drug for the treatment of epilepsy until the introduction of phenobarbital and phenytoin (6,7). The paper by Wuth presented here is noteworthy for two reasons. First, he described a rapid gold chloride method for the determination of bromide in urine and serum to screen for cases of bromide overdose. In the method, a sample of a protein free filtrate of serum is mixed with a colloidal gold chloride solution. The reaction between colloidal gold and bromide ions shifts the color of the solution from red to a red–brown. His gold chloride method is still in use today (8). Wuth and Faupel (9) in 1927 had previously described the preparation of a stable colloidal gold solution for use in the Lange protein spinal fluid test. With the bromide assay, Wuth found that 21% of 238 admissions to the psychiatric unit at Johns Hopkins in a six-month period had detectable levels of this drug in their serum. In 1929, the LaMotte Chemical Products Company of Baltimore, Maryland produced the "Wuth Bromide Comparator" test kit for serum bromide levels according to Wuth's colloidal gold procedure (10). A sample of a protein free filtrate of serum was added to a colloidal gold solution and the color read against test tubes containing known calibrators. The test kit was claimed to be accurate up to 19 mmol/L (1 500 mg/L), took only 15 minute to perform and cost $12.50.

Secondly, Wuth described the therapeutic and toxic serum levels found in patients on bromide therapy for epilepsy. Based on clinical information he established 15 mmol/L (1 200 mg/L) as the upper limit of therapeutic and results over 19 mmol/L (1 500 mg/L) as toxic. He further recommended that patient's on bromide therapy for the control of epilepsy be routinely monitored with serum levels. Wuth demonstrated with his case studies that the blood levels of bromide correlate with therapeutic efficacy and not with the dose administered. This fundamental principal in therapeutic drug monitoring (TDM) was advocated by Wuth almost 50 years before it became standard practice in clinical medicine (11).

References

(1) Hurst, D.L. and Hurst, M.J. (1984) Bromide psychosis: a literary case. Clinical Neuropharmacology. 7(3):259–264.
(2) Walter, J.J. (1989) Halogens *in Manual of Toxicologic Emergencies*. Noji, E.K. and Kelen, G.D. (eds), Year Book Medical Publishers, Chicago, pgs 684–702.
(3) McDanal, C.E., Owens, D., and Bolman, W.M. (1974) Bromide abuse: a continuing problem. American Journal of Psychiatry. 131(8):913–915.
(4) Bowers, G.N. and Onoroski, M. (1990) Hyperchloremia and the incidence of bromism in 1990. Clinical Chemistry. 36(8):1399–1403.
(5) Boyer, E.W., Kearney, S., Shannon, M.W., Quang, L., Woolf, A., and Kemper, K. (2005) Poisoning from a dietary supplement administered during hospitalization. Pediatrics. 109(3), E49 (electronics pages) March [www.pediatrics.org/cgi/content/full/109/3/e49] (Accessed January 2005).
(6) Joynt, R.J. (1974) The use of bromides for epilepsy. American Journal of the Diseases of Children. 128(3):362–363.
(7) Steinhoff, B.J. (1992) Antiepileptic therapy with bromides-historical and actual importance. Journal of the History of Neuroscience. 1(2):119–123.

(8) Bradley, C.A. (1986) Bromide in serum by spectrophotometry, in *Selected Methods of Emergency Toxicology*, Selected Methods of Clinical Chemistry, Vol 11, Frings, C.S. and Faulkner, W.R. (eds), American Association of Clinical Chemists Press, Washington, DC, pgs 51–52.

(9) Wuth, O. and Faupel, M. (1927) The significance of the H-ion concentration for the colloidal gold test. Bulletin of the Johns Hopkins Hospital. 40:297–303.

(10) [Anonymous] (1929) LaMotte Blood Chemistry Outfits for the General Practitioner's Routine Tests. (Product Catalog). LaMotte Chemical Products Company, Baltimore, MD, pg 11.

(11) Koch-Weser, J. (1972) Drug therapy. Serum drug concentrations as therapeutic guides. New England Journal of Medicine. 287(5):227–231.

JOURNAL OF THE AMERICAN MEDICAL ASSOCIATION, 1927, VOLUME 88; PAGES 2013–2017.

67

which is due to the fact mentioned that bromides in part replace chlorides. Thus, a sort of constant "saturation" of the body with bromides takes place, so that after a certain period in prolonged medication no more bromides are retained, and intake and excretion are balanced.[7] The chloride content of the blood is then diminished, the chlorides having been partly replaced by bromides.

A replacement of more than 40 per cent of the chlorides of the blood by bromides, according to Bernoulli,[8] is fatal. Intoxication symptoms generally appear, according to the experiences of Ulrich[9] gained by examination of the urine, when from about 25 to 30 per cent of the total halogens are represented by bromides; there exist, however, individual differences, a fact that must be borne in mind.

After this, it is easily understood that the action of the bromide medication depends not only on the bro-

RATIONAL BROMIDE TREATMENT

NEW METHODS FOR ITS CONTROL *

OTTO WUTH, M.D.
Associate in Psychiatry, Henry Phipps Psychiatric Clinic,
Johns Hopkins Hospital
BALTIMORE

Bromide treatment to be rational must, on the one hand, produce the desired effect of the drug and, on the other hand, avoid the danger of bromide intoxication. The foundations of bromide action, and consequently also those of a rational treatment, are based on the relations between chlorides and bromides—the chloride-bromide equilibrium or replacement—which therefore has to be discussed briefly.

Sodium chloride constitutes the greater part of the electrolytes of the body, and its ions are essential for the function of most cells. Since it is constantly excreted, mainly in the urine, it must be constantly replenished. The body maintains its chloride concentration with remarkable constancy. The excretion varies with the salt intake but lags somewhat behind in time. According to Borelli and Girardi,[1] with a steady income, equilibrium is reached within three or four days. If the supply of salt is stopped, excretion falls within three days to a lower level, but the body retains its normal salt content.

The excretion of chlorides can be hastened by the administration of bromides and iodides.[2] Conversely, the administration of chlorides hastens the elimination of these salts.[3]

If bromides are introduced into the body their excretion starts rapidly but proceeds very slowly;[4] so slowly, in fact, that even twenty days after medication has been stopped the excretion of bromides is not completed.[5] Hence, a retention of bromides takes place [6]

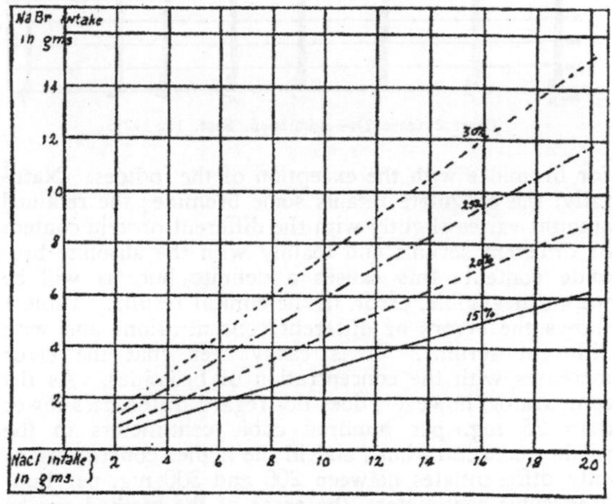

Chart 1.—Graphic illustration (from figures of Bernoulli) that the maintenance of a certain level of urine bromides to total urine halogens is dependent on the variation in sodium chloride intake as well as bromide intake.

mide intake but also on the chloride intake. That is to say, prescribing bromides without knowing the chloride intake or the bromide saturation is the same as letting a patient take as much or as little bromides as he chooses. The relations are clearly demonstrated in chart 1, which was constructed from Bernoulli's figures. Abscissa and ordinates of the chart give the intake of sodium chloride and sodium bromide; the curves give the urine saturation level. The fact is emphasized by Ulrich, that with equal doses of chlorides and bromides, bromide intoxication is produced in three weeks.

The methods for determining bromides in the blood or urine, i. e., in the presence of chlorides, are somewhat tedious and require a chemical laboratory outfit as well as some technical skill.

Walter[10] described a color reaction between gold chloride and bromides; his colorimetric method, however, according to Bieling and Weichbrodt, is practically useless, the limits of error are so great. Hauptmann's[11] modification gives better results but requires a colorimeter.

* From the Laboratory of Internal Medicine, Henry Phipps Psychiatric Clinic, Johns Hopkins Hospital.
1. Borelli and Girardi: Zentralbl. biochem. Bioph., Ref. 18: 159, 1915, cited by Sollmann.
2. Ellinger and Kotake: Arch. f. exper. Path. u. Pharmakol. 65: 87, 1911; Med. Klin. 1910, p. 1474. Laudenheimer: Zentralbl. f. Neurol., 1910, p. 461. Markwalder: Arch. f. exper. Path. u. Pharmakol., 1917, p. 81. Von Wyss: Arch. f. exper. Path. u. Pharmakol. 49: 186, 1906; 55: 266, 1906; 59: 196, 1908; Deutsche med. Wchnschr. 39: 345, 1913.
3. Sollmann, Torald: A Manual of Pharmacology, Philadelphia, W. B. Saunders Company, 1917.
4. Herzfeld and Gormidor: Zentralbl. biochem. Bioph. 13: 790, 1912.
5. Féré, Herbert and Peyrot: Compt. rend. Soc. de biol., 1892, p. 513. Mencki and Schoumow-Simanowsky: Arch. f. exper. Path. u. Pharmakol., 1894, p. 313. Pflaumer: Inaug. diss., Erlangen, 1896.
6. Frey: Ztschr. f. exper. Path. v Therap., 1910, p. 461. Von Wyss (footnote 2).

7. Fessil: München. med. Wchnschr., 1899, p. 1270. Laudenheimer (footnot 2). Von Wyss (footnote 2).
8. Bernoulli: Arch. f. exper. Path. u. Pharmakol. 73: 355, 1913.
9. Ulrich, A.: Schweiz. Arca. f. Neurol. u. Psychiat. 13, 1923.
10. Walter: Ztschr. f. d. ges. Neurol. u. Psychiat. 77, 1922; 99, 1925.
11. Hauptmann, A.: Klin. Wchnschr. 4, number 34, 1925.

I aimed to develop a method which would be simple enough for practical use and not require special equipment, and therefore devised a comparator consisting of tubes containing certain dyestuffs, the color shades of which correspond to the color changes of certain bromide concentrations after the addition of gold chloride.

The principles of the method involve the coagulation of the serum and the addition of gold chloride to the filtrate, producing color changes corresponding to the bromide concentration and varying from a yellow greenish brown to a red brown. The reaction is specific

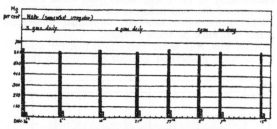

Chart 2 (case 1).—Admission, Sept. 19, 1925.

for bromides with the exception of the iodides. Naturally, the coagulate retains some bromide; the retained amount varies slightly with the different protein content of different serums and mainly with the absolute bromide content; this causes a definite but, as will be seen, a negligible, error, in the clinical results. Table 1 shows the errors of different concentrations and with different serums. It is easily seen that the error increases with the concentration of bromides. As the comparator, however, does not regard differences lower than 25 mg. per hundred cubic centimeters in the weaker concentrations, and in the higher concentrations only differentiates between 200 and 300 mg. per hundred cubic centimeters, the error of the method can be neglected. After all, I do not intend to make a quantitative chemical determination but rather an estimation which serves adequately for practical clinical use.

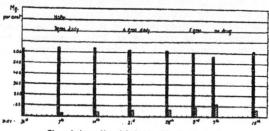

Chart 3 (case 2).—Admission, March 13, 1926.

Therefore, from this standpoint, it seems better to have an approximately exact method rather than none at all, and that the method meets this demand can be seen from the cases to be described later.

TECHNIC

The method of procedure is as follows: Ten cubic centimeters of blood is drawn from a vein of the patient and allowed to coagulate. To 2 cc. of serum, 4 cc. of distilled water and 1.2 cc. of 20 per cent trichloracetic acid are added. After standing for half an hour, the mixture is filtered through a small filter. The filtrate must be clear; if necessary, the solution is

refiltered. To each cubic centimeter of filtrate, 0.2 cc. of 0.5 per cent gold chloride (Merck) solution is added, and then compared in the comparator with the color tubes. On the color tubes are figures which indicate the milligrams of sodium bromide per hundred cubic centimeters of serum.[12]

TABLE 1.—*Errors with Different Concentrations and Different Serums*

Bromides Added to Serum, Mg. for 100 Cc.	Bromides Found in Filtrate by Comparator, Mg. per 100 Cc.	Error in Mg.	Error in Percentage
25	24.62, 24.75, 25.0	−0.38, −0.25, ±0	1.5, 1.0, 0
50	46.0, 46.0	−4.0, −4.0	8.0, 8.0
75	64.52, 67.45, 67.65	−10.48, −7.55, −7.35	13.9, 10.1, 9.8
100	91.85, 93.23, 98.0	−8.15, −6.77, −2.00	8.2, 6.8, 2.0
150	130.3, 130.7, 134.0, 132.3	−19.7, −19.3, −16.0, −17.7	13.1, 12.9, 10.7, 11.1
200	177.5, 175.9, 172.6	−22.5, −24.1, −27.4	11.3, 12.0, 13.7

Having described the mode of action of the bromides in its relation to the chlorides, and the method and apparatus for approximate bromide determination, I will demonstrate the practical application of the method in a series of cases.

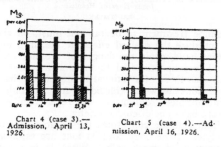

Chart 4 (case 3).— Admission, April 13, 1926.   Chart 5 (case 4).—Admission, April 16, 1926.

REPORT OF CASES

CASES[13] 1 and 2.—Varying doses of sodium bromide were given. The charts (2 and 3) were constructed by determining the total halogen amount of the serum by titration with silver nitrate; the bromides were then determined in the serum with the colorimetric method; the chlorides then could be calculated. The black bars represent the chlorides expressed in milligrams of sodium chloride per hundred cubic centimeters of serum. The shaded areas represent the bromides as sodium bromides per hundred cubic centimeters of serum. Both charts demonstrate how the bromides increase slowly with the increased doses of sodium bromides; they also show how the chlorides are diminished as the bromides increase, and finally how bromides are retained in the blood for quite a length of time after bromide administration has been stopped. The results obtained by this simple method check well with the previous results of those authors who have worked with more exact quantitative methods.

Of greater interest are the observations in pathologic cases.

CASE 3.—A woman, aged 23, of the schizophrenic reaction type, having taken bromides in large doses, amount not known, before coming to the hospital, was admitted, April 13,

12. The comparator with the necessary agents and pipets is manufactured by the La Motte Chemical Products Company, McCormick Building, Baltimore.

13. In the cases reported, all determinations were made with the colorimeter, with which the percentage of error due to serum absorption of the bromides varies from 0 to −13 per cent. The comparator does not aim to estimate closer than 25 mg. per hundred cubic centimeters of serum. Therefore the absorption error can be disregarded, as the comparator allows for variations from 33 per cent in the lower to 12.5 per cent in the higher concentrations. The absorption error therefore always falls within the limits of error of the comparator. Each comparator has been checked against standards with the colorimeter.

1926, with hallucinations, and very severe acne, but without neurologic disturbances.

This was a case of bromide intoxication, the blood containing 264.2 mg. of sodium bromide per hundred cubic centimeters of serum. The acne cleared up rapidly when bromide administration was stopped and sodium chloride solution was given by mouth. The antagonistic displacing behavior of the chlorides stands out plainly.

CASE 4.—A woman, aged 58, had been treated with bromides for some time, doses not known, until suddenly she fell into coma. The disturbance manifested itself in disorientation; drowsy, tearful hallucinations; limitation of the conjugate movements of the eyes; limited, somewhat masklike facial movements; difficulty in swallowing, and regurgitation through the nose; a speech defect suggesting the bulbar type; tremor of the arms; rigidity of the arms and legs to passive movements; hyperactive reflexes; absence of abdominal reflexes, and unstable gait. Because of these symptoms, the case was previously diagnosed as questionable encephalitis, cerebral arteriosclerosis, or pseudobulbar palsy.

When the patient was admitted, April 16, 1926, she had been for some time without bromides, so that the serum bromide content which, April 21, was still 124.4 mg. per hundred cubic centimeters of serum, must have been considerably higher before this date, and in the toxic zone. The condition cleared up rapidly after the administration of salt solution.

CASE 5.—A woman, aged 45, had been taking bromides at home for "nervousness," doses unknown, since the spring of 1925. She developed acute delirium, June 26, 1926, and when admitted, June 28, showed restlessness, disorientation, misinterpretations, visual hallucinations, tremor of the tongue and hands slurred speech, ataxic, reeling gait, and acne. The sodium bromide content of the serum was 339 mg. per hundred cubic centimeters. Treatment with about 150 grains of sodium chloride (10 Gm.) daily in water caused the ataxia to subside in five days; in ten days the speech was clear, and hallucinations were less vivid, and the sixteenth and seventeenth days were free of hallucinations for many hours. The twenty-third day the patient emerged from the delirium, oriented, and with full insight.

Chart 6 (case 5).—Admission, June 28, 1926.

This was a case of pronounced bromide intoxication with mental and neurologic symptoms. The chart demonstrates that it is sufficient to determine the absolute amount of bromides in the blood without chloride estimation.

CASE 6.—A man, aged 52, who had taken 80 grains (3.2 Gm.) of sodium bromide a day, from May 14, 1926, until May 29, 1926, became sleepy and drowsy, May 29. He was admitted to the hospital June 2, disoriented, delirious and talkative, with visual hallucinations and confabulations. There were thick speech, tremor of the tongue and fingers, absence of abdominal and cremasteric reflexes, and hypalgesia of skin. With chloride administration, the patient recovered completely in eighteen days.

Chart 7 (case 6).—Admission, June 3, 1926.

This patient also represented a case of severe bromide intoxication with delirium and neurologic symptoms, which yielded promptly to bromide elimination.

CASE 7.—A woman, aged 53, had had hypertension with headaches for twenty years. She had taken bromides for five years, amount not known. Now and then morphine had been taken, but the habit had not been formed; the last dose had been taken in February, 1926. There had been five confusional periods with hallucinations since November, 1925. When admitted, May 1, 1926, the patient was approximately oriented. She had visual and auditory hallucinations, tremor of the tongue and underactive knee jerks. Ankle jerks and abdominal reflexes were absent. No medication was given at first; later, salt solution was administered. The visual hallucinations stopped first; then orientation returned. After sixteen

days there were free intervals; on the twenty-eighth day the patient was free of mental symptoms and had good insight. She was discharged, May 30, 1926.

As this was one of our early cases, the blood was not tested for bromides until fourteen days after admission. The bromide content was then still 111.3 mg. per hundred cubic centimeters after fourteen days of bromide abstinence. The level on admission must therefore have been considerably higher.

CASE 8.—A woman, aged 61, had suffered from sleeplessness since October, 1926. She took approximately 90 grains (5.8 Gm.) of bromides from the end of October, 1926, to November 15. When admitted to the hospital, November 17, she was delusional, disoriented, and subject to misinterpretations; she had thick speech, difficulty in walking. muscular incoordination, blood pressure from 160 to 190, and albumin, + +, in the urine. The patient developed bronchopneumonia and died, November 27. The bromide content, November 18 (nine days before death), was 207 mg. per hundred cubic centimeters of serum.

Chart 8 (Case 7). — T w o weeks after admission.

This case is one of a bromide intoxication in a patient with cardiovascular disorder. It is of interest that the patient went through a confusional state with speech difficulty in July, 1926, which cleared up entirely with cessation of bromides.

CASE 9.—A man, aged 50, had had difficulty in sleeping since June, 1926. He took 45 grains (3 Gm.) of bromides a day from August 17 to September 17. Then he became delirious, and was admitted, Sept. 20, 1926. He was disoriented, with hallucinations, poor insight and memory, and restlessness. The pupils were irregular, and reacted sluggishly; the tongue protruded to the left with tremor. There was tremor of the lips and fingers; the Romberg sign was positive, the speech slurred, and the writing poor. The blood and spinal fluid (0.4 cc.) Wassermann reactions were positive; 14 cells; Pandy, + + + +; mastic and gold curve, negative. Septem-

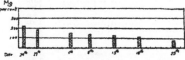

Chart 9 (case 9).—Admission, Sept. 20, 1926.

ber 28, the patient exhibited his first period of clearness; October 3, he became clear; October 6, there was a flareup of fear and hallucinations, which occurred also on October 11, 18 and 22. He was discharged, October 25, with no personality deterioration, normal memory and judgment. The speech was slightly slurred. The pupils were normal and the neurologic examination was negative.

This case represents a bromide delirium in a case of cerebrospinal syphilis. Without the bromide determination, we would not have recognized that there was a superimposed bromide intoxication and might have thought that the clinical picture was merely due to metasyphilitic disturbances.

CASE 10.—A woman, aged 51, feeling "run down," began, June 17, to take about 100 grains (7.5 Gm.) of bromides a day. She became weak and forgetful; August 5, she was forced to go to bed, because of delusions. August 13, the condition became worse, with disorientation, delusions, and thick speech. When admitted, Aug. 17, 1926, the patient

Chart 10 (case 10).—Admission, Aug. 17, 1926.

was disoriented, with confabulations, delusions, auditory hallucinations and no insight; test phrases were badly slurred, and abdominal reflexes were absent. August 23, the hallucinations and delusions persisted, although sometimes the patient was oriented. Salt was given by mouth. September 13, the patient was still disoriented to a large extent, with delusions and hallucinations. September 16, she felt better. During the next eight days she was only once delirious during the night;

there was growing insight. October 6, the mental condition was quite clear.

This was a case of bromide intoxication which presented mental symptoms of a delirium.

COMMENT

The series of cases shows how the clinical picture of bromide intoxication varies. The only sign may be a very severe acne; in other cases neurologic symptoms may occur which, together with euphoria, dulness and defective memory, suggest the clinical picture of a general paralysis; again, other cases show delirious states and some finally show an admixture of the bromide intoxication symptoms with the symptoms of the more fundamental disease or reaction type.

It is not my intention, however, to enter on the clinical aspects of the bromide intoxication. I wish only to draw attention to two facts. The first is the frequency of bromide intoxications, which still seems to be somewhat underestimated and is demonstrated by the fact that I was able in six months in 238 admissions to find fifty patients who had bromides in their blood and twenty who showed signs of bromide intoxication—a total of about 21 per cent of admitted patients who had been treated previously with bromides. The second point is that in most of these cases intoxication probably could have been avoided. This

TABLE 2.—*Urinary Chloride Excretion and Intake*

	Volume, Cc. (24 Hours)	NaCl., Gm.
F. S.	1,205	2.86
N. R.	220	2.68
J. S.	1,000	7.74
A. K.	470	6.61
J. T.	2,225	6.68
F. H.	1,775	9.05
C.	1,175	3.76
A. M.	1,350	7.80
S. T.	900	4.59
B.	1,100	5.28
C. H.	1,150	5.40
B.	2,360	13.9
K.	650	7.1
W. G.	1,275	7.56
A. M.	1,155	5.20
F. D.	2,065	15.84
A. B.	630	3.59
K. T.	825	8.2
W. G.	1,595	10.4
J. H.	1,950	7.8
H.	1,110	9.3
W. B.	1,000	13.2
M.	1,275	8.02
W. D.	1,720	6.88
G.	1,635	5.04

is seemingly contradicted by the fact that, in some cases, the doses that were taken were by no means abnormally high. But this contradiction is only apparent. An intoxication can doubtless be produced with the small customary doses of sodium bromide because the chloride intake may be too low; i. e., in an individual case the "normal" dosage of bromides is too high in relation to the patient's chloride balance. That this was the case in most of my patients can be seen by the high absolute bromide content of the serum. It has been demonstrated that equal doses of chlorides and bromides lead with certainty to an intoxication in about seventeen days to three weeks.[14] The routine dosage of bromides (from 45 to 75 grains [3 to 5 Gm.] a day) is based on the chloride intake of about 15 Gm. (225 grains) a day. But not all persons have a chloride intake of 15 Gm. Table 2, which shows the urinary chloride excretion and consequently also the

14. Toulouse: Gaz. d. hóp., 1900, p. 10; Rev. de psych., 1900; Gaz. d. hóp., 1904. Ulrich (footnote 9).

intake, taken from patients who had no other disease leading to a disturbance of chloride metabolism, such as pneumonia, cancer or nephritis, and who were on the relatively uniform hospital diet, demonstrates the enormous individual variations in chloride excretion and therefore intake. Taking into account the interaction of bromides and chlorides, it is evident that if these individual variations of chloride intake are not considered it is merely a matter of luck whether bromide treatment is successful or not, or whether it does or does not lead to intoxication. These facts justify a demand for a clinical method for bromide estimation, and the use of the method for patients under bromide therapy.

The ideal control would be to determine both the chloride and the bromide content of urine and blood. That, however, will be an impossible task for a practitioner and a very difficult one for institutions, such as state hospitals, which have to deal with a large number of patients under bromide treatment. Therefore, I have tried to present some simpler methods.

URINE TEST IN BROMIDE INTOXICATION

For the chemical diagnosis in a questionable case of bromide intoxication, it will be advisable first to try a simple test for bromides in the urine. If this is negative bromide intoxication can be excluded. This method is as follows:

To 25 cc. of urine about 1 Gm. of animal charcoal is added, mixed well, allowed to stand for a few minutes, and then filtered. To exactly 5 cc. of filtrate measured into a test tube, 1 cc. of 20 per cent trichloracetic acid and 1 cc. of 0.5 per cent gold chloride solution is added; a brown shade indicates the presence of bromides. In order to be able to recognize the color change more easily, it may be advisable for those not yet acquainted with the reaction to prepare the following bromide solution and to treat as follows: To 0.5 cc. of 0.1 per cent sodium bromide solution, 4.5 cc. of water, 1 cc. of 20 per cent trichloracetic acid and 1 cc. of 0.5 per cent gold chloride solution are added. (It must be borne in mind that the presence of iodides will be indicated by a precipitation, whereas in the absence of iodides no precipitation appears.)

If this test is positive, then it will be necessary and essential to determine the absolute bromide content of the serum with the comparator.

For the control of any prolonged bromide treatment it would, of course, be of value if possible to determine the patient's chloride balance in the urine before starting treatment. This control of the urine chlorides could prevent overdosing the patient from the onset. The salt intake, doubtless to a large extent, is determined by individual habit, and the estimation of the chlorides in the urine would inform one, for instance, whether the patient is taking 5 or 10 Gm. daily. In the former case, the bromides may not be prescribed in higher doses than 2 Gm. (30 grains) a day, as a patient otherwise would doubtless develop an intoxication within about twenty days; on the other hand, with a salt intake of 10 Gm. or more, the dose of 30 grains of bromides a day would probably be insufficient.

However, whether this preliminary chloride determination can be made or not, I should like to lay the main stress on the importance of determining the absolute bromide content of the serum. The importance of such a control for the prevention of bromide intoxication has been illustrated by the cases reported.

BROMIDE TREATMENT IN EPILEPSY

Finally, I will add a few words about the bromide treatment of epilepsy.

Ulrich was the first to control the bromide saturation or relative bromide content of his epileptic patients; that is to say, the percentage of total halogens in the urine represented by bromides. He found large individual differences as to the optimal individual saturation. Of forty-eight epileptic patients who had been kept free of convulsions for several years by bromide therapy, ten showed a relative bromide content of from 1 to 5 per cent; nineteen, from 5 to 10 per cent; five, from 10 to 15 per cent; eight, from 15 to 20 per cent, and six, from 20 to 25 per cent. Ulrich's results probably are the best ever obtained by bromide treatment; a status epilepticus has not occurred with his house patients, numbering from 300 to 400, since 1911. These results can be obtained only by a chemical study and control of the patient's urinary or blood bromides. That this method has not been taken up as readily as one would have expected, and as would have been justified, is due to the complexity and methodological difficulties of the Berglunds method, even in Bernoulli's modification, which is used by Ulrich. I hope I have eliminated this difficulty to a great extent by having given a relatively simple technic.

Although my work, to date, with this simplified technic has not been extensive (fifty cases), I feel justified in setting rough standards as to the safety limits of bromide content in the blood serum. I feel that it is not wise to exceed a limit of 125 mg. of sodium bromide per hundred cubic centimeters of serum for the average patient. It seems likely that the bromide tolerance of patients suffering from anemia, malnutrition, cardiorenal disease, and possibly also alcoholism and drug addiction, is lower, and therefore this arbitrary standard may be a little high in these particular cases. On the other hand, in some epileptic patients it is well known that certain factors tend to increase the frequency of convulsions (such as menstruation, unavoidable periods of overwork and excitements, or dietary alterations of salt balance), and under these circumstances it will probably be wise to allow this level of blood bromides to be exceeded. When the blood bromides increase beyond 150 mg. per hundred cubic centimeters of serum, I believe that the patient is liable to intoxication symptoms and therefore speak of bromide content above this figure as the "toxic zone."

As a matter of practical handling of epileptic patients, I recommend the following: It is wise to keep the patient slightly above the lowest level at which his convulsions have disappeared. This slightly increased level will guard against the effect of any sudden increase in salt intake. This will eliminate the possibility of the development of status epilepticus—an ever present serious danger for the epileptic patient. I feel that if convulsions, in the more severe cases, cannot be controlled by bromides up to from 170 to 200 mg. per hundred cubic centimeters of serum, there is no use to attempt to push bromide therapy—in fact, I should advise against maintaining this level for any length of time.

SUMMARY

There are two simplified methods for determining the extent of bromide intoxication: (1) a qualitative test for bromides in the urine, and (2) a comparator method for the clinical estimation of the bromide content of the blood serum.

The qualitative urine test is useful in (*a*) cases suspected of being bromide intoxications, and (*b*) puzzling delirious and organic neurologic cases, which not infrequently prove to be due to bromide intoxication.

In fact, with the simplicity of the method it deserves to be made a routine urine examination. With the frequency of bromide intoxications, it will well repay the little amount of work it requires.

The control of the bromide content in the serum seems essential when bromides are being used either as a sedative or in the bromide treatment of epilepsy.

The chemical control may help to prevent the mistaking of bromide intoxication for organic brain disease (organic deliriums, epileptic psychosis, general paralysis) in epileptic patients who have been uncontrollably dosed with bromides for a long period of time.

In both cases, whether bromides are used as a sedative or as an antiepileptic, the physician will be enabled by the chemical control to have a clear record of the case and also to be safeguarded against self-prescribed

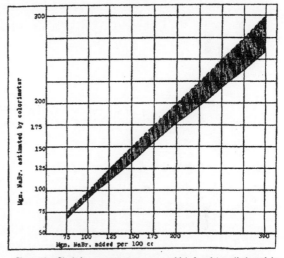

Chart 11.—Shaded area represents error which has been eliminated by standardizing the color tubes in the comparator, accordingly; broken line, sodium bromide in water compared with water standard; solid line, sodium bromide in serum compared with water standard.

changes on the part of the patient. He will be able to avoid insufficient doses on the one hand and intoxication on the other hand, and thus be in a position to render a bromide treatment more rational and consequently more successful.

A CORRECTION OF COMPARATOR STANDARD TO ELIMINATE ERROR OF SERUM ADSORPTION

Our previous experiments showed that serum adsorbed a certain amount of bromides—this amount increasing proportionally to the concentration of the bromides, and that the amount adsorbed is practically independent of the particular blood serum (between 1 and 4.1 per cent). This adsorption loss is indicated in chart 11 by the shaded portion, which represents the difference in the amount of bromides estimated from watery solution and from serum solutions of known quantities of bromides. The standard solutions in the comparator are corrected to eliminate this error.

COMMENTARY TO

7. Bratton, A. C. and Marshall, E. K., Jr. (1939)
A New Coupling Component for Sulfanilamide Determination. Journal of Biological Chemistry 128(2): 537–550.

In the early 1930s Gerhard Domagk in Germany discovered the first clinically effective antibiotic. The drug was a red azo industrial dye named 4'-sulfamyl-2,4-diaminoazobeneze and was sold under the trademark name Prontosil™ (1). Prontosil turned the patient's skin red, but it cured otherwise fatal streptococcal infections. It was effective in humans and animals but not *in vitro* in bacterial cultures. Studies in humans by numerous investigators including Colebrook and Kenny in England in 1935 (2) confirmed the clinical efficacy of Prontosil. Trefouel and co-workers in that same year at the Pasteur Institute determined that the active portion of the molecule was the sulfanilamide moiety and not the red dye (3). Biotransformation *in vivo* released the active compound sulfanilamide from the Prontosil molecule. Long and Bliss in the United States (US) soon reported on the clinical efficacy of sulfanilamide (4). When the son of the President of the US was cured of a life threatening streptococcus infection with sulfanilamide the front-page headline in the *New York Times* on December 17, 1936 read "Young Roosevelt Saved by New Drug" (5). This helped launch the era of antibiotics in the US.

Sulfanilamide was released at a time in history when safety and efficacy studies were not required prior to the drug being sold for use in humans. In the US, the Food and Drug Administration (FDA) laws required only that the product label list all the active ingredients. Eli Kennerly Marshall, Jr. at The Johns Hopkins Hospital whose paper is presented here was a dissenting voice in the rush to embrace this new wonder drug and the analogues that soon followed. Marshall produced reports on the sulfonamides that defined for the first time the pharmacokinetic characteristics of these drugs. He insisted that all new drugs must be carefully studied before being used in humans. His February 1937 paper in *Science* described for the first time a method for the measurement of sulfanilamide in biological fluids. This was almost five years after the drug was first used in humans. He used this method to study its metabolism in animals and humans (6). A month latter he reported on the blood levels in humans following standard oral dosing (7). He established that concentrations in the tissues are dependent on blood levels and recommended dosing intervals to maintain adequate levels to ensure clinical efficacy. He demonstrated that the drug passes the blood brain barrier and is present in cerebrospinal fluid. The method he developed was a colorimetric procedure that was based on a diazo dye formation with the free amino group on the benzene ring of the drug. The blue-purple color was read visually or with a Dubsocq colorimeter (8,9).

By the summer of 1937 Squibb, Merck, Calco, Lederle, Winthrop, Eli Lilly and Parke–Davis had sulfanilamide on the market without the benefit of any safety studies (10).

An editorial in the *Journal of the American Medical Association (JAMA)* with the title "Sulfanilamide-A Warning" claimed, "Many months of investigations of the pharmacology, toxicology, and clinical application of new preparations under carefully controlled conditions are needed to provide evidence of therapeutic value" (11). Four months latter Marshall, Cutting and Emerson published in the same journal a study on the acute toxicity of sulfanilamide in rabbits, dogs and mice along with the chronic toxicity in dogs and rats (12). Other researchers used Marshall's method to demonstrate that the drug passed into the human fetus following oral administration to the mother (13–14). Between September and October 1937 in the US 76 people died after taking an elixir of sulfanilamide. The newly released preparation contained diethylene glycol as a solvent. No animal toxicity studies were required or performed before release of the product (15).

The success of sulfanilamide led to the development of a wide range of structural analogues. In 1939 Marshall in *JAMA* referred to the release of these new drugs with out adequate experimental studies as "stupid and unscientific" (16). Marshall continued to refine his method and used it to study the pharmacokinetics and toxicity of the different sulfonamides (17–18). The paper presented here by Bratton and Marshall describes the final version of their assay. A protein free filtrate of blood is made with trichloroacetic acid (TCA) followed by the addition of nitrous acid to form a diazo link to the free amino group on the benzene ring. Sulfamate ions were used to remove free nitrous acid and the p-diazobenzenesulfonamide drug was coupled to N, N-dimethyl-1-naphthylamine to form a color complex. Goth substituted acetone for TCA and eliminated the need for filtration (19). The A.S. Aloe Company in St. Louis produced a test kit for sulfonamides that provided all the reagents for the Bratton–Marshall assay except for the acetone, in tablet form. An advertisement for the product claimed that the assay could be run on 200 μL of blood at the bedside in 7–8 minutes (20).

Marshall was elected President of the American Society for Pharmacology and Experimental Therapeutics in 1942 and died in 1966 (21). Today sulfonamides are most often measured in biologicals by high performance liquid chromatography (HPLC) (22).

The Bratton–Marshall assay however continues to find use in specialized applications. It has been used in pharmacogenomic studies when a sulfonamide is used to test for acetylator status (23). Laikind, Seegmiller and Gruber developed a novel application for the assay in 1986 (24). They used it to screen for elevated levels of succinyl-adenosine and ribosyl-4-(N-succinyl-carboxamide)-5-aminoimidazole in urine as an indicator of a rare neurological disorder due to a deficiency of the enzyme adenylosuccinate monophosphate lyase (EC 4.3.2.2). Urine specimens were collected onto filter paper strips and dried. The Bratton–Marshall assay was run on a simple water extract of the paper strips. Results were found to agree with HPLC.

References

(1) Domagk, G. (1935) Ein Beitrag zur Chemotherapie der Bakteriellen Infektionen. Deutsche Medizinische Wochenschrift. 61(7):250–253.

(2) Colebrook, L. and Kenny, M. (1936) Treatment of human puerperal infections, and of experimental infections in mice, with prontosil. The Lancet. 227(5884):1279–1286.

(3) Trefouel, J., Trefouel, J., Nitti, F., and Bovet, D. (1935) Activite du p-aminophenylsulfamide sur les Infections Streptococciques experimentales de la Souris et du Lapin. Comptes Rendus Societe de Biologie. 120:756–758.

(4) Long, P.H. and Bliss, E.A. (1937) Para-Aminobenzene-Sulfonamide and its derivatives. Experimental and clinical observations on their use in the treatment of beta-Hemolytic streptococcic infections: A preliminary report. Journal of the American Medical Association. 108(1):32–37.

(5) [Anonymous] The New York Times, Thursday, December 17, 1936, pgs 1 and 10.

(6) Marshall, E.K. Jr. and Cutting, W.C. (1937) Acetylation of para-aminobenzenesulonamide in the animal organism. Science. 85(2199):202–203.

(7) Marshall, E.K. Jr., Emerson, K., and Cutting, W.C. (1937) Para-aminobenzenesulfonamide. Absorption and excretion: method of determination in urine and blood. Journal of the American Medical Association. 108(12):953–956.

(8) Marshall, E.K. Jr. (1937) Determination of sulfanilamide in blood and urine. Proceedings of the Society for Experimental Biology and Medicine. 36(3):422–424.

(9) Marshall, E.K. Jr. (1937) Determination of sulfanilamide in blood and urine. Journal of Biological Chemistry. 122(1):263–273.

(10) [Anonymous] Council on Pharmacy and Chemistry. (1937) New and nonofficial remedies. Examination of Certain American Brands of Sulfanilamide. Journal of the American Medical Association. 109(5):358–359.

(11) [Editorial] (1937) Sulfanilamide — A Warning. Journal of the American Medical Association. 109(14):1128.

(12) Marshall, E.K. Jr., Cuting, W.C., and Emerson, K. (1938) The toxicity of sulfanilamide. Journal of the American Medical Association. 110(4):252–257.

(13) Speert, H. (1938) The passage of sulfanilamide through the human placenta. Bulletin of the Johns Hopkins Hospital. 38:337–339.

(14) Barker, R.H. (1938) The Placental Transfer of Sulfanilamide. New England Journal of Medicine. 219(2):41.

(15) Geiling, E.M.K. and Cannon, P.R. (1938) Pathological effects of elixir of sulfanilamide (diethylene glycol) poisoning. A clinical and experimental correlation: final report. Journal of the American Medical Association. 111(10):919–926.

(16) Marshall, E.K. Jr. (1939) An unfortunate situation in the field of bacterial chemotherapy. Journal of the American Medical Association. 112(4):352–353.

(17) Marshall, E.K. Jr. and Litchfield, J.T. (1938) The determination of sulfanilamide. Science. 88(2273):85–86.

(18) Marshall, K. Jr., Bratton, A.C., and Litchfield, J.T. (1938) The toxicity and absorption of 2-sulfanilamidopyridine and its soluble sodium salt. Science. 88(2295):597–599.

(19) Goth, A. (1942) A simple method for determining sulfonamides in blood. Journal of Laboratory and Clinical Medicine. 27(6):827–829.

(20) [Advertisement] (1943) Rapid Sulfonamides Test Kit, A.S. Aloe Company, St. Louis, MO. The American Journal of Medical Technology. 9(4), July.

(21) Harvey, A.McG. (1976) The Story of Chemotherapy at Johns Hopkins: Perrin H. Long, Eleanor A. Bliss, and E. Kennerly Marshall, Jr. The Johns Hopkins Medical Journal. 138(2):54–60.

(22) Bury, R.W. and Mashford, M.L. (1979) Analysis of trimethoprim and sulphamethoxazole in human plasma by high-pressure liquid chromatography. Journal of Chromatography. 163(1):114–117.

(23) Whelpton, R., Watkins, G., and Curry, S.H. (1981) Bratton–Marshall and liquid-chromatographic methods compared for determination of sulfamethazine acetylator status. Clinical Chemistry. 27(11):1911–1914.

(24) Laikind, P.K., Seegmiller, J.E., and Gruber, H.E. (1986) Detection of 5′-phosphoribosyl — 4-(N-succinylcarboxamide)-5-amino-imidazole in urine by use of the Bratton–Marshall reaction: identification of patients deficient in adenylosuccinate lyase activity. Analytical Biochemistry. 156(1):81–90.

J. Biol. Chem. 1939; 128(2): 537–550

A NEW COUPLING COMPONENT FOR SULFANILAMIDE DETERMINATION*

By A. CALVIN BRATTON and E. K. MARSHALL, Jr.

WITH THE TECHNICAL ASSISTANCE OF DOROTHEA BABBITT AND ALMA
R. HENDRICKSON

(*From the Department of Pharmacology and Experimental Therapeutics,
The Johns Hopkins University, Baltimore*)

(Received for publication, March 4, 1939)

The method proposed for the determination of sulfanilamide (1–3) has been widely used in estimating the drug in blood and urine both in experimental work and in controlling the dosage of the drug for patients. During the 2 years since the method has been in use, certain disadvantages have become apparent. The use of N, N-dimethyl-1-naphthylamine (dimethyl-α-naphthylamine) as the coupling component for the diazotized sulfanilamide is not entirely satisfactory on account of the necessity of a catalyst for rapid development of color in dilute solutions, the need of a large excess of the reagent, and the necessity of a certain amount of alcohol to keep the resultant azo dye in solution. A coupling component which can be obtained in the form of a crystalline salt of reproducible composition and which gives a soluble azo dye in acid solution appeared desirable. The other defect which was discovered in the method was that certain samples of dimethyl-α-naphthylamine did not give complete recovery of sulfanilamide added to normal blood. This was found to be due to the salts (mainly chloride) present in the blood filtrate catalyzing the destruction of the azo dye by the excess nitrite.

Modifications of our method by various authors offer no real advantages. In the main, these procedures have consisted in altering the amount of blood and reagents used (4), a purification of the coupling component (5), or a restatement of slight modifica-

* This investigation has been aided by a grant from the John and Mary R. Markle Foundation.

538   Sulfanilamide Determination

tions already described by the author (6, 7). Two important improvements in the method were described about a year ago (8); namely, the destruction of excess nitrite by ammonium sulfamate and the buffering of the diazotized solution before coupling with dimethyl-α-naphthylamine. The destruction of excess nitrite, by preventing the formation of nitroso compounds, allows the use of a much wider variety of coupling components than is otherwise possible. Previously, but unknown to us at the time, Hecht (9) determined N^4-sulfanilyl-N^1, N^1-dimethylsulfanilamide in urine and blood by coupling the diazotized compound with N-ethyl-1-naphthylamine after destruction of the excess nitrite with sulfamic acid or urea.

We decided that the ideal coupling agent for determination of sulfanilamide should exhibit rapidity of coupling, sensitivity, purity, and reproducibility, be unaffected in rapidity of coupling by changes of pH from 1 to 2, and that the azo dye formed should be acid-soluble and not affected in color by pH changes from 1 to 2. A number of compounds which appeared to be promising in these respects and which couple in acid solution[1] have been examined.

The rapidity of coupling (speed) was noted for diazotized 0.1 and 0.01 mg. per cent solutions of sulfanilamide buffered to pH 1.3; also the influence of pH on the rapidity of coupling with a 1 mg. per cent solution, the sensitivity with a 0.01 mg. per cent solution, the solubility of the azo dye with a 10 mg. per cent solution, and the effect of pH on the color of the dye in the same solutions used for determining the effect of pH on speed. Trichloroacetic acid was used for acidification, excess nitrite was destroyed with ammonium sulfamate, and pH was varied by adding excess acid or sodium dihydrogen phosphate. In Table I are summarized these preliminary tests on seventeen compounds. Two aqueous solutions of each coupling component were used, one of such a strength that 10 moles per mole of diazonium salt were used for the 10 mg. per cent solution of sulfanilamide and the other one-tenth as strong for use with the other dilutions of the drug. A minimal quantity of hydrochloric acid was used to dissolve coupling Compounds 1 to 3, 6 to 9, and 13; sodium

[1] Coupling in alkaline solution has certain disadvantages which have been cited (3).

A. C. Bratton and E. K. Marshall, Jr.   539

hydroxide was employed on Compounds 4, 10 to 12, 15, and 16; alcohol was necessary for Compound 17. Coupling Compounds 2 to 4, 7, and 10 to 17 were obtained from E. I. du Pont de Nemours and Company, Inc., through the courtesy of Dr. H. A. Lubs and Dr. D. E. Kvalnes. Compounds 1,[2] 2, 4, 7, 8, 15, and 17 were prepared or purified from the commercial base or salt before use, Compounds 1, 6, 8, and 9 were from the Eastman Kodak Company, and Compound 5 was synthesized by us.

On the basis of rapidity and sensitivity only five compounds (Nos. 2, 3, 5, 7, and 8) offer improvement over N,N-dimethyl-1-naphthylamine (No. 1). Of these, Compounds 7 and 8 are eliminated because they yield precipitates with diazotized 1 mg. per cent sulfanilamide. Compound 3 is eliminated because of the considerable influence of pH on rapidity of coupling, leaving only Compounds 2 and 5 for consideration. Compound 2 is very difficult to purify, and possesses no definite physical properties to aid in its characterization. It is to be expected that different batches of the compound would show considerable variation in purity, and the lack of reproducibility would render it unsuitable for use in the method. Compound 5, N-(1-naphthyl)ethylenediamine dihydrochloride may be readily prepared in a state of high purity. Its coupling is very rapid and uninfluenced by pH in the range of 1 to 2. This range of pH has no effect on the color of the dye, and the dye is more soluble in this range than that from any other coupler we have examined.

Preparation and Purification of N-(1-Naphthyl)Ethylenediamine Dihydrochloride

No adequate chemical description was given by Newman (10, 11) who first prepared this compound through the Gabriel synthesis. Therefore, the following modifications of Newman's synthesis and a rather detailed characterization of the compound seem justified.

β-(1-Naphthylamino)Ethylphthalimide—This was prepared by essentially the procedure of Newman, except that 2.5 moles of 1-naphthylamine to 1 mole of β-bromoethylphthalimide were used, and the product was recrystallized from glacial acetic acid.

[2] This sample of dimethyl-α-naphthylamine was a pure sample which had been aerated 30 minutes at 267° (3), and distilled at reduced pressure.

TABLE I

Suitability of Some Coupling Components for Sulfanilamide Determination

Compound	Speed for 0.01 and 0.1 mg. per cent solutions at pH 1.3	Color of dye, 10 mg. per cent solution	Influence on coupling speed, 1 mg. per cent solution			Sensitivity, 0.01 mg. per cent solution	Ppt., 10 mg. per cent solution	Effect on dye color, 1 mg. per cent solution		
			pH 1.0	pH 1.3	pH 1.6			pH 1.0	pH 1.3	pH 1.6
1. N,N-Dimethyl-1-naphthylamine	+	Purple-red	+	++	++	+	++	++	++	++
2. N-(1-Naphthyl)glucamine	+++	"	+++	+++	+++	++	++	+++	+++	+++
3. N,N-Di-(hydroxyethyl)-1-naphthylamine	++	Violet	+++	+++	+++	+	++	+++	+++	+++
4. Sulfonated N-ethyl-1-naphthylamine	+	Violet-red	+	+	++	+	++	+	++	++
5. N-(1-Naphthyl)ethylenediamine dihydrochloride	+++	Purple-red	+++	+++	++	+	0	+++	+++	+++
6. 2-Naphthylamine	0	Orange				++	++			
7. N-Ethyl-1-naphthylamine hydrochloride	++	Purple	−	−	−	++	++			
8. N-Methyl-1-naphthylamine acid sulfate	++	Violet	−	−	−	+	++			
9. 1-Naphthylamine	++	Violet-red				+	++			
10. 1-Hydroxyethylamino-5-naphthol	++	Purple-red				0	++			
11. 1-Amino-5-naphthol	++	Violet				0	++			
12. Phenyl J acid (N-phenyl-2-amino-5-naphthol-7-sulfonic acid)	++	Orange-red				0	++			

13. N,N-Diethyl-1-naphthylamine	Red	0	+	0	+
14. H acid (1-amino-8-naphthol-3,6-disulfonic acid)	"	+	++	0	+
15. Phenyl peri acid (phenyl-1-naphthylamine-8-sulfonic acid)	Violet	++	++	0	++
16. Tolyl peri acid (p-tolyl-1-naphthylamine-8-sulfonic acid)	Purple	+	+	0	+
17. N-Phenyl-1-naphthylamine	"	—	+++	—	+++

0 = color no greater than blank; — = precipitate formation; + = degree of color, speed, etc.

542 Sulfanilamide Determination

The yield was 68 to 78 per cent of theory. If desired, half the 1-naphthylamine in the synthesis may be replaced by an equivalent amount of sodium bicarbonate, resulting in a yield of 55 to 60 per cent.

The product is extensively soluble in hot glacial acetic acid, benzene, and ethyl acetate, moderately in alcohol. It is moderately soluble in cold ethyl acetate and benzene, fairly soluble in alcohol, and slightly soluble in ether. It crystallizes from acetic acid in greenish yellow, thin, irregular plates, m. p. 165.3–165.7°. (A second recrystallization from this solvent with the addition of activated charcoal yields a golden yellow product.)

Hydrolysis of β-(1-Naphthylamino)Ethylphthalimide—The use of ordinary hydrolytic agents, even fuming hydrochloric acid as employed by Newman, results in poor yields. Either of the following two methods may be employed.

(a) *By Fused Sodium Hydroxide*—β-(1-Naphthylamino)ethylphthalimide was ground intimately with an equal weight of solid sodium hydroxide, and the mixture was distilled rapidly at 10 to 20 mm. pressure. The mass melted at 160°, an oily yellow distillate passed over at 200–300°, and the distillation was discontinued when the residue began to char at 340°. The distillate was extracted with the calculated amount of 0.05 N hydrochloric acid (stronger acid yielded a turbid extract), and after the material was decolorized with charcoal, the base was liberated with excess sodium hydroxide, taken up in benzene, and dried over solid sodium hydroxide. Upon distilling off the solvent and taking up the base in a little alcohol, the calculated amount of hot alcoholic picric acid was added and the solution was cooled. The red-brown picrate was recrystallized from 6 N acetic acid (or 95 per cent alcohol), and precipitated as tiny red octahedra in 40 to 70 per cent yield. Its melting point is not sharp, and depends upon the rate of heating: placed at 222° into a fairly rapidly rising bath, m.p. 227–228° with decomposition; in a more slowly heated bath, m.p. 225–226° with decomposition.

$C_{18}H_{17}N_5O_7$. Calculated. C 52.02, H 4.13, N 16.87
Found. " 52.21, " 4.12, " 16.82

Smaller still charges (8 to 10 gm.) gave higher yields, probably due to the lower heat gradient in the mass. Larger batches (60

A. C. Bratton and E. K. Marshall, Jr.   543

to 70 gm.) gave lower yields because of a side reaction involving scission of the side chain to yield 1-naphthylamine, whose picrate may be obtained from the alcoholic picric acid mother liquor by addition of 5 volumes of water. In one experiment with 0.0994 mole of β-(1-naphthylamino)ethylphthalimide, 0.0398 mole of N-(1-naphthyl)ethylenediamine picrate and 0.0124 mole of 1-naphthylamine picrate were obtained.

Liberation of Base—The picrate was suspended in warm water, treated with sodium hydroxide in slight excess, and the liberated N-(1-naphthyl)ethylenediamine was taken up in benzene and dried over solid sodium hydroxide. The dihydrochloride may be prepared by bubbling dry hydrogen chloride into the benzene solution, or the solvent may be evaporated and the base distilled under reduced pressure.

(b) *By Hydrazine*—Refluxing β-(1-naphthylamino)ethylphthal-imide in alcoholic suspension for 2 hours with hydrazine hydrate, followed by addition of excess hydrochloric acid and continuation of refluxing for 1 hour, results in an 85 to 90 per cent yield of fairly pure N-(1-naphthyl)ethylenediamine dihydrochloride. The procedure was essentially that of the general method of Ing and Manske (12). Substitution of the hydrazine hydrate by an equivalent amount of hydrazine acid sulfate, sodium carbonate, and a minimal amount of water gave a more nearly pure product in slightly lower yield (70 to 80 per cent).

Properties of N-(1-Naphthyl)Ethylenediamine—The base is a straw-yellow, viscous liquid with an odor resembling that of the alkyl naphthylamines; the boiling point is 204° at 9 mm., about 320° with decomposition at 760 mm.; $n_D^{25} = 1.6648$; $d_4^{25} = 1.114$. The solubility in water is about 0.2 gm. in 100 cc. at 25°, more soluble in hot than in cold water; the pH of a saturated aqueous solution is 10.5. The base is readily soluble in the common organic solvents, except petroleum ether. It distils very poorly with steam, even from concentrated alkali.

Salts of Base—The dihydrochloride is prepared by introducing dry hydrogen chloride into a solution of the base in benzene or ether, or by dissolving the base in excess hot 6 N hydrochloric acid. Recrystallized from 6 N hydrochloric acid, it precipitates in long colorless hexagonal prisms. Use of activated charcoal is of advantage in obtaining a perfectly white preparation. We

544 Sulfanilamide Determination

were unable to dry the dihydrochloride without loss of a little hydrogen chloride, which resulted in a poor analysis. A mixture of the mono- and dihydrochlorides is perfectly satisfactory for use in the method, but it is, of course, advantageous to strive for a pure dihydrochloride in order that the melting point may serve in identification and control of purity. For this reason, the excess mother liquor should be removed by pressing between filter paper and the bulk of the remaining water should be removed *in vacuo* or by air drying. The last traces of water are removed by heating briefly at 110° and transferring while still warm to a vacuum desiccator. M.p. (placed at 184° in a fairly rapidly rising bath) 188–190°. If the dihydrochloride is distilled at reduced pressure, a product is obtained which melts at 231–232° with slight decomposition, and is probably the monohydrochloride. The dihydrochloride is easily soluble in 95 per cent alcohol, dilute hydrochloric acid, and hot water; it is rather difficultly soluble in cold water, acetone, and absolute alcohol.

$C_{12}H_{16}N_2Cl_2$. Calculated. C 55.58, H 6.23, N 10.82, Cl 27.37
Found.   " 55.52, " 6.20, " 10.41, " 26.81

Prolonged drying of the analytical sample at 110° raised the determined nitrogen to 11.60 per cent and lowered the halogen to 18.76 per cent.

The zinc chloride and mercuric chloride salts and the acid sulfate were prepared but the first is too soluble, and the melting points of the latter two too high, to be of use in purification or identification of the base.

Determination of Sulfanilamide in Blood and Urine

Reagents[3]—

1. A solution of trichloroacetic acid containing 15 gm. dissolved in water and diluted to 100 cc.

2. A 0.1 per cent solution of sodium nitrite.

3. An aqueous solution of N-(1-naphthyl)ethylenediamine dihydrochloride containing 100 mg. per 100 cc. This solution should be kept in a dark colored bottle.

4. A solution of saponin containing 0.5 gm. per liter.

[3] The reagents can be obtained from LaMotte Chemical Products Company, Baltimore.

A. C. Bratton and E. K. Marshall, Jr.   545

5. 4 N hydrochloric acid.

6. A solution of ammonium sulfamate, containing 0.5 gm. per 100 cc.

7. A stock solution of sulfanilamide in water containing 200 mg. per liter.   This solution can be kept for several months in the ice box.   The most convenient standards to prepare from the stock solution are 1, 0.5, and 0.2 mg. per cent.   To prepare these 5, 2.5, and 1 cc. of the stock solution plus 18 cc. of the 15 per cent solution of trichloroacetic acid are diluted to 100 cc.

Procedure for Blood[4]—2 cc. of oxalated blood are measured into a flask and diluted with 30 cc. of saponin solution, and after 1 or 2 minutes precipitated with 8 cc. of the solution of trichloroacetic acid.   The free sulfanilamide is determined in the filtrate as follows: 1 cc. of the sodium nitrite solution is added to 10 cc. of the filtrate.   After 3 minutes standing, 1 cc. of the sulfamate solution is added, and after 2 minutes standing, 1 cc. of the solution of N-(1-naphthyl)ethylenediamine dihydrochloride is added.   The unknown is compared with an appropriate standard which has been treated as above.   This comparison can be made immediately and no change in color is observed for 1 hour or more.   To determine the total sulfanilamide, 10 cc. of the filtrate are treated with 0.5 cc. of 4 N hydrochloric acid, heated in a boiling water bath for 1 hour, cooled, and the volume adjusted to 10 cc.   The subsequent procedure is as stated above for determining free sulfanilamide.

Procedure for Urine—Protein-free urine is diluted to contain about 1 to 2 mg. per cent of sulfanilamide and 50 cc. of the diluted urine plus 5 cc. of the 4 N hydrochloric acid are diluted to 100 cc. 10 cc. of the product of this second dilution are treated as a blood filtrate for free sulfanilamide, and 10 cc. heated without further addition of acid for total sulfanilamide.   If the urine contains protein, it is diluted and treated by the procedure for blood.

Photoelectric Colorimeter—When a photoelectric colorimeter is available, dilutions of blood of 1:50 or 1:100 can be used.   The blood is diluted with water (saponin is unnecessary), allowed to stand a few minutes, and precipitated with trichloroacetic acid solution, with a volume which is one-fifth that of the final mixture.

[4] Sample and reagent volumes can be proportionately reduced to give the minimal amount of filtrate necessary for an accurate color comparison.

546   Sulfanilamide Determination

This allows the use of 0.1 or 0.2 cc. samples of blood which are measured with washout pipettes. Determinations on urine or other body fluids are easily made after appropriate dilution. The reagent blank on distilled water is quite low, but increases with time if the solution is left in the light. For this reason, solutions to be read in the photoelectric colorimeter should be protected from light unless the reading is made immediately. Some reaction occurs between the trichloroacetic acid and the N-(1-naphthyl)ethylenediamine, since solutions acidified with hydrochloric acid do not show an increased color on exposure to light. The blood blank is extremely low and negligible for most purposes. With a 1:50 dilution of human blood the correction due to the blood blank varies from 0 to 0.03 mg. per cent. This blank can be easily determined by performing an analysis as usual except that water is substituted for the sodium nitrite solution. When small concentrations of sulfanilamide are to be determined or when a foreign dye such as prontosil or neoprontosil is present, this procedure is quite useful. The color of the normal urinary pigments can be conveniently corrected for by the same procedure.

When only a very small amount of blood is available, as in the case of small animals such as mice, a determination can be made with considerable accuracy on 0.02 cc. The adaptation is essentially that described by Marshall and Cutting (13), the dilution of blood being 1:200 or 1:400, depending on the concentration of sulfanilamide present. The proportion of reagents used is the same as in the other adaptations of the method. Centrifugation before filtration of the protein precipitate is useful in securing the maximum amount of filtrate.[5]

In using a photoelectric colorimeter a filter is essential. With dimethyl-α-naphthylamine the peak of the absorption of the azo dye formed occurs at 530 mμ (14). When N-(1-naphthyl)-ethylenediamine is used, the peak of absorption is shifted to 545 mμ, and the dyes from sulfapyridine and N[1]-ethanolsulfanilamide show the same absorption peak. In Fig. 1 is reproduced an absorption curve of the azo dye from sulfanilamide. We wish

[5] For these dilutions (1 : 200 or 1 : 400, a photoelectric colorimeter of high sensitivity, designed and constructed by Dr. Morris Rosenfeld of this department, was used. A description of this instrument will be published.

A. C. Bratton and E. K. Marshall, Jr.   547

to thank Dr. Elizabeth E. Painter of the Department of Physiology, Columbia University, New York, for these absorption data.

Blood Dilution and Recovery—When a 1:4 dilution of blood is used as suggested in Fuller's method (15) and in Proom's (6) adaptation of our method, a result 10 per cent too low may be obtained. Sulfanilamide was added to three samples of mixed human blood to give about 10 mg. per cent, and determinations made with various dilutions for precipitation. The filtrates were diluted when necessary to read with the photoelectric colorimeter.

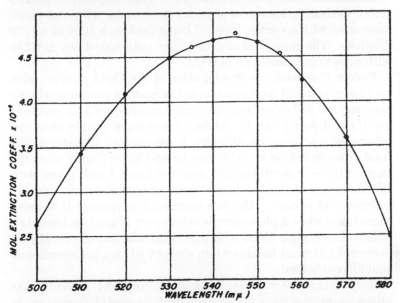

FIG. 1. Absorption curve of dye from coupling diazotized sulfanilamide with N-(1-naphthyl)ethylenediamine.

Average percentage recoveries were for the 1:4 dilution 90.0, for the 1:10 dilution 94.3, for the 1:20 dilution 96.9, and for the 1:50 dilution 99.5. Determinations on other samples of blood with added sulfanilamide gave similar recoveries for the 1:20 and 1:50 dilutions.

With acetylsulfanilamide recovery is not complete in a 1:20 dilution of blood (about 90 per cent) but is complete in a 1:50 dilution. A series of eight bloods to which varying amounts of acetylsulfanilamide were added gave recovery of 97.6 per cent

(96.2 to 99.2) with the dimethyl-α-naphthylamine reagent in a 1:50 dilution. Since experiments show that solutions of pure acetylsulfanilamide give theoretical results when hydrolyzed with hydrochloric acid and estimated by use of the new coupling component, no recoveries of the acetyl compound with the new reagent were made.

The accidental errors of the method can be best illustrated by some determinations made with the photoelectric colorimeter. Sulfanilamide was added to a sample of mixed human blood and fifteen determinations were made on the sample, 0.5 cc. of blood (measured with a syringe pipette) being used each time in a 1:50 dilution. The mean value in mg. per cent was 4.888 ± 0.039 with a maximum deviation of 0.113.

Protein Precipitation—Precipitation of the blood proteins with trichloroacetic acid has been adopted in place of p-toluenesulfonic acid for the following reasons: trichloroacetic acid of constant quality and purity can be obtained much more readily than can toluenesulfonic acid, no difficulty is experienced in determining total sulfanilamide in a trichloroacetic acid blood filtrate when the excess nitrite is destroyed (8), and the blood blank given with trichloroacetic acid is much less than that obtained when toluenesulfonic acid is used. The last mentioned advantage is of great importance when a photoelectric colorimeter is used, as the blank obtained on human blood with trichloroacetic acid can usually be neglected but must be taken into account when a toluenesulfonic acid filtrate is used.

Body Fluids Other Than Blood and Urine—No difficulty has arisen in estimating sulfanilamide and its acetyl derivative in other body fluids by the same procedure as used for blood. When tissues are to be analyzed, it appears desirable to extract the ground tissue in a Soxhlet apparatus with a limited amount of alcohol, dilute an aliquot portion of the extract with water, and proceed as in blood, with a photoelectric colorimeter. This is a simpler and less laborious method than the one previously used for tissues (16).

Sulfanilamide Derivatives—As previously indicated (3), our method can be used for determining diazotizable primary aryl amines containing either a free amino group or a blocked amino group which can be freed by hydrolysis. In the limited experi-

A. C. Bratton and E. K. Marshall, Jr.   549

ence which we have had in applying our method to the estimation of compounds other than sulfanilamide, three points of importance can be mentioned. With a very difficultly soluble substance, the recovery in the blood filtrate is generally not quantitative unless high dilutions (1:100 or greater) are used, so that either one must use such dilution or must resort to the original alcohol precipitation method (1). In the latter case, ammonium sulfamate is used to destroy excess nitrite and N-(1-naphthyl)ethylenediamine dihydrochloride is used as a coupling component. With certain derivatives, the azo dyes formed will not be acid-soluble and a certain amount of alcohol must be added with or just before the coupling component. With some substances buffering is necessary to obtain sufficient speed of coupling (*e.g.*, aniline).

In the determination of sulfapyridine (2-(sulfanilamido)-pyridine) with the present method, the following may be mentioned. To a sample of mixed human blood sulfapyridine was added to make about 10 mg. per cent. Precipitation in a 1:4 dilution gave 80.8 per cent recovery, in a 1:20 dilution 93.7 per cent, and in a 1:50 dilution 99.4 per cent. A number of other experiments indicate incomplete recovery (average 91 per cent) in 1:20 dilution with 5 to 10 mg. per cent in blood, but essentially complete with values below 5 mg. per cent. With a 1:50 or greater dilution, recovery is quantitative.

DISCUSSION

We have already discussed some of the so called modifications of our method which have been proposed. It remains to mention briefly other methods which have been suggested for the determination of sulfanilamide or allied compounds. Kühnau (17) has described a method for estimating N4-sulfanilyl-N1, N1-dimethylsulfanilamide by the color produced with dimethylaminobenzaldehyde, and Schmidt (18) one for sulfanilamide, with the color produced by sodium β-naphthoquinone-4-sulfonate. We have had no experience with either of these. Scudi (19) has described a diazotization procedure, followed by neutral coupling with chromotropic acid, while Doble and Geiger (20) used diphenylamine as an acid coupling agent. Neither of these latter methods appears to be as satisfactory as the method we have described.

With a new compound, a few preliminary trials should indicate

550   Sulfanilamide Determination

what slight modification of the method, if any, is necessary to determine it accurately. However, we must caution investigators against accepting results with a new compound until control recoveries from blood and urine have been made.

SUMMARY

In the determination of sulfanilamide by diazotization and coupling in acid solution, the use of N-(1-naphthyl)ethylenediamine dihydrochloride offers the following advantages over N,N-dimethyl-1-naphthylamine (dimethyl-α-naphthylamine): (1) reproducibility and purity, (2) greater rapidity of coupling, (3) increased sensitivity, (4) elimination of buffer, and (5) increased acid solubility of the azo dye formed. An improved synthesis and a complete characterization of the new coupling component is presented. Slight modification of previous technique and application to other primary aryl amines are described.

BIBLIOGRAPHY

1. Marshall, E. K., Jr., Emerson, K., Jr., and Cutting, W. C., J. Am. Med. Assn., 108, 953 (1937).
2. Marshall, E. K., Jr., Proc. Soc. Exp. Biol. and Med., 36, 422 (1937).
3. Marshall, E. K., Jr., J. Biol. Chem., 122, 263 (1937–38).
4. MacLachlan, E. A., Carey, B. W., and Butler, A. M., J. Lab. and Clin. Med., 23, 1273 (1938).
5. Stevens, A. N., and Hughes, E. J., J. Am. Pharm. Assn., 27, 36 (1938).
6. Proom, H., Lancet, 1, 260 (1938).
7. Kamlet, J., J. Lab. and Clin. Med., 23, 1101 (1938).
8. Marshall, E. K., Jr., and Litchfield, J. T., Jr., Science, 88, 85 (1938).
9. Hecht, G., Dermat. Woch., 106, 20 (1938).
10. Newman, H. E., Ber. chem. Ges., 24, 2199 (1891).
11. British patent 247,717, June 29, 1925.
12. Ing, H. R., and Manske, R. H. F., J. Chem. Soc., 2348 (1926).
13. Marshall, E. K., Jr., and Cutting, W. C., Bull. Johns Hopkins Hosp., 63, 328 (1938).
14. Gregersen, M. I., and Painter, E. E., Am. J. Physiol., 123, 83 (1938).
15. Fuller, A. T., Lancet, 1, 194 (1937).
16. Marshall, E. K., Jr., Emerson, K., Jr., and Cutting, W. C., J. Pharmacol. and Exp. Therap., 61, 196 (1937).
17. Kühnau, W. W., Klin. Woch., 17, 116 (1938).
18. Schmidt, E. G., J. Biol. Chem., 122, 757 (1937–38).
19. Scudi, J. V., J. Biol. Chem., 122, 539 (1937–38).
20. Doble, J., and Geiger, J. C., J. Lab. and Clin. Med., 23, 651 (1938).

COMMENTARY TO

8. Brodie, B. B. and Udenfriend, S. (1943)
The estimation of Atabrine in biological fluids and tissues. Journal of Biological Chemistry 151(1): 299–317.

W hen the Japanese in World War II took control of Indonesia they shut off the Allies' sole source of quinine, the drug of choice for the prevention and treatment of malaria. *Cinchona* tree plantations in Indonesia provided the bark from which quinine was extracted. In 1941, the world's demand for *Cinchona* bark exceeded 700 ton/year and 90% of that supply came from Indonesia (1). Synthetic antimalarials like quinacrine were the only option available. Quinicrine was a synthetic antimalarial drug discovered in Germany in 1932. Winthrop Chemical in the US made this drug and sold it under the trade name Atabrine™. In 1942 quinacrine became the official drug of choice for the US military as a substitute for quinine. Quinacrine however caused a number of adverse side effects and had a difficult dosing schedule. The drug caused gastrointestinal disturbances and turned the skin yellow. Preventative therapy required a 100 mg tablet three times a day. This, along with the side effects made compliance under military field conditions very difficult. In 1942 at the battle of Guadalcanal 8 500 soldiers or almost half of the 1st Marine Division was hospitalized with malaria (2). In that same year in Papua New Guinea 28 000 Australian and American soldiers were hospitalized with malaria. The battle casualties during that same period were 7 700 (3).

In 1942, the US National Research Council funded a major campaign to study antimalarial therapies and to develop new and more effective synthetic drugs. It was the largest government funded biomedical research program in the US in the first half of the 20th century (1). Bernard B. Brodie and Sidney Undenfriend, a 24-year-old graduate student, were part of a group funded by the Research Council. They were housed in a basement laboratory in Building D at the Goldwater Memorial Hospital on Welfare Island (now Roosevelt Island) in the East River in New York City (4). Brodie and Undenfriend developed methods to measure drug and metabolite levels; solved a major problem in the clinical pharmacology of antimalarials and helped introduce fluorescence technology into clinical chemistry and pharmacology.

Brodie and Undenfriend's paper on the development of a fluorometric assay for quinacrine and how they used it to study the pharmacokinetics of the drug is presented here. They recognized the importance of the interference of the drug's more polar water-soluble metabolites in the assay and developed an ethylene dichloride extraction procedure. Their selection of solvent excluded the interference of metabolites and improved the specificity of their assay. Parent drug was back extracted into acid and the fluorescence read at an excitation of 365 nm (blue) and emission above 500 nm (yellow). In a simpler version of the assay the solvent extraction was made acidic with glacial acetic acid and read directly. Tissue studies with this assay showed that after routine dosing the drug rapidly distributed into body stores such as soft tissues and blood cells. These reservoirs quickly saturated after a short loading dose and helped maintain adequate therapeutic plasma levels with a once a day maintenance dose (6). Toxic side effects were markedly reduced with the lower dose. It was shown that clinical efficacy correlated with the plasma levels and not the dose received. The three times a day empirical dosing schedule was quickly replaced by a once a day 100 mg tablet after a short loading dose. The Surgeon General of the Army adopted this dosing protocol as official policy in September 1943 (7). Quinacrine went on to become standard issue throughout the armed services and the tactical and strategic military problem of malaria was greatly reduced (4).

Brodie and Undenfriend along with others in their lab at the Goldwater Memorial Hospital expanded on the use of fluorometric methods for the estimation of drugs in biological fluids and tissues. In 1943, they published a fluorometric method for plasma quinine and quinidine, the cardiac antiarrhythmic drug also derived from *Cinchona* bark (8). This method was used to demonstrate the narrow therapeutic index of quinidine shortly after it was introduced into clinical practice in 1950 (9). In a single issue of the *Journal of Biological Chemistry* in April 1947, six papers were presented on the fundamental principals of drug assay and methods development (10–15). The second (11) and third papers (12) were based on fluorescence technology.

The fluorometric methods published in the 1940s from the Goldwater lab were performed with a quartz lamp filter fluorometer. Udenfriend in 1995 recounted how this early work with Brodie led him and Robert Bowman to develop the first commercial quartz spectrophotofluorometer (16). In their first description of the instrument published in *Science* in 1955 Bowman, Caulfield and Udenfriend credit Brodie with getting them started on the project to build the instrument that became the Aminco-Bowman spectrofluorometer (17). Udenfriend continued his research on the use of fluorescence technology and published a two-volume textbook on methods development and applications (18,19). These two volumes are still a standard reference work in the field of fluorescence technology. Brodie went on to become the director of the Laboratory of Chemical Pharmacology in the National Heart Institute at the National Institutes of Health. Today he is recognized as the founder of modern clinical pharmacology (5). In 1967 he received the Albert Lasker Award for Basic Medical Research. His citation read in part, "Probably no man has contributed more to the body of knowledge which makes possible the rational use of drugs in the treatment of many diseases than has Dr. Brody" (20).

References

(1) Slater, L.B. (2004) Malaria chemotherapy and the "kaleidoscopic" organization of biomedical research during World War II. Ambix. 51(2):107–134.

(2) Rocco, F. (2003) *The Miraculous Fever-Tree. Malaria and the Quest for a Cure that Changed the World.* Harper Collins, New York, NY, pg 291.

(3) Joy, R.J.T. (1999) Malaria in American troops in the South and Southwest Pacific in World War II. Medical History. 43(2): 192–207.

(4) Kanigel, R. (1986) *Apprentice to Genius. The Making of a Scientific Dynasty.* Macmillan, New York, pgs 13–30.

(5) Costa, E., Karczmar, A.G., and Vesell, E.S. (1989) Bernard B. Brodie and the rise of chemical pharmacology. Annual Review of Pharmacology and Toxicology. 29:1–21.

(6) Shannon, J.A., Earle, D.P. Jr., Brodie, B.B., Taggart, J.V., and Berliner, R.W. (1944) The pharmacological basis for the rational use of Atabrine in the treatment of Malaria. Journal of Pharmacology and Experimental Therapeutics. 81:307–330.

(7) Office of the Surgeon General of the Army. (1943) The drug treatment of malaria, suppressive and clinical. Circular Letter No. 153. Journal of the American Medical Association. 123(4):205–208.

(8) Brodie, B.B. and Undenfriend, S. (1943) The estimation of quinine in human plasma with a note on the estimation of quinidine. Journal of Pharmacology and Experimental Therapeutics. 78(2):154–158.

(9) Sokolow, M. and Edgar, A.L. (1950) Blood quinidine concentrations as a guide in the treatment of cardiac arrhythmias. Circulation. 1(4 Part 1):576–592.

(10) Brodie, B.B., Undenfriend, S., and Baer, J.E. (1947) The estimation of basic organic compounds in biological material. I. General principles. Journal of Biological Chemistry. 168(1):299–309.

(11) Brodie, B.B., Undenfriend, S., Dill, W., and Downing, G. (1947) The estimation of basic organic compounds in biological material. II. Estimation of fluorescent compounds. Journal of Biological Chemistry. 168(1):311–318.

(12) Brodie, B.B., Udenfriend, S., Dill, W., and Chenkin, T. (1947) The estimation of basic organic compounds in biological material. III. Estimation by conversion to fluorescent compounds. Journal of Biological Chemistry. 168(1):319–325.

(13) Brodie, B.B., Udenfriend, S., and Taggart, J.V. (1947) The estimation of basic organic compounds in biological material. IV. Estimation by coupling with diazonium salts. Journal of Biological Chemistry. 168(1):327–334.

(14) Brodie, B.B., Udenfriend, S., and Dill, W. (1947) The estimation of basic organic compounds in biological material. V. Estimation by salt formation with methyl orange. Journal of Biological Chemistry. 168(1):335–339.

(15) Josephson, E.S., Udenfriend, S., and Brodie, B.B. (1947) The estimation of basic organic compounds in biological material. VI. Estimation by ultraviolet spectrophotometry. Journal of Biological Chemistry. 168(1):341–344.

(16) Udenfriend, S. (1995) Development of the spectrophotofluorometer and its commercialization. Protein Science. 4:542–551.

(17) Bowman, R.L., Caulfield, P., and Udenfriend, S. (1955) Spectrophotofluorometric assay in the visible and ultraviolet. Science. 122(3157):32–33.

(18) Udenfriend, S. (1962) *Fluorescence Assay in Biology and Medicine.* Academic Press, New York.

(19) Udenfriend, S. (1969) *Fluorescence Assay in Biology and Medicine*, Vol 2. Academic Press, New York.

(20) [Anonymous] (1967) Lasker Awards citations. Journal of the American Medical Association. 202(7):599.

J. Biol. Chem. 1939; 151: 299–317
© 1943 The American Society for Biochemistry and Molecular Biology.
Reproduced with permission.

THE ESTIMATION OF ATABRINE IN BIOLOGICAL FLUIDS AND TISSUES*

By BERNARD B. BRODIE and SIDNEY UDENFRIEND

*(From the Research Service, Third (New York University) Medical Division, Goldwater
Memorial Hospital, New York, and the Department of Medicine,
New York University College of Medicine, New York)*

(Received for publication, August 24, 1943)

The following are two methods for the estimation of atabrine in biological fluids and tissues through the measurement of its fluorescence in an acidic environment. The methods have a high degree of specificity in that they exclude from the estimation the many fluorescent degradation products of atabrine as well as the naturally occurring fluorescent components of biological fluids and tissues.

The first method is a double extraction procedure. It is wholly satisfactory for the precise estimation of the concentration of atabrine in the plasma, whole blood, tissue, and urine of patients on the usual régimes of suppressive or definitive atabrine therapy. The procedure gives recoveries of added atabrine which average 98 per cent with amounts as low as 0.1 γ. Variation at this level is usually less than 5 per cent, and is minimized when larger amounts of atabrine are present. The precision of the estimation decreases when smaller quantities are present.

The second procedure involves only a single extraction. Its speed and simplicity recommend it for use when possible. However, the sensitivity of the measurement is somewhat less than that of the double extraction procedure.

Double Extraction Procedure

The method described below effects the isolation of the atabrine from the biological material by extraction of the free base with ethylene dichloride at pH 8.0. The latter phase is then washed with 2.5 N NaOH and the atabrine is returned as a salt to an aqueous phase of concentrated lactic acid.

General Considerations

Measurement of Fluorescence—The intensity of atabrine fluorescence is subject to many factors. Some of these relate to the activating energy

---

* The work described in this paper was done under a contract recommended by the Committee on Medical Research between the Office of Scientific Research and Development and New York University.

This work was the subject of a report submitted to the National Research Council, Division of Medical Science, on January 4, 1943.

300

and others to the environment in which atabrine occurs. It has usually been considered advisable to measure the fluorescence intensity of atabrine in a weakly alkaline solution with the 365 mμ band of the ultraviolet, because of the relatively high intensity of the fluorescence obtained under these conditions. However, it has been found that the fluorescence of atabrine in acid solution is also great when the 420 mμ band of the ultraviolet is utilized as the activating energy, and that the intensity may be enhanced without any sacrifice in the stability of the fluorescence by the presence of certain acids in high concentrations. The intensity of the fluorescence under the latter conditions appears to be close to the maximal obtainable.

It must be appreciated, however, that the intensity of the fluorescence to be measured may be low, since the concentration of atabrine is so low in certain of the biological fluids, more particularly plasma. The sensitivity and precision of the estimation will therefore be conditioned, to a considerable extent, by the sensitivity and the stability of the fluorometer used in the final assay of fluorescence. Several instruments with satisfactory characteristics are now available. The No. 12 Coleman electronic photofluorometer has been used in these studies, since it combines these essential characteristics with simplicity of operation.

A filter system has been selected which does not result in manifest fluorescence with extracts of tissues in the absence of added atabrine. This system consists of a 2 mm. No. 5113 Corning glass filter (Coleman B4) which is used to isolate the activating energy, and a Corning No. 3385 filter (Coleman PC9) which is used to limit the transmission of the resulting fluorescent light. Advantages which are derived from this combination are the large readings obtained with atabrine fluorescence in acid solution and the ability to augment these by the addition of a non-fluorescent solute. These are achieved in the method by making the final reading in a concentrated solution of lactic acid. An added advantage of the concentrated lactic acid may be its high viscosity, which appears to be one of the factors which operate in the enhancement of atabrine fluorescence.

It has been found that the fluorescence of atabrine is partly a function of the temperature of the medium. Frequent calibration of the instrument with the standard solution results in a diminished intensity of fluorescence unless progressive heating of the solution by the ultraviolet lamp is prevented. This end is achieved by keeping both standards and samples in a water bath at room temperature before and between readings. The bath must not be below room temperature, since otherwise the separation of the small amount of ethylene dichloride which is dissolved in the lactic acid may cause turbidity. The influence of temperature on fluorescence is so great that the manifest fluorescence may be increased 40 per cent by cooling the cuvette containing the samples and standards in an ice-salt bath. This procedure requires the addition of another ml. of lactic acid to prevent the separation of the ethylene dichloride and is not recommended as a routine procedure.

Solvents—Ethylene dichloride has been chosen as the organic solvent. It effects the extraction of atabrine in a highly efficient manner and at the same time does not remove many of its fluorescent degradation products from the biological sample. In addition, it has certain physical properties which recommend its use. The vapor pressure and water solubility are low and the specific gravity is high. The latter property is particularly important in that it minimizes the troublesomeness of emulsions which are commonly associated with organic solvent extractions of biological material. It must be appreciated, however, that, together with the other chlorinated hydrocarbons, ethylene dichloride is toxic to the human organism. All measurements or transfers of the solvent must, therefore, be made by automatic glass equipment.

Lactic acid is the acid of choice in the return of the atabrine to an aqueous phase, because of its efficient action in this procedure and because of the combination of properties which produce a large enhancement in the fluorescence of atabrine in acid solution (see above).

Most reagent grades of lactic acid contain fluorescent material of a foreign nature. This is routinely removed by extracting the lactic acid with ethylene dichloride in a separatory funnel. The lactic acid is then separated and shaken with a small quantity of charcoal. The charcoal is removed by filtering twice through a Buchner funnel. Small amounts of fluorescence due to foreign material in the ethylene dichloride do not usually constitute a hazard in the procedure. However, it is routine practice to remove such material with charcoal. The charcoal is removed in this case by filtering twice through a double thickness of filter paper which has been previously washed with purified ethylene dichloride. Care must be taken in both procedures to remove all carbon particles.

Standard Solutions—A strong solution of atabrine dihydrochloride (100 mg. per liter calculated as the free base) may be stored indefinitely in a refrigerator without deterioration. Concentrations of atabrine of 1 mg. per liter or less are used in the preparation of working standards. These are routinely prepared by diluting the concentrated atabrine solution with 0.2 M Na_2HPO_4. This precaution is essential, since the reversible adsorption of atabrine on glass surfaces in the absence of electrolyte would constitute a major error at the lower concentrations.

Procedure

Add 3 ml. of 0.2 M Na_2HPO_4 and 30 ml. of ethylene dichloride to 1 to 10 ml. of biological material[1] in a 60 ml. glass-stoppered Pyrex bottle (blood is first hemolyzed with 2 parts of water). Shake vigorously for 5 minutes, preferably on a shaking apparatus. Decant into a 50 ml. centrifuge tube and centrifuge for 10 minutes at moderate speed to break the emulsion. Remove the supernatant layer by aspiration. A solid gel sometimes forms in the ethylene dichloride which may be broken by vigorous stirring with a glass rod. A second centrifugation at high speed will then produce a clean separation of the two phases. Return the ethylene dichloride solution to the original rinsed out bottle, restraining the coagulum with a stirring rod. Add an equal volume of 10 per cent NaOH and shake for 3 minutes. Remove the major portion of the sodium hydroxide solution by aspiration and transfer the remainder of the contents of the bottle to a narrow test-tube. Centrifuge for 1 minute. Remove the supernatant layer by aspiration, wash the sides of the tube with water, and remove the water by aspiration. Pipette exactly 20 ml. of the ethylene dichloride into a prepared glass-stoppered bottle,[2] add 1 ml. of water and 10 ml. of 85 per cent lactic acid, and shake vigorously for 5 minutes. Transfer to a prepared,[2] narrow 35 ml. centrifuge tube and centrifuge for 1 minute at moderate speed. Transfer at least 8 ml. of the aqueous phase to a matched

[1] Tissue homogenates may be simply prepared by the use of an electrically driven homogenizer. A relatively inexpensive device is distributed by the Scientific Glass Apparatus Company, Bloomfield, New Jersey. 1 or 2 gm. of tissue are added to 2 or 4 ml. of water and ground to a fine emulsion in a few minutes. Cell fragmentation is the general rule.

It is essential, because of the distribution of atabrine in whole blood, to relate the chemotherapeutic activity of atabrine to its concentration in the plasma rather than to its concentration in whole blood. This conclusion is derived from the circumstance that the concentration of atabrine in leucocytes is about 400 times that of plasma, and that variations in whole blood levels are often only a reflection of changes in leucocyte count. The unequal distribution of atabrine in whole blood requires special precautions in the preparation of plasma for analysis. The technique used is as follows: Blood is drawn with adequate amounts of oxalate as the anticoagulant. It is immediately centrifuged at 1500 R.P.M. for 15 minutes, the upper portion of the plasma removed, recentrifuged for 1 hour at 1500 R.P.M., and then carefully separated from any solid residuum. This procedure is deemed advisable in order to remove any possibility of contaminating the plasma sample with leucocytes or leucocyte fragments.

[2] Small blanks equivalent to 1 to 2 γ of atabrine per liter may be obtained if the glassware has been previously exposed to air, presumably due to the accumulation of dust. The glassware for the second extraction and centrifugation must be free of water. It is rinsed with a small amount of ethylene dichloride just prior to use. These precautions are essential in the case of determinations of plasma concentrations, since these will usually range from 5 to 50 γ per liter with the usual suppressive or curative régimes of therapy.

cuvette and determine its fluorescence in relation to a properly prepared standard. No effort is made to read the galvanometer with a greater accuracy than a quarter of a division.

The sensitivity of the fluorometer is calibrated by an atabrine standard prepared in lactic acid. The standard used in the routine estimation of atabrine during ordinary atabrine administration to humans is prepared as follows: Add 1 ml. of solution containing 0.5 γ of atabrine to 10 ml. of 85 to 90 per cent lactic acid. A mixture of 10 ml. of lactic acid and 1 ml. of water is used for the blank setting of the instrument. The manifest fluorescence of atabrine is a linear function of its concentration in the range usually encountered when the measurement is made with a suitable fluorometer. Consequently, the calculation of atabrine concentration is by direct proportion.

Alternate Procedure—As noted below ethylene dichloride extracts of human blood and plasma contain little of the fluorescent degradation products of atabrine. Consequently, the alkali wash is omitted in the routine determination of atabrine concentration under ordinary conditions. The procedure is identical with that detailed above until one has produced a clean separation of the aqueous and ethylene dichloride phases after the initial extraction. A pipette is then carefully inserted below the coagulum and exactly 20 ml. of the ethylene dichloride are removed and placed in a prepared glass-stoppered bottle. 1 ml. of water and 10 ml. of 85 per cent lactic acid are added and the whole shaken vigorously for 5 minutes; the mixture is then transferred to a prepared narrow 35 ml. centrifuge tube and centrifuged for 1 minute at moderate speed. At least 8 ml. of the aqueous phase are transferred to a matched cuvette and the amount of fluorescence estimated as described in the above procedure.

Results

Recoveries of atabrine added in known amounts to whole blood and plasma were used to assay the precision of the method (Table I). These were consecutive runs performed over a period of several months in conjunction with the routine use of the method. They give, therefore, a fair appraisal of the precision which may be expected with the routine use of the method. Table I also contains a comparable series of results obtained with the alkali wash omitted. These data indicate that atabrine added to whole blood and plasma is recoverable with good precision. A limited series of analyses indicates that equally good results are obtainable when the procedure is applied to urine.

In Table II the results obtained when whole blood and plasma are analyzed with and without the alkali wash are compared. The blood of the latter series was obtained from patients during a course of atabrine therapy.

TABLE I

Recovery of Atabrine Added to Whole Blood and Plasma with and without NaOH Extraction. Double Extraction Method

10 ml. samples were used.

| | Washed | | | | | | Unwashed | | | | |
| | Whole blood | | | Plasma | | | Whole blood | | | Plasma | | |
Atabrine added	Atabrine found	Recovery	Atabrine added	Atabrine found	Recovery	Atabrine added	Atabrine found	Recovery	Atabrine added	Atabrine found	Recovery
γ	γ	per cent	γ	γ	per cent	γ	γ	per cent	γ	γ	per cent
1.0	1.02	102	1.0	1.02	102	2.00	1.93	97	0.50	0.49	98
	1.04	104		1.02	102		1.93	97		0.50	100
	1.04	104	0.5	0.50	100		1.90	95	0.30	0.29	97
	1.03	103		0.51	102		1.93	97		0.292	97
	1.03	103		0.50	100	1.00	0.98	98	0.20	0.18	90
	1.02	102		0.51	102		0.97	97		0.195	98
	1.03	103		0.51	102		1.01	101		0.20	100
0.5	0.48	96		0.50	100		1.03	103		0.19	95
	0.50	100		0.48	96		1.01	101		0.19	95
	0.51	102	0.2	0.18	90		0.99	99		0.197	99
	0.51	102		0.19	95		0.95	95		0.20	100
	0.51	102		0.21	105		1.04	104		0.20	100
0.25	0.26	104		0.20	100		1.00	100	0.10	0.10	100
0.3	0.30	100		0.195	98	0.50	0.50	100		0.105	105
							0.51	102		0.095	95
							0.49	98		0.095	95
							0.52	104		0.095	95
							0.47	94		0.095	95
							0.48	96		0.10	100
							0.53	106		0.102	102
							0.50	100		0.10	100
						0.20	0.20	100			
							0.195	98			
							0.20	100			

TABLE II

Comparison of Atabrine Estimation in Human Blood and Plasma with and without Alkali Extraction Step. Double Extraction Method

| Whole blood | | Plasma | |
Unwashed	Washed	Unwashed	Washed
γ per l.	γ per l.	γ per l.	γ per l.
273	273	42	37
332	327	70	70
297	300	66	60
123	121	94	100
307	303	42	42
277	266	63	57
		9	10

The small difference obtained with the two procedures in this series is the basis for the judgement that an alkali wash is not essential for most routine determinations on these fluids. Similar results have been obtained on the blood and plasma of dogs to which atabrine had been administered for several months. On the other hand, the urine of both dogs and humans contains appreciable amounts of fluorescent atabrine degradation products which are extractable by ethylene dichloride. It is necessary, therefore, to include the alkali wash in the procedure when urine is analyzed.

Analyses run on single samples of blood over a period of several days invariably give highly reproducible results. This indicates that atabrine in blood or plasma is quite stable when stored in a refrigerator.

Comment

The precision of the procedure is related to the absolute amount of atabrine contained in the sample rather than to its concentration. Samples of blood or plasma as large as 10 ml. need be used only when the concentration of atabrine is in the range of 30 γ per liter or less. Actually, good precision may be obtained at considerably lower concentrations than this with no larger volume of sample, provided special precautions are taken in the matching of cuvettes and the cleaning of glassware.

It might be thought that a reasonable procedure for obviating the alkali wash in all cases is to utilize a strongly alkaline medium during the initial extraction. Unfortunately, this is not a feasible procedure, at least for whole blood or plasma. Atabrine is quite unstable in dilute solutions when exposed to strong alkali and there is a consequent loss in precision. Secondly, extraction from strongly alkaline blood or plasma results in gel formation in the ethylene dichloride phase which is broken with difficulty.

Single Extraction Procedure

The single extraction procedure involves the measurement of atabrine by its fluorescence in the initial ethylene dichloride extract of the biological material. This measurement is made subsequent to the addition of acetic acid which serves to stabilize and enhance the fluorescence. The speed and simplicity of this method recommend it for most tissue analyses, even though it has somewhat lower sensitivity than the lactic acid procedure. The sensitivity of the method permits the estimation of atabrine down to 0.5 γ with good precision. The blank is negligible in urine and organ tissues, while in plasma and whole blood the blanks are equivalent to 2 and 4 γ of atabrine per liter. The method is therefore not recommended for ordinary use with plasma because of the low concentration of atabrine commonly observed in this fluid.

General Considerations

Measurement of Fluorescence—A Coleman glass filter No. B₂ (combination of Corning No. 5113, 2 mm.; No. 3389, 2 mm.) is used to isolate the activating energy and a Coleman Filter PC9 to limit the transmission of the fluorescent light. This filter combination diminishes the sensitivity obtaining with the Corning No. 5113 filter alone, but is used to exclude the large amount of fluorescence derived from normal components of biological material.

Fluorescence measurements taken directly on ethylene dichloride extracts are theoretically less specific than those obtained in a double extraction procedure. Actually, however, only a small difference has been observed between the two procedures when applied to blood and no difference when applied to urine of dogs and humans obtained during a course of atabrine therapy. The difference in the case of blood averages 5 per cent, and may be accounted for, in part, by the blank which is only present in the single extraction procedure.

Solvent—Ethylene dichloride is used as the organic extractor largely because of the reasons mentioned above. An additional consideration in the present procedure is that only small amounts of interfering substances are extracted from biological material with the solvent. This reagent is highly variable in the amount of foreign fluorescent material which it contains. However, the impurities may be easily removed with charcoal unless present in excessive amounts (see above).

Buffer—A borate buffer of pH 11.5 is used to adjust the samples to pH 9.5 to 10 before extraction with ethylene dichloride. This is prepared as follows: To 50 ml. of 0.6 M boric acid in 0.6 M KCl add 50 ml. of 0.6 M NaOH. A blood pH of 9.5 to 10 has been selected to minimize the extraction of interfering pigments which are extractable in significant amounts from human blood or plasma at a lower pH (8.0). The error due to the extraction of pigment at the higher pH is negligible except in the rare case.

Procedure

Add 1 part of borate buffer to 1 part of biological material in a 60 ml. glass-stoppered bottle.[3] (The blood is first hemolyzed with 2 parts of water.) Then add 15 ml. of ethylene dichloride and shake vigorously for 5 minutes, preferably on a shaking apparatus. Decant the mixture into a 50 ml. centrifuge tube and centrifuge for 10 minutes at a moderate speed to break the emulsion. Remove the supernatant aqueous layer by aspiration. A solid gel sometimes forms in the ethylene dichloride phase. This may be

---

[3.]The glassware in the single extraction procedure must be scrupulously clean. The bottles and centrifuge tubes are kept in calgonite solution and rinsed just prior to use. These precautions are necessary to prevent extraneous fluorescence due to dirt.

broken by vigorous stirring with a glass rod. A second centrifugation will then produce a clean separation of the ethylene dichloride. Return the ethylene dichloride solution to the rinsed out bottle, restraining the coagulum with a stirring rod. Add an equal volume of 10 per cent NaOH and shake for 3 minutes. Remove the major portion of the sodium hydroxide solution by aspiration and transfer the remainder of the contents of the bottle to a narrow test-tube. Centrifuge for 1 minute. Remove the supernatant layer by aspiration. Wash the side of the tube with water and repeat the aspiration. Pipette 10 ml. of the ethylene dichloride directly into a cuvette containing 1 ml. of glacial acetic acid. Slight turbidity of the ethylene dichloride does not introduce an error, since it clears in the presence of acid. The reading of fluorescence is made in relation to a properly prepared standard with the same precautions as noted above.

The standard used in calibrating the sensitivity of the instrument is prepared as follows: An aqueous solution of atabrine buffered as above is extracted with ethylene dichloride and handled in the same manner as the biological sample. Ethylene dichloride is used as the reagent blank. The computation of atabrine concentration is by direct proportion.

Alternate Procedure—The alkaline wash of the ethylene dichloride extracts may also be omitted when the present procedure is applied in a routine manner to human whole blood and plasma, since these contain little of the fluorescent degradation products of atabrine. The procedure is the same as the one described above until a clean separation of the ethylene dichloride and water phases after the initial extraction is obtained. A pipette is then carefully inserted below the coagulum and about 10 ml. of the ethylene dichloride are transferred directly to a cuvette containing 1 ml. of glacial acetic acid. The amount of fluorescence is estimated as described in the above procedure.

Results

Atabrine added to whole blood and urine is recovered with good precision as shown in Table III. The data in Table IV indicate that the estimation of atabrine concentration by the single extraction procedure yields results somewhat higher (averaging 5 per cent) than the double extraction procedure when these are applied to the blood and urine of patients during a course of atabrine therapy.

Comment

The single extraction procedure is not as sensitive as the double extraction procedure. However, it may be used to advantage for many routine purposes. The lessened sensitivity results largely from the presence of a

308   DETERMINATION OF ATABRINE

blank which is due to the presence of extraneous fluorescent material derived from the biological sample. The quantity of this material is small and may be neglected for most purposes when the concentration of atabrine is higher than 50 γ per liter.

TABLE III

Recovery of Atabrine Added to Whole Blood and Urine. Single Extraction Method

Blood			Urine		
Atabrine added	Atabrine recovered	Recovery	Atabrine added	Atabrine recovered	Recovery
γ	γ	*per cent*	γ	γ	*per cent*
1	1.00	100	3	2.96	99
	1.03	103		3.08	103
	1.02	102		3.03	101
	0.98	98	2	1.97	98
	1.01	101		1.98	99
	1.04	104		2.06	103
	1.02	102	1	1.02	102
	1.02	102		1.03	103
	1.04	104		1.08	108
	1.03	103			
	1.00	100			
	1.00	100			
	0.98	98			
	0.94	94			
	0.95	95			
	0.95	95			
	1.03	103			
	0.98	98			
0.5	0.48	96			
	0.51	102			
	0.49	98			
	0.53	106			
	0.48	96			
	0.49	98			
	0.48	96			

Appraisal of Specificity

The above methods appear to have adequate precision for most purposes. However, their general usefulness also depends upon the completeness with which extraneous fluorescent material is excluded in the analysis. This aspect of the problem is somewhat simplified by the absence of a significant blank in ethylene dichloride extracts as usually prepared from atabrine-free biological fluids. The degree to which the methods are specific in such a circumstance will depend upon the extent to which they exclude fluorescence due to the degradation products of atabrine in the final estimation of fluorescence.

The intensity of the fluorescence manifested by a solution containing an acridine is dependent in part upon the arrangement of the optical system and in part upon certain characteristics of the solution. The latter have been used to obtain additional evidence on the degree to which the recommended procedure is specific. It is not suggested that such an examination should be substituted for that utilized by Craig (1) but it does give im-

TABLE IV

Comparison of Single and Double Extraction Procedures

Biological sample	Atabrine	
	Single extraction	Double extraction
	γ per l.	γ per l.
Blood	824	794
	724	720
	933	828
	61	60
	840	819
	755	685
	106	104
	104	106
	111	104
	164	159
	588	520
	607	607
	213	205
	569	600
	448	411
Urines	1370	1410
	1930	1970
	283	274
	1990	2100
	3200	3110
	1430	1350

portant supplementary information on the specificity of the method through the use of different criteria. It also presents a general method which may be used to some advantage in a more general fashion, particularly in situations in which the amount of the fluorescent material available for study is small. The technique involves the measurement of the fluorescence manifested by a solution at a constant temperature but at various concentrations of HCl and NaOH. The variation in the intensity of the fluorescence under these conditions results from the operation of such factors as the pH, the ionic strength, the chloride and sodium concentration, and perhaps the viscosity of the solution.

The routine procedures recommend the use of lactic acid as the final extractor of the atabrine prior to the measurement of fluorescence, or the measurement of fluorescence in the ethylene dichloride phase itself. However, essentially all of the fluorescent material which is extractable by ethylene dichloride from plasma, whole blood, or urine at a pH of 8 to 10 can be returned to an aqueous phase with any one of a number of acids, including concentrated lactic acid and 0.1 N HCl. Consequently, if the fluorescent material remaining in the ethylene dichloride phase subsequent to washing with 2.5 N sodium hydroxide has the fluorescent characteristics of atabrine, then the probability that the method has the desired specificity has been strengthened.

Purified samples of atabrine dihydrochloride and a series of other acridines were used to construct curves in which fluorescence is related to the strength of acid or alkali in the solution. The atabrine curve, so constructed, has then been used as a standard of reference for comparison with similar curves which describe the fluorescent characteristics of the material contained in ethylene dichloride extracts of the blood and urine of patients and dogs receiving atabrine.

Procedure

Each test run started with a solution of the acridine in 0.1 N HCl at a concentration of 0.5 to 1.0 γ per ml. 1 ml. aliquots of the solution to be examined were placed in a series of fluorometer cuvettes and 10 ml. of various strengths of sodium hydroxide or hydrochloric acid were added. Fluorescence was then measured by means of the same filter system as was used in the double extraction procedure. A value of 1.0 was assigned to the fluorescence observed with the solution to which the 1.0 N hydrochloric acid had been added. The observations were performed at room temperature. A constant temperature for all samples in any run was assured by placing the cuvettes in a water bath for some time before and between readings of fluorescence. Moderate variations in the absolute temperature at which readings are made do not constitute a source of error provided the temperature is the same for all tubes in the series.

Four general types of solutions were examined. They were prepared as follows:

Samples of pure acridine solutions (Fig. 1) were prepared by the addition of a small amount of the acridine to 0.1 N HCl. The series of acridines selected for examination included atabrine, three relatively simple acridines, and two more complex acridines related somewhat to atabrine itself. The structural relationships of these six acridines are as follows: (1) 2-methoxy-6-chloroacridone-9, (2) 2-methoxy-6-chloro-9-aminoacridine, (3) 2-methoxy-6,9-dichloroacridine, (4) 2-methoxy-6-chloro-9(3-diamylaminopropyl)am-

inoacridine, (5) 2-methoxy-6-cyano-9(1-diethylamino-4-methylbutyl)am-
inoacridine, (6) 2-methoxy-6-chloro-9(1-diethylamino-4-methylbutyl)-
aminoacridine (*atabrine*).

Mixtures of pure acridines (Fig. 2) were prepared in order to examine

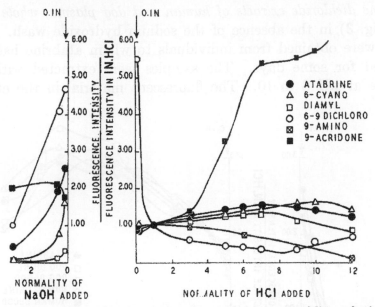

FIG. 1. Fluorescence characteristics of a series of pure acridine solutions

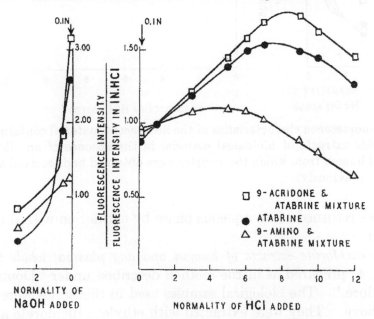

FIG. 2. Fluorescence characteristics of mixtures of pure acridines. The 9-acridone
and atabrine mixture consists of atabrine contaminated with a trace of 2-methoxy-6-
chloroacridone-9. The latter compound is too insoluble to make a solution of known
composition. The 9-amino and atabrine mixture consists of 10 per cent 2-methoxy-
6-chloro-9-aminoacridine and 90 per cent atabrine.

312   DETERMINATION OF ATABRINE

the fluorescent characteristics of atabrine contaminated by small quantities of other acridines. These mixtures were as follows: (1) 90 per cent atabrine and 10 per cent 2-methoxy-6-chloro-9-aminoacridine, (2) atabrine contaminated with a trace of 2-methoxy-6-chloroacridone-9.

Ethylene dichloride extracts of human and dog plasma, whole blood, and urine (Fig. 3) in the absence of the sodium hydroxide wash. Biological samples were obtained from individuals to whom atabrine had been administered for some days. The samples were extracted with ethylene dichloride at pH 8 to 10. The fluorescent material in the ethylene di-

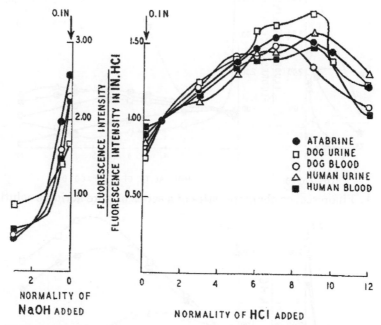

FIG. 3. Fluorescence characteristics of the fluorescent material contained in ethylene dichloride extracts of biological material in the absence of an alkaline wash. The dog and human from which the samples were obtained had received atabrine for several days previously.

chloride was returned to an aqueous phase by extraction with 0.1 N hydrochloric acid.

Ethylene dichloride extracts of human and dog plasma, whole blood, and urine (Fig. 4) prepared as in the method described under "Double extraction procedure." The biological samples used in this series were the same as those above. They were extracted with ethylene dichloride at pH 8 to 10; the ethylene dichloride was then extracted with an equal volume of 2.5 N sodium hydroxide and washed with water. The fluorescent material in the ethylene dichloride was then returned to an aqueous phase by extraction with 0.1 N HCl.

Results

The results on the pure acridine solutions are summarized in Table V and Fig. 1. The data in Table V indicate that at a constant strength of acid various acridines manifest quite different intensities of fluorescence.

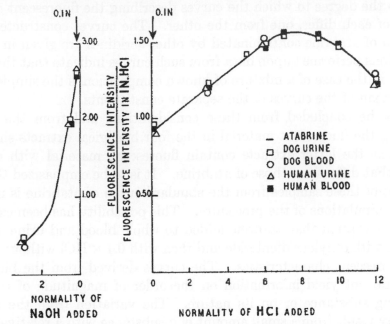

FIG. 4. Fluorescence characteristics of ethylene dichloride extracts of biological material after these extracts were washed with 2.5 N NaOH.

TABLE V

Relative Values for Intensity of Fluorescence Manifested by Six Acridines at a Concentration of 1 Mg. per Liter in 1.0 N HCl with Standard Filter Arrangements

Compound	Galvanometer reading
2-Methoxy-6-chloroacridone-9 (saturated solution).......	100
2-Methoxy-6-chloro-9-aminoacridine....................	400
2-Methoxy-6,9-dichloroacridine.......................	7
2-Methoxy-6-chloro-9(3-diamylaminopropyl)aminoacridine..................................	46
2-Methoxy-6-cyano-9(1-diethylamino-4-methylbutyl)-aminoacridine................................	26
2-Methoxy-6-chloro-9(1-diethylamino-4-methylbutyl)-aminoacridine................................	45

Furthermore, it is apparent from Fig. 1 that the fluorescence of each acridine varies in a systematic manner with variations in acid or alkali strength and that the variation is characteristic for each acridine. The curves relating fluorescence intensity to acid and alkali strength are highly reproducible.

It may be predicted from these results that the degree to which the fluorescence characteristics of a mixture of acridines will depart from the standard curve for each depends upon two factors. The first of these is the absolute magnitude of fluorescence of each of the acridines of the mixture at the standard hydrochloric acid concentration of 1.0 N. The second is the degree to which the curves describing the fluorescent characteristics of each differ, one from the other. The curves constructed from a solution of atabrine contaminated by other acridines are given in Fig. 2. Calculations performed upon data from such curves indicate that the curve observed in the case of a mixture of known composition is the simple arithmetical mean of the curves of the separate constituents.

It may be concluded from these considerations and from the curves describing the fluorescent material in the four biological extracts shown in Fig. 3 that the latter extracts contain fluorescent material with characteristics that differ from those of atabrine. It is to be emphasized that the departure of these samples from the standard curve of atabrine is not due to the manipulations of the procedure. This possibility has been excluded by the observation that atabrine added to whole blood and urine may be extracted with ethylene dichloride and then with 0.1 N HCl without change of its fluorescent characteristics. The curves derived from the biological samples do not yield information on the order of magnitude of the contaminating substance or on its nature. The variation from the normal curve may result from a small amount of a substance with a relatively high fluorescence or a large amount of a substance with a relatively low fluorescence.

The curves presented in Fig. 4 are from data which describe the fluorescent characteristics of material extracted from the same biological samples as were used in the experiments summarized in Fig. 3 with but one difference. This consisted in an alkaline wash of the initial ethylene dichloride extract, as in the analytical procedure. The data describing the characteristics of those extracts fall so closely along a curve which is characteristic of atabrine that it is unlikely for a significant amount of the fluorescence to be due to other material.

Similar results have been obtained from a series of samples of whole blood and urine derived from five patients and three dogs. It may be concluded from these data that those contaminants which cause departure from the standard curve in Fig. 3 can be removed by an alkaline wash.

Comment

Craig (1) has examined the solubility characteristics of the fluorescent material which is extractable by ethylene dichloride from blood and urine of dogs and humans during a course of atabrine therapy. The examination included the measurement of the distribution of the material and its manifest fluorescence in a system composed of ethylene dichloride and water-methyl alcohol mixtures. The examination was made at constant temperature and constant pH, the latter being obtained by the use of a cacodylate buffer. The distribution of the fluorescent material and the intensity of the fluorescence in each of the two phases of such a system depend upon the temperature, the pH of the aqueous phase, the dissociation constants of the fluorescent material, and the phenomenon of quenching. Ethylene dichloride extracts of blood and urine contained fluorescent materials, presumably degradation products of atabrine, which diverged in their combined solubility fluorescent characteristics from pure atabrine solutions. However, it was further demonstrated that these products are soluble in strong alkali and may be quantitatively removed by washing the ethylene dichloride extracts with 2.5 N NaOH. The residual fluorescent material was then found to have solubility characteristics which are identical with those of atabrine. It was in consequence of the latter finding that the alkali wash was incorporated in the analytical procedure.

Craig's work, together with that presented above and other unpublished studies,[4] clearly indicates that in the metabolism of atabrine degradation products are produced, some of which are fluorescent. However, the proposed method has a high degree of specificity, because it excludes these products in the final estimation, as evidenced by both the solubility and fluorescent characteristics of the material which is present in the final measurement. This result is achieved in part by the choice of ethylene dichloride as the organic solvent and in part by washing the ethylene dichloride extract with strong alkali. One of the groups of degradation products is insoluble in ethylene dichloride and is left behind in the initial extraction. The other group is extractable by ethylene dichloride, but can be removed from this solvent by a wash with 2.5 N sodium hydroxide. The latter group is quantitatively unimportant in the case of plasma, whole blood, and the tissues of humans and dogs. Consequently, the alkaline wash may be omitted for many routine purposes. However, most urine samples contain considerable amounts of these products and an alkaline wash is, therefore, essential even for routine determinations.

DISCUSSION

A method for the estimation of atabrine in biological fluids and tissue should be useful in several general circumstances. It will permit the detailed study of the general pharmacology of the drug as well as the study of its specific use in the suppression and treatment of malaria. Information on the plasma concentrations of atabrine in either of the latter conditions should prove as helpful in the quantitative control of such therapy as is

[4] Scudi, J., unpublished observations. Bush, T. M., Butler, T. C., and Greer, C. M., unpublished observation.

information on the plasma concentration of the sulfonamides in the control of sulfonamide therapy.

Two procedures have been described for determining atabrine. The general usefulness of the more simple single extraction procedure is limited somewhat by its lesser sensitivity. The inherent specificity of the procedure appears to be as great as that of the double extraction procedure except for the presence of a small blank which precludes its use at low atabrine concentrations.

The method of choice in any situation will depend on the concentration of atabrine in the biological sample to be analyzed as well as on the size of the sample available for analysis. The single extraction procedure is advised when the concentration of atabrine is at least 50 γ per liter and the amount of atabrine available for analysis is in excess of 0.5 γ. It should be noted, however, that special precautions must be taken to avert errors due to the presence of extraneous fluorescence when the single extraction procedure is used in this range of concentration.

The alkaline wash is only recommended when it is important that the measurement be wholly specific. Studies concerned with the pharmacology of the drug and its distribution and excretion fall in this general category. On the other hand, it is usually unnecessary to include this step in the routine estimation of plasma atabrine concentration when the latter datum is to be used in the appraisal of atabrine therapy. The error due to such an omission will usually be less than 10 per cent.

A method, somewhat similar to the double extraction procedure, has been recently developed elsewhere as the result of independent work (2). Extensive comparisons between this and the present methods have not been undertaken. However, it is known that the method has a sensitivity of the same order as the double extraction procedure and includes in the estimation a portion of the fluorescent degradation products of atabrine. The amounts of the latter compounds are not greatly in excess of those which are included in the double extraction procedure in the absence of an alkali wash of the ethylene dichloride. It would appear from this that it was suited for the routine of clinical therapy. However, its use for other purposes is problematic at the moment. Information is not available to indicate whether the method may be modified to make it wholly specific, as is the case for the present procedure. A disadvantage of the method stems from the fact that it is not capable of modification into a more simple single extraction procedure.

SUMMARY

Simple precise methods are described for the estimation of atabrine in biological fluids and tissues.

Atabrine is isolated from the biological material by an extraction of the free base with ethylene dichloride at a pH of about 8.   The ethylene dichloride extract is then washed free of degradation products with 2.5 N sodium hydroxide and the atabrine is returned as the salt to an aqueous phase of concentrated lactic acid.   The estimation of atabrine concentration is then made by its fluorescence in the lactic acid.

The second method for the estimation of atabrine involves the measurement of fluorescence in the ethylene dichloride phase.   The speed and simplicity of this procedure recommend it for routine use when the concentration of atabrine in the biological sample does not require excessive sensitivity on the part of the method.

These methods are specific for atabrine in that they do not include the degradation products of atabrine.

BIBLIOGRAPHY

1. Craig, L. C., *J. Biol. Chem.*, **150**, 33 (1943).
2. Masen, J. M., *J. Biol. Chem.*, **148**, 529 (1943).

COMMENTARY TO

9. Smith, T. W., Butler, V. P. and Haber, E. (1969)
Determination of Therapeutic and Toxic Serum Digoxin Concentrations by Radioimmunoassay. New England Journal of Medicine 281(22): 1212–1216.

W illiam Withering in his classic 1785 monograph on foxglove (*Digitalis purpurea*) described both its toxicity and its efficacy (1). He reported on the deaths of turkeys fed dried and chopped foxglove leaves mixed with their bran. Toxicity in humans included gastrointestinal and visual disturbances with some patients reporting that objects appeared yellow or green. He detailed 160 case reports on the use of foxglove in humans for a variety of diseases. Despite their potential for toxicity foxglove proved a great benefit to humanity and the drug quickly entered into standard medical practice. Vincent Van Gogh in 1890 painted two portraits of his personal physician Dr. Paul-Ferdinand Gachet. In one painting Gachet is holding a sprig of foxglove in his left hand, in the other painting, the plant is propped in a drinking glass (2).

In 1961 Rodensky and Wasserman (3) reported that up to 20% of hospitalized patients on digitalis drugs showed signs of toxicity. It would be another 8 years before methods became available for the determination of serum digitoxin and digoxin levels. These early methods would demonstrate the narrow therapeutic index of the cardiac glycosides and confirm the importance of therapeutic drug monitoring (TDM) in patient management. The therapeutic index for digoxin for example is 1.0–1.9 nmol/L (0.8–1.5 ng/mL). Most of these early methods for digoxin were based on the inhibition by cardiac glycosides of the membrane bound sodium–potassium adenosine triphosphatase (ATPase) pump. In 1965, Lowenstein (4) measured the inhibition of ^{86}Rb (rubidium) uptake into human red cells by plasma digoxin or digitoxin. The assay required solvent extraction of the plasma and took 24 hr to complete. Burnett and Conklin (5) in 1968 measured the inhibition of animal cerebral cortex ATPase by digoxin using a direct colorimetric phosphorus assay. Other researchers measured the competitive inhibition by tritiated ouabain on pig brain ATPase (6).

Oliver *et al.* were the first to describe an immunoassay for digitalis (7). The assay required solvent extraction of the serum followed by overnight incubation with iodinated digoxin derivative and rabbit antibody. Smith, Butler and Haber in their 1969 paper presented here provided the first practical method for cardiac glycosides in human serum. In this assay, serum digoxin is measured directly using tritiated digoxin and rabbit antibody developed by Butler and Chen (8). The assay used dextran-coated charcoal to separate free from bound tracer and was complete in 30 min. Smith, Butler and Haber (9) described in detail the antibody and Smith (10) reported on a similar assay for digitoxin. These methods were then used to confirm the therapeutic index and toxicity of the cardiac glycosides in a large number of patients (11–13).

Radioimmunoassay (RIA) procedures for digoxin and digitoxin based on the work of Smith, Butler and Haber became standard techniques in clinical chemistry by the mid-1970's. Gammill in 1976 in a survey of RIA procedures listed 28 different commercial test kits for digoxin that used either ^{3}H or ^{125}I as the tracer (14).

References

(1) Withering, W. (1785) An account of the foxglove, and some of its medical uses: with practical remarks on dropsy, and other diseases, in *Readings in Pharmacology*. Shuster, L. (ed), Little, Brown and Company, Boston, pgs 107–133, (1962).

(2) van Tilborgh, E.U.L. and van Heugten, S. (1990) *Paintings. Vincent van Gogh*. Arnoldo Mondadori Arte srl, Milan, pgs 269, 271.

(3) Rodensky, P.L. and Wasserman, F. (1961) Observations on digitalis intoxication. Archives of Internal Medicine. 108(2): 171–188.

(4) Lowenstein, J.M. (1965) A method for measuring plasma levels of digitalis. Circulation. 31(2):228–233.

(5) Burnett, G.H. and Conklin, R.L. (1968) The enzymatic assay of plasma digitoxin levels. Journal of Laboratory and Clinical Medicine. 71(6):1040–1044.

(6) Brooker, G. and Jelliffe, R.W. (1972) Serum cardiac glycoside assay based upon displacement of ^{3}H-Ouabain from Na–K ATPase. Circulation. 45(1):20–36.

(7) Oliver, G.C., Parker, B.M., Brasfield, D.L., and Parker, C.W. (1968) The measurement of digitoxin in human serum by radioimmunoassay. Journal of Clinical Investigation. 47(5):1035–1042.

(8) Butler, V.P. and Chen, J.P. (1967) Digoxin-specific antibodies. Proceedings of the National Academy of Sciences USA. 57(1):71–78.

(9) Smith, T.W., Butler, V.P., and Haber, E. (1970) Characterization of antibodies of high affinity and specificity for the digitalis glycoside digoxin. Biochemistry. 9(2):331–337.

(10) Smith, T.W. (1970) Radioimmunoassay for serum digitoxin concentration: methodology and clinical experience. Journal of Pharmacology and Experimental Therapeutics. 175(2):352–360.

(11) Smith, T.W. and Haber, E. (1970) Digoxin intoxication; the relationship of clinical presentation to serum digoxin concentration. Journal of Clinical Investigation. 49(12):2377–2386.

(12) Beller, G.A., Smith, T.W., Abelmann, W.H., Haber, E., and Hood, W.B. (1971) Digitalis intoxication. A prospective clinical study with serum level correlations. New England Journal of Medicine. 284(18):989–997.

(13) Smith, T.W. and Haber, E. (1971) The clinical value of serum digitalis glycoside concentrations in the evaluation of drug toxicity. Annals of the New York Academy of Sciences. 179:322–337.

(14) Gammill, J. (1976) Radioimmunoassay kits, in *Practical Radioimmunoassay*. Moss, A.J., Dalrymple, G.V., and Boyd, C.M. (eds), The C.V. Mosby Company, Saint Louis, MO, pgs 138–146.

New Engl. J. Med. 1969; 1212–1216

DETERMINATION OF THERAPEUTIC AND TOXIC SERUM DIGOXIN CONCENTRATIONS BY RADIOIMMUNOASSAY*

Thomas W. Smith, M.D., Vincent P. Butler, Jr., M.D., and Edgar Haber, M.D.

Abstract A sensitive (0.2 ng per milliliter), precise (standard deviation 3 to 4 per cent), and specific radioimmunoassay for serum digoxin concentration has been developed. Levels are determined by measurement of the extent to which digoxin in the patient's serum competes with tritium-labeled digoxin, added in vitro, for digoxin-specific antibody binding sites. Mean values in nontoxic patients with normal renal function receiving 0.25 or 0.50 mg per day were 1.1 and 1.4 ng per milliliter respectively; the ranges fell within relatively narrow limits. Patients with cardiac arrhythmias attributed to digoxin toxicity had a mean level of 3.3 ng per milliliter, and little overlap with the nontoxic group (p less than 0.001). A determination can be completed in one hour, and may provide useful information to the clinician faced with the difficult problem of evaluating his patient's state of digitalization.

S INCE the publication of Withering's remarkable "Account of the Foxglove" in 1785,[1] clinicians have sought better methods of judging proper thera-peutic dosage and diagnosing digitalis toxicity. Important advances have included the introduction of pure crystalline preparations facilitating standardization of dosage, and increased awareness of early electrocardiographic manifestations of digitalis excess.[2] Nevertheless, the low therapeutic ratio[3] of this class of drugs continues to result in an estimated rate of toxicity ranging from 7[4] to 22[5] per cent in hospitalized patients receiving digitalis glycosides.

The pharmacodynamics of the widely used cardiac glycoside digoxin have been well studied by the administration of tritium-labeled drug to patients.[6] The importance of renal function as the major determinant of excretion has been well

*From the Department of Medicine, Harvard Medical School, and the Cardiac Unit, Medical Service, Massachusetts General Hospital, Boston, and the Department of Medicine, Columbia University College of Physicians and Surgeons, New York City (address reprint requests to Dr. Smith at the Cardiac Unit, Massachusetts General Hospital, Boston, Mass. 02114).

Supported by the Myocardial Infarction Research Unit, under a contract (PH-43-67-1443), by research grants (HE-10608, HE-5196 and HEP-06664) from the United States Public Health Service, and by a grant from the New York Heart Association (Dr. Butler is the recipient of a research career-development award [1 K04-HE-11,315] from the United States Public Health Service).

Presented in part at the annual meeting of the American Society for Clinical Investigation, Atlantic City, N.J., May 5, 1969.

documented.[6] Clinically, however, prediction of total body digoxin stores is complicated by variation in absorption of oral doses[3,6] and variation in non-renal excretion.[7] Additional problems arise in hemo-dynamically unstable patients with fluctuating renal function, or when an accurate history of dosage cannot be obtained.

Since myocardial digoxin concentration has been found to bear a relatively constant ratio to serum levels in clinical studies using radioactively labeled digoxin, it has been suggested that the ability to measure serum digoxin concentration might be of clinical value.[8] To this end, we have developed a rapid, sensitive and specific radioimmunoassay for serum digoxin concentration and determined levels in groups of patients with and without evidence of digoxin toxicity.

MATERIALS AND METHODS

Materials

Digoxin. Crystalline digoxin was kindly supplied by Dr. Stanley Bloomfield, Burroughs-Wellcome and Company (USA), Incorporated, Tuckahoe, New York.

Tritiated digoxin. Digoxin, tritium labeled in the 12α position, specific activity 3.2 Ci per mM, was obtained from New England Nuclear Corporation, Boston.

Antidigoxin antibody was obtained as previously reported[9] by immunization of rabbits with a digoxin-protein conjugate. (Detailed characterization of this antibody is to be separately reported.[10])

Digitoxin. Crystalline digitoxin was donated by Wyeth Pharmaceuticals, Philadelphia.

Deslanoside. Crystalline deslanoside was donated by Sandoz Pharmaceuticals, Hanover, New Jersey.

Steroid compounds. Crystalline cholesterol, cortisol, dehydroepiandrosterone, 17-β estradiol, progesterone and testosterone were kindly supplied by Professor Lewis Engel, Massachusetts General Hospital and Harvard Medical School, Boston.

Patients

All the patients studied were adults hospitalized on the Medical or Surgical Services of the Massachusetts General Hospital. Ten (mean age of 62 years, and mean body weight of 146 pounds) had been on a stable oral maintenance digoxin dose of 0.25 mg daily for more than one week; 11 (mean age of 52 years, and mean body weight of 163 pounds) had received 0.5 mg orally per day for similar periods. Each of these 21 patients had normal blood urea nitrogen (BUN) (15 mg per 100 ml or less) or serum creatinine (1.0 mg per 100 ml or less) levels, and none had evidence clinically or by electrocardiogram of digitalis excess. Serum was obtained from these patients eight hours after the last dose of digoxin, and assayed as described below.

In a second group of 18 patients receiving digoxin there was evidence of toxicity according to the cri-

TABLE 1. *Criteria for Digoxin Toxicity.*

ONE OR MORE OF THE FOLLOWING DISTURBANCES OF IMPULSE FORMATION OR CONDUCTION IN A PATIENT RECEIVING DIGOXIN:

Supraventricular tachycardia (atrial or atrioventricular junctional) with block (12 patients)

Frequent or multifocal ventricular premature beats, ventricular bigeminy or ventricular tachycardia (7 patients)

Atrial fibrillation, with ventricular response <50 and ventricular premature beats (3 patients)

Sinus rhythm with 2d-degree or 3d-degree atrioventricular block (2 patients)

DISAPPEARANCE OF THE RHYTHM DISTURBANCE WHEN DIGOXIN WAS WITHHELD

teria listed in Table 1. Many in this group were acutely ill, and had received doses of digoxin varying from less than 0.25 mg orally per day to increments totaling 1.5 mg intravenously in a 24-hour period. Serum for assay was obtained from these patients at times varying from eight to 48 hours after the last administration of digoxin. The mean BUN for this group was 41 mg, with a range of 8 to 102 mg per 100 ml. Ten of the 18 had BUN values above 30 mg, and only four had values of 15 mg per 100 ml or less at the time of the serum digoxin determination.

Laboratory Methods

Serum was drawn by routine venipuncture, separated from formed elements by centrifugation and assayed immediately or stored for as long as five days at 4°C. Plasma obtained from heparinized blood gave identical results. Standards were prepared by the addition of gravimetrically determined amounts of crystalline digoxin to normal human serum. Determinations were run in duplicate. To 1 ml of serum in disposable plastic test tubes, 12 by 75 mm (Falcon Plastics, Los Angeles, California), was added, with thorough mixing, 3 ng of tritiated digoxin (Fig. 1). Antidigoxin antibody was then added in an amount sufficient to produce 37 to 45 per cent binding of the tritiated digoxin in the ab-

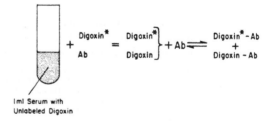

FIGURE 1. *Diagrammatic Representation of the Radioimmunoassay Method.*

Digoxin* indicates 12-α-tritiated digoxin, and Ab digoxin specific antibody. The amount of antibody-bound digoxin* (digoxin*-Ab) present at equilibrium is determined by the quantity of unlabeled digoxin present in the sample.

sence of unlabeled drug, and the mixture incubated at 25°C for 15 minutes. As shown in Figure 1, competition between labeled and unlabeled digoxin for antibody binding sites determines the amount of labeled digoxin-antibody complex present at equilibrium. Separation of bound from free labeled digoxin (Fig. 2) was achieved by the dextran-coated charcoal technic of Herbert et al.,[11] resulting in selective binding of free digoxin to the coated charcoal, which was then separated by centrifugation. The supernatant phase was added to 15 ml of liquid

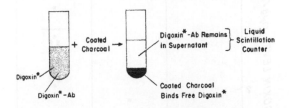

FIGURE 2. *Separation of Bound from Free Digoxin.**

Dextran-coated charcoal selectively binds free digoxin*; antibody-bound digoxin* in the supernatant after centrifugation is then counted.

scintillation medium[12] and heated in a water bath at 60°C for 10 minutes to complete denaturation of the antibody and release of labeled digoxin from antibody binding sites. Precipitated protein was separated by centrifugation, and the supernatant counted in an Ansitron liquid scintillation counter. Correction for quenching due to constituents of serum was made by recounting of each sample after addition of an internal standard of tritiated digoxin. Experiments quantitating the extent to which cardiac glycoside and steroid compounds competed with digoxin for antibody binding sites were carried out both in normal human serum and in phosphate-buffered saline (0.15M NaCl, 0.01M K_2HPO_4, pH 7.4), with the same results. Amounts of each steroid or steroid glycoside used were determined gravimetrically.

RESULTS

Standard Curve

Figure 3 represents a typical standard curve, derived from the addition of known amounts of digoxin to normal serum. Concentrations in each group of unknown samples were determined by comparison with a similar standard curve run simultaneously. Concentrations of 0.2 ng per milliliter, well below usual therapeutic levels, produced an easily measurable effect, and there was good resolution over the range of clinical interest. Twenty such curves run under similar conditions were virtually identical.

Precision

Replicate determinations were carried out on 10 aliquots of serum containing digoxin concentrations of 0.5, 2.0 and 10.0 ng per milliliter. Standard de-

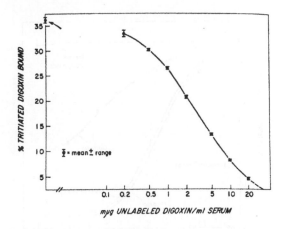

FIGURE 3. *Typical Standard Curve, as Run with Each Set of Unknowns.*

Sensitivity, precision and resolution in the range of clinical interest are demonstrated. The arrow on the vertical axis indicates binding of labeled digoxin in the absence of unlabeled digoxin.

viations for the final results were 3 to 4 per cent at each concentration level.

Specificity

Thirty serum samples chosen at random from patients not receiving cardiac glycosides all gave results within 2 per cent of the zero value; false-positive values were not encountered. Specificity was further defined by comparison of inhibition of antibody binding of tritiated digoxin by the closely related compounds digitoxin and deslanoside. Digitoxin differs from digoxin only in the absence of the C^{12} hydroxyl group, whereas deslanoside, with the identical steroid nucleus, has an additional glucose coupled to the terminal digitoxose sugar. Despite the marked similarity of structure, the small change in the steroid nucleus substantially reduced the binding of digitoxin as compared with digoxin (Fig. 4), whereas the addition of a glucose residue to the saccharide portion of the molecule had an almost negligible effect as reflected in the closely similar deslanoside curve. The antibody therefore appears to be quite specific for the structure of the steroid nucleus of digoxin. As might have been predicted from the relative lack of cross-reaction with digitoxin, physiologic steroids, including cholesterol, cortisol, dehydroepiandrosterone, 17-β estradiol, progesterone and testosterone, which differ from digoxin to a much greater extent, do not interfere with the assay even when present at levels 1000 times in excess of usual therapeutic digoxin levels (Fig. 4).

Serum protein binding of cardiac glycosides, although significant for albumin,[13] does not affect these results, nor does it interfere with the assay. This appears to be due to the fact that the antibody employed has association constants for digoxin and digitoxin that are several orders of magnitude higher than those of albumin.[10,13]

116

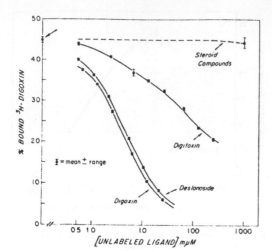

FIGURE 4. *Specificity: Comparison of the Ability of Related Compounds to Displace Tritiated Digoxin from the Antibody Binding Site.*

Steroid compounds (cholesterol, cortisol, dehydroepiandrosterone, 17-β estradiol, progesterone and testosterone) cause negligible displacement, even in concentrations in excess of physiologic levels. (Cholesterol concentration is limited by solubility in aqueous medium.) The arrow on the vertical axis indicates binding of labeled digoxin in the absence of unlabeled steroids or steroid glycosides.

Rapidity

It is possible to run an assay in one hour, rendering the method applicable to the assessment of patients requiring early management decisions.

Clinical Studies

Figure 5 includes in the left and middle columns individual serum digoxin concentrations for the nontoxic patients previously described; mean values are 1.1 and 1.4 ng per milliliter for patients receiving 0.25 mg and 0.50 mg per day respectively. The 18 patients who were digoxin toxic by the criteria listed in Table 1 had levels as shown in the right column of Figure 5. The mean value for this group was 3.3 ng per milliliter. The highest value, 8.7 ng per milliliter, occurred in a patient with simultaneous atrial and junctional tachycardias with atrioventricular dissociation who died shortly after this serum sample was obtained, and therefore represents an exception to the criterion of disappearance of rhythm disturbance when digoxin was withheld.

The difference between mean values for the toxic and nontoxic groups is statistically significant, with a p value of less than 0.001.

DISCUSSION

Therapeutic serum or plasma digitoxin concentrations, which are about 10 times higher than those of digoxin, have been measured by inhibition of red-cell [86]rubidium uptake,[14] double isotope-dilution derivative assay,[15] radioimmunoassay[16] and inhibition of Na-K-activated ATPase activity.[17] All these

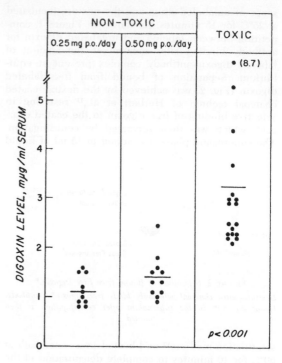

FIGURE 5. *Individual Serum Digoxin Concentrations.*

Mean values, denoted by the horizontal lines in each column, are 1.1 ± 0.3 (standard deviation) ng per milliliter (mμg/ml), range of 0.8 to 1.6, for nontoxic patients on oral doses of 0.25 mg per day, and 1.4 ± 0.4, range of 0.9 to 2.4, for the corresponding group on 0.50 mg per day. Values for toxic patients, as defined in Table 1, were 3.3 ± 1.5, with a range of 2.1 to 8.7.

methods require extraction of digitoxin from serum with organic solvents.

Digoxin has proved more difficult to determine because of the extremely low circulating levels present in patients given usual therapeutic doses. Serum concentrations have been detected by inhibition of red-cell [86]rubidium transport[14,18] and by displacement of tritiated digoxin from Na-K-activated ATPase.[19] However, extraction procedures are again required, and data defining specificity and rapidity of these methods have not been extensively documented. The remarkable specificity of the antibody used in the studies reported here obviates the need for extraction of digoxin from whole serum, and greatly enhances the rapidity and simplicity of the assay.

Useful data have been obtained by several groups, particularly Doherty[6] and Marcus and his co-workers,[20] by the direct administration of tritiated digoxin to patients with subsequent measurement of radioactivity in physiologic fluids, including blood or serum. Marcus et al.[21] studied normal volunteers on a steady-state oral dose of 0.5 mg of tritiated digoxin per day, and obtained a mean blood level of 1.4 ± 0.3 (standard deviation) ng per milliliter eight hours after the last dose. This value is in excellent

117

agreement with the data reported here for the group with normal renal function on 0.5 mg orally per day (Fig. 5). Grahame-Smith and Everest,[18] using a modified [86]rubidium transport inhibition assay, reported a mean plasma level of 2.36 ng per milliliter for nontoxic patients receiving 0.25 mg or more of digoxin per day. Direct comparison with the lower values reported in our study and by Marcus[21] is not possible because dosage, renal function and time of plasma sampling were not defined.

The interpretation of serum digoxin levels in the clinical context depends upon the assumption of a relatively constant serum-to-myocardial concentration ratio. Na[22] and K balance and thyroid state[6] have been shown to affect myocardial concentration of tritiated digoxin in animal experiments. However, within physiologic limits the ratio has been found to be relatively constant in human subjects,[8] and the serum level should allow useful inferences to be drawn regarding myocardial and total body stores.

The relative similarity of serum levels in the two groups of nontoxic patients summarized in Figure 5 bears comment. The group receiving 0.5 mg per day had a larger mean weight (by 17 pounds) and a younger mean age (by 10 years), suggesting a higher glomerular filtration rate and hence more rapid excretion of digoxin.[23] In addition, the clinical response of each patient probably results in dosage adjustments that tend to standardize the serum concentration at a desirable therapeutic level. For example, two of the patients receiving 0.25 mg per day had their doses decreased from 0.5 mg per day because of anorexia and mild nausea, and two of the group receiving 0.5 mg per day had their doses raised to this level from 0.25 mg per day because of inadequate control of the ventricular rate in atrial fibrillation.

Doses of digoxin received by patients in the toxic group were in most cases also in the range of 0.25 to 0.50 mg per day. The major factor leading to the higher serum digoxin concentrations therefore appears to be diminished renal function (mean BUN of 41 mg per 100 ml) compared with nontoxic patients, each of whom had a BUN of 15 mg per 100 ml or less or a serum creatinine level of 1.0 mg per 100 ml or less.

The rigid criteria for toxicity used in this study exclude many patients with early or subtle signs of digitalis excess such as occasional ventricular premature beats who might be expected to have serum digoxin levels closer to those of the nontoxic group. The upper limit of appropriate serum digoxin concentration is difficult to define, since we have seen a few patients receiving frequent increments of digoxin, usually for control of ventricular rate in atrial fibrillation or flutter, who had serum levels above those of the nontoxic control group but who gave no clinical evidence of toxicity.

In general, our experience has been that knowledge of the serum digoxin concentration is of con-

siderable clinical usefulness when weighed along with the many other factors that influence pharmacologic and toxic effects, including serum potassium, calcium and magnesium, acid-base balance, oxygen tension, other drugs that the patient has received, thyroid state, and nature of the underlying cardiac disease.[2] The rapidity and simplicity of the assay make it well suited to the early evaluation of problem patients, and the lack of need for administration of radioactive substances to the patient facilitates studies of the clinical pharmacology of the drug.

We are indebted to Miss Lynne M. Geever, R.N., for technical assistance.

REFERENCES

1. Withering W: An account of the foxglove and some of its medical uses, with practical remarks on dropsy and other diseases, Classics of Cardiology. Edited by FA Wilkins, TE Keys. New York, Dover Publications, 1941. pp 227-252
2. Irons GV Jr, Orgain ES: Digitalis-induced arrhythmias and their management. Progr Cardiovasc Dis 8:539-569, 1966
3. Moe GK, Farah AE: Digitalis and allied cardiac glycosides. Chap 31, The Pharmacological Basis of Therapeutics. Third edition. Edited by LS Goodman, A Gilman. New York, The Macmillan Company, 1965, pp 665-698
4. Sodeman WA: Diagnosis and treatment of digitalis toxicity. New Eng J Med 273:35-37 and 93-95, 1965
5. Rodensky PL, Wasserman F: The possible role of sex in digitalis tolerance. Amer Heart J 68:325-335, 1964
6. Doherty JE: The clinical pharmacology of digitalis glycosides: a review. Amer J Med Sc 255:382-414, 1968
7. Bloom PM, Nelp WB, Tuell SH: Relationship of excretion of tritiated digoxin to renal function. Amer J Med Sci 251:133-144, 1966
8. Doherty JE, Perkins WH, Flanigan WJ: The distribution and concentration of tritiated digoxin in human tissues. Ann Intern Med 66:116-124, 1967
9. Butler VP Jr, Chen JP: Digoxin-specific antibodies. Proc Nat Acad Sci USA 57:71-78, 1967
10. Smith TW, Butler VP Jr, Haber E: Characterization of antibodies of high affinity and specificity for the digitalis glycoside digoxin. Biochemistry (in press)
11. Herbert V, Lau K, Gottlieb CW, et al: Coated charcoal immunoassay of insulin. J Clin Endocr 25:1375-1384, 1965
12. Bray GA: A simple efficient liquid scintillator for counting aqueous solutions in a liquid scintillation counter. Anal Biochem 1:279-285, 1960
13. Lukas DS, De Martino AG: Binding of digitoxin and some related cardenolides to human plasma proteins. J Clin Invest 48:1041-1053, 1969
14. Lowenstein JM, Corrill EM: An improved method for measuring plasma and tissue concentration of digitalis glycosides. J Lab Clin Med 67:1048-1052, 1966
15. Lukas DS, Peterson RE: Double isotope dilution derivative assay of digitoxin in plasma, urine, and stool of patients maintained on the drug. J Clin Invest 45:782-795, 1966
16. Oliver GC Jr, Parker BM, Brasfield DL, et al: The measurement of digitoxin in human serum by radioimmunoassay. J Clin Invest 47:1035-1042, 1968
17. Burnett GH, Conklin RL: The enzymatic assay of plasma digitoxin levels. J Lab Clin Med 71:1040-1044, 1968
18. Grahame-Smith DG, Everest MS: Measurement of digoxin in plasma and its use in diagnosis of digoxin intoxication. Brit Med J 1:286-289, 1969
19. Brooker G, Jelliffe RW: Determination of serum digoxin by enzymatic isotopic displacement of H³ digoxin from Na - K ATPase. Fed Proc 28:608, 1969
20. Marcus FI, Kapiada GJ, Kapiada GG: The metabolism of digoxin in normal subjects. J Pharm Exp Ther 145:203-209, 1965
21. Marcus FI, Burkhalter L, Cuccia C, et al: Administration of tritiated digoxin with and without a loading dose: a metabolic study. Circulation 34:865-874, 1966
22. Harrison CE Jr, Wakim KG: Inhibition of binding of tritiated digoxin to myocardium by sodium depletion in dogs. Circ Res 24: 263-268, 1969
23. Ewy GA, Kapiada GG, Yao L, et al: Digoxin metabolism in the elderly. Circulation 39:449-453, 1969

COMMENTARY TO

10. Thompson, R. D., Nagasawa, H. T. and Jenne, J. W. (1974)
Determination of Theophylline and its Metabolites in Human Urine and Serum by High-Pressure Liquid Chromatography. Journal of Laboratory and Clinical Medicine 84(4): 584–593.

Theophylline (1,3-dimethylxanthine) is an anti-asthmatic drug that has been used in both adults and children since 1936. Therapeutic drug monitoring (TDM) of patients on theophylline therapy is well established as a means to avoid toxicity especially in neonates. The standard method for analysis of theophylline was for many years the ultraviolet (UV) spectrophotometric procedure of Schack and Waxler (1). Theophylline is measured at 277 nm following organic solvent extraction. The assay requires up to 2 mL of serum and suffers from interferences from barbiturates, dietary xanthines and theophylline metabolites. Numerous attempts were made to improve the procedure in order to avoid these problems (2–4). The first publication of a serum and urine theophylline assay by Thompson, Nagasawa and Jenne using liquid chromatography solved all of the problems of the UV assay. Thompson was a graduate student in 1974 finishing his Masters degree when he was given the assignment to develop an alternative to the Schack and Waxler method for monitoring theophylline in patient serums.

High-pressure liquid chromatography is the older term for this type of liquid chromatography that is now called high-performance liquid chromatography (HPLC). In 2004 one of HPLC's early advocates regarded it as the biggest revolution in analytical chemistry in the last 40 years (5). Some of the early pioneers in the development of HPLC include Horvath and co-workers (6), Snyder (7) and Kirkland (8).

In the 1974 HPLC paper for theophylline presented here the pressure used to drive the liquid phase through the column was provided by a compressed tank of helium. The 1.8 mm id column was packed with ion-exchange resin beads 13 μm in diameter. Serum sample volume was less than 10 uL and the analysis time was 24 minute. Metabolites were well separated from parent compounds and other drugs. The same assay was used to study the relationship between urine and serum metabolites in theophylline dosed patients (9). This paper by Thompson, Nagasawa and Jenne was the first HPLC method for serum theophylline. Within 2 years, 14 different HPLC methods were published for theophylline in biological fluids. The success of these theophylline assays helped establish HPLC as an essential tool in clinical chemistry.

References

(1) Schack, J.A. and Waxler, S.H. (1949) An ultraviolet spectrophotometric method for the determination of theophylline and theobromine in blood and tissues. Journal of Pharmacology and Experimental Therapeutics. 97(3):283–291.

(2) Jatlow, P. (1975) Ultraviolet spectrophotometry of theophylline in plasma in the presence of barbiturates. Clinical Chemistry. 21(10):1518–1520.

(3) Vasiliades, J. and Turner, T. (1976) A modified ultraviolet spectrophotometric method for the determination of theophylline in serum in the presence of barbiturates. Clinica Chimica Acta. 69(3):491–495.

(4) Schwertner, H.A., Wallace, J.E., and Blum, K. (1978) Improved ultraviolet spectrophotometry of serum theophylline. Clinical Chemistry. 24(2):360–361.

(5) Engelhardt, H. (2004) One century of liquid chromatography. From Tswett's columns to modern high speed and high performance separations. Journal of Chromatography B Analytical Technology and Biomedical Life Science. 800(1–2): 3–6.

(6) Horvath, C.G., Preiss, B.A., and Lipsky, S.R. (1967) Fast liquid chromatography: an investigation of operating parameters and the separation of nucleotides on pellicular ion exchangers. Analytical Chemistry. 39(12):1422–1428.

(7) Snyder, L.R. (1967) An experimental study of column efficiency in liquid–solid adsorption chromatography. Analytical Chemistry. 39(7):698–704.

(8) Kirkland, J.J. (1969) Controlled surface porosity supports for high speed gas and liquid chromatography. Analytical Chemistry. 41(1):218–220.

(9) Jenne, J.W., Nagasaw, H.T., and Thompson, R.D. (1976) Relationship of urinary metabolites of theophylline to serum theophylline levels. Clinical Pharmacology and Therapeutics. 19(3):375–381.

J. Lab. Clin. Med. 1974; 84(4): 583–593
Copyright © 1974 Mosby.
Reproduced with permission.

Determination of theophylline and its metabolites in human urine and serum by high-pressure liquid chromatography

RICHARD D. THOMPSON,* HERBERT T. NAGASAWA *Minneapolis, Minn.* and JOHN W. JENNE *Albuquerque, N. M.*

A high-pressure liquid chromatographic system has been developed for the quantitative determination of theophylline and its metabolic products in human urine. The procedure involves preliminary separation of the xanthines from the uric acids by anion-exchange chromatography followed by liquid chromatography at 725 p.s.i. on Aminex A-5 cation exchange resin. The 24-hour urinary excretion of theophylline, 3-methylxanthine, 1,3-dimethyluric acid, 1-methyluric acid, and uric acid has been quantitatively determined by this method in two human subjects who were being treated with oral aminophylline. By direct injection of diluted sera into the high-pressure column, serum theophylline levels, as well as the serum levels of its metabolite, 3-methylxanthine, and of uric acid were also determined in these two subjects without interference from caffeine and other methylated xanthines, or from hypoxanthine.

Theophylline 1,3-dimethylxanthine; 1,3-MX; Fig. 1), is a mild diuretic agent, a moderately-active central nervous system and myocardial stimulant, and a powerful bronchodilator which relaxes the smooth muscle fibers of the bronchi.[1] This latter property constitutes the basis of therapy for a large number of individuals with obstructive lung disease. Despite its widespread use in this regard, the metabolism of theophylline in humans has not received the attention that it deserves.

The pioneering work of Brodie, Axelrod, and Reichenthal[2] has established that the major urinary metabolite of theophylline in man was 1,3-dimethyluric acid (1,3-MU). Subsequently, Johnson,[3] Weinfeld and Christman,[4] and Cornish and Christman[5] have identified the remaining metabolites as 3-methylxanthine (3-MX) and 1-methyluric acid (1-MU). The conspicuous absence of 1-methylxanthine (1-MX) and 3-methyluric acid (3-MU) has been noted by these these investigators and will be confirmed herein.

Wide variations have been reported for the serum half-life values for

From the Medical Research Laboratories, Veterans Administration Hospital, and the Department of Medicinal Chemistry, University of Minnesota, Minneapolis; and the Pulmonary Disease Section, Veterans Administration Hospital, and the Department of Medicine, University of New Mexico, Albuquerque.

This work was supported by Program Grant 618/5968.1, Veterans Administration, and in part by a grant from Cooper Laboratories, Wayne, N. J.

Received for publication Feb. 27, 1974.

Accepted for publication June 19, 1974.

Reprint requests: Dr. H. T. Nagasawa, Veterans Administration Hospital, Minneapolis, Minn. 55417.

*Participant in a Health, Education, and Welfare Long-term Training Program of the United States Food and Drug Administration. Present address: United States Food and Drug Administration, Minneapolis, Minn. 55401.

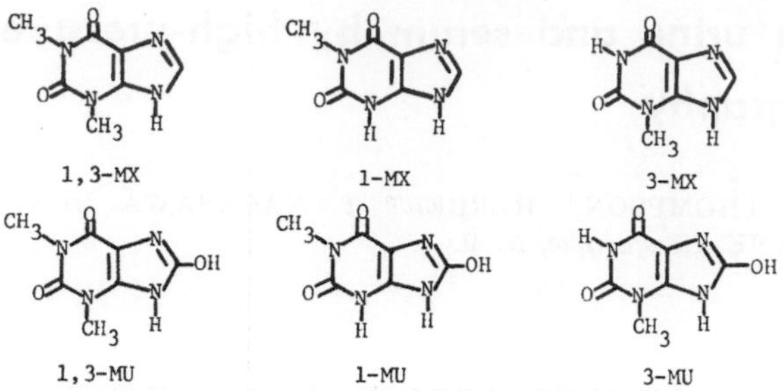

Fig. 1. Structures of theophylline (1,3-MX) and its possible metabolic products: 1-methylxanthine (1-MX); 3-methylxanthine (3-MX); 1,3-dimethyluric acid (1,3-MU); 1-methyluric acid (1-MU); and 3-methyluric acid (3-MU).

theophylline in patients on an essentially identical therapeutic regimen with aminophylline (theophylline ethylenediamine), and the suggestion has been made that these differences might relate to differences in the rate and/or pathway of metabolic degradation of theophylline.[6]

Direct application of the methods of Cornish and Christman[5] to the study of these variations in the metabolism of theophylline in man are hampered by the cumbersome procedures involved (anion-exchange chromatography followed by paper chromatography; then, ultraviolet spectrophotometry and colorimetry), and the paucity of authentic synthetic specimen of metabolites and possible metabolites for use as standards. These obstacles have all been overcome in the present study; and we wish to describe a rapid, high-pressure liquid chromatographic method for the separation, identification, and quantitation of the major urinary theophylline metabolites excreted by man. A simplified version of the method is directly applicable to the determination of *serum theophylline levels* without interference from its own metabolites, from hypoxanthine or uric acid, or from dietary caffeine.* Furthermore, the method of separation and detection, *viz.*, ion exchange chromatography followed by measurement of absorbtivity at 280 nm., precludes any interference from drugs such as ephedrine or phenobarbital which are commonly used in the therapeutic management of asthmatic patients, since these drugs have negligible absorption in this region of the spectrum. At this wavelength, theophylline and its metabolites all absorb near-maximally with log ϵ_{280} values equal to or greater than 3.88.

Materials

Apparatus. A schematic diagram of the high-pressure liquid chromatographic system employed in this study is shown in Fig. 2. The eluent delivery unit consisted of a 500 ml. high-pressure (0 to 750 p.s.i.) solvent cylinder (C) (Varian Aerograph, Walnut Creek, Calif.) pressurized by a tank of compressed helium (A) with the flow-rate controlled by a two-stage precision gas regulator (B). The liquid lines throughout the system, including the reference column (G), were constructed of Type 316, 1/8 inch outside diameter stainless-steel tubing with low dead volume stainless-steel Swagelok compression fittings (Crawford Fitting Company, Solon, Ohio).

*As caffeine is not metabolized to theophylline in significant amounts in man[5] and theophylline is separated from caffeine by this procedure, the serum method is applicable for either substance. However, 2 of the 5 known metabolites of caffeine are identical to the metabolites of theophylline; hence, dietary caffeine will interfere with the quantitative determination of the *urinary* metabolites of theophylline; and, conversely, theophylline administration will interfere with the determination of caffeine metabolities in urine.

586 *Thompson, Nagasawa, and Jenne* J. Lab. Clin. Med.
October, 1974

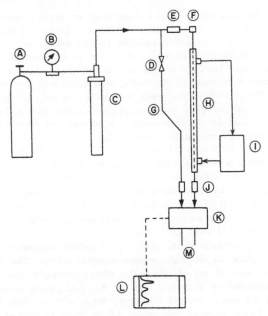

Fig. 2. Schematic diagram of the high-pressure liquid chromatographic system used for the quantitative determination of theophylline and its metabolites. See Materials section for details.

The analytical column (H) consisted of a 66.5 cm. length of Type 316, 3.2 mm. (outside diameter) and 1.8 mm. (inside diameter) precision-bore stainless-steel tubing equipped with an "on-column" sample injector (F) (Varian Aerograph) which utilized ¼ inch diameter laminated Tefseal septa (Hamilton Company, Whittier, Calif.) preceded by a check valve (E, 10 p.s.i.). The column (as well as the reference column) was terminated with a ⅛ inch to 1/16 inch reducing union (J) fitted with a 2.0 μ stainless-steel fritted disk. The flow-rate through the reference column (G) was adjusted using a regulating valve (D). A fibrous internal filter was also placed above the fritted disk on the analytic column to support the resin bed. The temperature of the column was controlled with water from a circulating constant temperature bath (I) which flowed through a ¾ inch (outside diameter) stainless-steel water jacket surrounding the column.

The detector (K) was a Pharmacia UV Monitor (Pharmacia Fine Chemicals, Piscataway, N. J.) operated at 280 nm., and its output was monitored with a 10 mV. recorder (L). Samples were introduced by means of a 10 μl Model CG(A) Pressure-Lok microsyringe (Precision Sampling Corporation, Baton Rouge, La.), and effluents were collected (or discarded) at the terminal end (M).

Reagents and reference compounds. All reagents used in this study were of analytic grade purity. The buffer eluent was prepared by dissolving 51.8 Gm. of ammonium dihydrogen phosphate ($NH_4H_2PO_4$) in 1 L. of boiled distilled water and adjusting the pH to 3.65 with 1 M H_3PO_4.

Uric acid, xanthine, hypoxanthine, and caffeine were purchased from Calbiochem, San Diego, Calif.; 7-methylxanthine from Pfaltz and Bauer, Inc., Flushing, N. Y.; and theophylline and theobromine from Nutritional Biochemicals Corporation, Cleveland, Ohio. 1-Methylxanthine was obtained from Cyclo Chemical Company, Los Angeles, Calif. This compound was found to be chromatographically impure and was, therefore, purified by recrystallization from hot dilute acetic acid. 1-Methyluric acid,[7, 8] 3-methyluric acid, 1,3-dimethyluric acid,[9, 10] and 3-methylxanthine[9, 11, 12] were synthesized by adaptation of published procedures in the literature. 1,7-Dimethylxanthine was a gift from Dr. Gertrude B. Elion of the Wellcome Research Laboratories, Research Triangle Park, N. C.

Ion-exchange resins. Dowex 2-X8 anion-exchange resin (Cl⁻ form) 50 to 100 mesh

(Bio-Rad Laboratories, Richmond, Calif.) was used to prefractionate the urine into a xanthine or uric acid fraction. This step was necessary because 3-MX and 1,3-MU, when present together as in urine, did not completely resolve under the conditions of the high-pressure liquid chromatography described here. This preliminary ion-exchange separation into xanthine and uric acid fractions also served as "clean-up" steps and rendered the urine samples essentially free from much of the endogenous substances that could interfere in the subsequent steps of the procedure. The resin was packed as an aqueous slurry into a 1 cm. (internal diameter) by 15 cm. glass chromatographic column with a Teflon stopcock and a reservoir at the top to hold 50 ml. of solvent. The column was filled with resin to a height of 5.5 cm. and the resin bed was capped with a plug of glass wool. The column was conditioned with 50 ml. of 6N HCl followed by a water wash to neutrality. To isolate xanthines from urine, the column was washed with 200 ml. of distilled water prior to use. For the isolation of uric acids from urine, the resin was equilibrated with 50 ml. of 0.01 N HCl. Xanthines are not retained by this column and are washed through with water, while uric acids are retained and require elution with acid (see below). A new column of this type was prepared for each of the urine samples analyzed and approximately 5 ml. of liquid was maintained above the resin bed when not in use.

Aminex A-5 cation exchange resin, 13 ± 2 μ (Bio-Rad Laboratories), was selected for use as the stationary phase in the high-pressure analytic column. The resin (1.8 Gm.) was allowed to equilibrate with 10 ml. of the buffer eluent overnight and then packed by the dynamic procedure described by Scott and Lee[13] at 650 p.s.i., using a stainless-steel reservoir 45 cm. in length by 4.5 mm. (internal diameter) fitted with ¼ inch to ⅛ inch reducing unions. The excess resin was removed from about 1.5 cm. from the top of the analytic column.

Standard solutions. Standard A (stock solution of methyl xanthines) was prepared by dissolving 10.0 mg. each of 3-methylxanthine and theophylline in approximately 35 ml. of 0.08 M $Na_2B_4O_7$ · 10 H_2O, adjusting the pH to 7 with 1 M H_3PO_4, then adding 5.8 Gm. of NaCl and diluting to 100.0 ml. with water. This solution was stable over a period of at least 4 months when stored under ambient conditions.

Standard B (stock solution of uric acids) prepared in the same manner as Standard A above, contained 10.0 mg. each of uric acid, 1-methyluric acid, and 1,3-dimethyluric acid in a final volume of 100.0 ml. The stability of this solution was also comparable to Standard A. Uric acid was added to Standard B to permit the quantitative analysis of this normal constituent both in urine and in serum.

Individual reference standard solutions for other uric acids and xanthines were similarly prepared, each at concentrations of 50 μg per milliliter and their chromatographic retention behavior was determined (Table III).

Methods

Serum and urine sampling. Blood and urine samples were obtained from two male subjects who were being treated as outpatients with maintenance doses of oral aminophylline for asthmatic conditions. The subjects did not consume coffee, tea, colas, or chocolate, i.e., were on a "xanthine-free" regimen, for at least 12 hours prior to sampling. Blood was collected immediately following an oral dose and at hourly intervals thereafter for a period of 4 and 6 hours, respectively. The serum was separated from the coagulated blood by centrifugation (2x) and a 3.0 ml. aliquot removed for theophylline determination by the ultraviolet procedure of Schack and Waxler.[14] The remainder of the serum was frozen and kept frozen until just prior to analysis the following day by the high-pressure liquid chromatographic procedure described below for serum.

Urine was collected at room temperature over a 24-hour period, the volume recorded, and a few drops of toluene added as preservative. Aliquots (100 ml.) from each urine sample were taken and quick-frozen for analysis the following day. The remainder was stored at –20° C.

For recovery studies (see below), serum and urine samples from healthy individuals not receiving medication of any kind and who were on a xanthine-free regimen were used.

Conditions for high-pressure liquid chromatography. The following instrumental parameters for optimal separation of theophylline metabolites using the Aminex A-5 resin

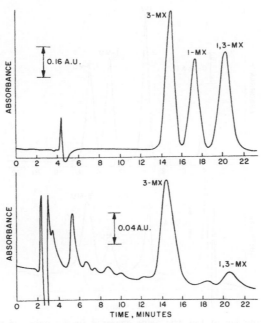

Fig. 3. Liquid chromatogram of the methylxanthines excreted in the urine of a subject receiving 300 mg. of oral aminophylline four times a day (lower), compared against a standard mixture of methylxanthines (upper). Conditions as described in the text.

in the high-pressure system were established experimentally: buffer eluent, 0.45 M $NH_4H_2PO_4$, pH 3.65; column temperature, 55° C.; column pressure, 725 p.s.i.; flow rate, 10.0 ml. per hour; and chart speed, 0.5 inch per minute. Typical chromatograms of standard mixtures of xanthines and uric acids determined under these conditions are present in Figs. 3 and 4 (upper curves).

Urine

FOR METHYLXANTHINES. Five milliliters of urine was adjusted to pH 7.1 to 7.2 with 0.1 N NaOH and transferred with the aid of 5 ml. of distilled water to the Dowex 2-X8 column previously equilibrated with distilled water. After the sample had passed into the resin bed, the column was eluted with 125 ml. of distilled water into a 500 ml. round-bottomed flask. The eluate was evaporated to dryness in vacuo at 40° C. on a rotary evaporator and the residue dissolved in 50.0 ml. of 1 M NaCl. (Note: Further dilution at this step, e.g., to 100 ml., may be necessary as 3-methylxanthine is excreted in large quantities in urine.) Seven microliters of this solution was injected into the liquid chromatograph at a detector absorbance range of 0.04. A calibration curve was prepared as follows: 6.0 ml. of Standard A was diluted to 25.0 ml. with 1 M NaCl, and 2.0, 4.0, 6.0, and 8.0 ml. aliquots were further diluted to 10.0 ml. with this solvent, 7 μl being injected. Peak areas were determined from the chromatograms by multiplying the peak heights by the width at half-height, and the area values were plotted against the nanograms of standard injected. The quantity (milligrams per 24 hours) of methylxanthines in the urine sample was calculated as follows:

$$\text{Nanograms methylxanthine} \atop \text{(from calibration curve)} \times \frac{\text{dilution factor}}{\mu\text{l urine injected}} \times \text{24-hour urine volume (in milliliters)} \times \frac{1}{1{,}000}$$

FOR URIC ACIDS. Two milliliters of urine was adjusted to pH 6.8 to 6.9 with 0.1 N NaOH and transferred with the aid of 5 ml. of 0.01 N HCl to the anion exchange column which had been previously equilibrated with 0.01 N HCl. After the sample had passed into the resin bed, the column was eluted with 90 ml. of 0.1 N HCl. The eluate was adjusted to pH 7.0, initially using 10 N NaOH with final adjustment with 1 N NaOH, transferred to a

126

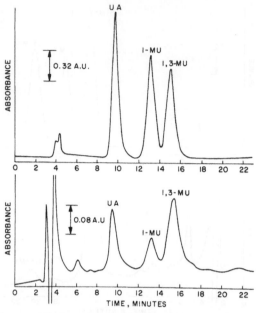

Fig. 4. Liquid chromatogram of the uric acids excreted in the urine of the subject described in Fig. 3 (lower), compared against a standard mixture of uric acids (upper). Conditions as described in the text.

100 ml. volumetric flask containing 5.8 Gm. of NaCl and then diluted to volume with water. Seven microliters of this solution was injected into the liquid chromatograph at a detector absorbance range of 0.08. Calibration curves were prepared from Standard B and the 24-hour uric acid levels in urine were calculated in the same manner as above for the methylxanthines.

Each of the respective components of Standards A and B exhibited linear detector response over at least a tenfold concentration range. The lower limit of detection was estimated to be 5 to 10 ng. for the methylxanthines and between 2 to 5 ng. for the uric acids.

Serum. One milliliter of serum was diluted with 2.0 ml. of 1 M NaCl, thoroughly mixed, and 7 μl of this diluted serum sample was injected directly into the liquid chromatograph at a detector absorbance range of 0.04. The serum concentrations (micrograms per milliliter) of 3-methylxanthine, theophylline, and uric acid were calculated using the appropriate calibration curve, taking account of sampling and dilution factors. [For routine applications, a replaceable "pre-column" maintained at 55° C. and consisting of a short length (10 to 15 cm.) of column packing placed above the analytic column is recommended. This is because repeated injections of diluted serum samples over a period of time reduced the column life.] The addition of NaCl to urine or serum samples resulted in much more symmetrical chromatographic peak shapes and is, therefore, recommended. Studies using NaCl concentrations over a range of 0.1 to 2.0 M indicated an optimal concentration of about 1 M. This phenomenon appears not to be due to a salt effect on the column resin, since incorporation of NaCl into the buffer *eluent* over a concentration range of 0.01 to 0.5 M resulted in severe loss of resolution of the standards.

Recovery studies. The efficiency of the preliminary ion-exchange chromatographic step for the separation of xanthines from the uric acids was evaluated by adding known amounts of Standard A or Standard B at two different levels (0.5 and 2.5 mg.) to 5 ml. aliquots of normal human urine. The recoveries, compared against like quantities of standards processed in the same manner as the calibration standards, ranged from 98 to 100 per cent, and duplicate samples were reproducible to within 5 ng. of each other.

The recoveries of 30 or 50 μg of theophylline and 3-methylxanthine added to normal

J. Lab. Clin. Med.
October, 1974

Table I. 24-Hour urinary excretion of theophylline and its metabolites by two male subjects receiving oral aminophylline (mg.)*

Subject	Theophylline	3-Methylxanthine	1-Methyluric acid	1,3-Dimethyluric acid
SS	18.0	91.4	225	920
JD	9.3	57.3	122	453

*Average of duplicate determinations. The lack of material balance here indicates that the patients did not diligently take their full 24-hour doses.

Table II. Serum levels of theophylline and 3-methylxanthine in two male subjects receiving oral aminophylline (μg/ml.)

Subject	Time following administration (hours)	Theophylline*	3-Methylxanthine
SS	0	7.3 (7.9)	3.3
	1	8.1 (8.0)	3.3
	2	11.5 (11.2)	4.0
	3	10.7 (10.7)	4.2
	4	8.1 (9.2)	3.8
JD	0	13.0 (12.9)	3.4
	1	18.5 (19.8)	3.1
	2	15.6 (17.2)	3.7
	3	16.7 (17.6)	3.4
	4	14.1 (15.0)	3.4
	5	13.8 (14.8)	3.4
	6	12.0 (11.7)	3.1

*Numbers in parentheses represent values determined by the Schack and Waxler procedure.[14]

human *serum* using identical serum/diluent volumes as described previously ranged from 99 to 102 per cent. This amount of theophylline added was in the concentration range normally found in serum or plasma after the administration of therapeutic doses.

Blank studies with normal urine or serum carried through the same procedure indicated that interference from endogenous 280 nm. absorbing components falling near the retention times of theophylline or its metabolites were negligible.

Results

The recovery studies with standards added to normal urine or serum indicated that theophylline and its metabolic products could be quantitatively determined in these biologic fluids by the high-pressure liquid chromatographic procedures herein described. Accordingly, the applicability of the method was evaluated in an actual clinical situation in two male subjects who were being actively treated with theophylline by following its metabolic disposition. Subject SS, age 50, weighing approximately 200 pounds and known to be a rapid deactivator of theophylline was receiving 400 mg. of aminophylline (theophylline ethylenediamine in tablet form) orally, six times daily; and subject JD, age 63, weighing approximately 150 pounds, was receiving 300 mg., four times daily (i.e., subject SS was being treated with twice the total daily amount of aminophylline as subject JD, or one and one-half times the daily dose on a per unit weight basis).

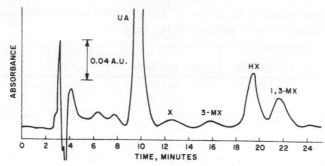

Fig. 5. Liquid chromatogram of a diluted serum sample from the subject described in Fig. 3. X = xanthine; HX = hypoxanthine. Conditions as described in the text.

Table III. Separation of some representative uric acids and xanthines by high-pressure liquid chromatography on Aminex A-5*

Compound	Retention time (min.)	Relative retention time†
Uric acid	10.2	1.00
3-Methyluric acid	11.3	1.11
Xanthine	13.3	1.30
1-Methyluric acid	13.7	1.34
3-Methylxanthine	14.8	1.44
7-Methylxanthine	15.2	1.49
1,3-Dimethyluric acid	15.6	1.53
3,7-Dimethylxanthine (theobromine)	17.0	1.66
1-Methylxanthine	17.3	1.69
Hypoxanthine	19.2	1.88
1,3-Dimethylxanthine (theophylline)	20.0	1.98
1,7-Dimethylxanthine	22.8	2.22
1,3,7-Trimethylxanthine (caffeine)	26.0	2.54

*Conditions as described in the text.
†Relative to uric acid.

The results of duplicate determinations of urine samples are shown in Table I. In addition to theophylline and its metabolic products, i.e., 3-methylxanthine, 1-methyluric acid, and 1,3-dimethyluric acid, the average 24-hour urinary uric acid levels were also determined for subject SS (385 mg.) and for subject JD (247 mg.). Examples of the chromatograms obtained are illustrated by the lower curves of Figs. 3 and 4. 1-Methylxanthine and 3-methyluric acid were *not* detected in the urine of these subjects.

The *serum* levels of theophylline and 3-methylxanthine determined in the same two male subjects are given in Table II. A representative liquid chromatogram appears as Fig. 5. As can be seen, this direct-injection method for serum samples assures the separation, identification, and quantitative analysis of 3-methylxanthine as well as of theophylline. Only 22.8 per cent of this metabolite was extractable from normal human serum by chloroform-isopropanol using the Schack and Waxler procedure. These results demonstrate the accuracy and superiority of the present method over the extraction/ultraviolet procedure. The relatively large peak eluted prior to theophylline in the serum samples has been tentatively identified from its relative retention, extraction behavior, and from "spiking" experiments, as hypoxanthine (HX).

Comparison of the serum theophylline levels determined by this high-pressure liquid chromatographic method with the extraction/ultraviolet procedure of Schack and Waxler (Table II) revealed that, whereas the values for subject SS were quite similar by both methods, the values obtained by the ultraviolet method for subject JD were 5 to 10 per cent higher. The reason for these differences is not immediately apparent, but does not appear to be due to 3-methylxanthine, since the concentrations of this metabolite in the serum were of the same order of magnitude for each subject. As can be seen from Fig. 5, the serum procedure also allows for the simultaneous analysis of serum uric acid levels, and these were determined in a number of representative samples. The amount of uric acid present in serum 3 hours following the administration of the aminophylline dose was thus calculated to be 7.1 mg. per 100 ml. for subject SS and 8.1 mg. per 100 ml. for subject JD. These values remained relatively constant over the time course, e.g., the average level for subject SS at 5 different times was 6.95 ± 0.06 mg. per 100 ml. The methyluric acids were not detected in any of the serum samples analyzed.

Discussion

The twofold differences exhibited by subject SS and subject JD in the 24-hour urinary excretion of theophylline and *each* of its metabolities (Table I) very likely reflect the twofold higher intake of aminophylline by subject SS. Yet the average serum concentrations of theophylline in this subject remained at only 62 per cent of the levels achieved by subject JD and never approached even the lowest serum values of the latter (Table II). The reason(s) for such individual differences in the physiologic disposition of theophylline remains a matter of conjecture[6] with the limited data yet available. However, the techniques described in the present communication should greatly facilitate a more thorough investigation of these phenomena. Indeed, quantitative knowledge of their magnitude is critical for the proper therapeutic management of patients receiving this drug,[6] and the present methods should also find application in the routine monitoring of serum or plasma theophylline concentrations to aid in the adjustment of initial and maintenance doses.

The relative retention data presented in Table III for a wide variety of xanthines and uric acids suggest an even broader application of this high-pressure liquid chromatographic procedure for their separation. Representative among the many possibilities that may be proposed are (1) the analysis of mixtures of methylated xanthines, (2) the quantitative analysis of caffeine and its metabolic products in urine, serum, or in in vitro systems, and (3) the simultaneous determination of uric acid, xanthine, and hypoxanthine in the sera of plasma of patients with gout.

We are indebted to Dr. Gertrude B. Elion for a sample of 1,7-dimethylxanthine, to Faye Rood and Siv Goulding for technical assistance during certain phases of this work, and to Dr. M. W. Anders for helpful discussions.

REFERENCES

1. Ritchie JM: Central Nervous System Stimulants. II. The Xanthines. *In:* The Pharmacological Basis of Therapeutics, ed. 4. Goodman LS and Gilman A, editors. New York, 1970, The Macmillan Company, pp. 358-370.
2. Brodie BB, Axelrod J, and Reichenthal J: Metabolism of theophylline (1,3-dimethylxanthine) in man. J Biol Chem **194:** 215-222, 1952.
3. Johnson EA: The occurrence of substituted uric acids in human urine. Biochem J **51:** 133-138, 1952.
4. Weinfeld H and Christman AA: The metabolism of caffeine and theophylline. J Biol Chem **200:** 345-355, 1953.
5. Cornish HH and Christman AA: A study of the metabolism of theobromine, theophylline, and caffeine in man. J Biol Chem **228:** 315-323, 1957.
6. Jenne JW, Wyze E, Rood FS, et al: Pharmacokinetics of theophylline. Application to adjustment of the clinical dose of aminophylline. Clin Pharmacol Therap **13:** 349-360, 1972.
7. Fischer E and Clemm H: Über 1-methyl und 1,7-dimethyl-harnsäure. Chem Ber **30:** 3089-3097, 1897.
8. Stein A, Gregor HP, and Spoerri PE: Preparation of 1-alkyluramil-7,7-diacetic acids. J Am Chem Soc **78:** 6185-6188, 1956.
9. Traube W: Der synthetische aufbau der harnsäure, des xanthins, theobromins, theophyllins, und caffeins aus der cyanessigsäure. Chem Ber **33:** 3035-3056, 1900.
10. Bergmann F and Dikstein S: The relationship between spectral shifts and structural changes in uric acids and related compounds. J Am Chem Soc **77:** 691-696, 1955.
11. Ukai T, Yamamoto Y, and Kanetomo S: Xanthine derivatives. I. Methylation of 4-aminouracil. J Pharm Soc Jap **74:** 674-677, 1954.
12. Bredereck H, von Schuh HG, and Martini A: Neue synthesen von xanthin, coffein, und theobromin. Chem Ber **83:** 201-211, 1950.
13. Scott CD and Lee NE: Dynamic packing of ion-exchange chromatographic columns. J Chromatogr **42:** 263-265, 1969.
14. Schack JA and Waxler SH: An ultraviolet spectrophotometric method for the determination of theophylline and theobromine in blood and tissues. J Pharmacol Exp Therap **97:** 283-290, 1949.

SECTION III

Enzymology

ALP (Alkaline Phosphatase, EC 3.1.3.1)

11. Bessey, O. A., Lowry, O. H. and Brock, M. J. **(1946)**
     A method for the rapid determination of alkaline phosphatase with five cubic millimeters of serum.

AST (Aspartate Aminotransferase, EC 2.6.1.1)

12. Karmen, A., Wroblewski, F. and LaDue, J. S. **(1955)**
     Transaminase activity in human blood.

CK (Creatine Kinase, EC 2.7.3.2)

13. Mercer, D. W. **(1974)**
     Separation of tissue and serum creatine kinase isoenzymes by ion-exchange column chromatography.

11. Bessey, O. A., Lowry, O. H. and Brock, M. J. (1946)
A Method for the Rapid Determination of Alkaline Phosphatase With Five Cubic Millimeters of Serum. Journal of Biological Chemistry 164(1): 321–329.

By the late 1940s serum alkaline phosphatase (ALP, EC 3.1.3.1) had become a well-established diagnostic marker for a wide variety of bone and liver disorders. Methods used were either modifications of H.D. Kay's original beta glycerophosphate assay (1–4) or King and Armstrong's phenyl phosphate procedure (5). Both methods required a protein precipitation step followed by a secondary color development of the phosphorus or phenol produced. They required 0.5 mL of serum and at least 1 to 2 hours to complete. Otto Bessey, Oliver Lowry and Mary Jane Brock set out to develop a more rapid ALP assay that could be run with pediatric samples. The results of their efforts are presented here in this 1946 paper. Their ALP method used 5 µL of serum, required no protein precipitation step and was complete in 35 min.

Bessey, Lowry and Brock used p-nitrophenyl phosphate (PNP), a self-indicating substrate. The product of the enzyme reaction with PNP is *p*-nitrophenol which has a strong absorbance at 405 nm and can be read directly in the serum–substrate mixture. Ohmori in 1937 (6), Fujita in 1939 (7) and King and Delory in 1939 (8) were actually the first to use PNP in an ALP assay. Lowry in 1990 wrote that because of this "…we received more credit than we deserved" (9). The Ohmori and Fujita papers unfortunately were published in journals with limited circulation at the time. King and Delory synthesized PNP and used it only to compare its rate of hydrolysis with other substrates.

A few years after the publication of this Bessey, Lowry and Brock paper, Eastman Kodak ceased production of PNP. Bessey and his group began to receive large numbers of requests from clinical laboratories for the substrate. They soon published a method for its synthesis (10). In 1950 on the train back from a scientific meeting, Lowry asked Dan Broida, the president of Sigma Chemical Co., to make PNP available as a product (11). In 1951 Sigma produced their first commercial test kit. It was Technical Bulletin No. 104 for the determination of ALP using PNP substrate. This product helped launch the widespread use of test kits into clinical chemistry. By 1974, the last pages of Technical Bulletin 104 listed over 77 different chemistry test kits available just from Sigma (12).

In the 1960s researchers began to measure the enzymatic formation of *p*-nitrophenol in a direct kinetic mode and this reduced the assay time to seconds (13–15). In 1979 McComb, Bowers and Posen published a book on ALP that detailed the history of the development of this and other assay methods for ALP (16). In 1983 the American Association of Clinical Chemists (17) and the International Federation of Clinical Chemistry (18) established the Bessey, Lowry and Brock ALP assay as an international reference method.

References

(1) Kay, H.D. (1930) Plasma phosphatase II. The enzyme in disease, particularly in bone disease. Journal of Biological Chemistry. 89(1):249–266.
(2) Jenner, H.D. and Kay, H.D. (1932) Plasma phosphatase III. A clinical method for the determination of plasma phosphatase. British Journal of Experimental Pathology. 13(1):22–27.
(3) Bodansky, A. (1933) Phosphatase studies III. Determination of serum phosphatase. Factors influencing the accuracy of the determination. Journal of Biological Chemistry. 101(1):93–104.
(4) Shinowa, G.Y., Jones, L.M., and Reinhart, H.L. (1942) The estimation of serum inorganic and "acid" and "alkaline" phosphatase activity. Journal of Biological Chemistry. 142(2):921–933.
(5) King, E.J. and Armstrong, A.R. (1934) A convenient method for determining serum and bile phosphatase activity. Canadian Medical Association Journal. 31(4):376–381.
(6) Ohmori, Y. (1937) Uber die Phosphomonoesterase. Enzymologia. 4:217–231.
(7) Fujita, H. (1939) Uber die Mikrobestimmung der Blut-Phosphatase. Journal of Biochemistry (Tokyo). 30(1):69–87.
(8) King, E.J. and Delory, G.E. (1939) CXLV. The rates of enzymic hydrolysis of phosphoric esters. Biochemical Journal. 33(8):1185–1190.
(9) Lowry, O.H. (1990) How to succeed in research without being a genius. Annual Review of Biochemistry. 59:1–27.
(10) Bessey, O.A. and Love, R.H. (1952) Preparation and measurement of the purity of the phosphatase reagent. Disodium *p*-nitrophenyl phosphate. Journal of Biological Chemistry. 196(1):175–178.
(11) Berger, L. (1993) Sigma diagnostics: pioneer of kits for clinical chemistry. Clinical Chemistry. 39(5):902–903.

(12) [Anonymous] *The Colorimetric Determination of Phosphatase*, Sigma Technical Bulletin No. 104, Revised October 1974, 29 pages. Sigma Chemical Co., St. Louis, MO. Sigma Product Labeling, Bulletin No. 104 (1993) (10–74). pgs 26–29.

(13) Garen, A. and Levinthal, C. (1960) A fine-structure and chemical study of the enzyme alkaline phosphatase of *E. coli*. 1. Purification and characterization of alkaline phosphatase. Biochimica Et Biophysica Acta. 38(March 11):470–483.

(14) Frajola, W.J., Williams, R.D., and Austad, R.A. (1965) The kinetic spectrophotometric assay for serum alkaline phosphatase. American Journal of Clinical Pathology. 43(3):261–263.

(15) Bowers, G.N. and McComb, R.B. (1966) A continuous spectrophotometric method for measuring the activity of serum alkaline phosphatase. Clinical Chemistry. 12(2):70–89.

(16) McComb, R.B., Bowers, G.N., and Posen, S. (1979) Measurement of alkaline phosphatase activity, in *Alkaline Phosphatase*, Plenum Press, New York, Chapter 7, pgs 289–372.

(17) Tietz, N.W., Burtis, C.A., Duncan, P., Ervin, K., Petitclere, C.J., Rinker, A.D., Shuey, D., and Zygowicz, E.R. (1983) A reference method for the measurement of alkaline phosphatase activity in human serum. Clinical Chemistry. 29(5): 751–761.

(18) Tietz, N.W., Rinker, A.D., and Shaw, L.M. (1983) International federation of clinical chemistry. IFCC methods for the measurement of catalytic concentration of enzymes. Part 5. IFCC method for alkaline phosphatase (orthophosphoric-monoester phosphohydrolase, alkaline optimum, EC 3.1.3.1). Clinica Chmica Acta. 135(3):339F–367F.

J. Biol. Chem. 1946; 164(1): 321–329

A METHOD FOR THE RAPID DETERMINATION OF ALKALINE PHOSPHATASE WITH FIVE CUBIC MILLIMETERS OF SERUM

By OTTO A. BESSEY, OLIVER H. LOWRY, AND MARY JANE BROCK

*(From the Division of Nutrition and Physiology, The Public Health Research Institute
of the City of New York, Inc., New York)*

(Received for publication, March 28, 1946)

The alkaline phosphatase of the serum increases early and markedly in rickets and returns completely to normal only after healing is complete. Because of this fact, serum phosphatase is the most satisfactory index now known for the detection of this deficiency. The phosphatase activity of serum is not strictly specific in this respect and has also proved clinically useful in a number of other pathological states; *e.g.*, Paget's disease, hyperparathyroidism, liver disease, etc.

In connection with nutritional studies on large groups of population, it became necessary to have a rapid method for the determination of this enzyme on small amounts of serum. By the use of a new substrate (*p*-nitrophenyl phosphate) a method has been devised which requires only 5 c.mm. of serum (0.005 ml.) and which permits 50 to 100 analyses to be made in 2 hours. The simplicity and speed of the method recommend it for macro- as well as microdeterminations and for either alkaline or acid phosphatase.

A number of methods have been described for the determination of the phosphatase content of serum and other biological materials, all of which depend upon the principle of measuring the rate of hydrolysis of various phosphate esters under specified conditions of temperature and pH. The two most widely used methods are those of Bodansky (1) and King and Armstrong (2) in which glycerol phosphate and phenyl phosphate respectively are employed as substrates. While these methods are satisfactory for many uses, they are rather time-consuming when large numbers of determinations are needed; furthermore, they require larger samples of serum than is convenient for the purpose of dietary surveys.

The substrate, *p*-nitrophenyl phosphate, was studied by King and Delory (3) and has been used for phosphatase estimations by Ohmori (4) and by Fujita (5). The compound is colorless, but upon splitting off the phosphate group, the yellow salt of *p*-nitrophenol is liberated (absorption maximum, 400 mμ). Hence the substrate is itself an indicator of the amount of splitting and thus a measure of phosphatase activity. It is only necessary to incubate serum with the buffered reagent, stop the reaction by dilution with alkali, and measure the amount of color developed. Since serum itself makes a small contribution to the color, the first colorimetric reading is followed by the addition of acid to the sample (converting the yellow sodium salt into colorless free nitrophenol) and a second colorimetric reading which furnishes a blank correction. This procedure is considerably simpler than those now in use. Furthermore, p-nitrophenyl phosphate is split by alkaline phosphatase 15 per cent faster than phenyl phosphate (3), 2 or 3 times more rapidly than glycerol phosphate, and 25 or 30 times faster than phenolphthalein phosphate (6). Because of this rapid splitting, and the high chromogenicity of the salts of p-nitrophenol, the reagent is well suited for adaptation to microprocedures.

Materials and Procedure

Reagents and Standards—

Reagent A. Dissolve 7.50 gm. (0.1 mole) of glycine and 95 mg. (0.001 mole) of $MgCl_2$ in 700 to 800 ml. of H_2O, add 0.085 mole (*e.g.* 85 ml., 1 N) of NaOH, and dilute to 1 liter.

Reagent B. Prepare 0.4 per cent disodium p-nitrophenyl phosphate (Eastman) in 0.001 N HCl. (At present the Eastman product contains about 50 per cent inert material; hence a double amount of this preparation should be used. If desired, the compound may be purified by recrystallization from hot 87 per cent alcohol.) If the pH is not 6.5 to 8.0, adjust with acid or base. To test for free nitrophenol, dilute 1 ml. with 10 ml. of 0.02 N NaOH and measure the light absorption at 415 mμ. If the extinction is greater than 0.08 (*i.e.* light transmission less than 83 per cent for a 1 cm. light path or 70 per cent for a 2 cm. light path), remove free phenol by extracting Reagent B two or three times with equal volumes of water-saturated butyl alcohol, and once with water-saturated ether, finally aerating off traces of ether. Store in the ice box. Reextract when Reagent B fails to pass the above test.

Reagent C (complete reagent). Mix equal parts of Reagents A and B. If necessary, adjust the pH to 10.3 and 10.4 with a little strong NaOH or HCl. Store in the refrigerator, or, better, store frozen. When 2 ml. plus 10 ml. of 0.02 N NaOH have an extinction (1 cm.) greater than 0.1, either discard or extract with butyl alcohol and ether as above and readjust the pH.

Standards. Prepare solutions containing 1, 2, 4, and 6 mM per liter of p-nitrophenol (Eastman), mol. wt. 139.1.

*Apparatus—*For 5 c.mm. serum volumes, (*a*) 5 and 50 c.mm. Lang-Levy constriction pipettes (Fig. 1); (*b*) 6 $\times$ 50 mm. serological tubes, Kimble No. 45060; (*c*) any spectrophotometer or photoelectric colorimeter adapted to 0.5 ml. volume measurements; *e.g.*, Beckman spectrophotometer or

Junior Coleman spectrophotometer (model 6)   (Adapters for the Coleman instrument may be obtained from Samuel Ash, 3044 Third Avenue, New York 56); (d) wire rack to hold 100 tubes; this may be made conveniently from ⅜ inch mesh wire screen.

Principle and Use of Lang-Levy Constriction Pipettes—These pipettes were originally described by Levy (7) working in the laboratory of Dr. Linderstrøm-Lang.   They are particularly useful for the easy, rapid, and precise measurement of volumes of 1 to 200 c.mm. (0.001 to 0.2 ml.), although larger pipettes are occasionally of value.   With 10 to 200 c.mm. pipettes the percentage accuracy of delivery compares favorably with conventional pipettes delivering 2 to 10 ml.; *i.e.*, approximately 0.1 per cent. Below 10 c.mm. the precision falls off somewhat, but even with 1 c.mm. volumes a precision of at least 1 per cent is obtained and may be much better with a well made pipette properly used.   These pipettes are easy to construct for any one familiar with the rudiments of glass blowing,[1] or they may be obtained from the Arthur H. Thomas Company, Philadelphia.

Drawings of Lang-Levy pipettes are shown in Fig. 1.   They are filled and emptied by the use of a small rubber tube such as is used with blood-diluting pipettes.   The pipette is dipped not more than 1 or 2 mm. into

[1] The following directions are for making an ordinary 50 c.mm. constriction pipette.   A 20 to 25 cm. length of either soft glass or Pyrex tubing, 4 or 5 mm. outer diameter, is heated in the middle and drawn down to a diameter of 1.5 to 2 mm.   This furnishes material for two pipettes.   2 cm. from that point where the tube is narrowed (Fig. 1, *a*) the slender portion is further drawn down to a diameter of 0.5 to 0.8 mm. This narrowest portion is bent 1 cm. below the 2 mm. portion (*b*) at an angle of about 45° from a straight line and is cut off with a diamond point 3 or 4 mm. below the bend (*c*).   The upper large end is fire-polished.   A 0.1 or 0.2 ml. graduated pipette is partially filled with water and placed horizontally on the table, a rubber tube is attached to the new pipette, its tip is touched to the tip of the graduated pipette, and 50 c.mm. of water are sucked into the new pipette.   A mark is made with the diamond point at the meniscus and just above this point the pipette is narrowed, for a distance of about 1 cm. or so (*d-d*), to an *inner* diameter of perhaps 1 to 1.5 mm., without thinning the wall.   Once again 50 c.mm. of water are drawn in, and the meniscus should now fall in this narrow portion.   If it does, a new mark is made with a diamond point and the actual constriction is made by heating just above this mark with as slender a flame as possible.   Without pulling or pushing, the glass will thicken where the flame strikes it, and heating is continued until the bore is 0.1 to 0.2 mm. at its narrowed point (*e*).   The constriction should be small enough to stop the meniscus from going by when moderate pressure is applied, but large enough so that undue pressure is not required to force the meniscus by.   The opening at the tip (*c*) should be a trifle smaller than the upper constriction so that the pressure which pushes the meniscus past the upper constriction will not cause the pipette to deliver too rapidly. The delivery time for a 50 c.mm. pipette should be 2 to 5 seconds.   With a little practice, pipettes within 1 to 2 per cent of the desired volume can easily be made. However, each pipette is subsequently calibrated by delivery of water into a weighing bottle containing moisture and weighing on a micro balance.

324   MICRODETERMINATION OF SERUM PHOSPHATASE

the liquid, and by sucking, liquid is pulled to just above the constriction. With gentle pressure the liquid level is blown down to the constriction, where surface tension stops the meniscus automatically. The pressure is not released until the tip of the pipette is removed from the remaining liquid. To deliver, the pressure is increased sufficiently to drive the meniscus past the constriction and this pressure is maintained until delivery is complete (2 to 5 seconds). With smaller pipettes the tip should be sufficiently constricted (by fire polishing) so that surface tension prevents air from following the liquid after it is delivered, since otherwise part of the liquid might be spattered. During delivery, the bend in the tip is used to keep the lower shaft of the pipette away from the wall of the tube into which the sample is being delivered. This is very important, since otherwise surface tension will cause the sample to run up the shaft of the pipette, and only part of the sample will be delivered into the tube, the rest clinging to the pipette. Similarly, the tip of the pipette must always touch a

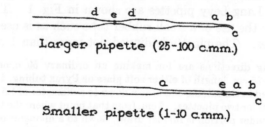

Larger pipette (25-100 c.mm.)

Smaller pipette (1-10 c.mm.)

Fig. 1. Lang-Levy pipettes. See foot-note 1

surface during delivery, since if the tip is free in the air, much of the sample adheres to the outside of the pipette.

Procedure with 5 C.mm. of Serum—Serum samples (5 c.mm.) are transferred to the bottom of 6 × 50 mm. tubes in a wire rack. (A simple method for collecting small samples of serum has been previously described (8).) The rack is immersed in a shallow pan of ice water and to each tube are rapidly added 50 c.mm. of ice-cold Reagent C, with a constriction pipette. Each tube is mixed by tapping with the finger. Care is taken not to warm the tube by so doing.

The whole rack of tubes is now immersed in a water bath at 38° at a depth sufficient to cover the bottom half of the tubes. After exactly 30 minutes the rack is again placed in the pan of ice water and 0.5 ml. of 0.02 N NaOH is added to each tube with sufficient force to mix the sample. (A syringe pipette is very convenient for this purpose.) This addition stops the reaction and dilutes the samples which are now transferred to colorimeter tubes and read at 400 to 420 mμ = R_1.

After the initial reading, 2 to 4 c.mm. of concentrated HCl are added with a 0.1 ml. graduated pipette (drawn out tip) and a second reading, R_2, is made.

R_1 and R_2 are converted into optical densities ($-$ log transmission or $2 - $ log per cent transmission) $= D_1$ and D_2. Then $D_1 - D_2 = D_c$, the corrected density.

Standards and blanks are provided by treating 5 c.mm. volumes of the standards and of distilled water exactly as though they were serum samples. The corrected densities (D_c) are used to construct a standard curve from which the serum values are calculated. Since sera and standards undergo the same dilution, it is unnecessary to take into account the exact volumes of the various pipettes. It is to be noted that D_c is the density corrected for possible residual absorption after acid addition but is not corrected for the D_c of the blank analysis. This second necessary correction is automatically provided by the standard curve. A "millimole unit" is defined as the phosphatase activity which will liberate 1 mM of nitrophenol per liter of serum per hour. Therefore, since the standard incubation time is only 30 minutes, the 1, 2, 4, and 6 mM standards are equivalent to sera with activities of 2, 4, 8, and 12 mM units. 1 such unit is approximately equal to 1.8 Bodansky units (see below). For adult sera, which have low phosphatase activity, the volumes of serum and reagent may be doubled without increasing the volume of alkali; this will nearly double the amount of color.

Procedure with 20 C.mm. of Serum—Volumes of serum, reagent, and 0.02 N NaOH are increased to 20 c.mm., 200 c.mm., and 2 ml., respectively. Otherwise the procedure is nearly identical with that described for 5 c.mm. of serum. The sample may be conveniently incubated directly in $\frac{3}{8}$ inch photocolorimeter tubes. 1 drop of 5 N HCl is added before the second reading.

Procedure with 0.1 Ml. of Serum—Use 0.1 ml. of serum, 1 ml. of reagent, and 20 ml. of 0.02 N NaOH. Because of the larger volumes, the tubes containing 1 ml. of reagent are placed in the water bath and allowed to come to temperature before the addition of serum. As each serum sample is added, the time is noted, and exactly 30 minutes later 20 ml. of NaOH are added.

DISCUSSION

Fig. 2 shows the differences in spectral absorption between *p*-nitrophenyl phosphate, *p*-nitrophenolate, and free *p*-nitrophenol. It is on these differences in absorption that the proposed method depends. Upon removing the phosphate group from *p*-nitrophenyl phosphate to form *p*-nitrophenolate, the absorption maximum is shifted from 310 to 400 mμ and

326   MICRODETERMINATION OF SERUM PHOSPHATASE

is nearly doubled in height. On converting the liberated nitrophenylate into free nitrophenol, by acidification, the absorption maximum is shifted back to 318 mμ and the absorption at 400 mμ is abolished. This latter reaction makes it possible to correct for the color contributed by the serum itself.

King and Delory (3) observed that the pH optimum of phosphatase with *p*-nitrophenyl phosphate as the substrate is more alkaline than with glycerol phosphate. The pH optimum with human serum was found to be 10.0 to 10.1, under the conditions described here (uncorrected glass electrode pH). Since alkaline phosphatase has a sharp pH optimum, devi-

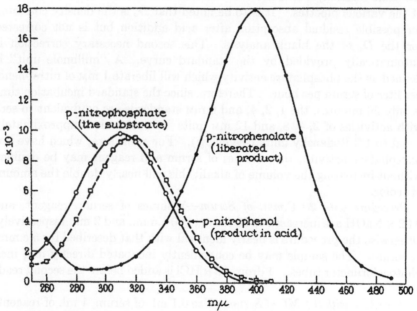

Fig. 2. Absorption curves of *p*-nitrophenol and *p*-nitrophenyl phosphate

ations of more than 0.1 pH unit will affect the readings significantly. For rat serum the optimum pH range is from 9.1 to 9.7. It is advisable to check the pH of the reagent against a standard buffer in the same pH range, such as 0.1 M sodium borate, pH 9.2. The reagent is well buffered; hence there is no danger of CO_2 from the air affecting the activity during the incubation, but the pH of the reagent itself should be rechecked occasionally. Since the buffer capacity is large, it is permissible in the case of sera having very high phosphatase to reduce the volume of serum to one-half or one-quarter without changing the other volumes. The phosphatase reagent of Bodansky (1) is less well buffered and considerably more care must be exercised with it to prevent pH changes from affecting phosphatase activity. The p-nitrophenyl phosphate reagent contains added magnesium, which, however, has less effect on the activity of the phosphatase than in the case of glycerol phosphate.

The degree of splitting in 30 minutes has been found to be proportional to the concentration of enzyme, but has not been found to be strictly linear with time for more than about 30 minutes. Therefore, it is desirable not to increase the period of incubation.

The blank correction for contribution of color from the serum is based on the finding that serum has very nearly the same absorption at 415 mμ in acid as in alkaline solution. Hemoglobin, if it is present as the result of

TABLE I

Comparative Results with Bodansky and Nitrophenyl Phosphate Procedures

Serum No.	Bodansky units	p-Nitrophenyl phosphate units	Ratio, p-nitrophenyl phosphate to Bodansky units
1	8.36	4.68	1.79
2	8.80	4.98	1.77
3	10.7	5.68	1.88
4	8.30	4.81	1.73
5	9.30	4.76	1.95
6	9.27	4.76	1.95
7	5.60	3.27	1.71
8	7.42	4.22	1.76
9	7.14	4.32	1.66
10	8.36	4.99	1.67
11	4.48	2.41	1.86
12	6.54	3.70	1.77
Average			1.79
Standard deviation			0.10

hemolysis, absorbs considerably less light at 415 than at 400 mμ, which is one reason for preferring the longer wave-length. Hemoglobin does not absorb quite the same amount of light in acid and alkaline solution; hence excessive hemolysis should be avoided.

Comparison with Other Methods—In Table I are shown comparative results of analyses of twelve children's sera made by the Bodansky procedure (1) on 0.1 ml. of serum and by the method described here on 5 c.mm. of serum. It would appear that the Bodansky units (mg. of P liberated per 100 ml. of serum per hour) bear a ratio of 1.79 to the mм unit of the present procedure (mм of substrate hydrolyzed per liter of serum per hour). We have found the same ratio for adult sera, but the activities were so low that accurate measurements by the Bodansky procedure were difficult to make, and hence a less regular correlation was found. In a personal communication, Dr. S. H. Jackson, Children's Hospital, Toronto, reported that the ratio of King-Armstrong units (mg. of phenol split from phenyl phosphate per 100 ml. of serum per hour) (2) to mM units is 7.3, with a variance of 8 per cent. With rat sera we have found in agreement with Fujita (5) that a much lower ratio is obtained. This observation deserves further investigation. The standard deviation of replicate determinations on the same serum by the proposed procedure with 5 c.mm. is 0.15 mM unit.

Huggins and Talalay (6) have recently introduced phenolphthalein phosphate as a phosphatase substrate. Like nitrophenyl phosphate it

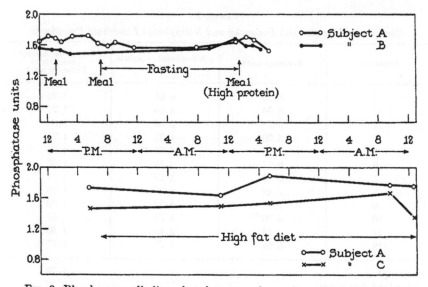

Fig. 3. Blood serum alkaline phosphatase under various dietary conditions

can function as its own indicator. It is, however, split only 3 or 4 per cent as fast as the latter and the color produced is stated not to be proportional to the enzyme concentration, presumably because there are two phosphate groups attached which may both have to be removed before color develops.

Illustrative Data—Fig. 3 gives the results of experiments to determine whether the alkaline phosphatase of human serum is influenced by the diet. Weil and Russell (9) observed that in the rat fasting for as little as 8 hours caused a marked decrease in serum alkaline phosphatase, confirming similar findings of Bodansky (10) for the dog. Gould (11) observed that a high fat diet produced an increase of 400 per cent in the alkaline phosphatase of rat serum. It is apparent (Fig. 3) that in the human subjects tested there were no significant changes in the serum alkaline phosphatase attributable to an 18 hour fast, a high protein meal, or a 40 hour period of very high fat intake. The blood samples were all obtained by finger puncture. Duplicate 10 c.mm. serum samples were analyzed. It seems possible to conclude that for human phosphatase studies it is unnecessary to consider the immediate dietary history.

SUMMARY

A method is described for the determination of serum alkaline phosphatase, which permits analysis of 5 c.mm. (0.005 ml.) samples of serum at the rate of 50 to 100 per 2 hours. The simplicity and speed of the method also recommend it for macro- as well as microdeterminations, and for either alkaline or acid phosphatase.

BIBLIOGRAPHY

1. Bodansky, A., *J. Biol. Chem.*, **101**, 93 (1933).
2. King, E. J., and Armstrong, A. R., *Canad. Med. Assn. J.*, **31**, 376 (1934).
3. King, E. J., and Delory, G. E., *Biochem. J.*, **33**, 1185 (1939).
4. Ohmori, Y., *Enzymologia*, **4**, 217 (1937).
5. Fujita, H., *J. Biochem.*, Japan, **30**, 69 (1939).
6. Huggins, C., and Talalay, P., *J. Biol. Chem.*, **159**, 399 (1945).
7. Levy, M., *Compt.-rend. trav. Lab. Carlsberg, Série chim.*, **21**, 101 (1936).
8. Lowry, O. H., and Hunter, T. H., *J. Biol. Chem.*, **159**, 465 (1945).
9. Weil, L., and Russell, M. A., *J. Biol. Chem.*, **136**, 9 (1940).
10. Bodansky, A., *J. Biol. Chem.*, **104**, 473 (1934).
11. Gould, B. S., *Arch. Biochem.*, **4**, 175 (1944).

COMMENTARY TO

12. Karmen, A. Wroblewski, F. and LaDue, J. S. (1955)
Transaminase Activity in Human Blood. Journal of Clincal Investigation 34(1): 126–133.

T he first serum enzyme assay introduced into routine clinical chemistry was Wohlgemuth's 1908 starch iodine method for amylase (EC 3.2.1.1) (1). Forty-five years latter, clinical enzymology consisted essentially of three serum assays; amylase, alkaline phosphatase (EC 3.1.1.1) and acid phosphatase (EC 3.1.3.2). Textbooks published in 1943 by Gradwohl (2), Bray in 1946 (3) and Caraway in 1960 (4) presented methods for only these three enzymes. All the assays were colorimetric end-point procedures. A paper by LaDue, Wroblewski and Karmen in *Science* in 1954 (5) and their 1955 paper presented here changed this forever. They demonstrated for the first time that elevated levels of serum aspartate aminotransferase (AST, EC 2.6.1.1 formerly called serum glutamate oxaloacetate transaminase, SGOT) could be used to help confirm suspected myocardial infarction (MI). This caused an explosive interest in the use of serum enzymes as markers in disease detection. This 1955 paper also helped create a second paradigm shift in clinical chemistry. The AST procedure by Karmen introduced into routine clinical chemistry the ultraviolet (340 nm) non-colorimetric optical technique for the measurement of serum enzymes.

Awapara and Seale in 1952 used a paper chromatography procedure to study the distribution AST in rat tissues (5). Two years latter LaDue, Wroblewski and Karmen used a similar paper technique and reported for the first time on the elevation of AST in patients following MI (6). In 1955 they reported that AST values were up to 20 times above normal in 74 of 75 patients following MI (7). The elevation of AST following MI was also confirmed experimentally in dogs (8). Wroblewski and LaDue in 1955 reported similar results for the enzyme, lactate dehydrogenase (LDH, EC 1.1.1.27) (9). Serum enzymes soon became a major topic within the scientific literature. Interest even entered into the popular press. In 1961 Wroblewski published an article in *Scientific American* called "Enzymes in Medical Diagnosis" (10).

AST catalyzes the transfer of an amino group from L-aspartate in the presence of 2-oxoglutarate to form oxaloacetate and L-glutamate. In the paper presented here the authors used paper chromatography to measure AST activity in serum. The method measured the formation of the amino acid glutamate by AST and took 36 hours to complete. In the Appendix to this paper written by Arthur Karmen, a new more practical method suitable for clinical use was introduced.

Otto Warburg in the 1930s discovered that enzymes that catalyzed the reduction or oxidation of the co-factors nicotinamide adenine dinucleotide (NAD) or nicotinamide adenine dinuleotide (NADP) could be measured at 340 nm with a spectrophotometer (11). This came to be known as Warburg's "optical assay." Karmen applied this concept to AST and replaced 36-hour paper chromatography with a 5-minute assay. In the assay, he added malate dehydrogenase (MDH, EC 1.1.1. 40) along with NADPH to the serum-substrate mixture. The oxaloacetate formed by AST was converted into malate by MDH with a stoichiometric oxidation of NADPH to NADP. The loss of hydrogen by NADPH resulted in a decrease in the absorbance of the reaction mixture that could be monitored kinetically at 340 nm. Karmen was one of the first to describe this "coupled" approach to a clinical enzyme assay. His UV method was quickly optimized (12,13) and colorimetric end-point versions of the coupled assay soon appeared (14). Growth of clinical enzymology was exponential following these reports. Barnett, Ewing and Skodon in 1976 evaluated the performance of 43 different commercial test kits available for AST in serum (15).

References

(1) Wohlgemuth, J. (1908) Uber Eine Neue Methode zur Bestimmung des Diastatischen Ferments. Biochemische Zeitschrift. 9: 1–9.

(2) Gradwohl, R.B.H. (1943) *Clinical Laboratory Methods and Diagnosis. A Textbook on Laboratory Procedures with Their Interpretation,* 3rd Edition. C.V. Mosby Company, St Louis, MO, Chapter III, pgs 149–360.

(3) Bray, W.E. (1946) *Synopsis of Clinical Laboratory Methods,* 3rd Edition C.V. Mosby Company, St Louis, MO, Chapter IV, pgs 190–242.

(4) Caraway, W.T. (1960) *Microchemical Methods for Blood Analysis.* Charles C. Thomas Publisher, Springfield, IL, pgs 94–100.

(5) Awapara, J. and Seale, B. (1952) Distribution of transaminases in rat organs. Journal of Biological Chemistry. 194(2): 497–502.

(6) LaDue, J.S., Wroblewski, F., and Karmen, A. (1954) Serum glutamic oxaloacetic transaminase activity in human acute transmural myocardial infarction. Science. 120(3117):497–499.

(7) LaDue, J.S. and Wroblewski, F. (1955) The significance of the serum glutamic oxalacetic transaminase activity following acute myocardial infarction. Circulation. 11(6):871–877.

(8) Agress, C.M., Jacobs, M.S., Glassner, H.F., Lederer, M.A., Clark, W.G., Wroblewski, F., Karmen, A., and LaDue, J.S. (1955) Serum transaminase levels in experimental myocardial infarction. Circulation. 11(5):711–713.

(9) Wroblewski, F. and LaDue, J.S. (1955) Lactic dehydrogenase activity in blood. Proceedings of the Society for Experimental Biology and Medicine. 90(1):210–213.

(10) Wroblewski, F. (1961) Enzymes in medical diagnosis. Scientific American. 205(2):99–102, August.

(11) Warburg, O. and Christian, W. (1936) Pyridin, der Wasserstoff-Ubertragende Bestandteil von Garungsfermenten (Pyridin–Nucleotide). Biochemische Zeitschrift. 287:291–328.

(12) Steinberg, D., Baldwin, D., and Ostrow, B.H. (1956) A clinical method for the assay of serum glutamic–oxalacetic transaminase. Journal of Laboratory and Clinical Medicine. 48(1):144–151.

(13) Henry, R.J., Chiamori, N., Golub, O.J., and Berman, S. (1960) Revised spectrophotometric methods for the determination of glutaminc–oxalacetic transaminase, glutamic–pyruvic transaminase, and lactic acid dehydrogenase. American Journal of Clinical Pathology. 34(4):381–398.

(14) Cabaud, P., Leeper, R., and Wroblewski, F. (1956) Colorimetric measurement of serum glutamic oxaloacetic transaminase. American Journal of Clinical Pathology. 26(9):1101–1105.

(15) Barnett, R.N., Ewing, N.S., and Skodon, S.B. (1976) Performance of "kits" used for the clinical chemical analysis of GOT (Aspartate amino-transferase). Clinical Biochemistry. 9(2):78–84.

TRANSAMINASE ACTIVITY IN HUMAN BLOOD

By ARTHUR KARMEN, FELIX WRÓBLEWSKI, AND JOHN S. LaDUE

(*From the Sloan-Kettering Institute, Department of Medicine, Memorial Center,
New York City, N. Y.*)

(Submitted for publication April 3, 1954; accepted July 15, 1954)

Enzymatic transamination consists of the enzyme catalyzed reversible transfer of the alpha amino nitrogen of an amino acid to an alpha-keto acid with the synthesis of a second amino acid and a second alpha-keto acid. Enzymes catalyzing different transamination reactions are found widely distributed in animal tissues and have been shown to change in activity in some tissues during disease (1–3). These observations prompted the present study to determine if transaminase activity could be demonstrated in human serum and blood cellular elements and, if so, to study any variations in activity of this enzyme in the blood of normal and diseased man.

METHODS AND MATERIALS

The two transaminases found most active in animal tissues are:

1. *"Glutamic-oxalacetic transaminase"*
   Aspartate + alpha-keto glutarate $\rightleftharpoons$ glutamate + oxalacetate
2. *"Glutamic-pyruvic transaminase"*
   Alanine + alpha-keto glutarate $\rightleftharpoons$ glutamate + pyruvate

When aspartate or alanine are incubated with alpha-keto glutarate and a source of enzyme, the rate of production of glutamate may be taken as a measure of transaminase activity. The amount of glutamate produced after a given incubation period under standardized conditions was measured by quantitative paper chromatographic analysis (4).

One-tenth molar solutions of 1-aspartate, 1-alanine, and alpha-keto glutarate were prepared in 0.06 M phosphate buffer and the pH of the solutions adjusted to pH 7.6. For serum transaminase determinations, 0.5 ml. of clear, non-hemolyzed serum, 1.5 ml. of 0.06 M phosphate buffer, pH 7.6, and 0.5 ml. of either the alanine or aspartate solutions were incubated for ten minutes at 37°C. At this time, 0.5 ml. of the alpha-keto glutarate solution was added and the incubation continued for 18 hours. For whole blood hemolysate transaminase determinations, equal volumes of blood and distilled water were shaken together for ten minutes, 1.0 ml. of the hemolysate was added to 1.0 ml. of the phosphate buffer, and the substrates added as above. The time of incubation of the hemolysate substrate mixture was three hours.

At the end of the incubation period, proteins were separated by adding 7.0 ml. of absolute ethyl alcohol, centrifuging for ten minutes, and washing the precipitate with 5 ml. of 70 per cent ethanol. The supernatant was evaporated to dryness over a water bath and the residue dissolved in 1.0 ml. of 0.06 M phosphate buffer. Aliquots of 0.05 ml. were then applied to Whatman No. 1 filter paper and chromatographed by the descending method for 18 hours, using phenol saturated with water as solvent and water saturated with phenol to saturate the atmosphere of the tank. The papers were then removed and dried in air at room temperature (5). The position of the amino acids was located by spraying the paper with a 0.1 per cent solution of ninhydrin in butanol and heating gently with an infra-red lamp.

The areas of paper corresponding to glutamate were cut out, rolled, and placed in test tubes. Elution of the amino acid from the paper and quantitative color development with ninhydrin were performed in one operation by adding the reagents and treating the paper as described in the procedure of Troll and Cannan (6). Areas of paper containing standard amounts of glutamate were analyzed concomitantly. Papers corresponding to incubation mixtures containing 0.5 ml. of serum, or 1.0 ml. of hemolysate were used as blanks.

RESULTS

Reliability of the method

No loss of glutamate was encountered in the incubation period with serum or in the process of paper chromatography. Satisfactory recovery of known quantities of glutamate added to serum and incubated for 18 hours, or applied to paper directly, was obtained (Table I).

Presence and properties of the enzyme in serum

Incubation of aspartate and alpha-keto glutarate or alanine and alpha-keto glutarate, without addition of serum, failed to form detectable quantities of glutamate after 18 hours of incubation. Incubation of 0.5 ml. of serum and any one of the substrates singly, similarly showed no formation of measurable glutamate. Thus, non-enzymatic transamination was inferred not to occur under the conditions of these experiments, and concen-

TABLE I

Recovery of added glutamate

Aliquot	Micromoles added	Micromoles recovered	% Recovery
A. From paper after phenol chromatography			
1.	0.405	0.410	101.1
2.	0.405	0.405	100.0
3.	0.405	0.415	102.3
4.	0.405	0.424	104.7
B. From serum after eighteen-hour incubation			
1.	8.00	8.10	101.1
2.	8.00	8.57	107.0
3.	8.00	7.60	95.0

tration of alpha-keto glutarate and amino group
donors in serum and hemolysates was inferred to
be negligible. Incubation of 0.5 ml. of serum with
aspartate and alpha-keto glutarate resulted in the
formation of from 4.4 to 15.0 micromoles of glutamate in 18 hours, depending on the serum sample tested.

The amount of glutamate produced was found
to be directly proportional to the time of incubation
when identical mixtures of serum, buffer, and substrates were incubated for varied intervals of time
(Figures 1 and 2). The rate of production of

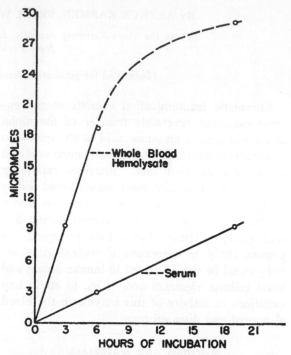

FIG. 2. RATE OF GLUTAMATE PRODUCTION *Via* TRANS-
AMINATION (WHOLE BLOOD HEMOLYSATE AND SERUM
COMPARED)

glutamate was seen to be directly proportional to
the quantity of serum, when 0.5 ml. and 1.0 ml.
samples of the same sera were incubated with
identical substrate mixtures (Table II).

Pyridoxal phosphate has been shown to act as
coenzyme in transamination reactions and O'Kane
and Gunsalus have determined the coenzyme saturated curve (7). Addition of a buffered solution
of pyridoxal phosphate in a concentration of ten
micrograms per ml. was found to have no measurable effect on the transaminase activity of the
serum. Addition of a boiled and filtered extract
of rat liver, used as source of possible activators,

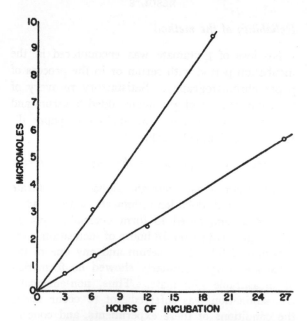

FIG. 1. TIME RATE OF GLUTAMATE PRODUCTION *Via*
TRANSAMINATION (TWO 0.5 ML. SERUM SAMPLES)

TABLE II

*The effect of enzyme concentration on rate of transamination
incubation period—twenty hours*

Sample	Micromoles glutamate produced by 0.5 ml.	Rate *micromoles/ ml./hr.*	Micromoles glutamate produced by 1.0 ml.	Rate *micromoles/ ml./hr.*
1.	5.6	0.56	10.4	0.52
2.	4.9	0.49	9.7	0.48
3.	6.6	0.66	11.0	0.55
4.	6.4	0.64	11.7	0.59
5.	5.9	0.59	12.3	0.61
6.	5.2	0.52	9.8	0.49

was similarly found not to affect the observed transaminase activity of serum.

Increased concentration of aspartate in a given serum incubation mixture was seen to cause a greater increase in the observed rate of glutamate production than an increase in the concentration of alpha-keto glutarate, demonstrating that complete saturation of the enzyme with substrate had not been achieved at these concentrations. These results are in essential agreement with those reported for transaminase preparations from pig heart muscle (7, 8).

The effect of pH on serum transaminase activity was studied by altering the composition of the buffer used. Phosphate buffer, 0.2 M, of several pH values was substituted for the 0.06 M buffer. The pH of each incubation mixture was determined before and after incubation. An increase in pH from 0.1 to 0.2 pH units was observed in each sample after the incubation period and the average of the pre- and post-incubation values was taken as the pH of the mixture. The finding of maximal activity between pH 7.0 and 8.0 (Figure 3) is in essential agreement with the results of Cohen (9) and others using pig heart muscle as source of transaminase.

No change in transaminase activity with time was noted in serum samples stored from ten minutes to 96 hours at room temperature, or for periods of from one hour to two weeks in the refrigerator (0 to 5°C.). The transaminase activity was not changed by freezing or lyophilization of

TABLE III

Distribution of transaminase activity in serum of healthy adults

Transaminase activity micromoles/ml./hr.	Aspartate-alpha-keto glutarate		Alanine-alpha-keto glutarate	
	No. of samples	%	No. of samples	%
0.20–0.39	0	0	2	5.1
0.40–0.49	7	7.8	11	28.2
0.50–0.59	22	25.0	16	41.1
0.60–0.69	24	27.2	5	12.5
0.70–0.79	15	17.1	2	5.1
0.80–0.89	5	5.7	1	2.5
0.90–0.99	9	10.2	1	2.5
1.00–1.09	5	5.7	1	2.5
1.36	1	1.1		
Total number tested	88	100	39	100

Mean transaminase activity micromoles/ml./hr.	Aspartate-alpha-keto glutarate	0.622 ± 0.191 Std. deviation
	Alanine-alpha-keto glutarate	0.525 ± 0.146 Std. deviation

the serum. No change in activity was noted in sera subjected to 56°C. for 25 minutes. Sera heated to 100°C. for ten minutes were found to have a decrease in activity to 10 per cent of the original transaminase activity.

No difference could be detected between transaminase activity in serum and in plasma from the same donors by using oxalate, citrate, or heparin as anticoagulants.

Transaminase activity in the blood of normal humans

The serum glutamic oxalacetic transaminase activity in 88 normal humans varied from 0.41 to 1.36 micromoles per ml. per hour with a mean activity of 0.622 ± 0.191 standard deviation. Serum glutamic pyruvic transaminase activity in 39 samples was found to be between 0.21 and 1.01 micromoles per ml. per hr. with a mean value of 0.525 ± 0.146 (Table III). The glutamic oxalacetic transaminase activity found in hemolysates ranged from 5.0 to 8.7 micromoles per ml. per hr. with a mean value of 6.86 ± 0.78 while the glutamic pyruvic transaminase in hemolysates varied from 1.6 to 3.3 micromoles per ml. per hr. with a mean value of 2.48 ± 0.36 (Table IV).

In each of 29 hemolysate samples, the ratio of

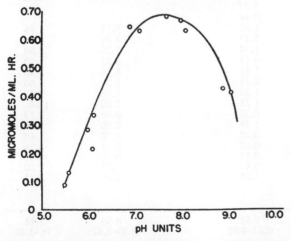

FIG. 3. EFFECT OF VARYING HYDROGEN ION CONCENTRATIONS ON SERUM TRANSAMINASE ACTIVITY

TABLE IV

Distribution of transaminase activity in whole blood hemolysates of healthy adults

Aspartate-alpha-keto glutarate			Alanine-alpha-keto glutarate		
Transaminase activity micromoles/ ml./hr.	No. of samples	%	Transaminase activity micromoles/ ml./hr.	No. of samples	%
5.0–5.9	3	10.3	1.5–1.9	3	10.3
6.0–6.9	12	41.4	2.0–2.5	12	41.4
7.0–7.9	12	41.4	2.5–3.0	11	38.0
8.0–8.9	2	6.9	3.0–3.3	3	10.3
Total No.	29	100.0	Total No.	29	100.0

Mean transaminase activity micromoles/ml./hr.	Aspartate-alpha-keto glutarate	6.86±0.78 Std. deviation
	Alanine-alpha-keto glutarate	2.48±0.36 Std. deviation

glutamic oxalacetic transaminase activity to glutamic pyruvic transaminase activity varied between 2.04 and 3.60 micromoles per ml. per hr. with a mean value of 2.70 ± 0.40. In 39 determinations of serum transaminase activity, this ratio was found to be 0.725 to 1.67 micromoles per ml. per hr. with a mean of 1.15 ± 0.23. No consistent relationship was noted between individual serum and corresponding hemolysate transaminase activities.

In no instance was transaminase activity absent in the sera of the normal humans tested or in any of the sera of hospitalized patients with various disease states tested. Increased activity was found in the sera of one patient with lymphomatous disease, one with extensive rhabdomyosarcoma, two with acute leukemia, one with acute hepatitis, and two patients with arteriosclerotic heart disease and recent myocardial infarction. Serum glu-

TABLE V

Transaminase activity in blood of hospitalized patients in micromoles per ml. per hr.

Clinical diagnosis	Serum transaminase glutamic		Hemolysate transaminase glutamic	
	Oxalacetic	Pyruvic	Oxalacetic	Pyruvic
Carcinoma of lung	0.54	0.41	8.86	2.07
Carcinoma of lung	0.51	0.38	1.73	2.54
Carcinoma of palate	0.89	0.70	6.25	2.60
Carcinoma of eyelid	0.75	0.47	8.05	1.80
Carcinoma of breast	0.80	0.73		
Fibromyoma of uterus	0.81	0.59	8.75	2.73
Lymphoma	0.46	0.51	11.3	2.70
Lymphoma	0.71	0.47	7.85	1.56
Lymphoma	0.52	0.39	6.45	2.53
Lymphoma	0.74	0.53	6.80	1.66
Lymphoma	1.06	0.70	6.35	2.00
Lymphoma	0.36	0.32	7.00	1.60
Lymphoma	0.47	0.37		
Lymphoma	0.71	0.48		
Rhabdomyosarcoma	1.70	1.60	9.13	7.46
Rhabdomyosarcoma	0.40	0.35	7.26	2.00
Chronic pulmonary tuberculosis	0.92	1.24	7.20	3.60
Chronic pulmonary tuberculosis	0.58	0.44	8.95	1.60
Acute leukemia	1.28	1.51	5.26	3.40
Acute leukemia	0.36	0.55	5.20	1.60
Acute leukemia	1.72	1.25	7.95	3.73
Acute hepatitis	1.97	1.75	7.20	10.4
Diabetes mellitus	0.46	0.35	8.46	2.73
Portal cirrhosis	0.99	1.28	8.70	3.50
Portal cirrhosis	0.49	0.38	3.10	1.40
Generalized arteriosclerosis	0.50	0.63	6.13	3.26
Cerebral hemorrhage	0.85	0.41	9.40	3.40
Arteriosclerotic heart disease with CHF	1.16	0.81	5.00	2.20
ASHD with anginal syndrome	0.95	1.25	8.65	3.27
ASHD with myocardial infarction— see Figure 5				
Average normal value	0.62	0.52	6.86	2.48
Normal range*	0.24–1.04	0.23–0.82	5.30–8.42	1.76–3.20

* Normal range represents the normal mean value plus or minus two standard deviations.

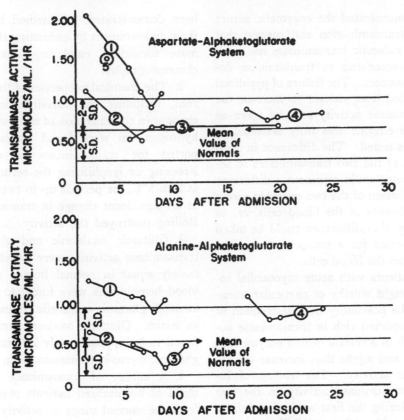

FIG. 4. SERUM TRANSAMINASE ACTIVITY IN PATIENTS WITH ADMISSION
DIAGNOSIS OF MYOCARDIAL INFARCTION

tamic oxalacetic transaminase activity and serum glutamic pyruvic transaminase activity was seen to vary together in most cases of marked departure from the normal range. Somewhat greater variation was found in the transaminase activity of the hemolysates from the same patients (Table V).

Figure 4 represents serial determinations of serum transaminase activity in five patients admitted to the hospital with the admitting diagnosis of acute myocardial infarction at various times after the onset of acute infarction. Of the five patients shown, the clinical picture subsequent to admission substantiated the diagnosis of transmural myocardial infarction in Nos. 1, 2, 4, and 5. Elevated values were found in patients Nos. 1 and 5, the only patients with acute infarctions studied during the first week after infarction. The finding of 2.02 units in patient No. 1 represents the highest value found in any of the samples tested and almost twice the value taken as the upper limit of normal.

DISCUSSION

The method of transaminase assay by quantitative paper chromatography of the glutamate produced was chosen because of its sensitivity and the simplicity of the equipment required. Other methods of measuring transamination reactions may be divided into two main types: Those utilizing either specific chemical or enzymatic decarboxylation of one of the products and subsequent manometric measurement of carbon dioxide evolved and those utilizing the high ultraviolet absorption of oxalacetate at wavelength 280 millimicra to follow the glutamic oxalacetic transaminase reaction by measuring a change in optical density as oxalacetate is produced or consumed. The high bicarbonate content of blood together with low transaminase activity makes a method depending on carbon dioxide evolution difficult to apply to serum. Low transaminase activity and high protein content together with the known instability of oxalacetate make the usual spectrophotometric assay difficult to apply to serum.

The results demonstrated the enzymatic nature of the observed transamination and suggest that serum glutamic oxalacetic transaminase has similar chemical characteristics to transaminase derived from other sources. The failure of pyridoxal phosphate or boiled liver extract to increase the measured transaminase activity may be taken as evidence that the enzyme was fully activated in the serum samples tested. The difference in comparative amounts of the two transaminases in serum and hemolysates could represent a difference in the rates of diffusion of the two enzymes across the cellular membranes of the blood cells, or, as seems more likely, this difference could be taken as suggestive evidence for a source of the serum enzymes apart from the blood cells.

The sera of patients with acute myocardial infarction were thought worthy of particular attention because of the possibility that destruction of cardiac muscle, reported rich in transaminase activity, might result in a release of this enzyme into the blood stream and might thus increase the serum transaminase activity. The finding of increased serum transaminase activity in the two patients studied during the first week after infarction is compatible with such a possibility.

SUMMARY

1. The presence of glutamic oxalacetic and glutamic pyruvic transaminase activity in human serum, plasma, and whole blood hemolysates has been demonstrated. A method is presented for their measurement by estimating the rate of glutamate formation employing quantitative paper chromatography.

2. The chemical properties of the enzyme in serum, including the variation in activity with changes in concentration of enzyme, substrate, and hydrogen ion were found similar to those reported for transaminases in animal tissues. Freezing or lyophilizing the serum, or storing it at 0 to 5°C. for periods up to two weeks resulted in no significant change in transaminase activity. Boiling destroyed the activity.

3. Glutamic oxalacetic and glutamic pyruvic transaminase activities were found to be approximately equal in normal human serum. Whole blood hemolysates were found to have ten times as much glutamic oxalacetic transaminase activity as serum. Glutamic oxalacetic transaminase was found to be approximately 2.7 times as active as glutamic pyruvic transaminase in hemolysates.

4. A survey of transaminase activity in the blood of hospitalized patients revealed departure from the normal range of activity in several disease states.

ACKNOWLEDGMENT

We wish to acknowledge gratefully the interest, advice and assistance of Drs. Aaron Bendich, Oscar Bodansky, Severo Ochoa, and Rulon Rawson.

APPENDIX

A NOTE ON THE SPECTROPHOTOMETRIC ASSAY OF GLUTAMIC-OXALACETIC TRANSAMINASE IN HUMAN BLOOD SERUM

By ARTHUR KARMEN

(*Department of Pharmacology, New York University College of Medicine*)

A spectrophotometric method was devised in which the transamination reaction (Reaction 1) is coupled to the reduction of oxalacetate to malate by reduced diphosphopyridine nucleotide (DPNH), in the presence of an excess of purified malic dehydrogenase (Reaction 2). The oxidation of DPNH, and therefore the transamination reaction, is followed by measuring the decrease in light absorption at wave length 340 mμ, at which the reduced pyridine nucleotides have an absorption peak.

(*1*) α-Keto glutarate + aspartate $\leftrightarrows$ L-glutamate + oxalacetate

(*2*) Oxalacetate + DPNH + H$^+$ $\rightleftarrows$ L-malate + DPN$^+$

METHODS

Materials. Aspartic acid, alpha-keto glutaric acid and reduced diphosphopyridinenucleotide were obtained commercially. Purified malic dehydrogenase, prepared from pig heart muscle by the method of Straub (10), was tested and found free of detectable transaminase or glutamic dehydrogenase activity.[1]

Experimental procedure. From 0.1 to 1.0 ml. of serum, 1.0 ml. of 0.1 M phosphate buffer, pH 7.4, 0.5 ml. of 0.2 M aspartate in buffer, pH 7.4, 0.2 ml. of DPNH (1 mgm. per ml.) and 0.1 ml. of a solution of purified malic dehydrogenase (50 micrograms of enzyme protein per ml.) were mixed and brought to a final volume of 2.8 ml. in a cuvette having a 1.0 cm. light path. The blank contained all reactants listed except DPNH. After 10 minutes, 0.2 ml. of 0.1 M alpha-keto glutarate in buffer, pH 7.4, was added. The optical density at wavelength 340 mμ was followed for five minutes, and the rate of decrease of

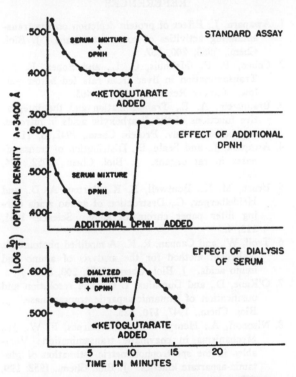

FIG. 5. SPECTROPHOTOMETRIC ASSAY OF SERUM TRANSAMINASE

Top and bottom curves, all components except α-keto glutarate present initially; α-keto glutarate added at time indicated by arrows. Middle curve: all components present initially, additional DPNH at time indicated by arrow. DPNH added at zero time in all cases. The increase in optical density when α-keto glutarate is added is due to the absorption of light by α-keto glutarate at this wavelength.

[1] I am indebted to Dr. Martin Schwartz for this preparation.

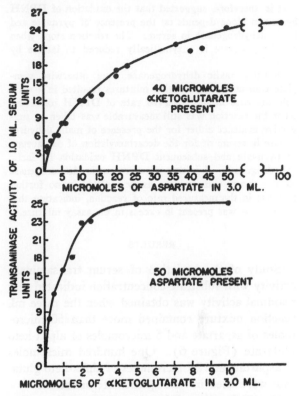

FIG. 6. SERUM TRANSAMINASE ACTIVITY AS A FUNCTION OF SUBSTRATE CONCENTRATION

optical density taken as the measure of the transaminase activity of the serum. The reaction was followed in a Beckman model DU spectrophotometer at room temperature. The activity is expressed as units per ml. of serum per minute. One unit equals a decrease in optical density of 0.001 under the conditions described.

Comments on procedure. When DPNH is added to serum without addition of substrates the optical density of the mixture decreases for six to seven minutes, indicating the oxidation of a finite quantity of DPNH. After this reaction has stopped, completion of the transamination system results in a steady decrease in optical density which is taken as a measure of the transaminase activity of the serum (Figure 5, top curve).

The rate or amount of DPNH oxidation by serum alone is not changed by the addition of malic dehydrogenase and either aspartate or alpha-keto glutarate. The amount oxidized is not appreciably affected by adding more DPNH (Figure 5, middle curve). Addition of 0.2 ml. of a solution of potassium pyruvate (4 mgm. per ml.), after this reaction has stopped, results in complete oxidation of the DPNH present at a measurable rate, demonstrating the presence of lactic dehydrogenase in serum. Dialysis of the serum against 0.1 M phosphate buffer reduces the amount of DPNH oxidized without changing the measurable transaminase activity of the serum (cf. Figure 5, top and bottom curves).

154

It is, therefore, suggested that the oxidation of DPNH by serum alone depends on the presence of pyruvate and lactic dehydrogenase in serum. The reaction stops when all the pyruvate is enzymatically reduced to lactate by DPNH.

Omitting malic dehydrogenase from otherwise complete transamination reaction mixtures resulted in a variable decrease in the observed rate of DPNH oxidation. That the reaction was still measurable was taken as suggestive evidence either for the presence of malic dehydrogenase in serum or for the decarboxylation of oxalacetate to pyruvate and subsequent DPNH oxidation by lactic dehydrogenase. Addition of more malic dehydrogenase to any serum transaminase assay resulted in no further increase in the measured rate of reaction, indicating that the enzyme was present in excess in the assay mixture.

RESULTS

Study of the variation of serum transaminase activity with substrate concentration indicated that maximal activity was obtained when the three ml. reaction mixture contained more than 50 micromoles of aspartate and 5 micromoles of alpha-keto glutarate (Figure 6). One hundred micromoles of aspartate and 20 micromoles of alpha-keto glutarate were used in subsequent determinations.

Proportionality of the observed rate of reaction to the amount of serum present was observed over a wide range of serum transaminase activities (Figure 7).

Transaminase activity was measured in the sera of 50 normal humans in addition to those deter-

mined by the chromatographic assay method. The values found ranged from a low of nine to a high of thirty-two units per ml. per minute with a mean value of 19.6. Conversion of these units to micromoles per ml. per hour, using the extinction coefficient for DPNH determined by Horecker and Kornberg (11) gives a mean value of 0.57, which falls within the range of 0.24 to 1.04 micromoles per ml. per hour found by the quantitative paper chromatographic assay.

ACKNOWLEDGMENT

The author wishes to express his appreciation to Dr. Severo Ochoa for invaluable help in performing this work.

REFERENCES

1. Awapara, J., Effect of protein depletion on the transaminating activities of some rat organs. J. Biol. Chem., 1953, 200, 537.
2. Cohen, P. P., Hekhuis, G. L., and Sober, E. K., Transamination in liver from rats fed butter yellow. Cancer Research, 1942, 2, 405.
3. Braunstein, A. E., Transamination and the integrative functions of the dicarboxylic acids in nitrogen metabolism. Adv. Protein Chem., 1947, 3, 11.
4. Awapara, J., and Seale, B., Distribution of transaminases in rat organs. J. Biol. Chem., 1952, 194, 497.
5. Brush, M. K., Boutwell, R. K., Barton, A. D., and Heidelberger, C., Destruction of amino acids during filter paper chromatography. Science, 1951, 113, 4.
6. Troll, W., and Cannan, R. K., A modified photometric ninhydrin method for the analysis of amino and imino acids. J. Biol. Chem., 1953, 200, 803.
7. O'Kane, D., and Gunsalus, I. C., The resolution and purification of glutamic-aspartic transaminase. J. Biol. Chem., 1947, 170, 425.
8. Nisonoff, A., Henry, S. S., and Barnes, F. W., Jr., Mechanisms in enzymatic transamination: Variables in the spectrophotometric estimation of glutamic-aspartate kinetics. J. Biol. Chem., 1952, 199, 699.
9. Cohen, P. P., Kinetics of transaminase activity. J. Biol. Chem., 1940, 136, 585.
10. Straub, F. B., Reinigung der Äpfelsäuredehydrase und die Bedeutung der Zellstruktur in der Äpfelsäuredehydrierung. Ztschr. f. physiol. Chem., 1942, 275, 63.
11. Horecker, B. L., and Kornberg, A., The extinction coefficients of the reduced band of pyridine nucleotides. J. Biol. Chem., 1948, 175, 385.

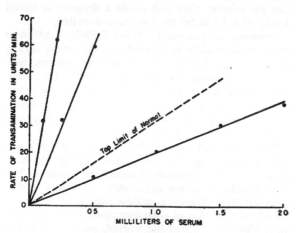

FIG. 7. TRANSAMINASE ACTIVITY AS A FUNCTION OF SERUM CONCENTRATION

Three different serum samples.

COMMENTARY TO

page 157 **13. Mercer, D. W. (1974)**
Separation of Tissue and Serum Creatine Kinase Isoenzymes by Ion-Exchange Column Chromatography. Clinical Chemistry 20(1): 36–40.

The measurement of isoenzymes in serum in order to improve the specificity of an enzyme assay for disease detection began with the work of Oscar Bodansky in 1937. He used taurocholate inhibition to differentiate bone, kidney and intestinal alkaline phosphatase (ALP, EC 3.1.3.1) in normal serum and in two patients with Paget's disease (1). Zone electrophoresis on paper was utilized throughout the 1950's to separate isoenzymes. Estborn in 1959 separated serum ALP isoenzymes using starch gel electrophoresis (2). Throughout the 50's and 60's chemical inhibition and electrophoresis were the principal methods for the identification of isoenzymes in serum.

In the early 1960's creatine kinase (CK, EC 2.7.3.2) was reported to be an early and sensitive marker of myocardial damage (3). The ultraviolet kinetic spectrophotometric procedure for total CK described by Oliver (4) as modified by Rosalki (5) was quickly adopted into routine use. CK in normal serum on agar gel or cellulose acetate electrophoresis showed three isoenzymes; CK1 (BB, brain), CK2 (MB, heart) and CK3 (MM, skeletal muscle and heart). Total CK was not specific for heart but electrophoresis with quantitative measurement of CK2 improved the sensitivity and specificity of the assay for detecting myocardial damage (6). The 1974 paper presented here by Mercer was a novel refinement in the analysis of a clinical isoenzyme. His method was eventually replaced with more rapid and specific antibody procedures for CK2 however for many years it remained a benchmark assay in the study of serum CK isoenzymes.

Wood in 1963 may have been the first to use an ion-exchange resin to separate and purify extracts of brain CK (7). Mercer packed 0.5×6.0 cm^2 mini-columns with diethylaminoethyl (DEAE) Sephadex®anion-exchange resin. A 3-step Tris sodium chloride buffer elution was used to isolate CK2. Varat and Mercer reported on the improved specificity of the column assay over electrophoresis in clinical studies (8). A number of modifications were reported (9–11) and in less than 4 years after Mercer's paper, numerous commercial test kits were made available for this column assay (12,13).

References

(1) Bodansky, O. (1937) Are the phosphatases of bone. kidney, intestine, and serum identical? The use of bile acids in their differentiation. Journal of Biological Chemistry. 118(2):341–362.

(2) Estborn, B. (1959) Visualization of acid and alkaline phosphatase after starch-gel elelctrophoresis of seminal plasma, Serum and Bile. Nature. 184(Suppl 21):1636–1637.

(3) Dreyfus, J.C., Schapira, G., Resnais, J., and Scebat, L. (1960) La creatine-kinase serique dans le diagnostic de L'infarctus myocardique. Revue Francaise D'Etudes Cliniques et Biologiques. 5:386–387.

(4) Oliver, L.T. (1955) A spectrophotometric method for the determination of creatine phosphokinase and myokinase. Biochemical Journal. 61(1):116–122.

(5) Rosalki, S.B. (1967) An improved procedure for serum creatine phosphokinase determination. Journal of Laboratory and Clinical Medicine. 69(4):696–705.

(6) Lott, J.A. and Stang, J.M. (1980) Serum enzymes and isoenzymes in the differential diagnosis of myocardial ischemia and necrosis. Clinical Chemistry. 26(9):1241–1250.

(7) Wood, T. (1963) Adenosine triphosphate–creatine phsophotransferase from Ox brain: purification and isolation. Biochemical Journal. 87(3):453–462.

(8) Varat, M.A. and Mercer, D.W. (1975) Cardiac specific phosphokinase isoenzyme in the diagnosis of acute myocardial infarction. Circulation. 51(5):855–859.

(9) Mercer, D.W. and Varat, M.A. (1975) Detection of cardiac-specific creatine kinase in sera with normal or slightly increased total creatine kinase activity. Clinical Chemistry. 21(8):1088–1092.

(10) Mercer, D.W. (1975) Simultaneous separation of serum creatine kinase and lactate dehydrogenase isoenzymes by ion-exchange column chromatography. Clinical Chemistry. 21(8):1102–1106.

(11) Nealon, D.A. and Henderson, A.R. (1975) Separation of creatine kinase isoenzymes in serum by ion-exchange column chromatography (Mercer's method, modified to increase sensitivity). Clinical Chemistry. 21(3):392–397.

(12) Hamlin, C. and Ackerman, E. (1978) Relative merits of two electrophoretic and two column-chromatographic kits for determining serum creatine kinase iosenzyme MB activity. Clinical Chemsitry. 24(11):2013–2017.

(13) McQueen, M.J., Mori, L., and Dey, E. (1978) Evaluation of the BMC column-chromatographic method for creatine kinase isoenzyme MB. Clinical Chemistry. 24(3):519–520.

Clinical Chemistry 1974; 20(1): 36–40
© 1974 American Association for Clinical Chemistry.
Reproduced with permission.

CLIN. CHEM. 20/1, 36–40 (1974)

Separation of Tissue and Serum Creatine Kinase Isoenzymes by Ion-Exchange Column Chromatography

Donald W. Mercer

I describe a simple, rapid anion-exchange column chromatographic technique for separating the creatine kinase (CK) isoenzymes in human serum and tissue. Extracts of CK-rich tissues (skeletal muscle, cardiac muscle, and brain) were used to determine optimum conditions for separating CK isoenzymes MM, MB, and BB. Samples, layered on mini-columns (0.5 × 6.0 cm) of DEAE-Sephadex A-50, were eluted stepwise with Tris-buffered sodium chloride (100, 200, and 300 mmol/liter). Column effluents were assayed by the Rosalki CK method. Distribution of total activity among the eluted fractions was tissue-specific and reproducible. Evaluation of sera from 71 patients with myocardial infarction and other diseases associated with elevated CK activity revealed isoenzyme patterns that resembled those of either cardiac muscle or skeletal muscle. Cardiac pattern (presence of MB isoenzyme) and clinical documentation of myocardial infarction were 100% correlated in the 35 patients so studied.

Additional Keyphrases: *diagnostic aids* • *isoenzyme separation* • *"kit" methods*

The effective use of CK[1] as a sensitive indicator of acute myocardial infarction has recently been diminished because of numerous reports of CK elevations owing to noncardiac conditions, such as chronic alcoholism (*1*), cardioversion (*2*), cerebrovascular disease (*3*), hypothyroidism (*4*), intramuscular injections (*5*) and surgical trauma (*6*).

Attempts to enhance the diagnostic specificity of CK assays have been stimulated by these reports. The chief aim of past work has been to separate serum CK isoenzymes MM, MB, and BB by conven-

tional electrophoretic techniques. Clinically significant separations of MM from MB have been reported (*7–10*). However, routine use of the electrophoretic technique is limited since laborious gel preparations and poor staining techniques remain an integral part of the electrophoretic procedure.

I describe here a simple, rapid, and quantitative anion-exchange chromatographic technique for CK isoenzymes that eliminates the difficulties associated with existing techniques.

Materials and Methods

Tissue and Serum Preparation

Human tissue used in this study was taken from autopsy material in which no gross anatomical changes were evident. Homogenates (1 g diluted to 10 ml) were prepared with Tris-hydrochloride buffer (50 mmol/liter, pH 8.0) containing sodium chloride (100 mmol/liter) and dithiothreitol (0.1 mmol/liter). After the homogenate was centrifuged (12 000 × *g*, for 10 min), the insoluble pellet was discarded and the supernate used in subsequent chromatographic experiments.

Sera with CK activity greater than six times normal were obtained from a general hospital population, and kept refrigerated until isoenzyme analysis, which usually was done within two days.

Enzyme Activity Analysis

CK was assayed with an ultraviolet kinetic test kit (Smith Kline Instruments, Palo Alto, Calif. 94304) based on the method of Rosalski (*11*). All assays were conducted either with the DSA 564-B (Beckman Instruments, Fullerton, Calif. 92634) or with the semi-automated "Eskalab" spectrophotometer (Smith Kline).

Electrophoresis

Isoenzyme electrophoresis was performed on polyacrylamide (7 g/dl), with the use of an analytical vertical-gel apparatus (Canalco, Rockville, Md. 20852). Samples were diluted 1:1 with sucrose (40

Biochemistry Section, Dept. of Pathology, Montefiore Hospital, Pittsburgh, Pa., 15213; and Dept. of Pathology, School of Medicine, University of Pittsburgh, Pa. 15213.

Presented at the 25th National Meeting of the AACC, New York, N.Y., July 16–20, 1973.

[1] Nonstandard abbreviations used: CK, creatine kinase (EC 2.7.3.2); Tris, tris(hydroxymethyl)aminomethane; DEAE, diethylaminoethyl; MM, skeletal-muscle isoenzyme of CK; MB, cardiac-muscle isoenzyme of CK; BB, brain isoenzyme of CK.

Received Sept. 20, 1973; accepted Oct. 24, 1973.

158

g/dl) and applied to the top of the polymerized gel (0.5 × 3.5 cm). Electrophoresis was performed at 2 mA per gel with the Tris-glycine buffer system as originally described by Davis (12). CK isoenzymes were detected on sliced 2.5-mm segments of the polyacrylamide gel. Gel segments were placed in separate test tubes containing 0.2 ml of the Tris-hydrochloride–sodium chloride–dithiothreitol buffer, and aliquots were removed and assayed for CK activity after they had soaked overnight at room temperature.

Column Chromatography

The mini-column consisted of a 12.5-cm Pasteur pipette (Chase Instruments, Lindenhurst, N.Y. 11757) filled with about 60 mg of the anion-exchanger "DEAE-Sephadex A-50" (Pharmacia, Piscataway, N.J. 08854). Column dimensions were 0.5 × 6.0 cm. The ion-exchanger was prepared for column packing by slowly mixing 5 g of dry DEAE-Sephadex with one liter of Tris-hydrochloride starting buffer (50 mmol/liter, pH 8.0) containing sodium chloride (100 mmol/liter). After sedimentation and decantation the ion-exchanger was resuspended in another liter of starting buffer. After repeating the sedimentation and decantation step, a slurry of about one part ion-exchanger to four parts starting buffer was transferred to a vacuum flask, where trapped air bubbles were removed under decreased pressure. The de-aerated slurry was poured into the column until the final height of the settled suspension was 6 cm. Before the sample was applied, 2 ml of starting buffer was passed through the column. (It was convenient to prepare large numbers of columns at the same time and to store them until needed; columns stored at room temperature for two weeks have performed satisfactorily).

Figure 1 is a diagrammatic representation of the chromatographic process. A sample volume of 1 ml, containing CK activity in the range of 600 to 2000 mU, was applied to the top of the column and sample effluent was collected in the first vial. After the mini-column had drained, the second collection vial was placed under it. Subsequent elution was stepwise with eluents of Tris-buffered sodium chloride:

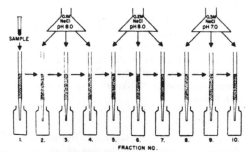

Fig. 1. Diagrammatic representation of the ion-exchange chromatographic procedure for separation of CK isoenzymes

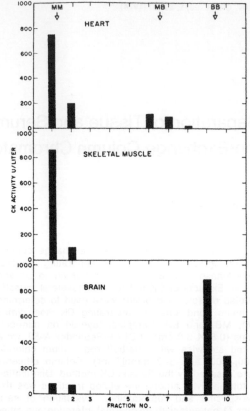

Fig. 2. Distribution of tissue CK isoenzymes MM, MB, and BB on DEAE-Sephadex columns

Three 1-ml fractions of Tris-hydrochloride starting buffer (50 mmol/liter, pH 8.0) containing sodium chloride (100 mmol/liter) were collected in vials 2, 3, and 4. Then three 1-ml fractions of Tris-hydrochloride (50 mmol/liter, pH 8.0) containing sodium chloride (200 mmol/liter, pH 8.0) were collected in vials 5, 6, and 7. Finally, three 1-ml fractions of Tris-hydrochloride (50 mmol/liter, pH 7.0) containing sodium chloride (300 mmol/liter) were collected in vials 8, 9, and 10. A flow rate of 0.5 ml/min was maintained. Total elution time was about 15 min. Aliquots (50 to 200 µl) of column effluents were assayed for CK activity.

Results

The ion-exchange chromatographic system was characterized by running extracts of CK-rich tissues through the column. Figure 2 depicts the behavior of the MM isoenzyme from skeletal muscle, the MB isoenzyme from cardiac muscle, and the BB isoenzyme from brain. Skeletal muscle extract exhibited CK activity assumed to be the MM isoenzyme in fractions 1 and 2. No activity was found in the other fractions. Cardiac extract exhibited about 90% of its CK activity in the MM isoenzyme region and about

10% in fractions 6 and 7. Activity in 6 and 7 was assumed to be the MB isoenzyme. MB isoenzyme was not found in the other tissues. Brain extract exhibited a predominant peak of CK activity in fractions 8, 9, and 10. Activity in these fractions was assumed to be the BB isoenzyme. All three of these tissue isoenzyme patterns have been reproduced many times with a number of different tissue preparations.

Evidence that these patterns are indeed real and not merely artifacts of the chromatographic system is shown in Figure 3. In this experiment, column fractions 1 and 2 from cardiac muscle (MM isoenzyme), column fractions 6 and 7 from cardiac muscle (MB isoenzyme), and column fractions 8, 9, and 10 from brain (BB isoenzyme) were rechromatographed on DEAE-Sephadex. As shown in Figure 3, the ion-exchange isoenzymes appeared as single peaks with unchanged elution characteristics.

Results of polyacrylamide-gel electrophoresis of the ion-exchange isoenzymes are shown diagrammatically in Figure 4. All three ion-exchange isoenzymes showed single bands of activity with characteristic migration rates similar to those reported for MM, MB, and BB by other investigators (7–10, 13). The uniqueness of the ion-exchange isoenzymes was also demonstrated by the results of a thermostability ex-

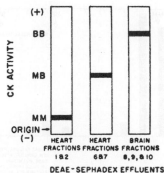

Fig. 4. Polyacrylamide gel electrophoresis of DEAE-Sephadex effluents

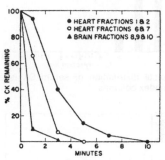

Fig. 5. Thermostability (56 °C) of DEAE-Sephadex effluents

periment at 56 °C (Figure 5).

Figure 6 shows typical ion-exchange patterns of sera obtained from patients with myocardial infarction and from patients with other diseases that also exhibit elevated CK. These patterns were obtained by applying 1 ml of serum, containing about 1000 mU of CK, to the column. A close relationship between serum isoenzyme patterns and those of cardiac and skeletal muscle (Figure 2) was observed. Although MM isoenzyme found in fractions 1, 2, and 3 was the predominant peak of activity in both patterns, only in sera of patients with myocardial infarction was the MB isoenzyme detected.

Table 1 relates diagnosis (of 71 patients in whose sera CK activity was more than six times normal) and serum isoenzyme distribution. Note that of the 35 patients having a known myocardial infarction, MM and MB isoenzymes were detected in each case. However, in the remaining patients who had different diagnoses, only the MM isoenzyme was detected, but not the MB isoenzyme. BB isoenzyme was not detected in this group of patients.

Results of quantitative analysis for MM and MB in 20 patients with myocardial infarction are shown in Table 2. Isoenzyme analysis was performed on specimens collected during peak activity of total CK without regard to specified post-infarct time periods. An average yield of 6% for MB isoenzyme was consistent with activities (approximately 10%) of

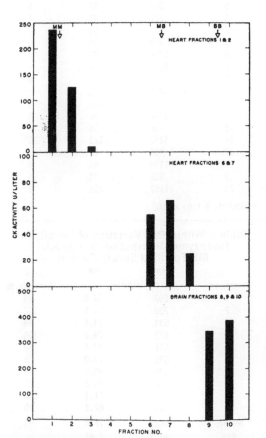

Fig. 3. Rechromatography of DEAE-Sephadex effluents

160

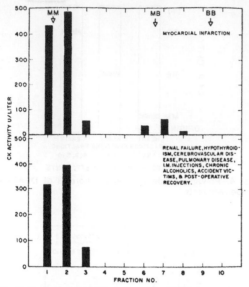

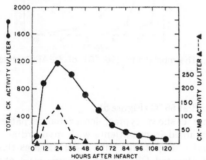

Fig. 6. Typical distribution of serum CK isoenzymes on DEAE-Sephadex columns

Fig. 7. Typical course of change in total CK and MB activities after myocardial infarction

MB found in cardiac tissue. Correlation between MB yields and total CK was not observed.

A typical time course for post-infarct activities of total CK and MB is shown in Figure 7. These activities paralleled one another until 36 h after infarct, when MB could no longer be detected.

Column reproducibility was evaluated by repeated analysis of a serum pool fortified with MB isoenzyme from an extract of heart tissue. A sample volume of 1 ml, containing 748 mU of CK, was applied to each of 10 ion-exchange columns. The results (Table 3) proved to be satisfactory, and excellent recoveries of total CK activity from the columns were also observed (Table 3).

Column capacity for MM isoenzyme was exceeded by CK samples with activities near 3000 U/liter. To prevent cross contamination of MB effluents by excess MM isoenzyme, I diluted high-activity CK samples with Tris-hydrochloride starting buffer to a CK activity near 1000 U/liter before they were applied to the column.

Table 1. Serum CK Isoenzymes in 71 Patients Exhibiting Sixfold Normal (or Greater) Serum CK Activity

Diagnosis	No. patients	CPK isoenzymes		
		MM	MB	BB
Myocardial infarction	35	35	35	0
Acute renal failure	4	4	0	0
Hypothyroidism	1	1	0	0
Cerebrovascular disease	8	8	0	0
Pulmonary disease	2	2	0	0
Chronic alcoholic	2	2	0	0
Muscular injections	1	1	0	0
Postoperative recovery	1	1	0	0
Muscle cramps	1	1	0	0
After cardioversion	1	1	0	0
Accident victims	6	6	0	0
Other diagnoses	9	9	0	0

Table 2. Quantitative Analysis of CK Isoenzymes in 20 Patients with a Diagnosis of Myocardial Infarction

	MM	MB	
Patient	U/liter		% MB[a]
1	1762	124	7
2	864	56	6
3	1246	50	4
4	837	103	12
5	654	42	6
6	1101	43	4
7	926	89	9
8	550	55	9
9	1218	46	4
10	1010	48	5
11	774	68	8
12	1054	62	5
13	1365	110	8
14	1247	42	3
15	796	16	2
16	1232	123	11
17	918	23	2
18	1139	96	9
19	857	78	9
20	1157	101	9

[a] (MB/MM + MB) × 100.

Table 3. Within-Run Variation of Results for Isoenzyme Distribution in a Special MB-Fortified Serum Control

	MM	MB	
Sample no.	U/liter		Recovery, %
1	592	77.0	89
2	700	73.5	103
3	594	74.4	89
4	672	79.2	100
5	661	77.4	99
6	635	77.0	95
7	599	70.0	89
8	631	76.6	95
9	680	71.3	100
10	609	67.3	90
Mean	638	74.4	95
SD	39.2	3.81	—

Discussion

Serum isoenzyme determinations have long been considered important diagnostic tests, especially when the tissue source of an elevated enzyme is not clinically apparent. But because complex electrophoretic techniques are used, isoenzyme determinations are considered to be highly specialized tests and are performed by only a few laboratories.

As an alternative to electrophoresis, selective inhibition techniques with urea, heat, and specific chemical inhibitors have been attempted. Results from these studies are useful if only one isoenzyme is involved, but yield equivocal results if more than one isoenzyme is present, as is frequently the case for serum.

Although ion-exchange chromatography has been used successfully to isolate and purify tissue isoenzymes (14–16), clinical application of this technique has been neglected. Recent attempts have been made to use ion-exchange as a method to separate cytoplasmic and mitochondrial aspartate aminotransferase (EC 2.6.1.1) in sera of patients with liver disease (17, 18). The results were clinically significant, but routine clinical use is limited, primarily because of the long elution time (2 to 4 h) and the long dialysis period (4 to 5 h) required before the serum is chromatographed.

The present ion-exchange method for separating CK isoenzymes is convenient for clinical use. The elution times usually associated with column chromatography have been eliminated by the use of minicolumns and stepwise elution techniques. Need for pre-treatment of sample (dialysis) has also been eliminated by setting the column's exchangeable chloride concentration equal to the chloride concentration of serum. Under these conditions, MM isoenzyme of skeletal muscle quickly passes through the ion-exchanger while the MB isoenzyme of cardiac muscle and BB isoenzyme of brain remain attached until a specified increase in chloride concentration decreases their ionic interaction with the ion-exchanger. Quantitative assay of column effluents by a technique usually used to assay total CK completes the CK isoenzyme analysis.

Clinical results for the present technique confirm previous observations made with electrophoretic techniques that MB is quite specific for the detection of myocardial infarction if specimens are collected and assayed between 12 and 36 h after the infarct (19). It is difficult to detect MB in late post-infarct specimens, either by electrophoresis or ion-exchange, because the activity becomes negligible. However, it appears that the ion-exchange technique is ideally suited for detecting trace activities of MB, because

one can collect and concentrate MB effluents from a series of ion-exchange columns on which the specimen in question has been run.

Studies directed toward the detection of MB in late post-infarct specimens and in other specimens in which total CK activity is normal or slightly elevated are now in progress.

References

1. Nygren, A., Serum creatine phosphokinase in chronic alcoholism. *Acta Med. Scand.* **182**, 383 (1967).

2. Mandecki, T., Leszek, G., and Kargul, W., Serum enzyme activities after cardioversion. *Brit. Heart J.* **32**, 600 (1970).

3. Acheson, J., James, D. C., Hutchinson, E. C., and Westhead, R., Serum creatine kinase levels in cerebral vascular disease. *Lancet* **i**, 1306 (1965).

4. Griffiths, P. D., Creatinephosphokinase levels in hypothyroidism. *Lancet* **i**, 894 (1963).

5. Meltzer, H. Y., Mrozak, S., and Boyer, M., Effect of intramuscular injections on serum creatine phosphokinase activity. *Amer. J. Med. Sci.* **259**, 42 (1970).

6. Dixon, S. H., Fuchs, J. C. A., and Ebert, P. A., Changes in serum creatine phosphokinase activity following thoracic, cardiac and abdominal operations. *Arch. Surg.* (Chicago) **103**, 66 (1971).

7. Smith, A. R., Separation of tissue and serum creatine kinase isoenzymes on polyacrylamide gel slabs. *Clin. Chim. Acta* **39**, 351 (1972).

8. Konttinen, A., and Hannu, S., Determination of serum creatine kinase isoenzymes in myocardial infarction. *Amer. J. Cardiol.* **29**, 817 (1972).

9. Elevitch, F. R., Isoenzymes. In *Fluorometric Techniques in Clinical Chemistry*. Little, Brown and Co., Boston, Mass., 1973, p 244.

10. Roe, C. R., Limbird, L. E., Warner, G. S., and Nerenberg, S. T., Combined isoenzyme analysis in the diagnosis of myocardial injury: application of electrophoretic methods for the detection and quantitation of the creatine phosphokinase MB isoenzyme. *J. Lab. Clin. Med.* **80**, 577 (1972).

11. Rosalki, S. B., An improved procedure for serum creatine phosphokinase determination. *J. Lab. Clin. Med.* **69**, 696 (1967).

12. Davis, B. J., Disc electrophoresis. Method and application to human serum proteins, *Ann. N. Y. Acad. Sci.* **121**, 404 (1964).

13. Dawson, D. M. and Fine, I. H., Creatine kinase in human tissues. *Arch. Neurol.* (Chicago) **16**, 175 (1967).

14. Wood, T., Adenosine triphosphate-creatine phosphotransferase from ox brain: Purification and isolation. *Biochem. J.* **87**, 453 (1963).

15. Wachsmuth, E. D., and Pleiderer, G., Biochemische Untersuchungen an kristallinen Isozymen der lactat Dehydrogenase aus menschlichen Organen. *Biochem. Z.* **336**, 545 (1963).

16. Takahashi, K., Ushikubo, S., Oimomi, M., and Shinko, T., Creatine phosphokinase isoenzyme of human heart muscle and skeletal muscle. *Clin. Chim. Acta* **38**, 285 (1972).

17. Schmidt, E., Schmidt, F. W., and Mohr, J., An improved simple chromatographic method for separating the isoenzymes of malic dehydrogenase and glutamic oxaloacetic transaminase. *Clin. Chim. Acta* **15**, 337 (1967).

18. Ideo, G., Franchis, R., Bellobuono, A., and Tornaghi, G., Aspartate aminotransferase isoenzymes in human serum in various liver diseases. *Enzymes* **12**, 529 (1971).

19. Wagner, G. S., Roe, C. R., Limbird, L. E., Rosati, R. A., and Wallace, A. G., The importance of identification of the myocardial-specific isoenzyme of creatine phosphokinase (MB form) in the diagnosis of acute myocardial infarction. *Circulation* **47**, 263 (1973).

14. Folin, O. and Wu, H. (1919)
A System of Blood Analysis. Journal of Biological Chemistry 38(1): 81–110.

COMMENTARY TO

T he Medical School at Harvard University planned to hold a dinner on November 16th, 1934 to celebrate Otto Folin's retirement after 25 years of service. A commissioned portrait would be presented to the University in his honor. Three weeks before the dinner, one of the invited speakers, Philip Shaffer, asked Folin how he would sum up his contributions to biochemistry. Folin replied in a letter to Shaffer dated October 21,

My portrait includes a colorimeter and a couple of volumetric flasks, and it might therefore fit in pretty well to say something about the introduction of colorimetry into biochemistry. This is about all that I can think of at the moment (1).

Folin died on October 25, 1934. The speeches and portrait were presented at a memorial service 1 month later (2). The colorimeter in the painting that Folin considered so important in summing up his major contribution to biochemistry was a Duboscq balancing method colorimeter (3). Folin's introduction of this instrument into biochemistry in 1904 created a technological advance in clinical chemistry that would not be surpassed until the development of the photoelectric colorimeter 35 years latter.

The Duboscq instrument was a color matching colorimeter first commercialized by Jules Duboscq in France in 1870 (4). It resembled from the outside a simple monocular microscope. Through the eyepiece the observer saw a split screen image. The left and right sides of the split image came from light that traveled up through two separate solid glass rods mounted below the eyepiece. Below each rod were glass bottom cups that sat on independently adjustable ring stands. In a typical colorimetric assay a standard was placed into the left hand cup and the unknown into the right hand cup. Each adjustable ring stand was linked to a mm scale that measured the depth of immersion of each glass rod. Reflected light passed up from a mirror below the ring stands through the solutions in the cups then up through the glass rods into the eyepiece. The color intensities of the two solutions seen through the eyepiece were made to match by varying the depth of the rods in their respective solutions. Color intensity was inversely proportional to the depth of the rods. The mm scale provided a quantitative measure of the depths of the rods. The concentration of the unknown could then be calculated from the simple formula,

$$\frac{\text{mm depth of the standard (concentration of the standard)}}{\text{mm depth of the unknown}}$$

Folin was the first to use the Duboscq colorimeter in a clinical chemistry procedure. He published methods for urine creatinine and creatine in 1904 based on the colorimetric alkaline picric acid reaction (5,6). Prior to the Duboscq, colorimetric clinical chemistry assays were read by holding the test tubes with the unknowns up against a white background and color matching them against tubes containing standards or up against colored glass filters previously matched against liquid standards for that assay. The Duboscq colorimeter replaced this with a quantitative measure of the difference in color intensity between samples. In addition, because of its optical design, it improved the ability to distinguish between weak differences in intensity.

Folin adopted his creatinine method for use with serum in 1914 (7). Colorimetric assays for glucose, urea, non-protein nitrogen, creatinine and uric acid read with the Duboscq colorimeter were integrated into a system of analytical assays by Folin and Hsien Wu in 1919. This paper is presented here. In 2002 the editors of the Journal of Biological Chemistry selected this paper as a "Classic Article" in biochemistry (8). Folin's Duboscq procedures were the methods routinely presented in clinical chemistry manuals and textbooks for the next 40 years (9–12). Folin's own textbook of methods, *Laboratory Manual of Biological Chemistry*, first published in 1916, went through five editions before his death (13).

Folin's reply to Shaffer in 1934 that he could think of only colorimetry as his major contribution to biochemistry was not the simple modest reply that it seemed.

References

(1)   Meites, S. (1989) *Otto Folin America's First Clinical Chemist*. American Association for Clinical Chemistry, Inc., Washington, DC, The letter from O. Folin to P.A. Shaffer dated October 21, 1934 is reproduced on pgs 355–356.

(2)   Shaffer, P.A. (1935) Obituary Otto Folin. Science. 81(2089):35–37.

(3)   Rosenfeld, L. (2002) Clinical chemistry since 1800: growth and development. Clinical Chemistry. 48(1):186–197, The Folin portrait is reproduced on pg 193.

(4)   Stock, J.T. (1994) The Duboscq colorimeter and its inventor. Journal of Chemical Education. 71(11):967–970.

(5)   Folin, O. (1904) Beitrag zur Chemie des Kreatinins und Kreatins im Harne. Hoppe-Seyler's Zeitschrift für Physiologische Chemie. 41:223–242.

(6)   Folin, O. (1904) Some metabolism studies. With special reference to mental disorders. American Journal of Insanity. 60(4):699–732.

(7)   Folin, O. (1914) On the determination of creatinine and creatine in blood, milk and tissues. Journal of Biological Chemistry. 17(4):475–481.

(8)   Simoni, R.D., Hill, R.L., and Vaughan, M. (2002) JBC centennial 1905–2005 100 years of biochemistry and molecular biology. Journal of Biological Chemistry. 277(20):e9.

(9)   Myers, V.C. (1924) *Practical Chemical Analysis. A Book Designed as a Brief Survey of This Subject for Physicians and Laboratory Workers*. 2nd Edition. C.V. Mosby Co., St Louis.

(10) Todd, J.C. and Sanford, A.H. (1931) *Clinical Diagnosis by Laboratory Methods. A Working Manual of Clinical Pathology*. 7th Edition. W.B. Saunders Company, Phildelphia.

(11) Peters, J.P. and Van Slyke, D.D. (1932) *Quantitative Clinical Chemistry Vol II Methods*. The Williams & Wilkins Company, Baltimore.

(12) Gradwohl, R.B.H. (1938) *Clinical Laboratory Methods and Diagnosis. A Textbook on Laboratory Procedures With Their Interpretation*, 2nd Edition. The C.V. Mosby Company, St Louis.

(13) Folin, O. (1923) *Laboratory Manual of Biological Chemistry With Supplement D*, 3rd Edition. Appleton and Company, New York.

J. Biol. Chem. 1919; 38(1): 81–110
© 1919 The American Society for Biochemistry and Molecular Biology.
Reproduced with permission.

A SYSTEM OF BLOOD ANALYSIS.

By OTTO FOLIN and HSIEN WU.

(From the Biochemical Laboratory, Harvard Medical School, Boston.)

(Received for publication, March 29, 1919.)

CONTENTS.

INTRODUCTION.

The main purpose of the research recorded in this paper has been to combine a number of different analytical procedures into a compact system of blood analysis, the starting point for which should be a protein-free blood filtrate suitable for the largest possible number of different determinations. It need scarcely be pointed out what a convenience and advantage it would be if one could take the whole of a given sample of blood and at once prepare from it a protein-free blood filtrate suitable for the determination of all or nearly all the water-soluble constituents, non-protein nitrogen, urea, creatinine, creatine, uric acid, and sugar.

In connection with our work on the problem we have also had in mind the desirability of reducing as far as practicable the amount of blood filtrate to be used for each determination, for by means of such reduction the total usefulness of the filtrate is correspondingly increased. There is no hard and fast limit as to the extent to which this reduction can be carried. It is doubtful, however, whether it is sound analytical practice regularly to use the smallest possible amount of material for each determination; whether, for example, blood filtrates corresponding to only 0.1 cc.

81

82 Blood Analysis

of blood should regularly be used for non-protein nitrogen determinations, because it may sometimes be advantageous or necessary to take no more. In this paper we deal chiefly with a semi microchemical scale of work representing only a moderate reduction of the quantities ordinarily taken for colorimetric work with the 60 mm. Duboscq colorimeter.

One of the main obstacles encountered in attempts to develop a definite system of blood analysis of the kind we have had in mind has been the determination of the uric acid. For several years we have had serious doubts as to the full trustworthiness of the uric acid results heretofore recorded in the literature; moreover, to be reasonably accurate the determination has required more blood (about 25 cc.) than can be obtained except in isolated special cases. A large share of the work involved in this research has therefore been a critical study of the uric acid determination; and a modification of the Folin-Denis-Benedict method has been developed which requires the filtrate from only 2 cc. of blood, and which we believe to be more dependable as well as more simple and convenient than the original method.

We have also satisfactorily solved the problem of how to make and keep standard uric acid solutions, and we have devised a new colorimetric method for the determination of sugar in blood.

The determinations included in this research, namely non-protein nitrogen, urea, creatinine, creatine, uric acid, and sugar, can all be determined in the filtrate obtained from 10 cc. of blood.

Preparation of Protein-Free Blood Filtrates.

The pivotal point in our projected general scheme of blood analysis was necessarily a searching review of the most promising methods which have been used for precipitating the blood proteins in connection with the various analytical procedures in common use. As a working principle or guide in this search we have first of all required that the procedure employed must permit the quantitative recovery of at least 10 mg. of uric acid and creatinine when added to 100 cc. of sheep, beef, or chicken blood, and that the total non-protein nitrogen must certainly be no higher than the figures obtained from a corresponding trichloroacetic acid filtrate representing a 10 per cent trichloroacetic acid concentration (in

O. Folin and H. Wu   83

the diluted unfiltered blood mixture)—or a corresponding 1.5 per cent m-phosphoric acid filtrate.

While we are not dependent on the urease method for the determination of the urea, we have, nevertheless, deemed it imperative that the blood filtrate must also be of such a character as readily to permit the use of the urease method for the determination of this important constituent. We do not claim to have exhausted this line of inquiry, for it is a laborious process to determine the merits and shortcomings of any particular reagent in connection with such a comprehensive program. None of the precipitation procedures described in recent years is free from serious shortcomings. Kahlbaum's phosphotungstic acid or sodium phosphotungstate (prepared by ourselves) met our requirements when used under certain very definite conditions, and for a time we concentrated our efforts on the standardization of these reagents and on the adaptation of the various analytical procedures to the blood filtrates obtained from them.

In connection with our work on sodium phosphotungstate we have discovered a new protein precipitant which probably has never before been used in blood analysis. We refer to it as new protein precipitant because so far as we have been able to learn it has never before been used in that capacity. This substance is tungstic acid. Tungstic acid, like sodium phosphotungstate or phosphotungstic acid, must be used in a definite way, but the necessary conditions are not difficult to find. Less than 1 gm. is used for the precipitation of the proteins from 10 cc. of blood, yet the precipitation is more complete than that produced by 10 gm. of trichloroacetic acid, and the filtrate obtained gives no trouble in connection with any of the determinatives so far investigated. Neither creatinine nor uric acid is carried down by the precipitate within the conditions to be described. As much as 20 mg. of uric acid may be added to 100 cc. of blood without incurring any loss by absorption.

The blood protein precipitation obtained by the help of tungstic acid is interesting. The precipitation is completed within a few seconds. When the mixture is shaken hard, the sound is almost like that of shaken mercury and the hardest kind of shaking will not produce more than a trace of foam. The precipitate is very fine, yet does not go through good filter paper and does not stop up the pores.   The filtration is slow, but the total amount of filtrate obtained is nearly as large as that obtained with trichloroacetic acid.   If the precipitated mixture is heated in a water bath for 2 or 3 minutes, the precipitate settles spontaneously.   With this modification, centrifuging can be substituted for the filtration as the supernatant liquid is water-clear and contains no more nitrogen than the unheated filtrate.   For the present we do not care to recommend this process except for quantities of blood so small that one cannot afford to filter.   The statement of Folin and Denis that no precipitation involving the use of heat is permissible is probably erroneous as has been pointed out by Bock, although it is true for the m-phosphoric acid precipitation and probably for many others.   Unless some compelling advantage is gained by the use of heat, precipitation in the cold does seem to be the safer process.

There are many other points of interest to be investigated in connection with tungstic acid as a precipitant, but as these have no direct bearing on the problem of this research, further consideration of them here is omitted.   As a precipitant for blood proteins we believe that tungstic acid will prove more useful than any other reagent yet proposed.

The precipitation of the blood proteins by means of our new reagent is made in the following manner.   Transfer a measured amount of blood into a flask having a capacity of fifteen to twenty times that of the volume taken.   Dilute the blood with 7 volumes of water and mix.   With an appropriate pipette add 1 volume of 10 per cent solution of sodium tungstate ($Na_2WO_42H_2O$) and mix.   With another suitable pipette add to the contents in the flask (with shaking) 1 volume of $\frac{2}{3}$ normal sulfuric acid.   Close the mouth of the flask with a rubber stopper and give a few vigorous shakes.   If the conditions are right hardly a single air bubble will form as a result of the shaking.   Much oxalate or citrate interferes with the coagulation and later with the uric acid determination.   20 mg. of potassium oxalate is ample for 10 cc. of blood.   Citrate, except in the minimum amount, is to be avoided. When a blood is properly coagulated, the color of the coagulum gradually changes from pink to dark brown.   If this change does not occur, the coagulation is incomplete, due, in every case we have encountered, to too much oxalate or citrate.   In such an

O. Folin and H. Wu   85

emergency the sample may be saved by adding 2 normal sulfuric acid drop by drop, shaking vigorously after each addition and allowing the mixture to stand for a few minutes before adding more, until the coagulation is complete. Pour the mixture on a filter large enough to hold the entire contents of the flask and cover with a watch-glass. If the filtration is begun by pouring the first few cc. of the mixture down the double portion of the filter paper and withholding the remainder till the whole filter has been wet, the filtrates are almost invariably as clear as water from the first drop. If a filtrate is not perfectly clear, the first 2 or 3 cc. may have to be returned to the funnel. (Filter papers of the following diameters will meet all ordinary needs: 11, 12½, 15, and 18½ cm.)

It will be noted that the precipitation of the blood proteins is not made in volumetric flasks. Our procedure is adapted to the full use of practically all of a given sample of blood, for by this system 7, 9, or 12 cc. can be utilized just as well as 5 or 10. For this work we have devised a special blood pipette,[1] a sketch of which is given in Fig. 1. This is simply a 15 cc. pipette, graduated from the long tip into 1 cc. portions. The lower part is more or less like that of a volumetric pipette, thus permitting one to draw the blood directly from small, narrow bottles. We find it convenient to use three such pipettes; one for the blood, one for the sodium tungstate solution, and one for the sulfuric acid. The water used for diluting the blood may be measured with a cylinder.

The preparation of protein-free blood filtrates by this new process is so simple that no one need go astray, provided that the sodium tungstate and the ⅔ normal sulfuric acid are correct. The only doubtful point is the quality of the sodium tungstate used. The acid is intended to set free the whole of tungstic acid with about 10 per cent excess (and to neutralize the carbonate usually present in commercial tungstates). A

[1] The pipettes are made for us by the Emil Greiner Co., New York.

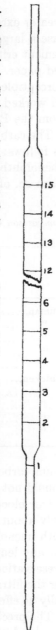

FIG. 1.

Blood Analysis

greater excess of sulfuric acid must not be used, for if this is the case a large part of the uric acid will be lost. A safe and convenient criterion is to test the blood filtrate obtained with Congo red paper. The reaction should be negative or at the most just perceptible. We have employed three different tungstates, and all worked equally well. The product we now use was obtained from the Primos Chemical Company, Primos, Pa.

The carbonate content of sodium tungstate is easily determined as follows: To 10 cc. of 10 per cent solution, add one drop of phenolphthalein and titrate with 0.1 normal hydrochloric acid. Each cc. of hydrochloric acid corresponds to 1.06 per cent of so-

TABLE I.

Comparison of Non-Protein Nitrogen in Blood Filtrates Obtained by Means of Trichloroacetic Acid and Tungstic Acid.

Source.	Mg. per 100 cc. blood.	
	Trichloroacetic acid.	Tungstic acid
Human...	35.7	31.5
"	32.4	28.2
"	35.4	33.0
"	42.0	42.0
"	50.7	48.6
Chicken...	66.5	54.5
"	48.0	42.5
"	53.2	44.4
"	50.0	44.8
"	51.6	47.6

dium carbonate. The amount of acid required for the titration should not exceed 0.4 cc.

Our blood filtrates are nearly neutral, 10 cc. of filtrate requiring only about 0.2 cc. of 0.1 normal sodium hydroxide when titrated with phenolphthalein as indicator. If the filtrates are to be kept for any length of time, more than 2 or 3 days, they need some preservation. One or two drops of toluene or xylene is adequate for the filtrate obtained from 10 cc. of blood. Xylene seems to be fully as effective as toluene as a preservative.

The precipitation process just described works equally well with any kind of blood which we have yet tried—human, beef, sheep, chicken, dog, and rabbit—and numerous comparisons with the tri-chloroacetic acid precipitation have shown that the non-protein nitrogen obtained by the process invariably tends to be lower than the figures given by trichloroacetic acid. The figures of Table I illustrate this point.

Determination of Non-protein Nitrogen.

The protein-free blood filtrates prepared by our new process lend themselves perfectly to nitrogen determinations by the direct Nesslerization process of Folin and Denis.[2] As a result of further experience with that method we are able to introduce certain modifications believed to represent improvements.

The acid digestion mixture is made as follows: Mix 300 cc. of phosphoric acid syrup (about 85 per cent H_3PO_4) with 100 cc. of concentrated sulfuric acid. Transfer to a tall cylinder, cover well to exclude the absorption of ammonia, and set aside for sedimentation of calcium sulfate. This sedimentation is very slow, but in the course of a week or so the top part is clear and 50 to 100 cc. can be removed by means of a pipette. (It is not absolutely necessary that the calcium should be thus removed, but it is probably a little safer to have it done.) To 100 cc. of the clear acid add 10 cc. of 6 per cent copper sulfate solution and 100 cc. of water. 2 cc. of this solution are substantially equivalent to 1 cc. of the acid mixture previously described by Folin and Denis. We prefer this diluted acid, first, because the objectionable viscosity of the undiluted reagent is practically eliminated, and, second, because we now use for a nitrogen determination only 5 cc. (instead of 10 cc.) of blood filtrate, and 1 cc. of acid (corresponding to 0.5 cc. of the undiluted acid reagent).

The micro-Kjeldahl digestion is made as before in test-tubes. While we still have an abundant supply of Jena test-tubes we no longer use them for this digestion because the Pyrex ignition test-tubes are very much better in nearly every respect. Test-tubes having a capacity of about 75 cc. (200 × 25 mm.) are suitable for this purpose, and if made of Pyrex ignition glass are almost as good as those of pure silica. These test-tubes should be graduated

[2] Folin, O., and Denis, W., Nitrogen determination by direct Nesslerization, J. Biol. Chem., 1916, xxvi, 473.

88   Blood Analysis

at 35 cc. and at 50 cc. on two sides, or by means of diamond marks going entirely around. The reason for this graduation is that we now Nesslerize in the digestion tube.

In micro-Kjeldahl digestions severe bumping is much more common than in ordinary macro-Kjeldahl digestions, but even in the latter the bumping phenomenon is often a source of serious difficulties. Glass beads, pumice stone, pieces of porcelain, etc., are used to remedy this trouble. For years ordinary quartz pebbles have been used in this laboratory, but at times these too have failed to prevent loss of a determination through sudden violent bumping. Occasionally a pebble may be hurled out of a 200 mm. test-tube by one intensive explosion. At other times no trouble at all is encountered. Langstroth,[3] who seems to have encountered very severe and persistent bumping, resorts to the device of holding the test-tube in as nearly a horizontal position as the contents in the tube will permit; but since it takes half an hour to boil off the liquid which ordinarily can be boiled off in less than 10 minutes, that remedy cannot be considered satisfactory. Langstroth seems to have concluded that such bumping is peculiar to blood filtrates obtained by means of *m*-phosphoric acid, but the phenomenon is quite general. Prolonged boiling of pure water in any glass vessel will lead to the most intense bumping. The most important cause of bumping is certainly the condition of the test-tube. A new test-tube does not cause bumping, but if the same one is used over and over again in one session, the bumping becomes progressively worse. The worst kind of a test-tube (or Kjeldahl flask) can be made as good as a new one by thoroughly drying it; also by rinsing with alcohol.

The reason why dry test-tubes cause less bumping and why dry pebbles tend to prevent bumping is manifestly the presence of very fine pores filled with air in the test-tube and in the pebbles. Until this air has been driven out by heat, localized formation of steam occurs and the boiling is smooth and even, but as these pores are gradually filled with the liquid the bumping begins. By keeping on hand a sufficiently large number of dry test-tubes so that no one need be used more than twice in one session the bumping phenomenon, in the presence of a fresh (that is a *dry*) quartz pebble or piece of granite, is almost entirely eliminated.

[3] Langstroth, L., Notes on Folin's direct Nesslerization method for the determination of nitrogen, *J. Biol. Chem.*, 1918, xxxvi, 377.

The Nesslerization process has also been simplified. The preliminary neutralization of the acid has been eliminated. The Nessler solution which we now use for all Nesslerizations is made as follows: The stock solution of mercuric potassium iodide can be made just as previously described. Dissolve 150 gm. of potassium iodide in 100 cc. of warm water, add 200 gm. of mercuric iodide, stir until the latter is dissolved, and dilute to a volume of about 1 liter; filter, if necessary, and dilute to a final volume of 2 liters. It is advantageous to make a large volume of this solution for a second sediment may form which takes a long time to settle.

The mercuric iodide obtainable from dealers frequently contains insoluble impurities (probably mercuric sulfide and mercurous iodide) which make it difficult to obtain a clear solution by the addition of potassium iodide. In such cases it is advisable to let the dissolved double iodide stand for 1 or 2 days and then filter, before diluting to volume.

Because of the difficulties encountered in obtaining high grade mercuric iodide, we have devised a new process for making the mercuric potassium iodide solution. This process is as follows: Transfer 150 gm. of potassium iodide and 110 gm. of iodine to a 500 cc. Florence flask; add 100 cc. of water and an excess of metallic mercury, 140 to 150 gm. Shake the flask continuously and vigorously for 7 to 15 minutes or until the dissolved iodine has nearly disappeared. The solution becomes quite hot. When the red iodine solution has begun to become visibly pale, though still red, cool in running water and continue the shaking until the reddish color of the iodine has been replaced by the greenish color of the double iodide. This whole operation usually does not take more than 15 minutes. Now separate the solution from the surplus mercury by decantation and washing with liberal quantities of distilled water. Dilute the solution and washings to a volume of 2 liters. If the cooling is begun in time, the resulting reagent is clear enough for immediate dilution with 10 per cent alkali and water, and the finished solution can at once be used for Nesslerizations.

The cost of the chemicals called for in this process of making Nessler's solution is less than when starting with mercuric iodide, and the disagreeable impurities present in many samples of mercuric iodide are avoided.

90   Blood Analysis

From the stock solution of mercuric potassium iodide, made according to either of the processes described above, we prepare the final Nessler solution as follows: From completely saturated caustic soda solution containing about 55 gm. of NaOH per 100 cc. decant the clear supernatant liquid and dilute to a concentration of 10 per cent. (It is worth while to determine by titration that a 10 per cent solution has been obtained within an error of not over 5 per cent.) Introduce into a large bottle 3,500 cc. of 10 per cent sodium hydroxide solution, add 750 cc. of the double iodide solution and 750 cc. of distilled water, giving 5 liters of Nessler's solution.

The Nessler solution so obtained contains enough alkali in 15 cc. to neutralize 1 cc. of the diluted phosphoric-sulfuric acid mixture and to give a suitable degree of alkalinity for the development of the color given by ammonia at a volume of 50 cc.

(In other Nesslerizations, as in urine analysis when there is no acid to be neutralized, 10 cc. of the Nessler reagent per 100 cc. of Nesslerized ammonia solution is the correct amount.)

Concise Description of Non-Protein Nitrogen Determination.— Introduce 5 cc. of the protein-free blood filtrate into a dry 75 cc. test-tube graduated at 35 cc. and at 50 cc. Add 1 cc. of the sulfuric-phosphoric acid mixture described on page 87. Add a dry quartz pebble and boil vigorously over a microburner until the characteristic dense acid fumes begin to fill the test-tube. This is usually accomplished in from 3 to 7 minutes. When the fumes are unmistakable, cut down the size of the flame so that the contents of the tube are just visibly boiling, and close the mouth of the test-tube with a watch-glass or a very small Erlenmeyer flask. Continue the heating very gently for 2 minutes from the time the fumes began to be unmistakable, even if the solution has become clear and colorless at the end of 20 to 40 seconds. If the oxidations are not visibly finished at the end of 2 minutes the heating must be continued until the solution is nearly colorless. Such cases are very rare; the oxidation is almost invariably finished within the 1st minute. Allow the contents to cool for 70 to 90 seconds and then add 15 to 25 cc. of water. Cool further, approximately to room temperature, and add water to the 35 cc. mark. Add, preferably with a pipette, 15 cc. of the Nessler solution described above. Insert a clean rubber stopper and mix. If the solution is turbid, centrifuge a portion before making the color comparison with the standard. The standard most commonly required is 0.3 mg. of N (in the form of ammonium sulfate) in a 100 cc. flask. Add to it 2 cc. of the sulfuric-phosphoric acid mixture, about 50 cc. of water, and 30 cc. of Nessler solution. Fill to the mark and mix. The unknown and the standard should be Nesslerized at approximately the same time. If the standard is set at 20 mm. for the color comparison, 20 divided by the reading and multiplied by 30 gives the non-protein nitrogen in mg. per 100 cc. of blood.

Determination of Urea.

Investigations on the most satisfactory method for the determination of urea have been pursued for the last 2 or 3 years (partly with the assistance of G. L. Foster and Guy Youngburg). Much of the work done on the subject has been an endeavor to find a direct Nesslerization process without the use of Merck's blood charcoal. Our attempts have not resulted in any thoroughly satisfactory method because very small amounts of ammonia cannot be Nesslerized in the presence of either amino-acids or peptones. Direct Nesslerization, even with the help of charcoal, cannot be made except at the expenditure of more blood filtrate than is actually used in the final stages of the determination, and a strictly economical use of the blood filtrate we have considered a fundamentally important point in our system of blood analysis. Direct Nesslerization has therefore been abandoned in connection with the determination of urea in blood. Extensive use has also been made of the permutit extraction after first decomposing the urea with urease, but this process has proved somewhat fallacious with bloods in which the total urea nitrogen is small, as in many normal bloods, so this process also has been abandoned. Since probably no other determination will be as useful and important to the clinician as the determination of the blood urea, we have considered it of the utmost importance to get a method which is as simple as possible, but above all reliable. In this connection we have had in mind not only the needs of well equipped hospital laboratories, but also the needs of private practitioners.

For the hydrolysis of the urea we make use of jack bean urease, or the autoclave; for the isolation of the ammonia produced we employ aeration or distillation; thus we have four combinations any one of which will give satisfactory results. The autoclave process is, of course, not advantageous for single urea determinations, but on the other hand is distinctly useful when it is a question of a large series of determinations, or when creatine determinations are also to be made, because the hydrolysis of the urea can then be accomplished simultaneously with the conversion of the creatine into creatinine. The chief merit of the autoclave process for decomposing urea in blood filtrates lies perhaps in the fact that by its help one is sure to get all the urea nitrogen; the values obtained may be too high, but not too low. Yet the results obtained by the autoclave process are as a matter of fact usually identical and rarely as much as 1 mg. per 100 cc. of blood higher than those obtained by the urea process.

Urease Decomposition.—For the decomposition of urea by means of urease we use exclusively jack bean powder extracts and not so called purified or concentrated urease preparation. It is doubtless possible to prepare such, but those obtainable in the market are usually less active than an equal weight of jack bean powder, and of course are much more expensive.[4]   An excellent urease solution can be prepared from jack bean powder in the following manner: Transfer to a 200 cc. flask or bottle about 3 gm. of permutit powder. Wash this by decantation, once with 2 per cent acetic acid, then twice with water. Add to the moist permutit in the flask 100 cc. of 30 per cent alcohol (35 cc. of 95 per cent alcohol mixed with 70 cc. of water). Then introduce 5 gm. of jack bean meal and shake for 10 minutes. Filter and collect the filtrate in three or four different small clean bottles. Set one aside for immediate use; it will remain serviceable at least 1 week at ordinary room temperature, if not exposed to direct sunlight. Put the others on ice where they will remain good for 3 to 5 weeks. The filtrate contains substantially the whole of the urease present in the jack bean powder and is very active. In the presence of a suitable phosphate mixture, 1 cc. added to 300 mg. of urea nitrogen at a volume of 200 cc. will yield 37 to 42 mg. of urea nitrogen

---

[4] The Arlington Chemical Co. supplies jack bean meal in a finer state of division than one can readily make by hand.

O. Folin and H. Wu   93

in 1 hour at 20°C.   In 18 hours all the urea will be decomposed. The use of permutit makes the extract free from ammonia (5 cc. containing less than 0.01 mg.), nor does more ammonia develop on standing.

Urease decompositions of urea are never dependable except in the presence of some buffer mixtures by which the reaction of the solution can be kept within certain limits.   The action of such mixtures is twofold.   They not only accelerate the decomposition of the urea, but also prolong greatly the acting period of the enzyme.   When urease solutions prepared as described above are added to urea dissolved in distilled water, it not infrequently happens that the enzyme acts for only a few minutes and then stops altogether, so that no more ammonia is obtained after 24 hours than after 15 minutes.   That the enzyme is only dormant and not entirely destroyed is shown by the fact that on adding phosphate mixture to the solution after 24 or even 48 hours standing, renewed urea decomposition begins and then continues for a long time.   The *Auxourease* found by Jacobi[5] to be present in blood serum represents probably nothing more or less than a preserving action of amphoteric serum proteins on the urease, action similar to that of phosphates.

In the course of our investigations on the determination of urea in blood filtrates by means of urease, it was accidentally found that other phosphates than those investigated by Van Slyke are equally good or better.   When the titratable acidity of *m*-phosphoric acid blood filtrates was neutralized with sodium bicarbonate, the urease action on (added) urea was surprisingly active and long sustained.   (The urea content of such blood filtrates can be determined conveniently both by the urease and by the autoclave processes.)   In consequence of this discovery, a series of experiments was made with pyro- and *m*-phosphates, and our observations have led to the conclusion that a solution containing 140 gm. of sodium pyrophosphate (u.s.p.) and 20 gm. of glacial phosphoric acid per liter is probably better than any of the phosphate mixtures investigated by Van Slyke.   We are at a loss for an explanation, for Van Slyke's mixtures cover the field sufficiently well from the standpoint of hydrogen and hydroxyl ion concentrations.

[5] Neumann, R., Über die Aktivierung der Soja-Urease durch menschliches Serum, *Biochem. Z.*, 1915, lxix, 134.

A thorough study of this subject has not been made, but it appears that the pyrophosphates are less injurious to urease than o-phosphates. One experiment may be cited.

Solutions containing mono- and disodium phosphate in the molecular ratios 1:1 and 1:2 were prepared. To 300 mg. of urea nitrogen in 200 cc. flasks were added (1) 5 cc. of phosphate 1:1, (2) 5 cc. of phosphate 1:2, and (3) 5 cc. of the pyrophosphate solution described above. To such mixtures were added water to 200 cc. and 1 cc. of urease solution (temperature 18°C.). Table II shows the results.

TABLE II.

Comparison of Effect of Different Buffer Mixtures on Rate of Hydrolysis of Urea by Action of Urease.

Buffer mixture.	Ammonia N.			
	15 min.	30 min.	1 hr.	19 hrs.
	mg.	mg.	mg.	mg.
Phosphate 1:1....................	5.7	9.6	21.4	266
" 1:2....................	5.7	9.5	19.8	180
Pyrophosphate....................	12.5	20.8	37.6	300

Determination of Urea by Urease Decomposition and Distillation.

Transfer 5 cc. of the tungstic acid blood filtrate to a *clean* and *dry* Pyrex ignition tube (capacity about 75 cc.). The graduated Pyrex tubes recommended for the non-protein nitrogen determination should never be used for urea determinations, because they have contained Nessler solutions and Nessler solutions leave behind films of mercury compounds which destroy the urease. If those tubes must be used, they should first be washed with nitric acid to remove the mercury films. Add to the blood filtrate two drops of the pyrophosphate solution described above or two drops of a molecular o-phosphate solution (⅓ molecular monosodium phosphate plus ⅔ molecular disodium phosphate). Then add 0.5 to 1 cc. of the urease solution described on page 92 and immerse the test-tube in a beaker of warm water and leave it there for 5 minutes. The temperature of the water is not very important but should not exceed 55°C. The warm water can perhaps scarcely be said to be essential, for the hydrolysis is very rapid at room temperature, but we nevertheless much prefer to use it.   If no hot water is used, continue the digestion for 10 to 15 minutes, or as much longer as is convenient.   The ammonia formed can be conveniently and quickly distilled into 2 cc. of 0.05 normal hydrochloric acid contained in a second test-tube.   The second test-tube should not be so heavy as the ordinary test-tubes and should be graduated at 25 cc.   A simple and compact arrangement for this distillation is indicated by Fig. 2.   The test-tube which serves as a receiver is held in place by means of a rubber stopper in the side of which has been cut a fairly deep notch to permit the escape of air (and some steam).   The rubber stopper serving as a holder for the receiver fits quite loosely to the delivery tube by means of which the two test-tubes are connected.   The delivery tube must, of course, be so adjusted as to reach below the surface of the hydrochloric acid solution in the receiver before the distillation is begun.[6]

Add to the hydrolyzed blood filtrate a dry pebble, 2 cc. of saturated borax solution, and a drop or two of paraffin oil; insert firmly the rubber stopper carrying both delivery tube and receiver, and boil moderately fast over a microburner for 4 minutes.   The size of the flame should never be cut down during the distillation, nor should the boiling be so brisk that the emission of steam from the receiving tube begins before the end of 3 minutes.   At the end of 4 minutes slip off the receiver from the rubber stopper and put it in the position shown in Fig. 2.   Continue the distillation for 1 more minute and rinse off the lower outside part of the delivery tube with a little water.   Cool the distillate with running water, dilute to about 20 cc., and add 2.5 cc. of the Nessler solution described on page 90.   Fill to the 25 cc. mark and compare in the colorimeter with a standard containing 0.3 mg. of N in a 100 cc. flask and Nesslerized with 10 cc. of the Nessler solution. The standard and unknown should always be Nesslerized as nearly simultaneously as practicable.

Calculation.—Multiply 20 (the height of the standard in mm.) by 15 and divide by the colorimetric reading to get the urea nitrogen per 100 cc. of blood.   The reasons for this calculation are, of course, to be found in the fact that the standard containing 0.3

[6] The distillation apparatus can be obtained from Knott Apparatus Co., Boston.

96   Blood Analysis

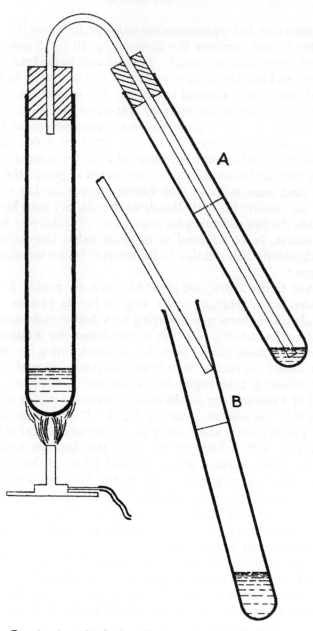

FIG. 2.   A, at beginning; B, toward end of distillation.

O. Folin and H. Wu   97

mg. of N is diluted to 100 cc., while the unknown, which corresponds to 0.5 cc. of blood, is diluted to only 25 cc.

It is even more important in this distillation than in the non-protein nitrogen digestion that the Pyrex test-tube should not be in a condition that leads to bumping. Dry the tube, or rinse it with alcohol, after each determination.

Borax, the alkali used in this distillation, is strong enough to set free the ammonia, yet is so weak that the blank ammonia which it gives with 5 cc. of urease solution is scarcely any greater than that obtained by the aeration process.

It will be noted that no condenser is used in connection with the microdistillation described above. Since ammonia can be quantitatively recovered by means of an air current, it would seem that it should be recovered as easily by means of a current of steam, especially since the first part of the distillate, containing probably 90 per cent of the ammonia, is automatically condensed just as in ordinary macro-Kjeldahl distillations. A few experiments made along this line indicate that condensers are indeed superfluous even in macro-Kjeldahl distillations.

The other three modifications for the determination of urea in the blood filtrates can be referred to very briefly, for they will be used only by those who are already familiar with the principles and practices involved.

Urea Determination by Means of Urease and Aeration.—The decomposition of the urea is made in the same kind of a Pyrex test-tube and in the manner already described. 1 or 2 cc. of 10 per cent sodium hydroxide are added and the ammonia is aspirated into a test-tube graduated at 25 cc. and containing 2 cc. of 0.05 normal hydrochloric acid. The only precaution which experienced investigators are likely to overlook is that the rubber tubing used for connections needs to be rinsed with water before being used the first time, and, later also, if the tubing has been idle for any length of time. The talcum powder with which the inner and outer surface of rubber tubing is coated is probably the source of the trouble in the case of new rubber tubing. It is probably contaminated with ammonia.

Urea Determination by Means of Autoclave Decomposition.—To 5 cc. of blood filtrate in a 75 cc. test-tube is added 1 cc. of normal acid; the mouth of the test-tube is covered with tin-foil, and the test-tube with contents is then heated in the autoclave at 150°C. for 10 minutes.

Allow the autoclave to cool to below 100°C. before opening. The ammonia is then distilled off exactly as in the first process described except that 2 cc. of 10 per cent sodium carbonate are substituted for the borax or it is removed by aeration in the usual manner. The autoclave process is of course only an adaptation of the process first recommended and then abandoned by Benedict for the determination of urea in urine. We are not prepared to say that in terms of per cent the results may not be as much too high in our blood filtrates as they were found to be in urine. An error of several per cent is, however, not at all important in the determination of the urea in blood. Whether one finds 15 instead of 14 mg. of urea nitrogen in human blood, or whether one obtains 3 instead of 2 mg. in 100 cc. of chicken blood is as yet of comparatively small consequence.

Determination of Creatinine and Creatine.

In this section we shall describe fairly obvious applications to our blood filtrate of Folin's colorimetric method for the determination of creatinine and creatine without thereby implying that the results so obtained are more accurate than the results which can be obtained by various other modifications which have been proposed during the past 2 or 3 years. The original methods as applied to blood were devised for the purpose of studying the absorption of creatinine and creatine and were adequate for that problem. Subsequent experience in many laboratories has shown that the method for the creatine gives results that are too high. The false step introduced in connection with the creatine determinations was undoubtedly the employment of picric acid as a protein precipitant, although at the time this seemed a peculiarly suitable process for securing the creatinine in concentrations then deemed necessary for reliable color comparisons. The process could perhaps be saved if it were worth while, for the cause of the high results is probably the formation of traces of hydrogen sulfide during the heating in the autoclave. The method is, however, now superfluous.

Determination of Preformed Creatinine.—Transfer 25 (or 50) cc. of a saturated solution of purified picric acid to a small, clean flask, add 5 (or 10) cc. of 10 per cent sodium hydroxide, and mix. Transfer 10 cc. of blood filtrate to a small flask or to a test-tube, transfer 5 cc. of the standard creatinine solution described below to another flask, and dilute the standard to 20 cc. Then add 5 cc. of the freshly prepared alkaline picrate solution to the blood filtrate, and 10 cc. to the diluted creatinine solution. Let stand for 8 to 10 minutes and make the color comparison in the usual manner, never omitting first to ascertain that the two fields of the colorimeter are equal when both cups contain the standard creatinine picrate solution. The color comparison should be completed within 15 minutes from the time the alkaline picrate was added; it is therefore never advisable to work with more than three to five blood filtrates at a time.

When the amount of blood filtrate available for the creatinine determination is too small to permit repetition, it is of course advantageous or necessary to start with more than one standard. If a high creatinine should be encountered unexpectedly without several standards ready, the determination can be saved by diluting the unknown with an appropriate amount of the alkaline picrate solution—using for such dilution a picrate solution first diluted with two volumes of water—so as to preserve equality between the standard and the unknown in relation to the concentration of picric acid and sodium hydroxide.

One standard creatinine solution, suitable both for creatinine and for creatine determinations in blood, can be made as follows: Transfer to a liter flask 6 cc. of the standard creatinine solution used for urine analysis (which contains 6 mg. of creatinine); add 10 cc. of normal hydrochloric acid, dilute to the mark with water, and mix. Transfer to a bottle and add four or five drops of toluene or xylene. 5 cc. of this solution contain 0.03 mg. of creatinine, and this amount plus 15 cc. of water represents the standard needed for the vast majority of human bloods, for it covers the range of 1 to 2 mg. per 100 cc. In the case of unusual bloods representing retention of creatinine, take 10 cc. of the standard plus 10 cc. of water, which covers the range of 2 to 4 mg. of creatinine per 100 cc. of blood; or 15 cc. of the standard plus 5 cc. of water by which 4 to 6 mg. can be estimated. By taking the full 20 cc. volume from the standard solution at least 8 mg. can be estimated; but when working with such blood it is well to consider whether it may not be more advantageous to substitute 5 cc. of blood filtrate plus 5 cc. of water for the usual 10 cc. of blood filtrate.

Calculation.—The reading of the standard in mm. (usually 20) multiplied by 1.5, 3, 4.5, or 6 (according to how much of the standard solution was taken), and divided by the reading of the unknown, in mm., gives the amount of creatinine, in mg. per 100 cc. of blood. In connection with this calculation it is to be noted that the standard is made up to twice the volume of the unknown, so that each 5 cc. of the standard creatinine solution, while containing 0.03 mg., corresponds to 0.015 mg. in the blood filtrate.

Determination of Creatine plus Creatinine.—Transfer 5 cc. of blood filtrate to a test-tube graduated at 25 cc. These test-tubes are also used for urea and for sugar determinations. Add 1 cc of normal hydrochloric acid. Cover the mouth of the test-tube with tin-foil and heat in the autoclave to 130°C. for 20 minutes or, as for the urea hydrolysis, to 155°C. for 10 minutes. Cool. Add 5 cc. of the alkaline picrate solution and let stand for 8 to 10 minutes, then dilute to 25 cc. The standard solution required is 20 cc. of creatinine solution in a 50 cc. volumetric flask. Add 2 cc. of normal acid and 10 cc. of the alkaline picrate solution and after 10 minutes standing dilute to 50 cc. The preparation of the standard must of course have been made first so that it is ready for use when the unknown is ready for the color comparison. The height of the standard, usually 20 mm., divided by the reading of the unknown and multiplied by 6 gives the "total creatinine" in mg. per 100 cc. blood.

In the case of uremic bloods containing large amounts of creatinine 1, 2, or 3 cc. of blood filtrate, plus water enough to make approximately 5 cc., are substitutes for 5 cc. of the undiluted filtrate.

The normal value for "total creatinine" given by this method is about 6 mg. per 100 cc. of blood.

Determination of Uric Acid.

The colorimetric method for the determination of uric acid in blood, like the colorimetric method for the determination of creatinine in urine, has furnished a tangible starting point for much important research. With the introduction by Benedict of potassium cyanide (or, as we prefer, sodium cyanide) for dissolving the silver urate, the uric acid method was materially simplified, and a new impetus was given to a widespread use of the process in researches of various kinds. For about 3 years doubts have been strong in this laboratory as to whether the method is really as reliable as it was at first believed to be, and in this laboratory at least we decided not to make further applications of it in research until these doubts could be removed. Our misgivings have proved in part unfounded and in part correct. Our fear that relatively large traces of uric acid are carried down with the blood proteins, during the coagulation process, have proved substantially groundless. On the other hand, it is certainly true that the precipitation of the uric acid from the concentrated blood filtrates by means of magnesia mixture and silver lactate (*i.e.*, essentially by Salkowski's process) is not quantitative, and the solubility of the silver urate is so large as to involve serious errors. On precipitating 0.1 mg. of uric acid from 10 cc. of solution an average loss of 50 per cent is encountered. By taking sufficiently large quantities of blood, 25 cc., the error due to the solubility of silver urate is of course largely eliminated, but the practical usefulness of the process is thereby much diminished. Another variable and uncontrollable source of error in the method is encountered during the concentration of the blood filtrates. If the total amount of water (acidified with acetic acid) to be boiled off does not exceed 100 cc. there is usually no destruction of uric acid, but when the volume is 200 to 400 cc. the losses, though variable, frequently amount to from 10 to 20 per cent, when starting with 0.1 or 0.2 mg. of uric acid. This source of error also can be eliminated almost wholly by taking aliquot portions of the blood filtrate (instead of the whole plus wash water). As a control method we have found the following process useful.

Heat about 160 cc. of water to boiling in a previously weighed beaker (capacity 500 cc.). Add 2 cc. of normal acetic acid, and add with pipettes 40 cc. of blood. Heat with constant stirring until the mixture is again boiling and continue the boiling for 2 minutes. Transfer beaker and contents to the scales, and add water until the total weight of the contents amounts to 200 gm. Mix and filter immediately. Transfer 100 cc. of the water-clear filtrate to an evaporating dish, add 1 cc. of 25 per cent acetic acid, and boil down as rapidly as possible to a volume of about 5 cc. Transfer the residue to a 15 cc. centrifuge tube, rinsing with 1 to 2 cc. of 0.1 per cent lithium carbonate solution. Cool. Add 2 cc. of Benedict's ammoniacal silver magnesia mixture, stir for 2 minutes, and centrifuge. Decant as completely as possible the supernatant liquid. To the residue in the tube add 2 cc. of 5 per cent sodium cyanide solution; stir, add 10 to 13 cc. of water, stir, and centrifuge again. Transfer the supernatant liquid to a 100 cc. volumetric flask, and make the color comparison in the usual manner.

As a control method the process outlined above is good, and if we were dependent on the method for regular use we should cut it down and introduce corrections for the solubility of silver urate.

Before describing the determination of uric acid in tungstic acid blood filtrates we wish to describe the preparation of a new standard solution of uric acid—a solution the keeping quality of which we now, after 18 months of constant use, consider much superior to any other as yet devised. The solvent is 10 per cent sodium sulfite, and the keeping quality of the solution depends on the fact that the sulfite keeps the solution free from dissolved oxygen. The solution is prepared as follows:

Make 1 to 3 liters of a 20 per cent solution of sodium sulfite, let stand over night, and filter. Dissolve 1 gm. of uric acid in 125 to 150 cc. of 0.4 per cent lithium carbonate solution and dilute to a volume of 500 cc. Transfer 50 cc., corresponding to 100 mg. of uric acid, to each of a series of volumetric liter flasks. Add 200 to 300 cc. of water, then 500 cc. of filtered 20 per cent sodium sulfite solution, and finally make up to volume, and mix well. Fill a series of 200 cc. bottles, and stopper very tightly with rubber stoppers. The solution in a bottle which is opened daily will keep for at least 3 to 4 months. Our records kept for one larger bottle so used show that no measurable loss of uric acid had occurred at the end of 6 months. In unopened bottles we expect the uric acid to keep for many years.

The surplus 20 per cent sulfite solution should be diluted to concentration of 10 per cent and should then be transferred to another series of small, tightly stoppered bottles. This sulfite is added to the unknown in order to offset the sulfite content of the standard.

O. Folin and H. Wu   103

Solutions Required for Uric Acid Determinations.

1. The standard uric acid sulfite solution already described (3 cc. used for each series of determinations).

2. A 10 per cent sodium sulfite solution, also described (2 cc. used for each determination).

3. A 5 per cent sodium cyanide solution, to be added from a burette (2.5 to 5 cc. used for each series of determinations).

4. A 10 per cent solution of sodium chloride in 0.1 normal hydrochloric acid (10 to 20 cc. used for each series of determinations).

5. The uric acid reagent prepared according to Folin and Denis. A still stronger reagent is obtained by heating the sodium tungstate (100 gm.) and the phosphoric acid (80 cc.) plus water (700 cc.) for 24 hours, instead of 2 hours; but the advantage gained, about 20 per cent, is not needed. Dilute the solution to 1 liter.

6. A solution of 5 per cent silver lactate in 5 per cent lactic acid (4 to 5 cc. needed for each determination).

In our new method for the determination of uric acid the latter is precipitated directly from the filtrate, without any previous concentration. 20 cc. of filtrate corresponding to 2 cc. of blood are used. In describing the process we shall have to introduce a slight variation from the way we actually do it. This variation is due to the fact that we use a larger centrifuge than most laboratories possess and by means of which we are able to use 30 cc. test-tubes for the precipitation. Using the small 15 cc. centrifuge tubes, it is necessary either to precipitate 10 cc. of filtrate in each of two tubes or to make the precipitation in two 10 cc. installments.

To 10 cc. of blood filtrate in each of two centrifuge tubes add 2 cc. of a 5 per cent solution of silver lactate in 5 per cent lactic acid, and stir with a very fine glass rod. Centrifuge; add a drop of silver lactate to the supernatant solution, which should be almost perfectly clear and should not become turbid when the last drop of silver solution is added. Remove the supernatant liquid by decantation as completely as possible. Add to each tube 1 cc. of a solution of 10 per cent sodium chloride in 0.1 normal hydrochloric acid and stir thoroughly with the glass rod. Then add 5 to 6 cc. of water, stir again, and centrifuge once more. By this chloride treatment the uric acid is set free from the precipitate. Transfer the two supernatant liquids by decantation to a 25 cc. volumetric flask. Add 1 cc. of a 10 per cent solution of sodium sulfite, 0.5 cc. of a 5 per cent solution of sodium cyanide, and 3 cc. of a 20 per cent solution of sodium carbonate. Prepare simultaneously two standard uric acid solutions as follows:

Transfer to one 50 cc. volumetric flask 1 cc. and to another 50 cc. flask 2 cc. of the standard uric acid sulfite solution described above. To the first flask add also 1 cc. of 10 per cent sodium sulfite solution. Then add to each flask 4 cc. of the acidified sodium chloride solution, 1 cc. of the sodium cyanide solution, and 6 cc. of the sodium carbonate solution. Dilute with water to about 45 cc. When the two standard solutions and the unknown have been prepared as described they are ready for the addition of the uric acid reagent of Folin and Denis. Add 0.5 cc. of this reagent to the unknown and 1 cc. to each of the standards, and mix. Let stand for 10 minutes, fill to the mark with water, mix, and make the color comparison.

Calculation.—In connection with the calculation it is to be noted (*a*) that the blood filtrate taken corresponds to 2 cc. of blood, (*b*) that the standard is diluted to twice the volume of the unknown, and (*c*) that the standard used contains 0.1 or 0.2 mg. of uric acid. The blood filtrate from blood containing 2.5 mg. of uric acid will be just equal in color to the weaker standard. 20 times 2.5 divided by the reading of the unknown gives, therefore, the uric acid content of the blood when the weaker standard is set at 20 mm.

The two standards recommended were adopted on the basis of the experience gained from the analysis of more than 150 different samples of human blood. About one-third of these bloods was from soldiers and most of the others were obtained from the State Wassermann Laboratory through the courtesy of Dr. Hinton. The bloods unfortunately do not cover the wider range occurring among hospital patients. A moderate number of blood samples have been obtained from the Massachusetts General Hospital, and these reveal that the uric acid may sink to as low as 1 mg. of uric acid per 100 cc. of blood. It seems hardly worth while to prepare a third and weaker standard regularly in order to provide for such low uric acid values. A standard corresponding to the color obtained from 1.25 mg. of uric acid per 100 cc. of blood can be prepared within a couple of minutes as follows:

O. Folin and H. Wu 105

Transfer 1 cc. of 10 per cent sulfite solution, 3 cc. of 20 per cent sodium carbonate, 2 cc. of the acidified sodium chloride, 0.5 cc. of the sodium cyanide solution, and 25 cc. of the weaker one of the two regular standard solutions already on hand. Dilute to 50 cc. and mix. Or, simply add 5 cc. of 20 per cent sodium carbonate to 25 cc. of the regular weaker standard, and dilute to 50 cc.

If a low uric acid value is expected, an alternate procedure is to dilute the unknown to a final volume of 10 cc. with corresponding reduction in the amount of the reagents used.

Special attention should perhaps be called to one small yet essential variation in the process for developing the blue uric acid

TABLE III.

Comparison of Old and New Methods For Determination of Blood Uric Acid.

Source.	Mg. per 100 cc. blood.		
	New method.	Old method.*	
		Without solubility correction.	With solubility correction.
Human..........................	2.6	2.2	2.4
"	3.5	3.4	3.6
Chicken........................	2.8	2.7	2.9
"	3.4	3.3	3.5
"	2.5	2.3	2.4
"	3.8	3.5	3.7

* Slightly modified.

color, a variation made necessary by the use of sodium sulfite. The uric acid reagent must invariably be added after, and not before, the addition of the sodium carbonate, because in acid solution the sulfite will itself give a blue color with the phosphotungstic acid.

It may also be worth while to mention that the peculiar increase in blue color obtained by the use of cyanide is not obtained in the presence of sulfite. Opinions will doubtless differ as to whether this is an advantage or disadvantage. The amount of color obtained from 2 cc. of blood is rather weak, and if we could conveniently have retained the intensifying effects of the cyanide we probably should have done so, though the fainter solutions can be read just as readily and accurately as the stronger ones obtained by means of the cyanide. The antifading effects of the cyanide are retained.

New Method for Determination of Sugar.

It was originally our intention to incorporate some adaptation of Benedict's picrate method for the determination of sugar to our tungstic acid blood filtrates. But a few exploratory experiments showed that an intense and stable color reaction can be obtained by the application of the phenol reagent of Folin and Denis to cuprous oxide. The color obtained from a given quantity of sugar is far more intense than that obtained by the alkaline picrate reaction; so that a small fraction of a mg. of dextrose (1 or 2 cc. of blood filtrate) is all that is required for a determination of the blood sugar. Some difficulties were encountered in trying to find the conditions under which the extent of reduction is strictly proportionate to the quantities of sugar used; but, by a systematic study of the various factors involved, these difficulties were overcome and a rapid and convenient process was obtained.

The copper solution used for reduction is a weakly alkaline copper tartrate solution. Qualitatively this solution is an extremely sensitive reagent for traces of sugar, yet is not affected by creatinine or uric acid in quantities corresponding to 50 mg. of each per 100 cc. of blood. We are therefore inclined to regard our method as more accurate than any method as yet proposed for the determination of sugar in blood.

The picrate methods,[7] whether we use Benedict's last modification or Myers' modification of Benedict's original method, in our hands give almost invariably results that are materially higher than the figures given by our new method. We are under the impression that the picrate methods are subject to sources of error similar to those encountered in Folin's original picrate method for blood creatine. The development of color in blood filtrates seems

[7] Lewis, R. C., and Benedict, S. R., A Method for the estimation of sugar in small quantities of blood, *J. Biol. Chem.*, 1915, xx, 61. Myers, V. C., and Bailey, C. V., The Lewis and Benedict method for the estimation of blood sugar, with some observations obtained in disease, *ibid.*, 1916, xxiv, 147. Benedict, S. R., A modification of the Lewis-Benedict method for the determination of sugar in the blood, *ibid.*, 1918, xxxiv, 203.

not to proceed at the same rate of speed as the color derived from a corresponding amount of dextrose. If the heating is interrupted at the end of 2 to 3 minutes the value obtained for the blood sugar will be nearly 50 per cent higher than when the heating is continued for 10 minutes or more. Such quantitative variations are not encountered in our process when equal amounts of dextrose in the form of pure sugar and of blood filtrate are heated, except that the reduced copper is, of course, more extensively precipitated and visible in the pure sugar solution. It need scarcely be stated that added sugar is quantitatively recovered by our method.

Solutions Needed for Determination of Sugar in Blood.

1. *Standard Sugar Solution.*—Dissol e 1 gm. of pure anhydrous dextrose in water and dilute to a volume of 100 cc. Mix, add a few drops of xylene or toluene, and bottle. If pure dextrose is not available, a standard solution of invert sugar made from cane sugar is equally useful. Transfer exactly 1 gm. of cane sugar to a 100 cc. volumetric flask; add 20 cc. of normal hydrochloric acid and let the mixture stand over night at room temperature (or rotate the flask and contents continuously for 10 minutes in a water bath kept at 70°C.). Add 1.68 gm. of sodium bicarbonate and about 0.2 gm. of sodium acetate, to neutralize the hydrochloric acid. Shake a few minutes to remove most of the carbonic acid and fill to the 100 cc. mark with water. Then add 5 cc. more of water (1 gm. of cane sugar yields 1.05 gm. of invert sugar) and mix. Transfer to a bottle; add a few drops of xylene or toluene, shake well, and stopper tightly. The stock solution made in either way keeps indefinitely. Dilute 5 cc. to 500 cc., giving a solution 10 cc. of which contain 1 mg. of dextrose or invert sugar. Add some xylene. Use 2 cc. for each determination.

2. *Alkaline Copper Solution.*—Dissolve 40 gm. of anhydrous sodium carbonate in about 400 cc. of water and transfer to a liter flask. Add 7.5 gm. of tartaric acid and when the latter has dissolved add 4.5 gm. of crystallized copper sulfate; mix, and make up to a volume of 1 liter. If the carbonate used is impure, a sediment may be formed in the course of a week or so. If this happens, decant the clear solution into another bottle.

3. *Phosphotungstic-phosphomolybdic Acid.*—Transfer to a large flask 25 gm. of molybdenum trioxide (MoO_3) or 34 gm. of ammonium molybdate $(NH_4)_2(MoO_4)$; add 140 cc. of 10 per cent sodium hydroxide and about 150 cc. of water. Boil for 20 minutes to drive off the ammonia (molybdic acid sometimes contains large amounts of ammonia as impurity). Add to the solution 100 gm. of sodium tungstate, 50 cc. of 85 per cent phosphoric acid, and 100 cc. of concentrated hydrochloric acid. Dilute to a volume of 700 to 800 cc.; close the mouth of the flask with a funnel and watch-glass. Boil gently for not less than 4 hours, adding hot water from time to time to replace that lost during the boiling. Cool and dilute to 1 liter. This solution is identical with the phenol reagent of Folin and Denis. For use in connection with the determination of blood sugar dilute 1 volume (100 cc.) of the reagent with one-half volume (50 cc.) of water and one-half volume (50 cc.) of concentrated hydrochloric acid.

4. *Saturated Sodium Carbonate Solution.*

The determination of blood sugar is carried out as follows: Heat a beaker of water to vigorous boiling. Transfer 2 cc. of the tungstic acid blood filtrate to a test-tube (20 m. $\times$ 200 mm.) graduated at 25 cc. The graduated test-tubes used as receivers when distilling off the ammonia in urea determinations (p. 95) are suitable for this work. Transfer 2 cc. of the dilute standard sugar solution to another similar test-tube. Add to each tube 2 cc. of the alkaline copper tartrate solution. Heat in the boiling water for 6 minutes. Remove the test-tube and add at once (without cooling), preferably from a graduated pipette, 1 cc. of the strongly acidified and diluted phenol reagent. This should be done as nearly simultaneously as possible; it is not advisable to use one standard for a set of more than four determinations. The purpose of the added hydrochloric acid in the reagent is to dissolve the cuprous oxide. Mix, cool, and add 5 cc. of saturated sodium carbonate solution. An intense blue color is gradually developed which will remain unaltered for several days. Dilute the contents of both test-tubes to the 25 cc. mark, and after at least 5 minutes make the color comparison in the usual manner.

The depth of the standard (in mm.) multiplied by 100 and divided by the reading of the unknown gives the sugar content, in mg., per 100 cc. of blood.

O. Folin and H. Wu 109

TABLE IV.

Sample Analyses of Protein-Free Blood Filtrates Obtained by Means of Tungstic Acid.

No.	Mg. per 100 cc. blood.					
	Total N.	Urea N.	Uric acid.	Preformed creatinine.	Total creatinine.	Sugar.
1	26	10	1.3	1.5	6.0	89
2	26	13	1.0	1.4	5.3	100
3	28	12	1.1	1.2	6.7	98
4	28	12	2.2	2.0	5.7	83
5	29	13	3.3	1.5	6.0	86
6	29	11	2.6	1.4	5.2	95
7	29	13	1.6	1.4	6.0	85
8	30	13	2.4	1.6	5.5	82
9	30	14	4.1	1.7	5.3	82
10	32	15	2.8	1.6	5.4	91
11	32	15	3.4	1.4	5.3	97
12	32	13	2.4	1.7	6.0	104
13	33	17	2.0	1.3	4.8	83
14	33	16	2.5	1.6	5.7	105
15	33	15	1.1	1.6	5.5	95
16	34	16	0.8	1.3	6.1	119
17	34	16	2.6	1.5	5.9	106
18	35	17	2.1	1.6	6.0	89
19	35	17	2.0	1.4	5.5	77
20	35	18	2.0	1.7	5.7	86
21	35	18	2.9	1.6	5.8	95
22	35	17	3.2	1.4	5.5	94
23	35	18	2.5	1.5	6.0	89
24	35	19	2.2	1.5	5.3	91
25	35	22	3.5	1.4	5.7	87
26	35	17	2.3	1.7	6.7	83
27	35	18	1.6	1.3	6.5	104
28	36	17	2.8	1.5	5.2	100
29	37	18	2.1	1.5	5.5	94
30	38	18	2.2	1.7	5.4	95
31	39	18	2.6	1.8	6.7	103
32	39	18	2.9	1.5	6.0	87
33	40	18	2.0	1.6	6.0	98
34	40	20	2.6	1.7	5.6	95
35	41	19	4.8	1.5	5.9	93
36	41	19	4.2	2.5	6.6	109
37	43	19	2.2	1.7	6.3	78
38	139	106	5.4	12.5	19.4	99
39	147	115	8.9	11.0	20.5	170
40	275	237	14.3	13.6	27.2	157

110   Blood Analysis

The copper solution is adjusted to give proportionate reductions with 0.12 to 0.4 mg. of dextrose.   This covers the range of hypoglycemic and hyperglycemic bloods.   But in extreme cases it is better to use 3 or 1 cc. of the filtrate, instead of 2 cc., adding water to the standard or to the unknown so as to equalize the concentration of the alkaline copper.

COMMENTARY TO

15. Van Slyke, D. D. and Neill, J. M. (1924)
The Determination of Gases in Blood and Other Solutions by Vacuum Extraction and Manometric Measurement. I. Journal of Biological Chemistry 61(2): 523–573.

Metabolic acidosis in diabetics was a common occurrence in the pre-insulin era of the early 1900s. Laboratory tests in these patients consisted mostly of a blood sugar and a spot test for ketones. The Scott-Wilson bedside test for ketones for example had the patient breath through a straw into a test tube that contained an alkaline solution of mercuric cyanide and silver nitrate. The presence of elevated ketones in the patient's breath turned the reagent cloudy. If the patient was comatose a glass rod was dipped into the reagent and held in front of the patient's mouth (1,2). Donald D. Van Slyke at the Rockefeller Institute for Medical Research in New York City reasoned that a more accurate assessment of acid–base balance could be made by measurement of the dissolved carbon dioxide in plasma from which the bicarbonate (HCO_3^-) level could be calculated. The problem was, there were no methods available for the measurement of carbon dioxide in plasma. Van Slyke invented an instrument to do this. His volumetric gas apparatus was the first instrument designed specifically for the clinical chemistry laboratory (3).

Van Slyke described his first volumetric blood gas apparatus in 1917 (4). Dissolved gas was released from 1.0 mL of a sample with acid in a closed glass chamber. The volume of gas was then measured in a graduated arm of the chamber. The Van Slyke and Neill manometric apparatus described in the paper presented here was published in 1924. The volumetric apparatus was modified in order to measure the pressure of the gas released from the serum. This improved the accuracy of the instrument. A mercury manometer was incorporated into the gas extraction chamber. Samples of serum could be tested for carbon dioxide, oxygen, and carbon monoxide. The paper by Van Slyke and Neill is 50 pages long and as presented here has been edited to include only the portions that describe the apparatus and its use for carbon dioxide. In 1927 a portable modification of the manometric apparatus was described (5). The mercury manometer and gas extraction bulb were attached to a rigid wooden frame. A motor was incorporated in order to facilitate the required shaking of the chamber during gas extraction and a light was placed behind the graduated arm of the manometer to facilitate readings. The unit measured at its base 28 cm (11 in.) × 50 cm (20 in.) and stood 100 cm (40 in.) tall. By 1927 and for the next 30 years many different laboratory supply houses manufactured and sold this instrument including Fisher Scientific (5,6).

The Van Slyke–Neill apparatus became the standard in the field of clinical chemistry for blood gas measurements. In addition it could be used to measure any analyte that could be converted into a gas. Peters and Van Slyke's textbook of methods in 1932 included manometric gas procedures for the blood gases and hemoglobin, sugars, urea, ammonia, nitrogen, calcium, potassium and lactic acid (7). Martin Hanke published a manual in 1939 on the use of the apparatus in clinical chemistry (8). Van Slyke and Plazin, the technician who helped design and build the apparatus, published their own procedure manual in 1961 (9). The apparatus was used for many years to measure enzyme activity levels in which a gas was consumed or released, for example for catalase (EC 1.11.1.6) (10). Brix in 1981 modified the Van Slyke–Neill apparatus for use with 25 to 100 µL of serum or plasma (11).

Van Slyke's discoveries in acid–base balance made with his manometric blood gas apparatus were described in the Peters and van Slyke textbook, *Quantitative Clinical Chemistry*, *Volume I Interpretations* published in 1931 (12). Peters died in 1955, and a planed complete revision of the original text became impossible. Williams & Wilkins reprinted Chapters 12 and 18 on acid–base and blood gases from the original Volume I text as a separate book. In the Publisher's Preface to this reprint it was stated that, "…the present volume may relieve the wear and tear on original copies now held together by string and tape and kept under lock and key" (13).

References

(1) Scott-Wilson, H. (1911) A method for estimating acetone in animal liquids. Journal of Physiology. 42(5–6):444–470.
(2) Abrahamson, E.M. (1940) *Office Clinical Chemistry. A Laboratory Guide for the Practitioner and Hospital.* Oxford University Press, London, pgs 229–230.
(3) Rosenfeld, L. (1999) *Four Centuries of Clinical Chemistry.* Gordon and Breach Science Publishers, Amsterdam, pgs 169–172.
(4) Van Slyke, D.D. (1917) Studies of acidosis II. A method for the determination of carbon dioxide and carbonate in solution. Journal of Biological Chemistry. 30(2):347–368.

(5) Van Slyke, D.D. (1927) Note on a portable form of the monometric gas apparatus, and on certain points in the technique for its use. Journal of Biological Chemistry. 73(1):121–126.

(6) [Advertisement] (1954) Journal of Clinical Investigation. 33(2), February.

(7) Peters, J.P. and Van Slyke, D.D. (1932) *Quantitative Clinical Chemistry Vol II Methods*. The Williams & Wilkins Company, Baltimore, MD.

(8) Hanke, M.E. (1939) *Manometric Methods. Detail of Procedure for the Manometric Gas Apparatus of Van Slyke and Neill*. The University of Chicago, Chicago, IL.

(9) Van Slyke, D.D. and Plazin, J. (1961) *Micromanometric Analyses*. The Williams & Wilkins Company, Baltimore, MD.

(10) Kirk, J.E. (1963) A rapid procedure for catalase determination in blood and tissue samples with the Van Slyke manometric apparatus. Clinical Chemistry. 9(6):763–775.

(11) Brix, O. (1981) A modified Van Slyke apparatus. Journal of Applied Physiology. 50(5):1093–1097.

(12) Peters, J.P. and Van Slyke, D.D. (1931) *Quantitative Clinical Chemistry Vol I Interpretations*. The Williams & Wilkins Company, Baltimore, MD.

(13) Peters, J.P. and Van Slyke, D.D. (1963) *Hemoglobin and Oxygen; Carbonic Acid and Acid–Base Balance*. The Williams & Wilkins Company, Baltimore, MD.

J. Biol. Chem. 1924; 61(2): 523–573

THE DETERMINATION OF GASES IN BLOOD AND OTHER SOLUTIONS BY VACUUM EXTRACTION AND MANOMETRIC MEASUREMENT. I.*

By DONALD D. VAN SLYKE AND JAMES M. NEILL.

(*From the Hospital of The Rockefeller Institute for Medical Research.*)

(Received for publication, July 3, 1924.)

CONTENTS.

I. Principles of Construction and Use of the Manometric Apparatus.

The method reported in the present paper resembles that previously described by Van Slyke (1917) and by Van Slyke and Stadie (1921) in that the gases are extracted from solution by shaking the latter in a relatively large free space in a Torricellian vacuum, and in that for subsequent measurement the volume of the extracted gas is reduced and the pressure increased. In the former

*A preliminary note on the apparatus described in this paper was published in the Proceedings of the National Academy of Science (Van Slyke. D. D., *Proc. Nat. Acad. Sc.*, 1921, vii, 229).

524   Blood Gases. I

"volumetric" method, however, the pressure was brought always to atmospheric, and the gas volume was read on a scale; while in the present "manometric" method the volume is brought to an arbitrarily chosen size, and the amount of gas is determined from the pressure exerted on a manometer.

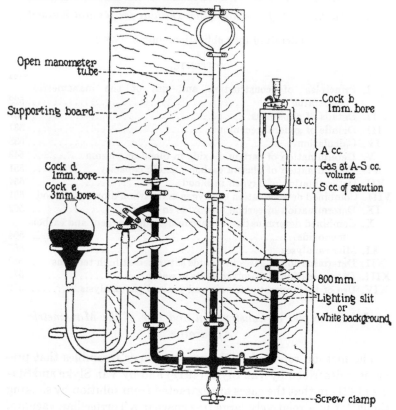

FIG. 1. Apparatus, with solution in position for extraction of gases. Manometer is of open type.

Under the conditions of the former method, when all measurements were made at atmospheric pressure, the error in volume reading might be 10 or even 100 times that of the pressure measurement.   The advantage of the present method is that it permits the analyst to choose the magnitude of both the volume and the pressure which he measures.   He may accordingly fix them within such limits that the errors in measuring both are of the same order

D. D. Van Slyke and J. M. Neill   525

of magnitude. Thereby the percentage error in gas measurement becomes the minimum attainable with given absolute errors in pressure and volume.

The apparatus consists of a short pipette with the upper stem closed by a stop-cock, the lower connected with a glass tube. The

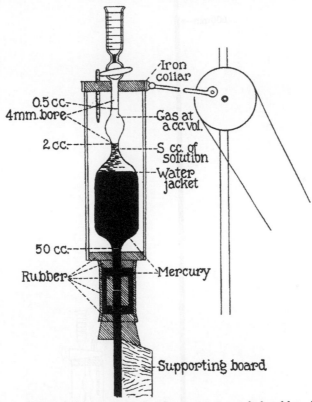

FIG. 2. Extracting chamber, showing mercury-sealed rubber joint at bottom and attachment of Stadie shaker at top. Gas is at 2 cc. volume for pressure reading.

latter descends 800 mm., then turns at a right angle to connect with a leveling bulb and a mercury manometer, which may be either open at the upper end (Fig. 1) or closed (Fig. 3). The pipette is calibrated at two points to hold a cc. of gas for pressure measurement and A cc. of total volume, respectively, as shown in Figs. 1 and 2.

526 Blood Gases. I

For analysis the sample of blood or other solution is introduced into the chamber over mercury, together with the reagents to free the desired gases from combination. A Torricellian vacuum is

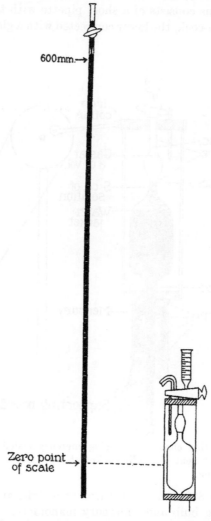

FIG. 3. Apparatus with closed manometer.

obtained, as in the previous "volumetric" apparatus, by lowering the leveling bulb, and the gases are extracted from solution by 2 or 3 minutes shaking. The gas volume is then reduced

D. D. Van Slyke and J. M. Neill   527

to a cc. by admission of mercury through cock e, and the reading p_1 is made on the manometer. The gases are either ejected or are absorbed by proper reagents, and the reading p_2 is taken, with the same gas volume. The partial pressure P of the gas at a cc. volume is then $P = p_1 - p_2$ mm. of mercury, from which the gas volume at 0°, 760 mm. may be calculated. The calculation is simpler than with the constant pressure apparatus, because the barometric pressure is not a factor. No corrections are required for vapor tension or for capillary attraction in the manometer tube, since these factors are the same at both readings. It is assumed that the temperature in the apparatus is the same when the p_1 and p_2 readings are taken. The time interval between the two readings is so brief (1 to 4 minutes) that sufficient constancy is attained (within 0.1°C.) by the insulation afforded by the water jacket. When the open manometer is used, constancy of barometric pressure over the same time interval is also assumed.

The total volume A of the chamber is a matter of convenience, but it is desirable to have it so large that the greater part of the dissolved gases shall be extracted. For analysis of 1 cc. of blood convenient magnitudes are $A = 50$, $a = 2$, $S = 3.5$ to 7.0 cc. (see Fig. 1). At 20° 0.01 cc. of gas under these conditions gives a reading of $P = 3.9$ mm. For minute amounts of gas an additional a mark at 0.5 cc. is desirable. At this volume 0.01 cc. of gas exerts about 16 mm. pressure.

For the most precise CO_2 determinations we have employed an analysis chamber in which all the dimensions are twice those given above, and in which as much as 3 cc. of blood could be analyzed for CO_2 with an error not exceeding 1 part in 500. For micro analyses of 0.2 cc. blood samples we have used both the 50 cc. chamber and a 10 cc. chamber, with results approximating 1 per cent accuracy. The different chambers can all be used on the same manometer.

II. Details of Apparatus.

The *extraction chamber* is simpler than that of the former "volumetric" apparatus; it is calibrated at only three points (*e.g.*, 0.5, 2, and 50 cc.), and has no cock at the bottom. The lack of necessity for a long narrow graduated tube for measurement of varying volumes of gas makes the chamber short and convenient to shake. The short length of space a also simplifies the problem of drainage.

The *mercury seal*, shown in Fig. 2, around the rubber joint at the bottom of the chamber was found necessary. Even thick pressure tubing, exposed to the air, permits the diffusion, during extraction *in vacuo*, of sufficient amounts of air to affect pressure readings as accurate as those that can be made on the present apparatus.

For details of the *mechanical shaker*, see Stadie (1921). The driving shaft should not exceed 8 or 9 cm. in length. Too long a shaft causes wobbling of the chamber sideways during the shaking.

Stop-cock d offers a convenient means of expelling the air which gradually diffuses through the rubber tubing connected with the leveling bulb and collects below this cock.

The screw clamp at the bottom of the manometer permits withdrawal of the mercury.

The *open manometer tube* is of 4 mm. inner diameter, and is expanded into a bulb at the top to prevent loss of mercury when the leveling bulb is raised high to expel solutions from the analyzing chamber. At the bottom it is contracted as shown in Fig. 1, to 1.5 mm. bore to minimize the tendency of the mercury to oscillate between the manometer and the analysis chamber when cock e is closed.

Alternative Closed Manometer Tube.—Instead of ending at the top with an open bulb as in Fig. 1, the manometer tube may be closed with a cock at the top, as shown in Fig. 3. The air is expelled before the manometer is used, so that there is no atmospheric pressure on the mercury surface in the tube. Consequently all the readings are about 800 mm. higher from the laboratory floor than in the open manometer. The closed manometer is of advantage from the standpoint of comfort, inasmuch as the operator does not need to bend down to take the zero readings.

The closed manometer tube is contracted to 1 to 1.5 mm. a little below the cock at the top, as shown in Fig. 3, to lessen the force with which the rising mercury strikes the cock.

D. D. Van Slyke and J. M. Neill   529

Slight amounts of moisture find their way into the closed manometer as the mercury flows back and forth from the analysis chamber. Some error from the vapor pressure of this moisture would be probable, as the manometer tube is not protected against sudden temperature changes. To absorb the water vapor, 2 or 3 drops of concentrated sulfuric acid are admitted through the cock at the top and permitted to flow down the tube for about 10 cm. Mercury is then forced up through the cock, leaving behind enough sulfuric acid to wet the upper end of the manometer, but not enough to flow down and interfere with readings of the mercury meniscus. The acid is renewed occasionally in this manner.

The closed manometer has the theoretical advantage over the open manometer in that the former is not affected by change in barometric pressure during the time interval between the p_1 and p_2 readings. In the 1 to 4 minute intervals, however, this is a negligible factor. The closed manometer is obviously not quite so simple as the open one. The latter has been used for most of the analyses thus far performed, but in comparisons of the two, identical results have been uniformly obtained.

For *the manometer scale* we at first stretched a steel meter tape divided into millimeters behind the tube, or sunk a brass scale into the board. The most satisfactory readings, however, have been made with the scale, about 600 mm. long, etched on the manometer tube itself. The marks are 1 millimeter apart, and are in the form of semicircles, passing around the tube from the middle of the front to the back. The centimeter and half-centimeter marks are complete rings. By sighting the meniscus across these marks error from parallax is avoided.

A good background is provided by a strip of white paper or of mirror glass behind the manometer. A still more satisfactory background is given by an illuminated strip of frosted glass, as shown in Fig. 1. From the board behind the manometer a section about 12 mm. wide is cut out. The rear of the aperture is covered by the strip of frosted glass, which is illuminated from behind by two electric lights, each in a tubular bulb of frosted glass 30 cm. long and 3 cm. wide. In order to avoid heating the manometer, the lights are turned on only at the moment a reading is made. The switch for them is conveniently attached to the same board with the apparatus. A practised observer, with the

530 Blood Gases. I

help of a reading glass, can approximate an accuracy of 0.1 mm. in his readings, either with the etched manometer tube or with a scale behind the tube, but the attainment of such accuracy is considerably easier with the etched tube.

It is advisable, before each reading, to tap the manometer tube with the finger. There may be a slight lag on the part of the mercury in adjusting itself to final pressure conditions, and a consequent error in the reading unless this precaution is taken.

Range of the Apparatus.—As stated before, for maximum accuracy the analyst chooses such sizes of a and of the blood or other sample analyzed, that the percentage errors in measuring the gas volume of a cc. and the resultant pressure of P millimeters are about equal. By so doing it is possible with the larger apparatus to approach a constancy of 2 *pro mille*. The precision of the manometer readings, however, enables one to determine a wide range of gas magnitudes with accuracy sufficient for most purposes (*e.g.* 1 per cent), in a single apparatus. In the 50 cc. apparatus, with a volumes of 0.5 and 2.0 cc., we have analyzed satisfactorily liquids containing from 0.2 to over 100 volumes per cent of gas (O_2 or CO_2), and have used blood samples varying from 0.2 to 2.0 cc.

Calibration of Apparatus.—The volume from the *bottom* of cock b down to the a mark is determined by weighing the water delivered by the technique described in the paper[1] by Van Slyke and Stadie (1921). Since in the present apparatus only two a volumes are used, calibration is simpler than in the case of the former "volumetric" apparatus with its graduated scale, and correction curves are unnecessary.

The volume between cock b and A mark is similarly determined. When A is 50 cc., an accuracy of 0.2 cc. is sufficient.

The cup at the top of the chamber is graduated in 0.5 cc. divisions *with the zero mark at the point where the bottom of the cup joins the capillary.* When solutions are measured into the cup, the capillary is filled with mercury.

Simplified Construction of Apparatus.—The apparatus may be constructed in any laboratory with details slightly less convenient, but not essentially less accurate than those above outlined. The manometer tube and the two tubes connecting with the manom-

[1] Van Slyke and Stadie (1921), p. 4.

D. D. Van Slyke and J. M. Neill   531

eter on the right and left, instead of being fused together, may be joined to a central + tube with very heavy pressure tubing. The rubber joints are reinforced against stretching under mercury pressure by wrapping them firmly with adhesive tape. Instead of the millimeter scale etched on the glass, a metal meter stick or metallic tape may be fastened to the board behind the manometer tube. The analysis chamber may be made from a sturdy 50 cc. pipette by fusing a cock on the stem. The cock, although conveniently 3-way, as in Fig. 1, may be 2-way, particularly if suction is available for drawing off used solutions. The Stadie shaking device needs to be well constructed, and requires the services of a mechanic. The values of A and a in a laboratory-made apparatus will probably be such that Tables II and III for calculation will not be applicable. A table for each apparatus will need to be calculated by Equations 6 and 7 with the values for $\frac{1}{1+0.00384t}$ and α' from Table I.

TABLE II.

Factors for Calculation of Volumes Per Cent of Blood Gases from Pressures in 50 Cc. Apparatus. *

Temperature.	Factors for CO_2.			Factors for O_2, CO, and N_2.				
	Sample =0.2 cc. S=2.0 " a=0.5 " i=1.03	Sample =1 cc. S=3.5 " a=2.0 " i=1.014	Sample =2 cc. S=7.0 " a=2.0 " i=1.014	Sample =0.2 cc. S=2.0 " a=0.5 " i=1.00	Sample=1 cc. S=3.5 "		Sample=2 cc. S=7 "	
					a=0.5 cc. i=1.00	a=2.0 cc. i=1.00	a=0.5 cc. i=1.00	a=2.0 cc. i=1.00
°C.								
15	0.335	0.2725	0.1483	0.312	0.0623	0.2493	0.0317	0.1251
16	33	11	70	10	21	85	15	46
17	31	0.2697	59	09	19	78	14	42
18	30	83	49	08	17	68	12	37
19	28	69	39	07	15	59	11	32
20	27	55	29	07	13	50	09	28
21	26	40	19	06	10	41	08	24
22	24	26	10	05	08	32	06	19
23	23	13	01	03	06	23	05	15
24	22	00	0.1391	02	04	14	03	10
25	20	0.2588	82	01	02	06	02	06
26	18	75	73	00	00	0.2398	01	02
27	17	62	64	0.299	0.0598	90	0.0299	0.1198
28	16	49	56	98	96	82	98	93
29	14	37	49	97	93	74	96	89
30	13	26	41	96	92	66	95	85
31	12	15	33	95	90	58	94	81
32	11	04	25	94	88	50	92	77
33	10	0.2493	18	93	86	42	91	73
34	08	82	10	92	83	33	90	69

*If calibration of an apparatus shows a value of a significantly different from the 0.500 or 2.000 cc. in the column heading, the factors in the column are corrected by multiplying them by $\frac{a}{0.500}$ or $\frac{a}{2.000}$.

V. Determination of CO_2 in Blood or Plasma.

Introduction of Sample and Reagents.—The apparatus having been cleaned with dilute lactic acid as previously described, a drop of octyl alcohol is drawn into the capillary above cock b, and 2.3 cc. of CO_2-free water are put into the cup for each cc. of

544   Blood Gases.   I

blood or plasma to be added.   Stop-cock *b* is closed, with *e* open
and the leveling bulb at the level shown in Fig. 1.   The blood
sample is delivered beneath the layer of water in the cup from a
pipette.   Preferably a pipette with cock and rubber tip is used,

TABLE III.

*Factors for Calculation of Millimols of Blood Gases per Liter from Pressures
in 50 Cc. Apparatus.*

Temperature.	Factors for CO₂.			Factors for O₂, CO, and N₂.				
	Sample =0.2 cc. S=2.0 " a=0.5 " i=1.03	Sample =1 cc. S=3.5 " a=2.0 " i=1.014	Sample =2 cc. S=7.0 " a=2.0 " i=1.014	Sample =0.2 cc. S=2.0 " a=0.5 " i=1.00	Sample=1 cc. S=3.5 " a=0.5 cc. i=1.00	Sample=1 cc. S=3.5 " a=2.0 cc. i=1.00	Sample=2 cc. S=7 " a=0.5 cc. i=1.00	Sample=2 cc. S=7 " a=2.0 cc. i=1.00
°C.								
15	0.1493	0.1216	0.0662	0.1388	0.02780	0.1113	0.01396	0.0558
16	85	11	56	84	70	09	90	56
17	80	04	51	80	61	05	85	54
18	73	0.1198	47	75	51	01	80	52
19	67	91	42	70	41	0.1097	75	50
20	61	85	38	65	31	93	70	48
21	54	78	34	60	21	89	65	46
22	46	72	29	55	11	85	60	44
23	39	66	25	50	02	81	55	42
24	33	61	21	45	0.02692	77	50	40
25	28	55	17	40	83	74	45	38
26	22	49	13	35	73	70	41	36
27	15	43	09	31	64	67	36	34
28	09	38	05	26	55	63	31	32
29	03	33	02	22	47	59	27	30
30	0.1398	28	0.0598	18	38	55	22	29
31	92	23	95	13	29	52	18	27
32	86	18	91	09	20	48	14	25
33	81	13	88	04	11	44	09	24
34	74	08	85	00	02	41	05	22

See foot-note to Table II.

as described under "Measurement of samples."   If an ordinary
pipette is used, the delivery is made with the pipette tip resting
on the bottom of the cup, and during the delivery cock *b* is partially opened, so that most of the blood flows directly through *b*

into the chamber below, leaving a minimum amount accumulated in the cup to be washed in.   After the delivery of the sample this residue of blood in the cup is run into the chamber below followed by the water layer, which washes the last traces of blood with it. Finally, 0.2 cc. of CO_2-free 1 N lactic acid per cc. of blood or plasma is added.   The lactic acid may be conveniently measured by counting the drops; the drop number per cc. having been established.   Stop-cock b is finally sealed with a drop of mercury.

Liberation of Carbon Dioxide.—The leveling bulb is lowered until the surface of the mercury (not of the water) has fallen to the A mark at the bottom of the chamber.   Stop-cock e is then closed. The reaction mixture is shaken for 3 minutes.

Measurement of Pressure of Total Gases.—Mercury is readmitted by stop-cock e with the precautions previously discussed for CO_2 determinations under "Adjustment of gas volume," until the gas volume in the analysis pipette is reduced to 2,0 cc.   Cock e is closed, the manometer is tapped with the finger, and the height of the mercury column is read (p_1 $mm.$).

Absorption of CO_2.—With blood, the variable amount of O_2 extracted with the CO_2 makes determination of the latter by absorption imperative.   With plasma, although the air accompanying the CO_2 is sufficiently constant to be allowed for by calculation with fair accuracy, it is our custom to use absorption to determine the CO_2 when precision is desired.

The manner in which the aborbent alkali solution is added is important.   The gas-free alkali solution (see "Gas-free reagents") must be run into the mixture of gases present without absorption of any gas other than CO_2.   It was at first our custom before admitting the 1 N alkali, to open cock e, so that, with the leveling bulb at the level shown in Fig. 1, the pressure in the chamber rose to nearly atmospheric.   This practice, however, permitted the absorption of enough oxygen by the gas-free alkali solution to cause, in whole blood analyses, an error of about + 0.2 volume per cent in the CO_2 determination.

To avoid this error, the 1 N NaOH solution is added under reduced pressure as follows.   After measuring the pressure of the total extracted gas mixture the pressure is diminished so that the mercury level in the chamber is lowered somewhat below the position shown in Fig. 2, and a space of several cc. is left between the

water meniscus and the lower a mark. Cock e is then closed.
2 cc. of air-free 1 N NaOH are measured into the cup, with the
delivering tip resting on the bottom of the cup so that the solution will not stream through and absorb air. 1 cc. of the alkali
is then permitted to flow gradually through cock b into the
chamber. Presumably because of the relatively uninterrupted
motion of the gas molecules at the reduced pressure, absorption
of CO_2 is extremely rapid.   If 30 seconds are taken for running in
the alkali we find absorption is complete.

That no measurable amounts of oxygen or air are absorbed
with this technique we have proved by repeated comparisons
of the manometer lowering (c values) caused by addition of 1.0
cc. of solution when the a space has been made free of all gases
save water vapor, and also when it contained amounts of CO_2-
free air greater than the amounts of oxygen obtained in blood
analyses. The c values obtained in both cases were identical,
indicating that no absorption of air occurred.

A simpler technique may be used for analyses of serum or water
solutions of carbonates. The CO_2 may be absorbed with 0.2
cc. of 5 N NaOH measured with a pipette into the cup of the
apparatus.   This solution has solubility coefficients for O_2 and N_2
only one-tenth those of water, and may be used without being
previously freed from air. It is added with the contents of the
chamber at but slight negative pressure (cock e open, leveling
bulb as in Fig. 1).   When the CO_2 is absorbed the solution rises
and washes the concentrated alkali out of the upper tube of the
chamber with sufficient completeness to prevent error from vapor
tension lowering. With whole blood, there is so much oxygen
left in the top of the chamber that the blood solution does not rise
high enough to rinse out the alkali; consequently, the technique
described previously must be used, with air-freed 1 N NaOH
added under diminished pressure.

Measurement of Pressure p_2 after Absorption of CO_2.—Absorption
of CO_2 being completed, the solution in the chamber is lowered,
if necessary, until its meniscus is a little below the a mark.   The
leveling bulb is placed in the position shown in Fig. 1, mercury is
readmitted from cock e until the solution's meniscus is again on
the a mark, and the p_2 reading is noted on the manometer.   The
CO_2 pressure P_{CO_2} is

$$P_{CO_2} = p_1 - p_2 - c$$

where c is the correction discussed on page 537. The CO_2 content of the solution is calculated from P_{CO_2} by multiplying with the factor obtained from Table II or III, or from a similar table derived from Equation 6 or 7. Cock b is sealed with mercury during the p_2 reading.

Sources of Error.—Aside from measurements of the sample and of the pressure and volume of the extracted gas there are certain minor sources of error, which, if not considered, may exert appreciable effect.

The ratio of unextracted gas to that extracted and measured is indicated by the term $\dfrac{S}{A-S}\alpha'$ in the equation by which the results are calculated. It is almost negligible when α' is as small (*viz.* 0.02 to 0.03) as in the cases of H_2, N_2, and O_2. At room temperature α'_{CO_2}, however, is about 1 (see Table I), so that when extraction has reached maximum completion about the same concentration of CO_2 exists per cc. of solution as per cc. of the supernatant gas phase. It is essential, consequently, that the value S, of total solution in the apparatus, be measured with an accuracy of about 0.05 cc. in order to keep errors from variation in S below 1 part per 1,000 in the final result.

Less accuracy in the measurement of A is required. In the 50 cc. apparatus an error of 1 cc. in A is required to introduce an error of 1 part per 1,000 in the final result.

A possible source of error may be encountered by the immediate analysis of samples chilled in the ice box, unless equilibration in the analysis chamber is continued for a time sufficient to allow the reaction mixture to reach the temperature recorded in the water jacket. *E.g.*, if the liquid in the chamber is at 19°C., and the water bath at 20°C., an error of 0.1 per cent is introduced into the calculated CO_2 content, chiefly because of the effect of temperature on the solubility coefficient of CO_2.

A constant slight error in the calculation of the CO_2 in blood and plasma analyses from the factors in Table I lies in the fact that the α' values used are those of pure water instead of the lower values of the diluted blood solution actually extracted. According to experiments by Van Slyke, Hastings, and Neill by the technique reported on page 568 of this paper, the solubility coefficients of the reaction mixtures used in plasma and blood analyses (1 cc. of blood or plasma + 2.5 cc. of water) are at 20°C., 0.850 and 0.838,

548   Blood Gases.   I

respectively, instead of 0.878, the value for water. The difference causes the CO_2 results calculated to be 1.5 parts per 1,000 too
high for plasma, and 2 parts per 1,000 too high for whole blood.
The effect is so slight, however, that it has not seemed to justify
the compilation for this paper of additional tables for blood and
plasma analyses.

Careful measurement of volume of the NaOH absorbent solution introduced is essential because of its effect on the c correction.
In the different 50 cc. machines in use in our laboratory, with S
= 3.5 cc., the value of correction c for addition of 1 cc. of alkali
ranges between 2.0 and 4.0 mm. Consequently, an error of 0.1
cc. in the volume of the absorbent introduced in analysis of a 1
cc. blood sample, may result in an error of 0.4 mm. in the manometer reading, and an error of approximately 0.1 volume per cent
of CO_2 in the results. For accuracy in measuring the alkali
solution in the cup, mercury should just fill the capillary below the
cup when the solution is placed in it.

The c correction is determined in a blank analysis in which S
cc. (for 1 cc. of blood S = 3.5) of water, made alkaline with 2 or
3 drops of 1 N NaOH, are extracted in the apparatus, p_1 and p_2
being read before and after addition of 1 cc. of 1 N NaOH, or
0.2 cc. of 5 N NaOH, as above described.

$$c = p_1 - p_2$$

Determination of Correction i for Reabsorption of CO_2.—Factor i
is determined by analysis of standard Na_2CO_3 solutions made up in
CO_2-free water and protected from access of atmospheric CO_2.
The value of i is calculated by rearrangement of Equation 7 in the
form

$$(8) \qquad i = \frac{1}{P} \times \frac{[CO_2]\,(cc.\ sample)}{0.0587\,a} \times \frac{1 + 0.00384\,t}{1 + \frac{S}{A - S}\alpha'}$$

where $[CO_2]$ = millimols of CO_2 per liter of the standard
solution.

Na_2CO_3 was prepared by heating "reagent" $NaHCO_3$ with the
precautions customary in preparing carbonate for the standardization of acids. Three different preparations of Na_2CO_3 were made

TABLE IV.

Estimated Corrections for Dissolved O_2 and N_2 in Blood.

Blood.	Determined.	Sought.	Correction to subtract.	
			vol. per cent	mm per l.
Venous.........................	Total O_2	Combined O_2	0.1 (O_2).	0.04 (O_2).
Arterial.......................	" "	" "	0.2 "	0.09 "
Saturated with air at 20°, 760 mm.................	" "	" "	0.5 "	0.22 "
Venous.........................	Total $O_2 + N_2$	" "	1.3 $O_2 + N_2$.	0.57 ($O_2 + N_2$).
Arterial.......................	" + "	" "	1.5 "	0.62 "
Saturated with air at 20°, 760 mm.................	" + "	" "	1.9 "	0.85 "
Venous.........................	" + " or CO + $O_2 + N_2$.	Total O_2 or CO + O_2.	1.2 (N_2).	0.53 (N_2).
Arterial.......................	" "	" "	1.2 "	0.53 "

550 Blood Gases. I

from two different lots of $NaHCO_3$. The carbonate prepared was titrated against 0.1 N HCl which had been standardized gravimetrically. Two different solutions were made from each Na_2CO_3 preparation. As a check on the constancy of the i correction with different amounts of CO_2, solutions were prepared of 15, 30, and

TABLE V.

Determination of Reabsorption Correction i by Analysis of Standard Na_2CO_3 Solution.

$A = 100.0$ cc.; $S = 5.00$ cc.; $a = 5.004$ cc.; $c = 1.5$ mm. 2 cc. of standard Na_2CO_3 solution for each determination.

Date.	Na_2CO_3 preparation No.	Solution.		p_1	p_2	$P_{CO_2} = p_1 - p_2 - 1.5$	Temperature.	i*
		No.	Concentration.					
1922			*mM*	*mm.*	*mm.*	*mm.*	*°C.*	
Nov. 27	B	B I	30.00	266.9	56.3	209.1	22.6	1.014
				266.0	55.9	208.6	22.6	1.016
	B	B II	29.91	267.2	57.5	208.2	22.5	1.014
				268.1	58.1	208.5	22.5	1.013
				266.0	57.0	207.5	21.4	1.013
Dec. 10	C	C I	48.85	397.9	53.0	343.4	24.7	1.015
				394.3	51.1	341.7	23.8	1.016
	C	C II	30.00	255.5	44.6	209.4	22.7	1.013
				254.4	43.6	209.3	22.5	1.012
Dec. 12	D	D I		248.6	37.9	209.2	22.5	1.013
				248.8	37.9	209.4	22.5	1.012
	D	D II	15.00	141.1	35.9	103.7	20.9	1.014
				140.1	34.9	103.7	20.6	1.012
Reagents.....................				31.7	30.1	0.1		
Average...								1.014

*From Equation 8.

45 millimolar concentration. Duplicate analyses were made on 2 cc. of each solution. Analyses were also made on different days as a further check on the constancy of the reabsorption correction. Results are given in Table V.

Since i is an empirical factor, it may vary slightly for different apparatus. It is also essential for the maintenance of the constancy of i that when the CO_2 volume is being reduced to a cc. after extraction, about the same time should be taken for the process in all analyses. If the time is materially prolonged i is somewhat increased.

Constancy of Results in CO_2 Determinations on Blood.—Duplicate analyses of 1 cc. samples of blood determined in the 50 cc. apparatus with $a = 2.0$ cc. usually agree within 0.2 volume per cent or 0.1 mM.

In experiments requiring the greatest precision we have used 2 or 3 cc. samples in the 100 cc. apparatus, with $a = 4$ or 5 cc. With 2 and 3 cc. samples our average variation in a series of duplicate analyses has invariably been below 0.05 mM. Table V represents a typical series of results with 2 cc. samples of blood and plasma saturated at different CO_2 tensions and determined in an apparatus whose $A = 100$ cc. and $a = 5.0$ cc.

D. D. Van Slyke and J. M. Neill 573

the solution is shaken 1 minute to absorb the CO_2. The gas volume is brought back to 2 cc. and p_2 is measured.

Similarly, to absorb O_2, 1 cc. of air-free hydrosulfite solution is introduced, and is shaken for 2 minutes. p_3 is then read on the manometer.

The c corrections for the effect of the added 1 cc. portions of alkali and hydrosulfite are found in blank determinations in the gas-free apparatus. The results are calculated as follows:

$$\text{Per cent } CO_2 = \frac{p_1 - p_2 - c}{p_1 - p_0}$$

$$\text{Per cent } O_2 = \frac{p_2 - p_3 - c}{p_1 - p_0}$$

For this calculation based on pressure changes only, it is assumed that during the few minutes required for the analysis the temperature in the water jacket remains constant within 0.1°C.

Results by the above method check within usually 0.1 volume per cent those by the Haldane apparatus. They are somewhat more consistent if obtained with the 100 cc. apparatus, with pressure measurements at 4 cc. of gas volume, than with the 50 cc. apparatus.

The methods here presented were in part developed with the technical assistance of John Plazin, who constructed the first apparatus and performed many of the analyses with which the details of technique were developed.

BIBLIOGRAPHY.

Austin, J. H., Cullen, G. E., Hastings, A. B., McLean, F. C., Peters, J. P., and Van Slyke, D. D., J. Biol. Chem., 1922, liv, 121.
Conant, J. B., J. Biol. Chem., 1923, lvii, 401.
Poulton, E. P., J. Physiol., 1919-20, liii, p. lxi.
Stadie, W. C., J. Biol. Chem., 1921, xlix, 43.
Van Slyke, D. D., J. Biol. Chem., 1917, xxx, 347.
Van Slyke, D. D., J. Biol. Chem., 1921, xlviii, 153.
Van Slyke, D. D., and Cullen, G. E., J. Biol. Chem., 1917, xxx, 289.
Van Slyke, D. D., and Stadie, W. C., J. Biol. Chem., 1921, xlix, 1.

COMMENTARY TO

16. Malloy, H. T. and Evelyn, K. A. (1937)
The Determination of Bilirubin With the Photoelectric Colorimeter.
Journal of Biological Chemistry 119(2): 481–490.

This chapter by Helga Tait Malloy and Kenneth A. Evelyn is noteworthy for two reasons. First, it simplified and made practical the van den Berg serum bilirubin assay. Secondly, their bilirubin assay because of its wide popularity helped introduce the filter photometer into clinical chemistry.

Van den Bergh was the first to apply Erhlich's diazo reagent (sodium nitrite and sulfanilic acid) to the measurement of bilirubin in serum (1). Diazo reagent combines with bilirubin to form a blue azobilirubin. Van den Bergh and Muller (2) discovered that color develops within 30 seconds in the absence of alcohol and a second increase in color occurs with the addition of alcohol. The color that develops rapidly in the absence of alcohol is the direct or conjugated (water soluble) bilirubin. The Van den Bergh diazo procedure measured total bilirubin after protein precipitation of the serum with alcohol. Numerous modifications included the addition of the diazo reagent before or after protein precipitation (3). Malloy and Evelyn in the paper presented here greatly simplified this assay. They measured total bilirubin (termed indirect in the paper) by mixing diluted serum, diazo reagent and 50% methanol. The 50% methanol prevented protein precipitation and allowed full color development of the azobilirubin. Direct or conjugated bilirubin was measured in the same way except that water was substituted for the 50% methanol. This is essentially the form of the assay as performed today except that the direct and total are run in the same tubes and the direct is read at 3 minute before the addition of the 50% methanol (4).

Malloy and Evelyn developed their improved bilirubin assay on a newly introduced filter photometer designed by Evelyn in 1936. The decade of the 1930s was a period of major growth in the description and commercialization of filter photometers. Most were based on the use of either one or two photoelectric cells, glass interference filters, a cuvette or test tube chamber and an ammeter readout. The first commercial filter photometer designed specifically for clinical chemistry was the Sheard and Sanford Photelometer™ (5). Methods for hemoglobin, Folin and Wu's sugar and Folin's creatinine were adapted to this instrument. The patent and the manufacturing rights were assigned to the Central Scientific Co. in Chicago. Sheard and Sanford also assigned all royalties to the American Society of Clinical Pathologists.

Malloy and Evelyn developed their bilirubin assay on the Evelyn photometer. In his first paper on the description of the instrument published in 1936 Evelyn stated that, "the color filter technique can be refined to such an extent that it not only improves the accuracy of existing procedures, but renders entirely new ones possible" (6). Evelyn latter described cuvette adapters that allowed less than 200 μL of colored reaction product to be read (7–8). Methods developed on this photometer by Evelyn and Malloy and co-workers appeared in rapid succession including procedures for bilirubin in bile and meconium (9), oxyhemoglobin, methoglobin and sulfhemoglobin in blood (10) and ascorbic acid in urine (11).

Filter photometers soon replaced the Duboscq visual colorimeter and remained the mainstay of colorimetric clinical chemistry assays for many years. In 1941 William S. Hoffman wrote *Photelometric Clinical Chemistry* (12). In his book, 23 different methods were presented for use with the filter photometer. Included were methods for bilirubin, sulfanilamides, total cholesterol, sodium and potassium. In 1948 eleven different manufacturers offered filter photometers for clinical use (13). By the late 1950s filter photometers were mostly replaced by diffraction grating instruments like the Coleman Jr. and the quartz prism grating spectrophotometer by Beckman called the Model DU. The Malloy and Evelyn serum bilirubin assay was adopted onto both of these instruments in 1959 (14).

References

(1) Van den Bergh, A.A.H. and Snapper, J. (1913) Die Farbstoffe des Blutserums I. Eine Quantitative Bestimmung des Bilirubins im Blutsesum. Deutsche Archiv für Klinische Medizin. 110:540–561.

(2) Van den Bergh, A.A.H. and Muller, P. (1916) Uber eine Direkte und Eine Indirekte Diazoreaktion auf Bilirubin. Biochemische Zeitschrift. 77:90–103.

(3) Godfried, E.G. (1935) CLXII. The technique to be selected for the determination of Bilirubin in blood by the diazo-method. Biochemical Journal. 29(6):1337–1339.

(4) Meites, S. (1982) Bilirubin, direct-reacting and total, modified Malloy–Evelyn method, in *Selected Methods of Clinical Chemistry*, in *Selected Methods for the Small Clinical Chemistry Laboratory*, Vol 9. Faulkner, W.R. and Meites, S. (eds), American Association for Clinical Chemistry, Washington, DC, pgs 119–124.

(5)   Sanford, A.H., Sheard, C., and Osterberg, A.E. (1933) The photelometer and its use in the clinical laboratory. American Journal of Clinical Pathology. 3(6):405–420.

(6)   Evelyn, K.A. (1936) A stabilized photoelectric colorimeter with light filters. Journal of Biological Chemistry. 115(1):63–75.

(7)   Evelyn, K.A. and Cipriani, A.J. (1937) A photoelectric microcolorimeter. Journal of Biological Chemistry. 117(1):365–369.

(8)   Evelyn, K.A. and Gibson, J.G. (1938) A new type of absorption cell for the photoelectric microcolorimeter. Journal of Biological Chemistry. 122(2):391–394.

(9)   Malloy, H.T. and Evelyn, K.A. (1938) Oxidation method for bilirubin determinations in bile and meconium with the photoelectric colorimeter. Journal of Biological Chemistry. 122(3):597–603.

(10) Evelyn, K.A. and Malloy, H.T. (1938) Microdetermination of oxyhemoglobin, methemoglobin, and sulfhemoglobin in a single sample of blood. Journal of Biological Chemistry. 126(2):655–669.

(11) Evelyn, K.A., Malloy, H.T., and Rosen, C. (1938) The determination of ascorbic acid in urine with the photoelectric colorimeter. Journal of Biological Chemistry. 126(2):645–654.

(12) Hoffman, W.S. (1941) *Photelometric Clinical Chemistry*. William Morrow & Company, New York.

(13) Snell, F.D. and Snell, C.T. (1948) Photoelectric filter photometer, in *Colorimetric Methods of Analysis Including Some Turbidimetric and Nephelometric Methods, Vol I Theory–Instruments–pH*, Third Edition, Chapter XI, D. Van Nostrand Company, Inc., New York, pgs 89–104.

(14) Hogg, C.K. and Meites, S. (1959) A modification of the Malloy and Evelyn procedure for the micro-determination of total serum bilirubin. American Journal of Medical Technology. 25(5):281–286.

J. Biol. Chem. 1937; 119(2): 481–490

THE DETERMINATION OF BILIRUBIN WITH THE PHOTOELECTRIC COLORIMETER

By HELGA TAIT MALLOY* AND KENNETH A. EVELYN*

(*From the Department of Medicine, McGill University Clinic, Royal Victoria Hospital, Montreal, Canada*)

(Received for publication, August 10, 1936)

The main source of error in the determination of bilirubin in serum by the diazo reaction has been the loss caused by adsorption on the protein precipitate. In addition, the lack of an accurately matching artificial standard and the sensitivity of the azobilirubin color to changes in pH have made accurate colorimetric determinations impossible, without the aid of some type of objective photometer such as the spectrophotometer, the Pulfrich photometer, or the photoelectric colorimeter (1–4).

In this paper we shall describe a method for the quantitative determination of both direct and indirect bilirubin in serum, in which protein precipitation and consequent loss of bilirubin have been eliminated. By a slight modification of the method, a quantitative study of the behavior of the direct reaction has also been made possible. Artificial standards have been eliminated in the colorimetric determinations by the use of the photoelectric colorimeter (5) with a specially selected light filter.

Selection of Color Filter

The spectrophotometric curve of the rose-mauve color of the azobilirubin solutions obtained by the method to be described below has a single broad absorption band at 540 $m\mu$ (Fig. 1). We have therefore chosen a filter which transmits a narrow spectral band in the vicinity of 540 $m\mu$, so that light which has passed through the filter is readily absorbed by solutions of azobilirubin. This filter has the further advantage of being unaffected by the

* Aided by grants from The Banting Research Foundation, Toronto, Canada.

482   Determination of Bilirubin

presence of the yellow serum pigments whose absorption at 540 $m\mu$
is negligible.   The only other interfering color is that due to the
possible presence of hemoglobin in the serum.   Although this
pigment does absorb light in the vicinity of 540 $m\mu$, any error

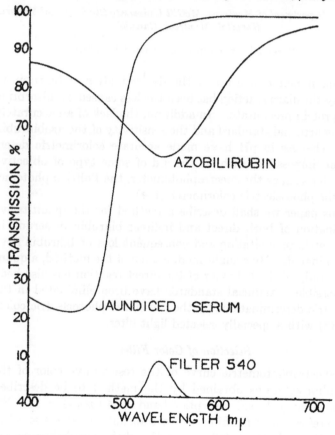

FIG. 1. Spectrophotometric curves of azobilirubin, Filter 540, and
jaundiced serum.

from this source is obviated by the use of a blank tube in the
initial adjustment of the instrument.

Indirect Reaction

Van den Bergh (6) assumed that indirect bilirubin in serum
would only react with diazo reagent in the presence of alcohol,

H. T. Malloy and K. A. Evelyn 483

after the proteins had been removed by precipitation. Our experiments, however, have proved that complete coupling of indirect bilirubin will take place in the presence of serum proteins provided the alcohol concentration is of the order of 50 per cent. This concentration can be achieved without protein precipitation if the serum is first diluted with water, and this procedure forms the basis for our technique for indirect bilirubin. This not only eliminates loss of bilirubin on the protein precipitate, but also provides a buffer substrate sufficient to stabilize the pH-sensitive color of the azobilirubin.

Method

Reagents—

1. Solution A, 1.0 gm. of sulfanilic acid dissolved in 15 cc. of concentrated HCl, and diluted to 1 liter with water.

2. Solution B, 0.5 per cent sodium nitrite.

3. Diazo reagent freshly prepared by adding 0.3 cc. of Solution B to 10 cc. of Solution A.

4. Hydrochloric acid for blank tubes (diazo blank), 15 cc. of concentrated HCl in 1 liter of water.

5. Absolute methyl alcohol. The use of absolute methyl alcohol is recommended, since it yields clearer solutions than 95 per cent ethyl alcohol. The amount of color produced is, however, unaltered.

Procedure—Two colorimeter tubes are set up as follows: Tube 1, indirect blank, 5 cc. of absolute methyl alcohol and 1 cc. of diazo blank solution; Tube 2, indirect sample, 5 cc. of absolute methyl alcohol and 1 cc. of diazo reagent.

1 cc. of serum or plasma is diluted to 10 cc. with distilled water, and 4 cc. of the diluted material are added to each tube. The contents are mixed by inversion, care being taken to handle both tubes in the same way, so that any turbidity which may result from too vigorous shaking will be the same in both tubes. If bubbles form they are best removed by gentle tilting and rotation of the tubes.

Tube 2 is read in the colorimeter with Filter 540[1] 30 minutes

[1] This filter is one of the set of eight filters which have been selected for use with the photoelectric colorimeter. Any of these filters may be obtained from the Rubicon Company, 29 North 6th Street, Philadelphia. The complete colorimeter may also be obtained from the Rubicon Company.

after addition of the serum, Tube 1 being used for the initial adjustment of the galvanometer. If the galvanometer reading is less than 10, it is advisable for the sake of greater accuracy to dilute both tubes with 10 cc. of 50 per cent methyl alcohol and read again immediately. In this case, the final answer in mg. per 100 cc. must be multiplied by 2.

Calculation—The bilirubin concentration in mg. per 100 cc. of serum is obtained from the formula

$$X = \frac{2 - \log G}{6.72} \times 100$$

where X is mg. of bilirubin per 100 cc. of serum, and G is the galvanometer reading.

For routine determinations, a calibration curve may be made from this formula from which the values for X may be read directly (Fig. 2).

Calibration—Since the concentration of azobilirubin in a solution is proportional to the negative logarithm of the light transmission, the following formula is valid

$$C = \frac{2 - \log G}{K_1} \tag{1}$$

where C is the concentration of azobilirubin (expressed as bilirubin) in mg. per cc. of colored solution, G is the galvanometer reading, and K_1 is a constant.

Conversion of the bilirubin concentration in terms of mg. per cc. of colored solution to mg. per 100 cc. of serum is made by means of the formula

$$X = \frac{2 - \log G}{K_1} \times \frac{V}{A} \times 100 \tag{2}$$

where X is mg. of bilirubin per 100 cc. of serum, V is the volume of colored solution, and A is the amount of serum used.

Thus in the method described above

$$X = \frac{2 - \log G}{K_1} \times \frac{10}{0.4} \times 100 \tag{3}$$

Calibration of the instrument consists essentially in determining the value of K_1 by obtaining the galvanometer readings for known

H. T. Malloy and K. A. Evelyn 485

concentrations of bilirubin. For this purpose about 10 mg. (accurately weighed) of pure bilirubin[2] were dissolved in 100 cc. of chloroform. Portions were withdrawn from this solution and diluted with ethyl alcohol to a final concentration of 0.01 mg. per cc. Varying amounts of the alcoholic solutions were placed in a series of colorimeter tubes to which were added 1 cc. of diazo reagent and sufficient ethyl alcohol to make the final volume 10 cc. The tubes were read at 5 minute intervals until the color was at a maximum. The known values for C and G were then substituted

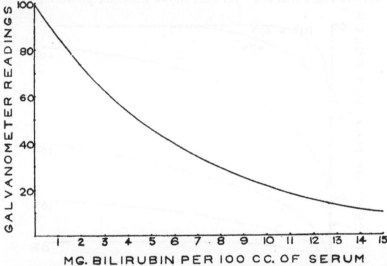

Fig. 2. Calibration curve. This curve has been checked at frequent intervals for nearly 2 years, and no variation greater than ±1 per cent has been found.

in Equation 1, and the value of K_1 was found to be 168.0. Since all photoelectric colorimeters of this type are interchangeable with respect to calibration, we recommend that this value of K_1 be generally adopted to insure uniformity of results.

[2] We have tested both Eastman Kodak and Hoffmann-La Roche bilirubin, but have used the latter for calibration, since the yield of color is 10 per cent higher. Moreover, van den Bergh and Grotepass (7) state that they have found Hoffmann-La Roche bilirubin to be identical with the chemically pure bilirubin prepared by Professor Hans Fischer.

486 Determination of Bilirubin

Substituting K_1 in Equation 3 above we obtain

$$X = \frac{2 - \log G}{6.72} \times 100 \text{ mg. bilirubin per 100 cc. serum}$$

The value 6.72 we have termed K_2.

Results

Effect of Alcohol Concentration upon Color Development—The curve in **Fig.** 3 marked 0 per cent shows the color produced when

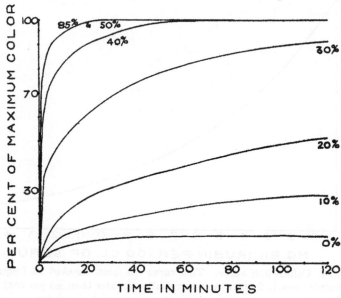

FIG. 3. Effect of alcohol concentration on development of color from bilirubin.

diazo reagent, without alcohol, is added to serum. By definition, this must be due to direct bilirubin only. The addition of alcohol in increasing amounts causes more and more of the indirect bilirubin to take part in the reaction, until at concentrations above 40 per cent the indirect reaction is complete. An alcohol concentration of 50 per cent was chosen for the method, since it affords an adequate margin of safety, yields perfectly clear solutions, and produces maximum color in a shorter time than 40 per cent.

H. T. Malloy and K. A. Evelyn 487

Recovery of Bilirubin Added to Serum—Table I shows that bilirubin added to serum is recovered with a maximum error of

TABLE I

Recovery of Pure Indirect Bilirubin Added to Normal and Jaundiced Sera
0.4 cc. of serum was used in each case. The figures in the second column represent the amount of bilirubin present in this 0.4 cc., as measured directly on the instrument. The first five experiments were made on normal sera. (Normal values obtained by our method are between 0.2 and 0.8 mg. per 100 cc.)

Bilirubin per 100 cc. serum	Bilirubin in 0.4 cc. serum	Bilirubin added to 0.4 cc. serum	Total bilirubin present	Amount of bilirubin measured	Percentage recovery
mg.	*mg.* $\times 10^{-3}$	*mg.* $\times 10^{-3}$	*mg.* $\times 10^{-3}$	*mg.* $\times 10^{-3}$	
0.25	1.0	4.4	5.4	5.5	101.8
0.5	2.0	8.8	10.8	10.6	98.1
0.5	2.0	50.0	52.0	50.0	96.1
0.75	3.0	34.0	37.0	36.4	97.7
0.75	3.0	80.0	83.0	80.0	96.3
1.0	4.0	43.6	47.6	48.0	100.8
1.0	4.0	88.0	92.0	95.0	103.4
1.7	6.8	31.2	38.0	38.4	101.1
2.7	10.8	30.0	40.8	40.0	98.0
3.2	12.8	50.0	62.8	63.4	101.0
4.0	16.0	40.0	56.0	54.0	96.5
4.0	16.0	50.0	66.0	67.4	102.0
8.0	32.0	40.0	72.0	73.2	101.9
16.1	64.0	20.0	84.0	83.6	99.7

TABLE II

Comparison of Determinations Made on Same Sera by Different Techniques
All the figures represent mg. per 100 cc. of serum. All the final colorimetric measurements were made on the photoelectric colorimeter.

Technique	Experiment 1	Experiment 2	Experiment 3	Experiment 4	Experiment 5
Van den Bergh and Grotepass (7) (indirect reaction)	2.0	4.2	3.1	7.8	12.4
Thannhauser and Andersen (8)....	3.3	4.9	4.6	9.7	12.8
Jendrassik and Czike (9)..........	3.4	4.9	5.2	10.5	15.1
Malloy and Evelyn (indirect reaction)............................	4.8	6.5	9.0	15.2	22.8

±4 per cent, the average error being only 2 per cent. Since the calibration curve used in these experiments was made with known amounts of pure indirect bilirubin in alcoholic solution, the results prove that the presence of serum proteins does not interfere with the diazo reaction of bilirubin, or with the quantitative determination of the resulting azobilirubin by means of the photoelectric colorimeter.

Duplicate determinations on serum will usually agree within ±1 per cent, with a maximum variation of ±2 per cent.

Comparison with Other Methods—The results of duplicate analyses on the same sera by different methods are shown in Table II. All the final colorimetric measurements were made on the photoelectric colorimeter so that any discrepancies which occurred could only have been due to differences in the preliminary treatment of the serum.   From Table II we conclude that (1) precipitation of proteins before addition of the diazo reagent, as recommended by van den Bergh and Grotepass (7), causes a large and variable loss of bilirubin (30 to 60 per cent); (2) this loss can be decreased but not eliminated by the technique of Thannhauser and Andersen (8), in which the diazo reagent is added before the proteins are precipitated; (3) the addition of caffeine sodium benzoate, as suggested by Jendrassik and Czike (9), is a further slight improvement, but the recovery of azobilirubin is still seldom more than 70 per cent, as compared with our method.

Direct Reaction

Van den Bergh and Grotepass (7), have recommended a quantitative method for the direct reaction, but, since 25 per cent alcohol is used, the method measures not only direct bilirubin but also a fraction of the indirect (see Fig. 3).   The results obtained are therefore too high, and since the amount of direct bilirubin in any serum bears no constant relationship to the amount of indirect bilirubin, the error involved is variable.

For routine determinations in which a clinical interpretation is required, we recommend that the direct reaction be carried out in the accustomed manner.   For a more quantitative study of the behavior of the direct reaction, we have, however, adopted the following procedure: Two tubes are set up as in the method for the indirect reaction except that 5 cc. of water are substituted for 5 cc. of methyl alcohol.   Readings are made on the photoelectric colorimeter at 10, 30, 60, and 120 minutes, and the corre-

H. T. Malloy and K. A. Evelyn   489

sponding bilirubin concentrations, obtained from the calibration curve, are plotted to show development of color with time.

Curve C of Fig. 4 shows the slow development of color from a serum of the "delayed" type, in which the true end-point of the reaction is not attained for several hours. Curve A shows the rapid development of maximum color typical of sera of the "prompt" type. Curve B shows a reaction intermediate between these extremes, which corresponds to what is usually termed the "biphasic" reaction.

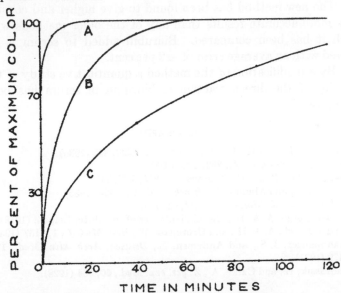

Fig. 4. Behavior of the direct reaction in different sera; Curve A "prompt," Curve B "biphasic," and Curve C "delayed."

In addition, the final galvanometer reading is used to determine the concentration of direct bilirubin in the serum. Sera have been found to differ not only in the shape of their color development curves, but also in the ratio of the amounts of direct and indirect bilirubin. The clinical significance of these two variables will be discussed elsewhere.

SUMMARY

1. A method has been described for the accurate photoelectric determination of both direct and indirect bilirubin in serum, in

which protein precipitation and consequent loss of bilirubin by adsorption have been eliminated.

2. The interfering effect of yellow serum pigments in the color determination has been overcome by the use of a specially selected light filter, which also eliminates the necessity for artificial color standards.

3. It has been shown that all the bilirubin in serum will react with diazo reagent even in the presence of serum proteins, provided a sufficiently high concentration of alcohol (50 per cent) is present.

4. The new method has been found to give higher and consistently more accurate results than any of the older methods with which it has been compared. Bilirubin added to serum is recovered with an average error of ±2 per cent.

5. By a modification of the method a quantitative study of the behavior of the direct reaction of bilirubin in serum has been made possible.

BIBLIOGRAPHY

1. Heilmeyer, L., and Krebs, W., *Biochem. Z.*, **223**, 352 (1930).
2. Peter, J. R., *Biochem. Z.*, **262**, 432 (1933).
3. Theil, A., and Peter, O., *Biochem. Z.*, **271**, 1 (1934).
4. Krupski, A., and Almazy, F., *Biochem. Z.*, **279**, 424 (1935).
5. Evelyn, K. A., *J. Biol. Chem.*, **115**, 63 (1936).
6. van den Bergh, A. A. H., Der Gallenfarbstoff im Blute, Leipsic (1918).
7. van den Bergh, A. A. H., and Grotepass, W., *Brit. Med. J.*, **1**, 1157 (1934).
8. Thannhauser, J. S., and Andersen, E., *Deutsch. Arch. klin. Med.*, **137**, 179 (1921).
9. Jendrassik, L., and Czike, A., *Z. ges. exp. Med.*, **60**, 554 (1928).

COMMENTARY TO

17. Gornall, A. G., Bardawill, C. J. and David, M. M. (1949)
Determination of Serum Proteins by Means of the Biuret Reaction.
Journal of Biological Chemistry 177(2): 751–766.

Biuret is the chemical product that forms when urea is heated to 180°C. In this reaction two molecules of urea condense to form a bi-urea or biuret molecule. Biuret reagent in the presence of copper ions forms a violet color complex. Peptide bonds in proteins will also link with copper ions to form a similar violet complex. This has been termed the biuret reaction. Biuret reagent for the measurement of total protein is simply copper salt in alkaline solution. In the biuret reaction, cupric ions (Cu^{2+}) in alkaline solution bind to peptide bonds which shifts the copper to the cuprous form (Cu^{1+}). This shift in valance produces a quantitative colorimetric change in alkaline solution from blue to blue-purple (1). The biuret reaction is the basis of the most widely used assay for the measurement of total protein in serum in clinical chemistry.

Kantor and Gies applied the biuret reaction to the measurement of protein in urine in 1911. They soaked the biuret reagent into filter paper and ran a spot test to screen for elevated urine protein (2). Riegler in 1914 was the first to apply the biuret reaction to measure proteins quantitatively in biological fluids (3,4). He precipitated urine proteins with acid, dissolved the precipitated proteins in alkali then added copper sulfate to form a purple color. George Kingsley in 1939 developed a total protein, albumin and globulin assay in serum based on the biuret reaction (5). Serum total protein was measured by adding 100 μL of a sample to 4 mL of 10% sodium hydroxide followed by 0.5 mL of a 1% solution of copper sulfate. The blue to purple color was read in a filter photometer after a 25-minute incubation. In 1942 Kingsley described a one-step biuret reagent (6). He premixed the copper sulfate in 14% sodium hydroxide. Weischselbaum added sodium potassium tartrate and potassium iodide to the single reagent (7). This and similar one-step formulations for the biuret reagent turned out to be unstable due to the eventual precipitation of the copper ions as $Cu(OH)_2$.

Gornall, Bardawill and David solved the stability problem in 1949. Their paper presented here is a careful and detailed study of the biuret reaction applied to serum. In it they also solved through step-wise experimentation the stability problem. They balanced the copper sulfate concentration with the sodium potassium tartrate, and added potassium iodide to provide a single biuret reagent that was stable at room temperature for up to one year. Gornall 28 years latter reviewed the development of this reagent by him and his co-workers and confirmed that the potassium iodide was essential for the stability (8). In 1978 Burkhardt and Batsakis reported on the results of a nation wide quality control survey in which 83% of 2234 reporting clinical chemistry laboratories used the biuret reaction for serum total protein (9).

References

(1) Cannon, D.C., Olitzky, I., and Inkpen, J.A. (1974) Proteins, in *Clinical Chemistry Principles and Techniques*, 2nd Edition, Henry, R.J., Cannon, D.C., and Winkelman, J.W. (eds), Harper & Row, Publishers, Hagerstown, MD, pgs 411–421.

(2) Kantor, J.L. and Gies, W.J. (1911) Additional experiments with the biuret reagent, Proceedings of the American Society of Biological Chemists, Fifth Annual Meeting, December 28–30, 1910, Journal of Biological Chemistry, 9:xvii–xviii.

(3) Riegler, E. (1914) Eine Kolorimetrische Bestimmungsmethod des Eiweisses. Zeitschrift für Analytische Chemie. 53: 242–245.

(4) Rosenfeld, L. (1982) Colorimetry and Photometry, in *Origins of Clinical Chemistry. The Evolution of Protein Analysis*. Academic Press, New York, Chapter 9, pgs 128–129.

(5) Kingsley, G.R. (1939) The determination of serum total protein, albumin, and globulin by the biuret reaction. Journal of Biological Chemistry. 131(1):197–200.

(6) Kingsley, G.R. (1942) The direct biuret method for the determination of serum proteins as applied to photoelectric and visual colorimetry. Journal of Laboratory and Clinical Medicine. 27(6):840–845.

(7) Weichselbaum, T.E. (1946) An accurate and rapid method for the determination of proteins in small amounts of blood serum and plasma. American Journal of Clinical Pathology Technical Section. 16(3):40–49.

(8) Gornall, A.G. and Manolis, A. (1977) Comments on a new biuret reagent. Clinical Chemistry. 23(6):1184–1185.

(9) Burkhardt, R.T. and Batsakis, J.G. (1978) An interlaboratory comparison of serum total protein analyses. American Journal of Clinical Pathology. 70(3):508–510.

J. Biol. Chem. 1949; 177(2): 751–766

DETERMINATION OF SERUM PROTEINS BY MEANS OF THE BIURET REACTION

By ALLAN G. GORNALL, CHARLES J. BARDAWILL,* AND MAXIMA M. DAVID

(*From the Department of Pathological Chemistry, University of Toronto, Toronto, Canada*)

(Received for publication, August 2, 1948)

In the course of an investigation of the biochemical changes following experimental liver injury we felt the need of a simple, rapid, and accurate method for determining the protein fractions in small amounts of serum. Among the simpler procedures known, the biuret reaction seemed to offer the most encouraging possibilities.

Variations and improvements in the application of the biuret reaction to clinical chemistry can be traced in the works of Autenrieth (1), Hiller (2), Fine (3), Kingsley (4), and Robinson and Hogden (5). Kingsley (6) simplified the technique by adding serum directly to a "one piece" reagent. Efforts have been made to increase the stability of such biuret reagents with ethylene glycol (7), tartrate (8), and citrate (9).[1]

We began our investigation with Kingsley's (6) method and report briefly on the two main difficulties encountered in its use. The first is that the total protein (TP) reagent and, to a lesser extent, the albumin (ALB) reagent are not sufficiently stable. The length of time they remain so depends upon the technique of their preparation. One consequence of this variable stability is a difficulty in duplicating calibration curves with different lots of reagent. Errors may arise when results with a new reagent are read from an old calibration curve. Serious errors occur if a reagent is used after the separation of any black deposit gives evidence of deterioration.

A second difficulty has been that total protein estimations made with the TP reagent and read, as prescribed, from calibration curves prepared with the ALB reagent have tended to be too low. Recorded in Table I are the results of a number of analyses in which Kingsley's biuret procedure has been compared with the Kjeldahl method[2] on both normal and ab-

---

* Medical Research Fellow, National Research Council, Canada.

[1] We have had access only to an abstract of this paper. Even with 10 times the amount of citrate there stated the reagent has been found unstable.

[2] Kjeldahl digestion with copper selenite catalyst (10) heated 15 to 20 minutes after clearing. Distillation into boric acid and titration with standard acid (11); protein = nitrogen (corrected for non-protein nitrogen) × 6.25. Values so obtained were 99+ per cent of the results with a digestion time of 3 hours or longer.

752   SERUM PROTEIN DETERMINATION

normal human and dog sera. It will be noted that, in such a comparison, the results for total protein by Kingsley's method are invariably and considerably low, while the results for albumin are in reasonably good agreement. Kingsley (12) reported low values for total protein by his method in patients suffering from chronic liver disease, but good agreement with the Kjeldahl method in normal individuals or persons suffering from a variety of other diseases. We have noted the discrepancy in any serum when a result for total protein with the TP reagent is read from a calibration curve prepared with the ALB reagent. The error, it is felt, must be

TABLE I

Results of Comparison of Analyses by Kingsley and Kjeldahl Methods
The values are measured in gm. per cent.

Serum		Diagnosis	Total protein			Albumin	
				Kingsley biuret			
			Kjeldahl nitrogen	Total protein reagent	Albumin reagent	Kjeldahl nitrogen	Kingsley biuret
Human	A. G.	Normal	7.49	7.10	7.45		
	C. B.	"	7.01	6.60	7.10		
	B. B.	"	7.04	6.50	7.10		
	J. D.	"	6.70	6.30	6.65		
	McK.	Lupus erythematosus	7.38	6.30		3.45	3.04
	M. S.	Cirrhosis	6.00	5.00		3.17	3.07
Dog No.	1	Normal	6.32	5.92			
	2	"	6.55	6.05		3.68	3.36
	5	"	6.36	6.17		3.94	4.00
	14	"	5.95	5.30			
	15	"	6.34	5.95			
	1	CCl₄ liver injury	5.40	5.00	5.53	3.26	3.52
	10	" " "	6.32	5.92	6.30	3.38	3.22

attributed to differences in behavior of the two reagents, prepared and stored at different strengths, one diluted to the strength of the other only at the time of analysis. Note, for example, that in Table I, sixth column, where total protein is determined by adding diluted serum to the ALB reagent, agreement with the Kjeldahl determination is satisfactory.

Our experience with Mehl's (7) method has been more limited. The reagent is stable but not easily duplicated, especially if the glycol contains any reducing substances as impurity. Certain minor disadvantages of the method are not encountered in the procedures which follow.

Weichselbaum (8) has described a biuret reagent stabilized with sodium potassium tartrate and potassium iodide. It is characterized by its high content of copper and low concentration of alkali. When a biuret reaction is developed with protein, the change appears to be simply an intensification of the blue color of the reagent, though spectrophotometric analysis reveals the presence of the reddish violet complex with an absorption maximum at about 545 mμ. In a foot-note, this author describes a second, "dilute" reagent as having the advantage of added sensitivity when the analyst is using a spectrophotometer or photoelectric colorimeter. This reagent has a fifth as much copper, and the reddish violet color developed upon addition of protein is clearly apparent.

It has been our experience that the biuret reagents of Weichselbaum possess the advantages claimed by their originator with respect to stability and optical clarity when mixed with clear sera. Differences between the two reagents, objections to the first, and a paucity of information concerning the behavior of either led us, however, to reinvestigate in some detail the factors affecting optimal sensitivity, stability, and practical usefulness of biuret reagents stabilized with tartrate. Our experiments, which have led to certain modifications in the reagent and in its application to the determination of protein fractions in serum or plasma, are described in the following section.

Study of Biuret Reagents Stabilized with Tartrate

Concentration of Copper—The concentration of cupric sulfate ($CuSO_4\cdot 5H_2O$) in the final biuret reaction mixtures of different investigators has varied from about 0.06 to 0.75 gm. per 100 ml., most commonly in the case of serum protein estimations, from 0.1 to 0.2 per cent. Mehl has related the optical density of reaction mixtures to copper concentration for a reagent stabilized with ethylene glycol. Comparable data for a biuret reagent containing tartrate are reported in Figs. 1 and 2.

It will be noted in both Figs. 1 and 2 that, at the protein levels stated, color development increases with increasing concentrations of copper sulfate, rapidly at first, then more slowly, and finally (Fig. 1) reaches a stage in which further additions of copper effect only a very slight, uniform, and linear increase in optical density of the copper-protein complex. This last rise in the curve has been found to result from the change in tartrate to copper ratio, because (Fig. 2), when this ratio is held constant, the curve becomes essentially horizontal over a wide range of copper concentrations. It will be noted that at protein levels up to 0.04 per cent (4 per cent in the undiluted serum) a reagent yielding a final copper sulfate concentration of 0.06 per cent would be adequate. For sera containing about 7 per cent protein, the reaction mixture should contain not less than 0.09 per cent copper sulfate and, for levels up to 12 per cent protein, a concentration of 0.12 per cent is desirable. With the serum dilution used,

754   SERUM PROTEIN DETERMINATION

an increase in copper above this level makes no apparent contribution to the biuret reaction. Its chief effect is the somewhat undesirable one of increasing the optical density of the reagent "blank."

Concentration of Sodium Hydroxide—The importance of sodium hydroxide in the biuret reaction has long been recognized and was studied in some detail by Rising and Johnson (13). Before the development of stabilized reagents, the biuret reaction was generally carried out in a

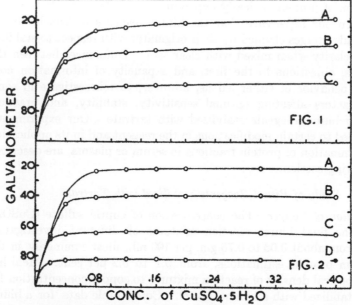

FIGS. 1 AND 2. Effect of copper sulfate concentration upon color development in biuret reaction mixtures containing 2.4 per cent sodium hydroxide and protein concentrations of Curve A, 0.127 per cent; Curve B, 0.077 per cent; Curve C, 0.039 per cent, and Curve D, 0.016 per cent. In Fig. 1, the sodium potassium tartrate concentration is constant at 0.48 per cent. In Fig. 2, the tartrate concentration is in each case 3.5 times that of the copper sulfate present.

medium containing about 3 per cent alkali. It was necessary to separate a precipitate of cupric hydroxide before color comparisons could be made. Kingsley (6) found that very high concentrations of alkali (12 to 17 per cent) prevented cupric hydroxide precipitation and by this means obtained fairly stable "one piece" biuret reagents. With the introduction of reagents containing glycol or tartrate, the need for such high levels of sodium hydroxide was obviated. Mehl obtained best results with 2 to 4 per cent alkali in the presence of glycol. Weichselbaum concluded that 0.4 per cent sodium hydroxide was optimal for reagents containing tartrate.

GORNALL, BARDAWILL, AND DAVID   755

We have investigated the effect of alkali concentration in a reaction mixture containing 0.12 per cent copper sulfate and 0.48 per cent sodium potassium tartrate and observed the color development shown in Fig. 3.

It will be noted that with 0.2 per cent alkali color development is poor, but that from a minimum of 0.4 per cent to about 7 per cent the results

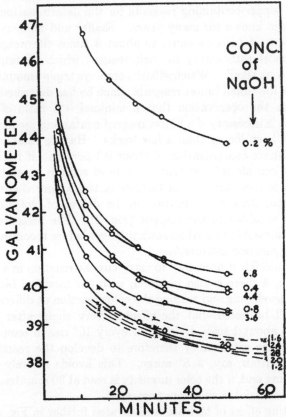

FIG. 3. The effect of sodium hydroxide upon color development in biuret mixtures containing 0.12 per cent copper sulfate, 0.48 per cent sodium potassium tartrate, and 0.077 per cent protein.

are not markedly different. With the final copper sulfate and tartrate concentrations here employed, it would appear that optimal biuret color development occurs with about 1.2 to 2.8 per cent alkali. It has been observed also that an increase in copper concentration tends to shift this optimal alkalinity toward higher values. From the shape of the curves shown it can be noted that the higher alkali concentrations favor a somewhat more rapid development of maximal color.

With low concentrations of alkali, a reaction mixture will remain clear for several days. When levels above about 2 per cent are used, the mixture may show, but only after 12 to 24 hours, a slight flocculent coagulum. This precipitation, which occurs when alkali alone is added to serum, has no significance for readings made at 30 minutes.

Effects of Tartrate—The use of sodium potassium tartrate as a stabilizing agent in the copper-containing reagents for the determination of reducing sugars has been known for many years. Shaffer and Somogyi (14) found that an amount of tartrate equal to about 3 times the weight of copper sulfate was most satisfactory in their reagent which contained approximately 3 per cent alkali. Weichselbaum employs triple amounts of tartrate in both of the low alkali biuret reagents which he has described.

We confirm the observation that a minimal 3:1 ratio of tartrate to copper sulfate is necessary if a biuret reagent containing moderate amounts of alkali is to keep longer than a few weeks. Having established a minimal copper sulfate concentration of about 0.1 per cent, it follows that the reagent must contain 0.3 per cent or more of sodium potassium tartrate. The addition of such amounts of tartrate causes a very slight retardation of, but a rather distinct diminution in the degree of, color development when protein is added to the reagent (Fig. 4). The presence of tartrate does make it advisable to wait somewhat longer before recording the optical density of the reaction mixture.

Weichselbaum felt it necessary to carry out the reaction in a warm water bath. In Fig. 4 there is also recorded, with our modified biuret reagent, the effect of developing and reading the color reaction at different temperatures. It will be noted that the effect is very slight after 30 minutes, amounting to about 1 scale division for every 10° rise in temperature. It would seem quite satisfactory therefore to develop the reaction at room temperature, within, say, a 5° range. This avoids entirely the complication of heating and, if the color intensity is read at 30 minutes, the changes are so slow that strict timing is not essential.

The inhibiting effect of tartrate is illustrated further in Fig. 5. Here we record, for different concentrations of copper sulfate and a constant amount of sodium hydroxide, the optical density obtained after 30 minutes in the presence of increasing ratios of sodium potassium tartrate to copper sulfate. It will be noted, first, that the greater the concentration of tartrate, for any fixed level of copper, the less the color development when protein is added to that reagent. A comparison of identical tartrate to copper ratios, for different copper levels, shows that the readings are virtually the same. One can conclude, then, that the effect of tartrate depends not on its actual concentration, but upon the relative amounts of tartrate and copper.

It will already be clear that the behavior of a biuret reagent is affected by its content of copper, alkali, and tartrate. An attempt is made to illustrate the interrelationships of all three of these factors in Fig. 6. Here a comparison is made of the color developed upon addition of serum to reagents containing two levels of copper, each with 3 and 10 times the

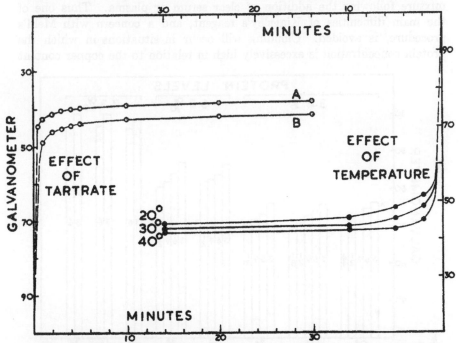

Fig. 4. Rate and degree of color development in biuret reaction mixtures. O, Curve A, without tartrate, Curve B, with 0.48 per cent sodium potassium tartrate; sodium hydroxide 6 per cent; protein 0.065 per cent. ●, the effect of temperature on a reaction mixture containing 0.12 per cent copper sulfate, 0.48 per cent tartrate, and 2.4 per cent alkali; protein 0.063 per cent.

amount of tartrate, and all four at alkali levels of 0.4, 2.4, and 4.8 per cent. All concentrations shown are those in the final reaction mixture.

From a consideration of Fig. 6 the following statements can be made: (1) In all cases the optical density is greater with the lower tartrate to copper ratio. (2) At the 0.4 per cent alkali level, 0.12 per cent copper sulfate is better than 0.32 per cent for either tartrate ratio. (3) With 2.0 to 2.5 per cent sodium hydroxide, there is no difference between copper levels of 0.12 and 0.32 per cent at a 3:1 tartrate ratio, but 0.12 per cent copper is still somewhat better for a 10:1 ratio. (4) When the alkali concentration is about 5 per cent, the reagent with 0.32 per cent copper

240

758   SERUM PROTEIN DETERMINATION

is somewhat better for a 3:1 tartrate ratio, but 0.12 and 0.32 per cent copper sulfate are virtually identical in reagents containing 10 times these amounts of tartrate.

Another important property of biuret reagents stabilized with tartrate is the absence of any serious tendency to develop turbidity in the reaction mixture following the addition of clear serum or plasma. Thus one of the main difficulties of Kingsley's reagent, and a concern with Mehl's procedure, is avoided. Clouding will occur in situations in which the protein concentration is excessively high in relation to the copper content

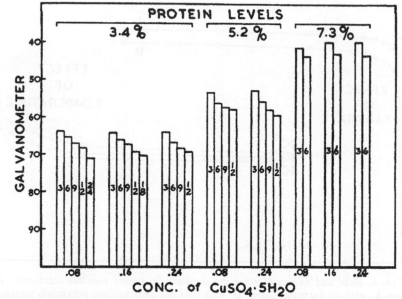

FIG. 5. Effect of the tartrate to copper sulfate ratio, indicated within the columns, on the degree of color development. Serum protein levels and final copper concentrations as shown. Sodium hydroxide 2.4 per cent.

of the reagent. We feel safe, however, in asserting that turbidity will not arise at final 1 in 100 dilutions of sera containing up to 15 per cent protein when the copper sulfate level is 0.12 per cent and the tartrate ratio 4:1. Tartrate is without effect on the precipitate produced by 3 per cent alkali after 12 to 24 hours.

Use of Potassium Iodide—Shaffer and Somogyi observed that, in tartrate-containing copper reagents for the determination of reducing sugar, potassium iodide served to prevent "autoreduction" and separation of cuprous oxide. At room temperature, deterioration of the reagent was prevented by 1 gm. of potassium iodide per liter. For maximal stability during periods of heating, however, it was recommended that 5 gm. per liter be added. Weichselbaum has incorporated 5 gm. of potassium iodide per liter in his reagents.

We have studied the effect of omitting potassium iodide entirely and, with more than 100 different preparations of the reagent, have encountered cuprous oxide precipitation in only one instance. This reduction, and

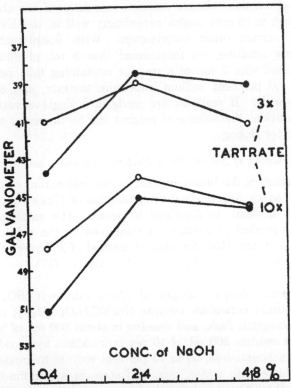

FIG. 6. The interrelationships of copper, tartrate, and alkali concentrations in the biuret reagent. O, 0.12 per cent copper sulfate, ●, 0.32 per cent copper sulfate.

indeed the so called autoreduction, we feel bound to regard as due to some contamination, or impurities in the chemicals used. It is well known that potassium iodide favors the reoxidation of reduced copper and may therefore mask the presence of such impurities. It would seem that potassium iodide can, with reasonable safety, be omitted from the reagent. If, under conditions of preparation or use, any tendency to reduction should be noted, 1 gm. of potassium iodide can be included in each liter of reagent. The presence of this amount of iodide has no detectable effect upon the rate, degree, or quality of biuret color production.

Choice of Biuret Reagent—It is plain that there are many biuret reagents, differing slightly in composition, which will work with comparable accuracy and satisfaction if they meet the conditions defined above. The data presented have led us to prefer a reagent containing 0.15 per cent copper sulfate, 0.6 per cent sodium potassium tartrate, and 3.0 per cent sodium hydroxide (with 0.1 per cent potassium iodide optional). When this reagent is mixed in a proportion of 8:2 with 1 in 20 dilutions of serum, the final reaction mixture contains four-fifths of the above concentrations. Such a reagent works exceedingly well in the Evelyn photoelectric and certain other colorimeters. With instruments that are somewhat less sensitive, we recommend that 3 ml. of diluted protein solution be used with 7 ml. of a reagent containing 0.25 per cent copper sulfate, 1.0 per cent sodium potassium tartrate, and 3.5 per cent sodium hydroxide. If readings are made in a single cuvette requiring 10 ml. of solution, the volume of reagent can be increased to allow 2 to 3 ml. of fluid for rinsing.

Determination of Serum Total Protein, Albumin, and Globulins

An adaptation of the biuret method to the estimation of protein fractions, separated by Kingsley's (6) modification of Howe's procedure (but at a 1 in 20 dilution), is described in detail. The same general technique can be applied to protein fractionations by the sulfite method of Campbell and Hanna (15), the alcohol method of Pillemer and Hutchinson (16), or the modified sulfate method of Majoor (17).

Reagents—

Biuret reagent. Weigh 1.50 gm. of cupric sulfate ($CuSO_4 \cdot 5H_2O$) and 6.0 gm. of sodium potassium tartrate ($NaKC_4H_4O_6 \cdot 4H_2O$); transfer to a *dry* 1 liter volumetric flask, and dissolve in about 500 ml. of water. Add with constant swirling 300 ml. of 10 per cent sodium hydroxide (prepared from stock, carbonate-free, 65 to 75 per cent sodium hydroxide solution). Make to volume with water, mix, and store in a paraffin-lined bottle. This reagent should keep indefinitely but must be discarded if, as a result of contamination or faulty preparation, it shows signs of depositing any black or reddish precipitate.

Globulin precipitants. (*a*) Sodium sulfate, 22.6 per cent; (*b*) ethyl ether. *Sodium chloride*, 0.9 per cent.

Procedure

Select three colorimeter cuvettes (or test-tubes) and mark them Cuvettes B (blank), T (total protein), and A (albumin).

Into Cuvette B pipette 2.0 ml. of 22.6 per cent sodium sulfate solution. This "blank" will serve for all the protein analyses being carried on at any one time.

Into a centrifuge tube measure, with an Ostwald pipette, 0.5 ml. of serum and add 9.5 ml. of 22.6 per cent sodium sulfate. Stopper the tube and mix thoroughly by inversion (not by shaking). *At once* transfer 2.0 ml. of the mixture to Cuvette T.

To the remaining serum-sulfate mixture add 3 ml. of ether, mix for 30 seconds, cap, and centrifuge. Slant the tube and transfer 2.0 ml. of the aqueous phase to Cuvette A.

Now into each of the three cuvettes pipette 8.0 ml. of biuret reagent and mix thoroughly by swirling. Allow these to stand for 30 minutes at room temperature (20–25°).

Using a photoelectric colorimeter, or spectrophotometer, transmitting maximally at 540 mμ, adjust it to 100 per cent transmission with the "blank" Cuvette B in position. Replace Cuvette B with Cuvettes T and A in turn and record the percentage transmission (or optical density) of each.

Obtain the concentration of total protein and of albumin in the serum by reference to the calibration curve; the total protein minus the albumin gives the globulin concentration.

Calibration Curve—Pipette 5.0 ml. of clear, normal serum into a stoppered, graduated cylinder, dilute to 50 ml. with 0.9 per cent sodium chloride, and mix.

Prepare in duplicate a series of nine cuvettes (or test-tubes) and into them pipette successively 2.0, 1.85, 1.70, 1.55, 1.40, 1.20, 1.00, 0.80, and 0.60 ml. of 0.9 per cent saline. Then, in the same order, add 0, 0.15, 0.30, 0.45, 0.60, 0.80, 1.00, 1.20, and 1.40 ml. of the diluted serum.[3]

Now pipette into each cuvette 8.0 ml. of biuret reagent, mix, and read after 30 minutes as above.

Kjeldahl nitrogen determinations can be carried out on the remaining diluted serum. The total nitrogen content of 100 ml. of serum, corrected for non-protein nitrogen from a separate determination on the original serum, gives protein nitrogen which, multiplied by 6.25, is taken as the standard protein concentration.

Plot the transmission or density values observed against 0.15, 0.30, 0.45, 0.60, 0.80, 1.0, 1.2, and 1.4 times the protein concentration of the standard serum.

Results

The calibration curves we have obtained using the reagent and the method described above obey satisfactorily the laws of Lambert and Beer.

[3] For work with 3.0 ml. samples use 0, 0.24, 0.45, 0.66, 0.9, 1.2, 1.5, 1.8, and 2.1 ml. of diluted serum, made to 3.0 ml. with saline. Plot the transmission values against 0, 0.16, 0.3, 0.44, 0.6, 0.8, 1.0, 1.2, and 1.4 times the protein concentration of the standard serum.

762   SERUM PROTEIN DETERMINATION

A slight loss of blue color must occur in the reaction because a proportion of the copper enters the protein "biuret" complex, but the effect on the extinction coefficient is scarcely apparent even at serum protein concentrations around 10 per cent.   In Fig. 7 a typical calibration curve is shown, plotted on semilogarithmic paper.   With the Evelyn colorimeter our K value $((2 - \log G)/\text{concentration})$ has averaged about 580 with a range in any one curve of ± 10.

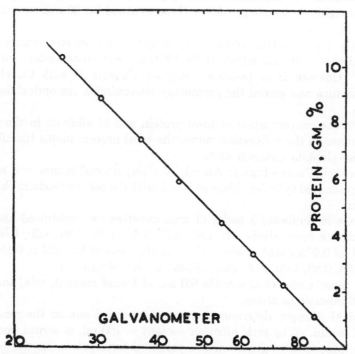

FIG. 7. Calibration curve for protein estimations with the biuret reagent, plotted on semilogarithmic graph paper.

Total protein and albumin analyses made with the modified biuret reagent have been compared with values calculated from Kjeldahl nitrogen determinations on the same sera; the results are shown in Table II.   It can be seen that the biuret and Kjeldahl results can be expected to agree within about 0.1 gm. per cent of protein.   The observed differences are statistically insignificant.

DISCUSSION

In our experience, the use of tartrate-stabilized biuret reagents offers a highly satisfactory method for estimating the proteins of serum or plasma.

The values obtained depend ultimately upon a Kjeldahl nitrogen determination, hence upon an ability to perform such an analysis accurately, and with the assumption of the factor 6.25. At present there appears to be no practical method of standardization superior to nitrogen analysis.

TABLE II

Results of Comparison of Total Protein and Albumin Analyses by Biuret and Kjeldahl Methods

The values are measured in gm. per cent.

Serum		Diagnosis	Total protein		Albumin	
			Kjeldahl	Biuret	Kjeldahl	Biuret
Human	A. G.	Normal	7.49	7.35		
	C. B.	"	7.01	7.05		
	B. B.	"	7.04	7.05		
	J. D.	"	6.70	6.60		
	Pooled	"	7.45	7.35		
	H. G.	"	7.62	7.80		
	F.	Pyelonephritis	4.50	4.50	1.40	1.40
	R.	Pneumonia	7.12	6.95	3.79	3.81
	G.	Rheumatoid arthritis	6.80	6.70		
	C.	Cirrhosis	6.97	7.05		
	L.	Obstructive jaundice (serum bilirubin, 25 mg. %)	5.81	5.82		
	M.	Obstructive jaundice	5.91	6.07		
	E.	Lipoid nephrosis	3.85	3.65		
	T.	Lupus erythematosus			3.25	3.20
Dog No.	11	Normal	5.67	5.70		
	15	"	5.84	5.85		
	17	"	6.60	6.60		
	10	CCl₄ liver injury	6.08	5.95	3.35	3.24
	12	" " "	7.20	7.15		
	16	" " "	6.45	6.40		
	17	" " "			2.72	2.88
	48	Uremia (non-protein N = 200 mg. %)	5.06	5.15		

The procedure described for total protein determinations involves redissolving the globulins precipitated by sodium sulfate or other agents. No difficulty has been experienced with this technique, nor has the presence of sulfate any effect upon the rate or degree of color development. Sodium sulfite and methanol (at a final concentration of 2.1 per cent) may cause a very slight increase in color intensity, but for practical purposes the effect is negligible.

It is pertinent to comment briefly on a few special points.

Interfering Factors—There are practically no substances other than protein in biological fluids which give the biuret reaction, certainly none that cause significant interference. The pigment, bilirubin, absorbs light very weakly at 540 mμ. We confirm Kingsley's (6) observation that serum levels as high as 25 mg. per cent introduce (at the dilution employed) an error not greater than 0.1 gm. per cent protein. Although ammonium ion is a disturbing factor in the biuret reaction, the amount present when a mixture of ammonium and potassium oxalates is used as anticoagulant will not cause any significant error. With the method described, ammonium sulfate cannot be used as a globulin precipitant, but its interference can be minimized (*a*) by analysis of the separated precipitate or (*b*) by using a reagent containing 8 to 10 per cent alkali.

Careful separation of the serum or plasma should avoid the disturbing effect of hemolysis, and, if the patient is in the fasting state, the specimen will almost always be clear. There are certain diseases and situations, however, which give rise to lipemic sera and it is then necessary to vary the technique somewhat if the results are to have any significance. Such sera yield a somewhat cloudy reaction mixture and clearing must be effected before readings can be made. For this purpose we have used Kingsley's (6) method of adding 3 ml. of ether. The phases are mixed by flicking the tube for a definite period, say 10 or 20 seconds; the mixture is then centrifuged in a capped tube and the aqueous phase transferred to a new cuvette for reading at 30 minutes. This technique introduces a dilution error which in our hands has amounted to about 3 per cent. It is necessary for each worker to establish this correction, not only by measuring the change in volume, but by noting the optical density of a clear reaction mixture before and after treatment with ether. The use of petroleum ether (ligroin, b.p. 70–90°) will avoid the dilution error, though occasionally the two phases may not separate quite so readily.

Age of Serum—Wokes and Still (18) have reported that the intensity of the biuret color reaction increases rather strikingly as plasma ages. Our own experience is not in accord with their findings. If serum, or oxalated plasma, has been removed within a reasonable time ($\frac{1}{2}$ to 2 hours) and stored under conditions in which it remains clear, no significant change has been observed in the protein concentration determined by the biuret reaction at $\frac{1}{2}$, 2, 4, 24, or 168 hours.

Biuret Reaction with Different Proteins—Although the exact nature of the "biuret" complex is still unknown, it is apparent that the accurate determination of serum protein fractions must presume a constant, average number of groupings which combine to give the reaction. Evidence supports the view that in this respect the biuret reaction is a more reliable quantitative method for protein estimation than one depending on the content of some particular amino acid. Autenrieth (1) and Robinson and Hogden (5) regarded the color development by serum albumin and globulins as virtually identical. Fine (3), Kingsley (12), and Wokes and Still (18) have reported apparent differences in behavior of these two fractions in the biuret reaction.

We have determined the K values $((2 - \log G)/C)$, which are comparable to extinction coefficients, for the proteins listed in Table III. The protein concentration (C) in each case has been calculated from a Kjeldahl nitrogen determination, corrected for non-protein nitrogen, and with the factor 6.25. The K values differ from the 580 previously mentioned because the reagent used for this experiment was of slightly different composition. It will be noted that, among the proteins tested, all the albumins and

TABLE III

K Values for Different Proteins, Comparable to Extinction Coefficients

Nature of protein	$K = \dfrac{2 - \log G}{C}$
Human serum albumin	520
" " globulins	514
" "abnormal" serum globulin (pptd. with 13.5% Na_2SO_3)	515
" serum (pooled)	520
Dog " "	517
Egg albumin	513
Casein	458
Zein	479
Gelatin (animal)	386

globulins show essentially identical values. The results with casein and zein differ to some extent, while gelatin shows a rather striking difference, one which was apparent visibly as a bluer tint.

The authors gratefully acknowledge the valued suggestions made by Dr. Andrew Hunter during the preparation of this report.

SUMMARY

Among the methods of preparing stable, "one piece" biuret reagents, the use of sodium potassium tartrate appears most promising.

The tartrate-stabilized reagent devised by Weichselbaum has been investigated in some detail and has been modified to contain less copper, more alkali, and less or no potassium iodide.

A simple procedure for the determination of serum total protein, albumin, and globulins is described.

766   SERUM PROTEIN DETERMINATION

The biuret reaction is approved as a simple, rapid, yet highly satisfactory and accurate method for the determination of the protein fractions in serum or plasma.

BIBLIOGRAPHY

1. Autenrieth, W., *Münch. med. Wochschr.*, **62**, 1417 (1915); **64**, 241 (1917).
2. Hiller, A., *Proc. Soc. Exp. Biol. and Med.*, **24**, 385 (1927).
3. Fine, J., *Biochem. J.*, **29**, 799 (1935).
4. Kingsley, G. R., *J. Biol. Chem.*, **131**, 197 (1939).
5. Robinson, H. W., and Hogden, C. G., *J. Biol. Chem.*, **135**, 707 (1940).
6. Kingsley, G. R., *J. Lab. and Clin. Med.* **27**, 840 (1942).
7. Mehl, J. W., *J. Biol. Chem.*, **157**, 173 (1945).
8. Weichselbaum, T. E., *Am. J. Clin. Path., Tech. Suppl.*, **10**, 40 (1946).
9. Zecca, A. M., *Semana méd.*, Buenos Aires, **2**, 709 (1947); *Chem. Abstr.*, **42**, 1621 (1948).
10. Campbell, W. R., and Hanna, M. I., *J. Biol. Chem.*, **119**, 1 (1937).
11. Ma, T. S., and Zuazaga, G.; *Ind. and Eng. Chem., Anal. Ed.*, **14**, 280 (1942).
12. Kingsley, G. R., *Proc. Am. Soc. Biol. Chem., J. Biol. Chem.*, **140**, p. lxix (1941).
13. Rising, M. M., and Johnson, C. A., *J. Biol. Chem.*, **80**, 709 (1928).
14. Shaffer, P. A., and Somogyi, M., *J. Biol. Chem.*, **100**, 695 (1933).
15. Campbell, W. R., and Hanna, M. I., *J. Biol. Chem.*, **119**, 9 (1937).
16. Pillemer, L., and Hutchinson, M. C., *J. Biol. Chem.*, **158**, 299 (1945).
17. Majoor, C. L. H., *Yale J. Biol. and Med.*, **18**, 419 (1946).
18. Wokes, F., and Still, B. M., *Biochem. J.*, **36**, 797 (1942).

page 251

COMMENTARY TO

18. Natelson, S. (1951)
Routine Use of Ultramicro Methods in the Clinical Laboratory. Estimation of Sodium, Potassium, Chloride, Protein, Hematocrit Value, Sugar, Urea and Nonprotein Nitrogen in Fingertip Blood. Construction of Ultamicro Pipets. A Practical Microgasometer for Estimation of Carbon Dioxide.
American Journal of Clinical Pathology 21(12): 1153–1172.

C linical chemistry has had a long history of commitment to the development of assays that require less and less sample volume. This has been driven by the need to reduce or eliminate venipunctures in special needs patients such as pediatric and neonatal patients, in diabetic patients who require daily multiple glucose assays and in burn patients. Rapid developments in instrumentation and methodologies over the years however quickly dates any attempt to define terms such as "micro" or "ultramicro". In 1955 a method for calcium that used 200 µL of serum was called a "micromethod" (1). Today, automated instruments routinely use 5 µL or less of sample and protein and DNA automatic array pipettes dispense 0.005 µL of sample.

The commitment to develop reduced sample volume assays in clinical chemistry began with Ivar C. Bang at the University of Lund in Sweden. In 1916 he published a book of clinical chemistry methods that described the collection of finger stick drops of blood on washed and dried pieces of filter paper (2). The filter paper was then weighed and used as the source of sample in various assays. Bang died at the age of 49 in 1918 and is regarded as the founder of clinical microchemistry (3,4).

Otto Folin recognized the need for reduced sample volume assays in order to decrease repeated venipunctures. In 1928 he published a new method for blood sugar that required only 100 µL of sample compared to his original Folin-Wu sugar method which used 1 000 µL of blood (5–7). Earl J. King in 1937 in *The Lancet* described 11 different basic clinical chemistry assays that each required only 200 µL of blood or serum (8). The assays were read with small sized cups and plungers retrofitted into the standard Duboscq colorimeter. In 1942 he presented in the same journal a method for blood sugar that required only 50 µL of sample (9). By 1947 all of his methods were adopted for use with the Klett-Summerson filter photometer (10). King's textbook of methods, *Micro-Analysis in Medical Biochemistry* was published in 1946 and went through four editions over the next 18 years (11).

Samuel Natelson at the Rockford Memorial Hospital in Rockford, IL developed a system of assays and techniques designed especially for the pediatric and neonatal patient. He prepared a booklet in 1950 on the physiology of the immature infant that included methods applicable for the neonate. The booklet was published in 1952 (12). The 1951 paper presented here presents the details of his system of clinical chemistry methods for the newborn, premature infants and the burn patient. Natelson describes the preparation of glass capillary tubes 1.5 mm in internal diameter, coated with anticoagulant and used for heel and fingertip collection of blood. These tubes are still sold today as the Natelson Micro-Collection Tubes (13). Sodium, potassium, chloride and total protein methods are presented in this paper that use a single 40 µL sample of plasma for the four assays. The Kopp-Natelson manometric gasometer invented by Natelson measured plasma carbon dioxide in 30 µL of sample. He describes the preparation of constriction pipettes with internal diameters of 0.5 mm along with methods for calibration using liquid mercury. Tables of data are presented comparing results between venous and capillary samples.

Natelson expanded on reduced sample volume assays and published his book, *Microtechniques of Clinical Chemistry* in 1957. The book went through three editions (14). In 1959 he published a special review article in the journal *Analytical Chemistry* titled "Microanalysis in the Modern Analytical Laboratory of Clinical Chemistry" (15). Natelson continued to publish new methods and techniques for use with neonates over the next 20 years. By 1976 he had developed a wide range of clinical chemistry methods that required only 1.0 to 10 µL of sample per test (16). Other authors published textbooks, manuals and papers devoted to pediatric methods during the 1950s and 1960s. All of them are indebted to the work by Natelson in clinical microchemistry (17–24).

References

(1) Harrison, H.E. and Harrison, H.C. (1955) A micromethod for the determination of serum calcium. Journal of Laboratory and Clinical Medicine. 46(4):662–664.

(2) Bang, I.C. (1916) Methoden zur Mikrobestimmung Einiger Blutbestandteile. J.F. Bergmann, Wiesbaden, Germany.

(3) Van Slyke, D.D. (1958) Ivar christian bang. Scandinavian Journal of Clinical Laboratory Investigation. 10(Supplement 31): 18–26.

(4) Schmidt, V. (1986) Ivar christian bang (1869–1918), Founder of modern clinical microchemistry. Clinical Chemistry. 32(1): 213–215.

(5) Folin, O. (1928) A new blood sugar method. Journal of Biological Chemistry. 77(2):421–430.

(6) Folin, O. (1929) Supplementary note on the new ferricyanide method for blood sugar. Journal of Biological Chemistry. 81(2):231–238.

(7) Folin, O. and Malmros, H. (1929) An improved form of Folin's micro method for blood sugar determination. Journal of Biological Chemistry. 83(1):115–120.

(8) King, E.J., Haslewood, G.A.D., and Delory, G.E. (1937) Micro-chemical methods of blood analysis. The Lancet. 229(5928): 886–892.

(9) King, E.J., Haslewood, G.A.D., Delory, G.E., and Beall, D. (1942) Micro-chemical methods of blood analysis. Revised and extended. The Lancet. 239(6181):207–209.

(10) King, E.J. and Delory, G.E. (1947) Photoelectric colorimeters, in Recent Advances in Clinical Pathology. Dyke, S.C. (ed), The Blakiston Company, Philadelphia, pgs 200–209, Chapter 20.

(11) King, E.J. (1946) *Micro-Analysis in Medical Biochemistry*. Grune & Stratton, New York.

(12) Natelson, S., Crawford, W.L., and Munsey, F.A. (1952) *Correlation of Clinical and Chemical Observations in the Immature Infant. A Working Manual for Physicians and Chemists*. Rockford Health Department, Rockford, IL.

(13) VMR International Catalog. (2003–2004) VWR International Inc., West Chester, PA 19380, pg 1517.

(14) Natelson, S. (1957) *Microtechniques of Clinical Chemistry*. Charles C. Thomas Publishers, Springfield, IL.

(15) Natelson, S. (1959) Microanalysis in the modern analytical laboratory of clinical chemistry. Analytical Chemistry. 31(3): 17A–30A, March.

(16) Natelson, S. (1976) The analytical laboratory of neonatalogy, in *Clinical Chemistry*, ACS Symposium Series 36, Forman, D.T. and Mattoon, R.W. (eds), American Chemical Society, Washington, DC, pgs 95–152.

(17) Kaplan, S.A. and del Carmen, F.T. (1956) Quantitative ultramicro-analysis for the clinical laboratory. Pediatrics. 17(6): 857–869.

(18) Knights, E.M., MacDonald, R.P., and Ploompuu, J. (1957) *Ultramicro Methods for Clinical Laboratories*. Grune & Stratton, New York.

(19) Saifer, A., Gerstenfeld, S., and Zymaris, M.C. (1958) Rapid system of microchemical analysis for the clinical laboratory. Clinical Chemistry. 4(2):127–141.

(20) Caraway, W.T. (1960) *Microchemical Methods for Blood Analysis*. C.C. Thomas Publisher, Springfield, IL.

(21) Wilkinson, R.H. (1960) *Chemical Micromethods in Clinical Medicine*. C.C. Thomas Publishers, Springfield, IL.

(22) van Haga, P.R. and de Wael, J. (1961) Ultramicro methods, in *Advances in Clinical Chemistry*, Vol 4. Sobotka, H. and Stewart, C.P. (eds), Academic Press, New York, pgs 321–350.

(23) Meites, S. and Faulkner, W.R. (1962) *Manual of Practical Micro and General Procedures in Clinical Chemistry*. C.C. Thomas, Springfield, IL.

(24) O'Brien, D., Ibbott, F.A., and Rodgerson, D.O. (1968) *Laboratory Manual of Pediatric Micro-Biochemical Techniques*, 4th Edition, Hoeber Medical Division, Harper & Row, Publishers, New York.

Am J Clin Pathol. 1951; 21: 1153–1172
© 1951 American Society of Clinical Pathologists.
Reprinted with permission.

KOPP-NATELSON MICROGASOMETER

is a patented ultra-micro precision instrument especially adapted to the determination of CO₂ and O₂ in blood by modification of the Van Slyke macro manometric method.

Reprinted from The American Journal of Clinical Pathology
Vol. 21, No. 12, December, 1951
Printed in U.S.A.

ROUTINE USE OF ULTRAMICRO METHODS IN THE CLINICAL LABORATORY

Estimation of Sodium, Potassium, Chloride, Protein, Hematocrit Value, Sugar, Urea and Nonprotein Nitrogen in Fingertip Blood. Construction of Ultramicro Pipets. A Practical Microgasometer for Estimation of Carbon Dioxide*

SAMUEL NATELSON, Ph.D.

From the Department of Biochemistry of the Laboratories of the Rockford Memorial Hospital, Rockford, Illinois

For more than a year a routine procedure has been in use in this hospital for analysis of blood from the fingertip or heel for certain chemical constituents. These ultramicro procedures are accurate, simple and rapid. They have been invaluable in following routinely the levels of many constituents of the blood in newborn or premature infants and in several subjects with severe burns, in whom the few available veins were reserved for administration of fluids. We have also used these procedures in patients with severe edema, extensive thrombosis or excessive obesity, in whom the performance of daily venipuncture was difficult. Figure 1 illustrates the scheme of analyses.

PROCEDURES

I. Drawing of the Sample

The tube in which the sample is drawn (Fig. 2) is made from ordinary pyrex tubing having a bore of 1.5 mm. and an outer diameter of 2.8 mm. Heat the tubing in an ordinary gas flame and draw out to a tip suitable for blood sampling (approximately a bore of 0.7 mm.). Break at the tip and 11 cm. from the tip with an ampoule file. Fire-polish the ends of the tube and clean by immersing in dichromate-sulfuric cleaning mixture and then flush with distilled water. Shake out excess water and dry in oven at 100 C. Just before use, draw up and expel the anticoagulant twice so as to coat the walls of the tube.

Puncture the site from which blood is to be drawn, with the tip of a Bard-Parker blade and wait until a large drop has collected. Allow the blood to enter the tube to a height of 5 to 7 cm. (approximately 0.1 ml.). This is done by holding the tube almost in a horizontal position to allow the blood to enter by capillarity. It is not advisable to aspirate the blood inasmuch as air bubbles may be introduced. If insufficient blood has been obtained, the tube may be touched to a second drop and the blood will still rise in the tube.

II. Hematocrit Reading

Hold the tube at an angle of approximately 30° with the wide end down and warm the wide end gently in the flame of a bunsen burner, alcohol lamp or

---

* Demonstrated at the Sixth National Chemical Exposition, September 5–9, 1950, Chicago, Illinois.
Received for publication, April 27, 1951.

cigarette lighter. Simultaneously heat DeKhotinsky cement or ordinary red sealing wax until soft and press against the large opening of the tube. While adhering cement is still soft, press the end of the tube gently on a flat surface holding the tube vertically. This is done to force a small amount of the cement up into the tube. The cement may be reheated if it has become too hard. Allow the tube to cool and the cement to harden. If an oxygen torch is available the wide end may be sealed with the torch without the use of cement.

After cooling, place the tube in an ordinary test tube and centrifuge until the length of the cell column remains constant. For a ⌗1 International Centrifuge this will require a speed of 2000 r.p.m. for ten minutes. Cut away one side of the cement with a sharp knife or spatula and measure the length of the whole blood column and red cell column with a set of calipers. If a set of calipers is not available a centimeter ruler with divisions 1 mm. apart will yield sufficiently accurate results.

$$\frac{\text{Length of red cell column}}{\text{Length of whole blood column}} \times 100 = \text{hematocrit reading}$$

III A. Sodium, Potassium, Chloride and Protein

Break the tube after scratching it with an ampoule file at a point slightly above the level of plasma. Remove 0.04 ml. of plasma and wash into approximately 1 ml. of distilled water. Make up to 2 ml. with distilled water. This solution will be referred to as the "plasma solution."

The sample (0.04 ml.) is measured by means of narrow-tipped constriction pipets described below. If constriction pipets are not available the plasma is first transferred to a small test tube. This is done by scratching the cut tube just above the erythrocyte level, inverting the tube into a small test tube and then breaking the tube with a downward motion. The plasma will then drop into the test tube. Any plasma which is held in the tube by capillarity is shaken or blown into the tube. The erythrocytes with a small amount of plasma, which is now contained in the cut tube sealed on one end with cement, are reserved for sugar and urea analysis as described below. The 0.04-ml. volume is now measured with an ordinary Sahli pipet* (0.02 ml.) into two portions. One portion of 0.02 ml. is made up to 1 ml. for estimation of sodium and potassium and the second 0.02-ml. portion is made up to 1 ml. for estimation of chloride and protein.

1. Sodium and potassium. Remove 1.0 ml. of plasma solution (0.02 ml. plasma) by means of a transfer pipet. To this add, with mixing, 4 ml. of lithium solution (0.5 mg. per ml.) as internal standard. Use approximately half of this solution for potassium and the remainder for sodium, employing a nozzle with a 0.3-mm. bore in the Perkin-Elmer† flame photometer.[12] For flame photometers in which

---

* Central Scientific Co., Chicago, Illinois; ⌗41065.

† Perkin Elmer Co., Norwalk, Conn. The sprayer supplied with this instrument normally cannot be used for the procedure described. Using a 0.3 to 0.4-mm. orifice for the fluid and a 1-mm. orifice for the compressed air, adjust the angle and relative position of the two so that 2 ml. sprays through in approximately 18 to 20 seconds at 10 pounds of air pressure.

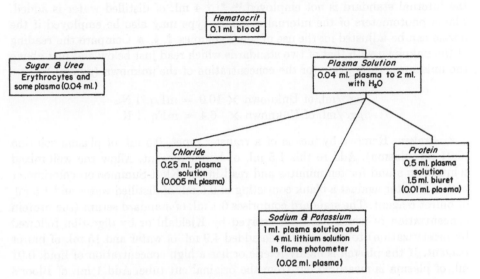

FIG. 1. Scheme of analysis for estimation of hematocrit value, Na, K, Cl, protein, sugar and urea on 0.1 ml. of whole blood drawn from fingertip or heel.

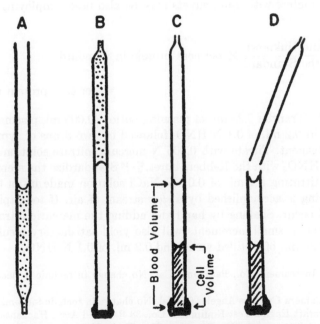

FIG. 2. Sequence of steps in estimating hematocrit value and preparing blood sample for further analysis. A. Blood suspended in bore pyrex tube of 1.5 mm. B. Inverted tube sealed with cement. C. After centrifuging, hematocrit reading is taken. D. Tube broken to allow sampling of plasma.

the internal standard is not employed,[14, *, †] 4 ml. of distilled water is added. Flame photometers of the internal standard type may also be employed if the nozzle can be adjusted for the use of small volumes. ‡, §, ⌗ Compare the reading of the unknown with that of two standards which read just below and just above the unknown. Interpolate for the concentration of the unknown (μg./ml.).

$$\mu\text{g. /ml. of Unknown} \times 10.9 = \text{mEq./l Na}$$
$$\mu\text{g. /ml. of Unknown} \times \ \ 6.4 = \text{mEq./l K}$$

2. Protein. Remove by means of a transfer pipet, 0.5 ml. of plasma solution (0.01 ml. plasma). Add to this 1.5 ml. of biuret reagent. Allow the well-mixed solution to stand for ten minutes and read in the Klett-Summerson colorimeter with a 54 filter against a blank consisting of 0.5 ml. of distilled water and 1.5 ml. of biuret reagent. The standard comprises 0.1 ml. of standard serum (the protein concentration of which has been assayed by Kjeldahl or by digestion followed by nesslerization) to which has been added 4.9 ml. of water and 15 ml. of biuret reagent. If the plasma is highly icteric or has a high concentration of lipid, 0.01 ml. of plasma is measured out from the original cut tube. Add 1 ml. of Bloor's reagent (3 parts of alcohol to 1 of diethyl ether) mix, centrifuge, decant the Bloor's reagent and add 1.5 ml. of biuret reagent and 0.5 ml. of water; mix until the precipitated protein is dissolved, and read as above. Other colorimeters or spectrophotometers with 2-ml. cuvets may·be also used, employing the 540-mμ wave length.

$$\frac{\text{Reading of the Unknown}}{\text{Reading of the Standard}} \times \text{per cent protein in Standard}$$

$$= \text{per cent protein in Unknown}$$

3. Chloride. Transfer 0.25 ml. of plasma solution (0.005 ml. plasma) to a 3-ml. test tube. Add 0.2 ml. of 0.1 N HNO_3 followed by two drops of sym. Diphenylcarbazone Reagent. Titrate with 0.005 N mercuric nitrate solution made up in 0.2 per cent HNO_3 with the Rehberg buret.[¶, 16] Standardize the mercuric nitrate solution by titrating 0.1 ml. of 0.01 N NaCl solution made up in 0.2 per cent HNO_3. Stirring is accomplished by a *slow* stream of air. If too rapid, excessive foaming will occur. Shaking by hand and adding the mercuric nitrate from the Rehberg buret in small increments will also yield satisfactory results. A blank consists of 0.25 ml. of distilled water and 0.2 ml. of 0.1 N HNO_3.

---

* Beckman Instrument Co., Pasadena, Cal. No change in technic is needed for 2-ml. volume.

† Fearless Camera Co., Los Angeles 64, Cal. No change in technic for 2-ml. volume.

‡ Janke Aircraft Engine Test Equipment Co., 38 Railroad Ave., Hackensack, N. J.

§ Process and Instruments Co., Brooklyn, New York, Flame Photometer Model 1b.

⌗ Patent Button Co., Barclay Instruments Division, Waterbury, Conn.

¶ Macalaster Bicknell Co., Cambridge, Mass; Emil Greiner Co., New York City; Arthur H. Thomas Co., Philadelphia, Pa.; Micrometric Instrument Co., Cleveland, O.; Microchemical Specialties Co., Berkeley, California.

$$\frac{2 \times (\text{Titration of Unknown} - \text{Blank})}{\text{Titration of Standard} - \text{Blank}} \times 100 = \text{mEq./l. chloride}$$

If the mercuric nitrate is exactly 0.005 N, the buret will read directly in mEq./l. of chloride. The titration will then be 100 divisions (0.1 ml.) for a chloride value of 100 mEq./l.

IIIB. Alternative Procedure for Na, K, Cl and Protein

When only certain of the constituents are being followed or when a specimen has an unusually high hematocrit value with relatively little plasma and one is limited to only certain of the procedures, then the samples are measured directly from the cut tube with the ultramicro pipets having the narrow tips, which are described below.

1. Sodium and potassium. Measure 0.01 ml. of plasma for sodium or potassium determinations into approximately 0.2 ml. of water, add 2 ml. of lithium solution and dilute to 2.5 ml. If the internal standard is not used, omit the lithium solution. Proceed as above.

2. Protein. Wash 0.01 ml. of plasma into approximately 0.2 ml. of water, add 1.5 ml. biuret reagent and make to 2.0 ml. and proceed as above.

3. Chloride. Wash 0.01 ml. of plasma from the cut tube into 0.2 ml. of water. Add 0.2 ml. of 0.1 N HNO_3 and proceed as above. In this case twice the amount of plasma is being used and the calculations become:

$$\frac{\text{Titration of Unknown} - \text{Blank}}{\text{Titration of Standard} - \text{Blank}} \times 100 = \text{mEq./l. chloride.}$$

IV. Sugar and Urea

Transfer the residuum of cells and serum in the original tube in which the blood was drawn to a 3-ml. test tube. This is done by scratching the tube just above the sealing wax with an ampoule file, holding the tube inverted into a small test tube and breaking off the sealed end. The mixture of cells and remaining plasma will fall into the test tube. Blow or shake any of the mixture which remains in the narrow tube into the test tube. This mixture may be used for the estimation of sugar and urea. If sufficient plasma is available it is preferable to do these analyses on the plasma rather than on this mixture.

Draw up 0.04 ml. of the mixture of plasma and cells or plasma alone and wash into 0.5 ml. of water. Make up to 2 ml. with tungstic acid, mix well. Centrifuge for five minutes at 2000 r.p.m. Mix a 1-ml. aliquot of the supernatant (equivalent to 0.02 ml. of mixture or plasma) with 1 ml. of urea-diacetyl reagent. To a 0.5-ml. aliquot (equivalent to 0.01 ml. of mixture or plasma) add 0.5 ml. of alkaline copper reagent. Heat both tubes in a boiling water bath for ten minutes. Do not allow the spray from the boiling water bath to enter the tubes. It is best to stopper the tubes loosely with corks. Cool tubes to room temperature in running water (3 minutes).

A. Urea. Read the color in the urea tubes in the Klett-Summerson colorimeter, with a 44 filter using a mixture of 1 ml. of water and 1 ml. of urea-diacetyl

1158   NATELSON

reagent as blank. Results are compared to the standard curve prepared by running 2.5 μg. and 5 μg./ml. standards.[13] These are best read in the 4-cm. length cuvets described.[13] In the Coleman Spectrophotometer a 475 mμ wave length is used.

B. Sugar. The tube with the alkaline copper reagent is diluted to 2 ml. with phosphomolybdic acid (1 ml.). After standing for a few minutes the color is read in the Klett-Summerson colorimeter with a 42 filter against a blank containing 0.5 ml. of water, 0.5 ml. alkaline copper reagent and 1 ml. of phosphomolybdic acid, treated similarly to the unknown. Prepare a sugar standard by heating 0.5 ml. of dilute sugar standard with 0.5 ml. alkaline copper reagent and then diluting to 2 ml. with phosphomolybdic acid as described for the unknown.

$$\frac{\text{Reading of the Unknown}}{\text{Reading of the Standard}} \times 100 = \text{mg. sugar per 100 ml. blood}$$

For higher readings a 54 filter may be used but with this filter a standard curve (not a straight line) must be prepared.

V. Nonprotein Nitrogen

Wash 0.01 ml. of plasma into 0.2 ml. of water and add 0.5 ml. of 15 per cent trichloracetic acid solution. Mix, centrifuge for five minutes at 2000 r.p.m. Pour supernatant into a second tube and transfer a 0.5-ml. aliquot of the supernatant to a Pyrex test tube (13 × 100 mm.) constricted to approximately 8-mm. bore at a 2-ml. mark. Add 0.2 ml. of the digestion mixture and place in a steel rack on a hot plate at an angle of 30° from the horizontal. Set the hot plate to medium heat until the solution becomes clear. Total heating time should be from 15 to 25 minutes. Cool, add approximately 0.5 ml. of water, 1 ml. of Nessler's Reagent against the wall of the tube, and dilute to the 2-ml. mark. Read in the Klett-Summerson colorimeter with the 44 filter. Blank comprises 0.2 ml. of water and 0.5 ml. of trichloracetic acid from which a 0.5-ml. aliquot has been taken to be treated like the unknown. The value of the reading of the unknown is compared to the standard obtained by mixing 0.5 ml. of the dilute nitrogen standard (8 μg./ml.) with 0.2 ml. of the digestion mix, 0.3 ml. of water and 1 ml. of Nessler's Reagent. A separate blank is made for the standard containing 0.8 ml. of water, 0.2 ml. of acid digestion mix and 1 ml. of Nessler's Reagent. This blank is also used to estimate the amount of nitrogen picked up from the trichloracetic acid and the air by the blank which was digested. A standard curve is prepared by analyzing stands containing 2, 4, 6, 8, and 10 μg./ml. as described above. The curve obtained is a straight line passing through the origin when results are read as color density.

$$\frac{\text{Reading of the Unknown}}{\text{Reading of the Standard}} \times 56.8 = \text{mg. N per 100 ml.}$$

VI. Estimation of Concentration of CO_2*

Collect the sample under oil, allow to clot, ream and centrifuge. If fingertip blood is used, the finger is punctured and the punctured surface is immersed in a

---

* SCIENTIFIC INDUSTRIES, INC. 132 Front Street, New York 5, N. Y. manufactures an instrument of the design described.

test tube full of oil and the blood is allowed to drop into the tube. Light mineral oil is employed. The tube used is a standard 1-ml. pyrex centrifuge tube the wide end of which is flared. Thus a funnel is produced with a diameter of 1 to

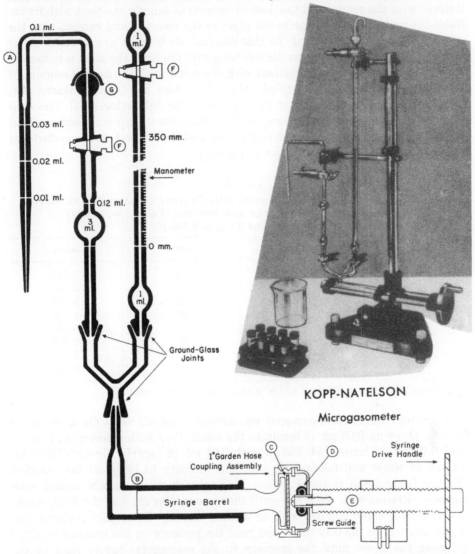

KOPP-NATELSON

Microgasometer

FIG. 3. Ultramicro manometric gasometer. A. 0.5-mm. bore pyrex capillary tube joined to 1-mm. bore pyrex capillary tube. B. 10-ml. syringe and side tube sealed on, as shown. C. Metal blank (from electric junction box). D. Ball bearing for rotating screw without turning coupling. E. Machined screw ¾″ D, 20 threads per inch, recessed and tapped at one end to receive machine screw. F. Vacuum stop-cock. G. Ball-and-socket ground-glass joint.

1.5 cm. at its widest end, leading into 1-ml. centrifuge tube. This tube may be centrifuged in the usual trunnion cups.

In introducing the standard, lactic acid, water or NaOH solutions into the instrument, air bubbles are avoided by having mercury in the test tube containing the standard, water and alkali and caprylic alcohol covering the lactic acid. Place the tip of the pipet (Fig. 3A) into the liquid to be drawn up, advance the screw until a globule of mercury protrudes from the tip or drops off into the tube. Retreat with the screw until the desired amount of sample has been withdrawn. Raise the test tube and immerse the pipet in the mercury and retreat with the screw to form a mercury seal. In this manner, air bubbles are avoided. After measuring the lactic acid, lower the test tube until the tip of the pipet is immersed in the caprylic alcohol layer. Retreat with the screw until the desired amount of caprylic alcohol has been sampled. After the serum has been measured out, lower the tube so that the tip of the pipet is in the light mineral oil. Draw up approximately 0.005 ml. of mineral oil. Now immerse the tip into mercury held in a test tube. Advance the screw so that most of the mineral oil is expelled and then draw up the required amount of mercury.

TABLE 6

CORRECTION FACTORS FOR ESTIMATION OF CO_2 CONTENT (VOLUME/100 ML. SERUM)
CALCULATED FROM VAN SLYKE AND SENDROY. FACTOR $=$ (0.5353)
$(1/1 + 0.00384\ T)\ (1 + 0.0601\alpha')$

TEMP. C.	FACTOR*	TEMP. C	FACTOR
17	0.536	25	0.513
18	0.533	26	0.510
19	0.529	27	0.508
20	0.526	28	0.506
21	0.524	29	0.504
22	0.522	30	0.502
23	0.518	31	0.500
24	0.516	32	0.497

Values for $(1/1 + 0.00384\ T)$ and α' may be found in reference 18.

Observing the above-mentioned precautions, analysis for CO_2 is made as follows. Draw up 0.03 ml. of serum to the mark. Plug with mercury and follow with 0.03 ml. of lactic acid. Follow with 0.01 ml. of caprylic alcohol. Draw up 0.1 ml. of water and follow with sufficient mercury to bring all the reaction mixture beyond the stopcock. Close the stopcock which is now sealed with mercury. Retreat with the screw until the mercury has reached the 3-ml. mark. Let stand for three minutes with occasional shaking. Bring the meniscus (caprylic alcohol) to the 0.12 mark and read the pressure on the manometer (P_1). Advance the screw until the mercury in the manometer barely rises to the manometer stopcock, open the reaction chamber stopcock. Advance the mercury in the pipet until a globule of mercury protrudes from the tip. Draw up 0.1 ml. of 1 N NaOH followed by sufficient mercury to bring all of the NaOH into the reaction chamber. Close the stopcock and evacuate to the 3-ml. mark as before. Promptly bring the meniscus to the 0.12-ml. mark. Read the pressure again (P_2). Clean the instrument by expelling the aqueous contents. Draw up 0.1 ml. of water and bring to the 3-ml. mark leaving the stopcock open. Expel the water. Repeat with 0.1 ml. of lactic acid followed by 0.1 ml. of water.

Calculation. The aqueous layer is well in the 3-ml. bulb. For this reason the correction which has to be subtracted for depression of the mercury owing to the addition of 0.1 ml. of NaOH, is small and of the order of 0.2 mm. This is determined by adding all the reagents for estimating CO_2 but substituting water for the unknown and taking a reading. Add 0.1 ml. of NaOH solution and take a second reading. The difference is the correction C.

$$(P_1 - P_2 - C) \times \text{factor (Table 6)} = CO_2 \text{ concentration}$$
$$\text{(volumes per 100 ml. serum.)}$$

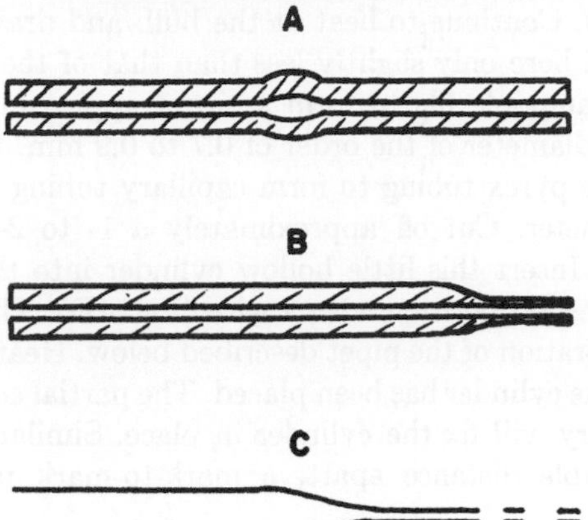

FIG. 4. Construction of constriction-type pipets. A. 0.5-mm. bore capillary tube heated and blown. B. Pipet drawn from A. C. 4 mm.-bore tubing drawn to less than 0.5 mm. O.D. D. Glass cylinder in place; constriction-pipet. E. Mark-to-mark constriction-pipet.

APPENDIX

I. Calibration of Test Tubes

2-ml. and 1-ml. calibrated pyrex test tubes are available under the label "Volumetric Flask Test Tube Shape". However, they are easily prepared from 3-ml. test tubes by pouring in 27.2 Gm. of mercury and scratching with a diamond pencil at the mercury meniscus for 2-ml. capacity. The marking of these tubes is best done with the electric vibrating tools sold in hobby shops. This mercury poured from tube to tube allows rapid calibration of many tubes. In this manner tubes can be calibrated for any capacity.

For NPN a pyrex tube (13 x 100 mm.) is heated at a predetermined 2-ml. mark and allowed to collapse until the diameter at the 2-ml. mark is of the order of 8 mm. The 2-ml. mark is now accurately determined with mercury and a scratch made there with a diamond pencil.

IIA. Preparation of Constriction Type of Pipets

Select pyrex capillary tubing of 0.5-mm. bore. Seal at one end by drawing out in an oxygen torch. Approximately 15 cm. from one end, heat and blow out a small bulb (Fig. 3A). Continue to heat at the bulb and draw out. A narrow tip is formed of internal bore only slightly less than that of the main tube. Scratch at the tip with an ampoule file to form a tip of approximately 2 to 3 cm. in length and of outer diameter of the order of 0.7 to 0.9 mm. (Fig. 4B). Draw out ordinary 3-mm. bore pyrex tubing to form capillary tubing of slightly less than 0.5 mm. outer diameter. Cut off approximately a 1- to 2-mm. length of this capillary (Fig. 4C). Insert this little hollow cylinder into the rear of the pipet and push into the desired position with a wire (Fig. 4D). This position is determined from the calibration of the pipet described below. Heat in flame to softness at the point where the cylinder has been placed. The partial collapsing of the walls of the heavy capillary will fix the cylinder in place. Similarly by inserting two cylinders at a suitable distance apart, a mark-to-mark pipet is constructed (Fig. 4E).

IIB. Calibration of Constriction Pipets

Draw approximately 100 mg. of mercury into the pipet. Mark with marking crayon, the point to which the meniscus of mercury has been drawn. Draw the mercury back to a uniform part of the tube and measure its length with calipers or centimeter ruler. Now blow the mercury out into a tared watch glass and weigh. Assuming the weight of the mercury is 100 mg. and its length L then for 0.01 ml. (135.5 mg.) its length should be $135.5/100 \times L$ in the uniform part of the tube. The difference between this calculated length and the measured length is the length which has to be added to the mark made when the mercury extended to the tip, since the meniscus was in the uniform part of the tube. Measuring from the original mark, this difference is added and a new mark made to which the constriction cylinder is pushed. After the constriction cylinder is sealed in place the pipet is calibrated by delivering from it 0.1 N NaCl standard and titrated in the Rehberg buret with standardized 0.005 N mercuric nitrate. The pipet may also be standardized colorimetrically by delivering from it phenolsulfonphthalein (PSP 6 mg./ml.), diluting to 2 ml. and comparing the color with PSP diluted 1:200, in the Klett-Summerson colorimeter employing the 54 filter.

For mark-to-mark pipets the length of a mercury column weighing 135.5 mg. is determined as above, placing the constriction cylinders this distance apart. Make the tip at a suitable point by blowing out a bulb and drawing out as described. In this pipet the tip is made as close as possible to one of the constriction cylinders to minimize holdback (Fig. 4E). This pipet is calibrated in the same manner as the constriction type.

III. Construction of Overflow Pipets

Determine the distance in the 0.5-mm. bore capillary which contains 0.01 ml. (135.5 mg. Hg) or any desired volume, as described above. Make two small bulbs separated by this distance and draw out and cut at both ends (Fig. 5A and B). Slip a collar, 5 mm. in length cut from soft rubber tubing, over this tubing. Make a holder by flaring 8-mm. bore pyrex tubing and sealing a 2-mm. bore tubing as shown in Figure 5C. This 8-mm. bore tubing may be connected to the 2-mm.

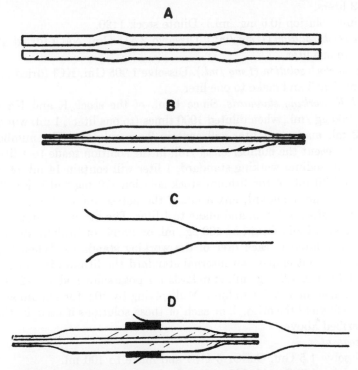

FIG. 5. Construction of overflow-type pipet. A. 0.5-mm. bore capillary heated and blown. B. Pipet drawn from A at both ends. C. Pipet holder. D Assembled pipet; B and C joined with rubber-collar cut from soft rubber tubing.

bore tubing by means of a one-holed rubber stopper. Assemble the pipet by slipping the holder over the rubber collar as shown in Figure 5D. This pipet is calibrated by weighing the mercury, titrating the 0.1 N NaCl solution or reading the color of dye it will deliver, as described above for the constriction pipets.

The pipets are washed by aspirating through them successively 10 per cent ammonium hydroxide, water and methyl alcohol.

IV. Reagents

All reagents are of analytical reagent grade. The solvent is distilled water unless otherwise specified.

Anticoagulant.

   a. Heparin (3 mg./ml.); Dissolve 300 mg. heparin and make to 100 ml.

   b. Ammonium oxalate. Saturated ammonium oxalate (2.5 per cent approximately) is employed.

   c. Ammonium citrate (pH 7.0). Dissolve 3 grams of citric acid dihydrate in 50 ml. Add concentrated NH_4OH until neutral to litmus. Make the volume to 100 ml.

Stock lithium solution (10 mg./ml.). Dissolve 200 grams of lithium nitrate and make to 2 liters.

   Lithium solution (0.5 mg./ml.). Dilute stock 1:20.

Sodium stock solution (1 mg./ml.). Dissolve 2.541 Gm. NaCl (dried at 100 C.) and make to one liter.

Potassium stock solution (1 mg./ml.). Dissolve 1.905 Gm. KCl (dried at 100 C.), add 30 Gm. NaCl and make to one liter.

Na and K working standards. Since 1 ml. of the stock K and Na solutions contains 1000 μg./ml., when diluted 1000 times (to one liter), 1 ml. will contain 1 μg./ml., 2 ml. made to 1 liter will contain 2 μg./ml., etc. The number of ml. used will represent the number of μg./ml. in the solution made to 1 liter. Thus, for 14 μg./ml. sodium working standard, 1 liter will contain 14 ml. of the stock standard and 40 ml. of the lithium stock solution (10 mg./ml.). For 3 μg./ml. working potassium standard, mix 3 ml. of the potassium stock solution, 40 ml. of the lithium stock solution and make to 1 liter. The lithium content of each of the working standards should be 400 μg./ml. or 40 ml. of lithium stock solution (10 mg./ml.) diluted in each liter of the working standards. When the instrument used does not employ an internal standard the lithium is omitted. Make 0.25, 0.5, 1.75, 1.0, 1.25 μg./ml. standards for potassium and 11, 12, 13, 14, 15 and 16 μg/ml. standards for sodium. Multiplying by 10.9 for sodium and 6.4 for potassium will yield the mEq./l. of each of these solutions if used in the procedures described above.

Biuret reagent.

   a. Dissolve 1.5 Gm. $CuSO_4 \cdot 5\ H_2O$ and make to 100 ml.

   b. Dissolve 160 Gm. NaOH and make to 1 liter. Keep bottle well-stoppered.

   On the day of the test, mix 10 ml. of the copper sulfate solution (*a*) with 50 ml. of the NaOH solution (*b*).

Mercuric nitrate 0.005 N. Dissolve 0.834 Gm. $Hg(NO_3)_2 \cdot \frac{1}{2}\ H_2O$ in 100 ml. of distilled water containing 2 ml. concentrated HNO_3 and make to 1 liter.

s-diphenylcarbazone (0.1 per cent). Dissolve 100 mg. s-diphenylcarbazone (Eastman Kodak Co.) and make to 100 ml. with methyl alcohol. Warm to dissolve. Keep in dark brown dropping-bottle in refrigerator.

Sodium chloride 0.1 N. Dissolve 5.845 Gm. of NaCl (dried at 100 C.) add 2 ml. concentrated HNO_3 and make to 1 liter with distilled water. Keep bottle well stoppered to avoid evaporation. Dilute 1:10 with 0.2 per cent nitric acid for 0.01 N.

Tungstic acid.

   a. Make 4.17 ml. of concentrated H_2SO_4 (sp. gr. 1.84) to 1 liter (0.15 N).

b. Dissolve 22 Gm. sodium tungstate dihydrate and make to 1 liter (2.2 per cent).

Mix equal parts of 0.15 N H_2SO_4 (a) and 2.2 per cent sodium tungstate (b) on the day of the test.

Alkaline copper reagent.

a. Dissolve 50 Gm. $CuSO_4 \cdot 5 H_2O$ and make to 1 liter.

b. Dissolve 70 Gm. Na_2CO_3, 22 Gm. $NaHCO_3$, 26 Gm. sodium tartrate and make to 2 liters.

Mix one part of the 5 per cent copper sulfate (a) and nine parts of the alkaline tartrate (b) on the day of the test.

Phosphomolybdic acid.

a. Dissolve 300 Gm. sodium molybdate in 500 ml. distilled water, add 2 drops of bromine and make to 1 liter.

b. Add 135 ml. concentrated sulfuric acid to 500 ml. of distilled water and make to 1 liter (25 per cent by weight).

In a 1-liter volumetric flask add 500 ml. of clear supernatant 30 per cent sodium molybdate (a), 225 ml. 85 per cent phosphoric acid and 150 ml. of 25 per cent sulfuric acid (b) Aspirate or blow air through the solution to remove the bromine, add 75 ml. glacial acetic acid and make to 1000 ml. with distilled water.

Sugar stock standard (1 mg./ml.). 1.000 Gm. glucose is dissolved and made to 1 liter with 0.25 per cent benzoic acid.

Dilute sugar standard. Dilute stock 1:50 with 0.25 per cent benzoic acid to make 20 μg./ml.

Urea–diacetyl reagent.

a. Make 5 ml. of diacetyl (Eastman Kodak Co.) to 100 ml. with 95 per cent alcohol. Keep in refrigerator.

b. Add 750 ml. 85 per cent phosphoric acid to 1000 ml. distilled water, add 250 ml. concentrated H_2SO_4 and mix.

On the day of the test, measure 1 ml. of the 5 per cent diacetyl solution (a) into a 25 ml. volumetric flask and make to mark with the acid mix (b). Use within one hour, keeping the solution in the refrigerator when not in use.

Urea stock standard (1 mg.N/ml.). Dissolve 2.144 Gm. urea (dried in a desiccator) and make to 1 liter. Keep in refrigerator under toluene.

Urea dilute standard.

a. 5 μg./ml. (dilute Stock 1:200.)

b. 2.5 μg./ml. (dilute Stock 1:400.)

Add one drop of chloroform to each and keep in refrigerator.

Ammonium stock standard (1 mg.N/ml.). Dissolve 2.360 Gm. $(NH_4)_2SO_4$ (dried in a desiccator) and make to 500 ml.

Dilute nitrogen standard (8 μg./ml.). Dilute 2 ml. of the ammonium stock standard to 250 ml. in a volumetric flask.

Digestion mix for NPN. Dissolve 25 mg. HgO, 25 Gm. K_2SO_4 in 300 ml. of water to which has been added 100 ml. concentrated H_2SO_4 and make to 1 liter.

Trichloracetic acid (15 per cent). Dissolve 15 Gm. of trichloracetic acid and make to 100 ml.

1166 NATELSON

Nessler's reagent.

a. Dissolve 500 Gm. NaOH and make to 1 liter. Allow carbonate to settle.

b. Dissolve 51 Gm. KI in 200 ml. of water. Add 16.2 Gm. HgO and shake until dissolved. Add 160 ml. of 50 per cent NaOH (*a*) and make to 1 liter.

CO_2 standard (50 volumes per 100 ml.) Dissolve 1.191 Gm. Na_2CO_3 (anhydrous, dried at 100 C.) and make to 500 ml. Keep under light mineral oil. This solution will tend to dissolve CO_2 from the air if exposed. Transfer 2 ml. of this solution to a small test tube, add 2 ml. of mercury and cover with mineral oil.

Lactic acid solution. Make 90 ml. lactic acid (85 per cent) to 1 liter. Transfer 5 ml. of this solution to a test tube and cover with 2 ml. of caprylic alcohol.

Sodium hydroxide (1 N). Dissolve 40 Gm. NaOH and make to 1 liter. Transfer 5 ml. of this solution to a test tube add 5 ml. of mercury and stopper well.

Distilled water (for CO_2). Add 5 ml. distilled water to test tube, add 5 ml. of mercury and keep well-stoppered.

DISCUSSION

It is conventional to refer to procedures performed on milligram quantities as micro procedures and those performed on microgram quantities as ultramicro procedures.[8] The procedures described are, for the most part, ultramicro procedures.

The scheme of analysis as shown in Figure 1 permits of wide variation in the type of analysis to be performed. The sample drawn can be used for many of the analyses normally performed in the clinical laboratory but not shown here. For example, salicylate and sulfa levels have been followed serially in infants that have received these drugs. Sampling of the plasma directly with ultramicro pipets has advantages when individual analyses are to be performed. Occasionally the hematocrit value is so high, in cases of dehydration, that very little plasma is obtained. In these cases sampling by means of the ultramicro pipets permits of analysis of some of the constituents which are considered of greatest importance in the particular case.

In comparison with the usual methods, the ultramicro methods (Table 1) for all constituents, with the exception of urea, give results that are fairly close. It is well known that the urea content of cells is approximately 20 per cent lower than that of plasma.[1, 23] It is thus to be expected that analysis of the residuum in the tube comprising all the cells and some plasma will give lower results. However, these results are still of value for clinical purposes. As can be seen from Table 2 the results obtained for urea compare well with those obtained by the micro procedures when done on the same plasma. Urea estimation is recommended on the mixture only when a high hematocrit concentration limits the amount of plasma available.

Nonprotein nitrogen content of erythrocytes is as much as 40 per cent higher than that obtained on the plasma. For this reason the cell plasma residue cannot be used for this determination. The procedure is therefore described for plasma. Since tungstic acid will bring down certain of the amino acids (arginine, histidine, tryptophane), trichloracetic acid is used as the protein precipitant. The tri-

TABLE 1

COMPARISON OF RESULTS OBTAINED BY ULTRAMICRO PROCEDURES WITH THOSE OF MACRO PROCEDURES ON BLOOD OF SAME PATIENTS. VENOUS SERUM WAS USED FOR MACRO PROCEDURES AND CAPILLARY PLASMA FOR ULTRAMICRO PROCEDURES

PATIENT NO.	MACRO RESULTS							ULTRAMICRO RESULTS						
	Hem*	Na	K	Cl	Urea	S	Pr	Hem	Na	K	Cl	Urea	S	Pr
1	40	141	4.6	100	11.0	94	6.5	41	143	4.7	99	9.0	95	6.5
2	35	136	3.9	94	23.2	120	5.5	36	135	4.0	95	19.1	125	5.4
3	32	128	5.5	86	50.1	85	4.5	32	130	5.5	87	44.6	86	4.5
4	69	148	6.1	110	16.2	70	6.0	71	149	6.0	110	13.6	75	6.2
5	52	145	5.1	107	18.3	69	7.1	54	143	4.9	107	15.5	70	7.1
6	39	139	4.5	103	9.0	87	6.7	40	136	4.6	102	7.5	91	6.5
7	38	140	4.7	101	10.0	91	6.5	38	139	5.0	100	9.2	95	6.4
8	39	138	4.4	102	11.1	92	6.8	37	139	4.1	103	9.2	92	6.7
9	37	145	2.9	107	29.2	141	5.0	39	146	3.1	106	26.5	144	5.1
10	33	138	4.6	109	9.8	95	5.6	36	136	4.3	109	8.0	98	5.7
11	45	147	5.1	108	20.1	84	4.2	44	148	4.9	107	18.9	80	4.1
12	32	143	4.7	99	18.2	124	5.6	34	140	4.9	101	15.1	120	5.6
13	29	136	2.5	92	15.3	75	4.8	28	141	2.1	92	12.8	81	5.0
14	42	138	4.5	104	15.6	250	6.8	42	138	4.7	103	14.1	255	6.9
15	49	165	3.6	123	34.6	164	5.3	48	168	3.0	125	31.2	161	5.2
Averages........	40.7	141.8	4.5	103	19.4	109	5.8	41	142	4.3	103	17.0	111	5.8

Sugar and urea carried out on a mixture of cells and plasma after approximately 60 per cent of the plasma originally present in the blood had been removed.

* Hem. indicates hematocrit reading in per cent; S, sugar; Pr, protein. Values for Na, K, Cl are in mEq./l.; for sugar and urea, in mg./100 ml.; for protein, in grams per 100 ml.

1168   NATELSON

chloracetic acid must be assayed for nitrogen content since many samples tested
contained sufficient nitrogenous substances to interfere in the ultramicro procedure.

From Table 1 it can be seen that the results obtained for sugar by both the
macro and the ultramicro method are sufficiently close to be within experimental
error of the macro method. The ultramicro method was performed on the mixture of cells and residual plasma of capillary blood while the macro method was
done on venous blood. Cells and plasma contain amounts of sugar which are
within 5 per cent of each other. Since capillary blood contains somewhat higher
concentrations of sugar than venous blood the accuracy of the ultramicro method
is not sufficient to show this difference consistently.

The hematocrit value as obtained by the ultramicro method was generally
higher than that found by the macro procedure. This might be accounted for by
the fact that sedimentation in the narrow tube (1.5-mm. bore) is slower than in

TABLE 2

COMPARISON OF RESULTS OF ULTRAMICRO PROCEDURES WITH THOSE OF MACRO PROCEDURES
ON 15 CONSECUTIVE ANALYSES ON POOLED SERA, TO SHOW
CLOSENESS OF AGREEMENT

		Sugar[8]	Urea[18]	NPN[15]	Na[12]	K[12]	Hem*[22]	Protein[7]	Cl[17]
		Mg. per 100 ml.			mEq./l.		%	Gm./100 ml.	mEq./l.
Average	Macro	92	12.1	32	139	4.6	41.0	6.4	103
	Ultramicro	91	11.8	36	139	4.5	41.5	6.4	103
Standard deviation	Macro	±1.3	±0.4	±0.3	±2.0	±0.10	±0.7	±0.15	±1.2
	Ultramicro	±1.9	±0.6	±0.6	±2.4	±0.13	±0.7	±0.20	±0.9

* Hematocrit value: All samples of oxalated blood were centrifuged simultaneously at
2000 r.p.m. for 10 minutes in International Centrifuge, size #1.

the wider tube. However centrifuging for longer periods did not seem to affect
the results appreciably. Actually lower results were expected on capillary blood
since oxygenated cells are smaller than cells having a lower oxygen concentration,
owing to a shift of water outside the cells during oxygenation.

The precision obtainable under routine conditions is compared in Table 2 for
macro and ultramicro procedures. Only for chloride estimation is the ultramicro
procedure more regularly reproducible. This is probably because titration with
the Rehberg buret does not require fractionation of drops and the length of the
buret employed for titration of 0.01 ml. of serum is approximately twice the length
(0.2-ml. or 200 scale divisions) of the buret used in the macro titration in our
laboratory.

The most satisfactory anticoagulant tested was heparin. This substance had
the least tendency to distort the cells and was most efficient in preventing clotting.
Ideally, the anticoagulant used should not be devoid of sodium but should have
a concentration of sodium of the order of that found in the plasma (3.2 mg./ml.).
If heparin (3 mg./ml.) is prepared from the solid state, an excess of concentra-

tion of sodium over that of the plasma is obviously impossible. The 3 mg./ml. solution of heparin used* contained 0.24 mg. of sodium/ml. Potassium concentration was negligible. Since the amount of heparin used is small (of the order of 0.003 ml. to wet the walls of the tube), interference in the estimation of sodium and potassium was within experimental error for the macro procedure (Table 1). The ammonium oxalate and ammonium citrate used, while more readily available, would not keep the blood from clotting for as long a period of time as the heparin. Samples taken with these anticoagulants could not be used for the estimation of NPN. Both of these anticoagulants cause cell distortion.

It is advisable that the cement be applied to the wide end of the tube in which the blood is drawn, at the bedside. A downward snap of the wrist will then drive the blood to the bottom of the tube. In this manner the last of the blood sample entering the tube is exposed to the action of the anticoagulant. With heparin the sample can then remain for several hours without clotting.

The accuracy of the ultramicro methods depends essentially on the accuracy with which the original sample is pipetted. The ultramicro pipets for measuring volumes of the order of 0.01 ml. need an automatic stop point in order to attain a precision of 1 per cent. Two types have been in general use: the constriction type[10, 11] and the overflow type. The conventional constriction type is made by heating a capillary tube until the tube partly collapses, to form ideally a point of lesser diameter. After the liquid has been drawn beyond the point of constriction it is blown forward gently. Increased resistance will be experienced at the point of constriction. The tip is now wiped and the contents expelled by blowing more vigorously. In actual practice it is difficult to achieve a sharp constriction. The operator must decide which point to choose since the constriction is usually a gradual one. For this reason narrow-walled tubing is usually chosen so as to permit production of sharp constrictions. This results in fragile pipets.

The constriction type of pipet described above is easily made from heavy-walled pyrex tubing. The site of constriction is definite since the upper edge of a little cylinder acts as the constriction. The site of constriction is at a point rather than broad and gradual.

The overflow type of pipet depends upon the same principle used in many automatic burets. The fluid is allowed to overflow an orifice drawn to capillary dimensions and the excess is discarded. Capillary forces hold the liquid at the zero point and at the tip of the pipet. When blown forward, a constant amount is delivered.

Ultramicro pipets achieve the greatest precision when washed out with the fluid into which the sample is to be measured. In measuring samples from a micropipet, one must never draw the wash-water above the mark since plasma adhering to the walls above the mark will be washed in.

Of the ultramicro pipets described, the overflow type has the greatest precision (Table 3). However because this pipet is wasteful of plasma we have used the constriction type for routine work. The mark-to-mark constriction type is particularly useful for the falling-drop method of protein determination. For

* Glogau & Co., Chicago, Illinois.

1170   NATELSON

this procedure pipets which deliver 0.005 ml. are used. They need not be calibrated with great accuracy, since the same pipet is used for delivery of the standard and the plasma.

The device for estimation of concentration of CO_2 (Fig. 3) is familiar to those accustomed to the use of the Van Slyke-Neill apparatus. No pipet is needed for sampling, the pipet being contained in the instrument. The volume is measured

TABLE 3

COMPARISON OF RESULTS OBTAINED WITH CONSTRICTION TYPE PIPET (WASHOUT) AND AND MOHR TYPE PIPET OF 0.1-ML. CAPACITY (KIMBLE #37020, BLUE LINE EXAX), MEASURED BY DELIVERING 0.1 N NaCl AND TITRATING WITH STANDARDIZED MERCURIC NITRATE SOLUTION (0.005 N) IN 15 CONSECUTIVE DETERMINATIONS

	MOHR PIPET*	CONSTRICTION PIPET†
Mean volume (ml.)	0.0991	0.00850
Standard deviation	±0.0011	±0.00009
Per cent standard deviation from average	1.11	1.06

* Titrated in Machlet buret, 5-ml. capacity.

† Titrated in Rehberg buret, 0.2-ml. capacity. Errors include those attributable to procedure and titration.

TABLE 4

COMPARISON OF VALUES OBTAINED ON 10 CONSECUTIVE SERA FOR CO_2 CONTENT (VOL. PER 100 ML.) CLOTTED VENOUS BLOOD UNDER OIL

NO.	MACRO	*ULTRAMICRO
1	40.1	39.7
2	61.3	60.7
3	75.5	75.9
4	115.0	113.2
5	26.1	25.7
6	16.4	16.5
7	50.8	50.0
8	56.6	55.1
9	57.5	55.2
10	65.4	64.1
Averages	56.5	55.6

* Calculated with the Van Slyke-Sendroy factors shown in Table 6.

with precision because the screw permits careful control. Tables 4 and 5 indicate that in routine use the instrument gives results which are as precise and accurate as those of the macro device. This instrument lends itself to direct sampling of blood from the fingertip. This is useful in estimation of the CO content of the blood in cases of suspected poisoning. In contrast to the macro instrument, mercury losses are minimal because of the small volumes of mercury needed for the instrument.

Estimation of CO_2 content of serum on ultramicro quantities has been limited to instruments which measure the volume of CO_2 evolved[5, 9, 18] after the method of Van Slyke.[19] These procedures have the disadvantage that other gases present in serum such as oxygen and nitrogen are measured. Experimentation by the author over many months indicated that one of the difficulties experienced with ultramicro volumetric measurements of CO_2 over an aqueous medium is that the volume of the liberated bubble of gas constantly decreases with time, owing to redissolving of CO_2 brought to atmospheric pressure, following liberation under vacuum. Thus, unless readings are taken quickly, erroneous results are obtained.

The instrument described above has the advantages of the Van Slyke-Neill[20] apparatus while measuring the CO_2 concentration of only 0.03 ml. of serum. The CO_2 is measured under reduced pressure (approximately $\frac{1}{8}$ atmosphere). Only the gas soluble in alkali is measured. This allows for specificity in analyzing for CO_2. Similarly the apparatus can be used for oxygen and CO estimation. It is of interest to note that recent studies seem to indicate that the spectrophotometric method for oxygen estimation are not as accurate as the gasometric methods.[2]

TABLE 5

COMPARISON OF RESULTS OBTAINED WITH VAN SLYKE-NEILL APPARATUS AND THE ULTRAMICRO APPARATUS IN 15 CONSECUTIVE DETERMINATIONS ON A SODIUM CARBONATE STANDARD (50 VOLUME PER 100 ML. SERUM) STANDARDIZED BY TITRATION WITH STANDARD SULFURIC ACID

	MACRO*	ULTRAMICRO*
Average	49.4	48.7
Standard deviation	1.6	1.3
Per cent standard deviation of average	3.2	2.6

* Calculated with the Van Slyke-Sendroy factors.[21]

Our experience with the procedures described herein have clearly shown that these procedures are practical, can be easily learned by the average technician, and when routinely used require no greater skill than that required for the macro procedures.

SUMMARY

Procedures are described for the estimation of sodium, potassium, chloride, protein, hematocrit value, nonprotein nitrogen, sugar and urea on one sample of blood drawn from a site of puncture of the heel or fingertip. A new micro-gasometer is described for the estimation of CO_2 on 0.03 ml. of serum. Directions are given for the construction of simple ultramicro pipets to be used in these procedures. Data are given comparing the results obtained with the ultramicro methods with those obtained by the standard macro procedures.

The accuracy and precision achieved with the ultramicro methods are comparable to those obtained with the macro procedures carried out under routine conditions. The methods described are simple and often easier to perform than the macro procedures.

1172   NATELSON

REFERENCES

1. BERGLUND, H.: Nitrogen retention in chronic interstitial nephritis and its significance. J. A. M. A., **79:** 1375–1382, 1922.
2. CHASTONAY, J. L.: Oxygen capacity of the blood and its measurement. Schweiz. Ztschr. f. Tuberk., **7:** 117–128, 1950.
3. FOLIN, O.: Two revised copper methods for blood sugar determinations. J. Biol. Chem., **82:** 83–93, 1929.
4. FOX, C. E.: Stable internal standard flame photometer. Analyt. Chem., **23:** 137–142, 1951.
5. FÜRST, V., JR., AND MØRSTAD, O.: Microgasometric determination of CO_2 with the Scholander-Roughton syringe analyzer. J. Clin. Lab. Investigation, 1: 258–262, 1949.
6. GLICK, D.: Techniques of Histo- and Cytochemistry. New York: Interscience Publishers, 1949, 525 pp.
7. KINGSLEY, G. R.: The direct biuret method for the determination of serum proteins as applied to photoelectric and visual colorimetry. J. Lab. and Clin. Med., **27:** 840–845, 1942.
8. KIRK, P. L.: Quantitative Ultramicroanalysis. New York: John Wiley and Sons, 1950, 310 pp.
9. LAZAROW, A.: In Medical Physics. Edited by GLASSER, O. Vol. II, p. 498, Year Book, Chicago, 1950.
10. LEVY, M.: A micro Kjeldahl estimation. Compt. rend. d. trav. du Lab., Carlsberg, serie chim., **21:** 101–110, 1936.
11. LOWRY, O. H., AND HUNTER, T. H.: The determination of serum protein concentration with a gradient tube. J. Biol. Chem., **159:** 465–474, 1945.
12. NATELSON, S.: The routine use of the Perkin-Elmer flame photometer in the clinical laboratory. Am. J. Clin. Path., **20:** 463–472, 1950.
13. NATELSON, S., BEFFA, C. E., AND SCOTT, M. L.: A rapid method for the estimation of urea in biologic fluids. Am. J. Clin. Path., **21:** 275–281, 1951.
14. PROEHL, E. C., AND NELSON, W. P.: The flame photometer in determination of sodium and potassium. Am. J. Clin. Path., **20:** 806–813, 1950.
15. RAPPAPORT, F.: Rapid and Microchemical Methods for Blood and C S F Examinations. New York: Grune and Stratton, Inc., 1949, 404 pp.
16. REHBERG, P. B.: A method of micro titration. Biochem. J., **19:** 270–278, 1925.
17. SCHALES, O., AND SCHALES, S. S.: A simple and accurate method for the determination of chlorides in biological fluids. J. Biol. Chem., **140:** 879–884, 1941.
18. SCHOLANDER, P. F., AND IRVING, L.: Micro blood gas analysis in fractions of a cubic millimeter. J. Biol. Chem., **169:** 561–569, 1947.
19. VAN SLYKE, D. D.: A method for the determination of carbon dioxide and carbonates in solution. J. Biol. Chem., **30:** 347–368, 1917.
20. VAN SLYKE, D. D., AND NEILL, J.: The determination of gases in blood and other solutions by vacuum extraction and manometric measurement. J. Biol. Chem., **61:** 523–573, 1924.
21. VAN SLYKE, D. D., AND SENDROY, J. JR.: Carbon dioxide factors for the manometric blood gas apparatus. J. Biol. Chem., **73:** 127–144, 1927.
22. WINTROBE, M. M.: A simple and accurate hematocrit. J. Lab. and Clin. Med., **15:** 287–289, 1929.
23. WU, H.: Separate analyses of the corpuscles and plasma. J. Biol. Chem., **51:** 21–31 1922.

COMMENTARY TO

19. Huggett, A. St. G. and Nixon, D. A. (1957)
Use of Glucose Oxidase, Peroxidase, and o-Dianisidine in Determination of Blood and Urinary Glucose. The Lancet 270(6991): 368–370.

The enzyme glucose oxidase (GO, EC 1.1.3.4) catalyzes the following reaction,

$$\text{(i) glucose} + O_2 + H_2O \rightarrow \text{D-glucuronic acid} + H_2O_2$$

Keilin and Hartree in 1948 were the first to use GO in an enzymatic assay for glucose in biological fluids (1). They added catalase (EC 1.11.1.6) to the above reaction mixture and converted the hydrogen peroxide to oxygen and water. The production of oxygen was measured with a manometer. Froesch and Renold (2) in 1956 used GO to improve the specificity of the standard Somogyi–Nelson copper reduction method for glucose (3). They measured glucose in samples incubated with GO then subtracted those values from matching samples assayed without GO.

Albert S. Keston from the New York University College of Medicine was the first to use GO to develop a direct quantitative colorimetric assay for glucose in biological fluids. The method was presented at a meeting of the American Chemical Society (ACS) in April 1956 (4). Urine samples were mixed with GO and the hydrogen peroxide produced coupled with horseradish peroxidase (HRP, EC 1.11.1.7) in the following reaction,

$$\text{(ii) } H_2O_2 + o\text{-dianisidine} \rightarrow H_2O + \text{oxidized } o\text{-dianisidine}$$

The oxidation of o-dianisidine produced a brown chromophore that was read at 480 nm. Six months latter Joseph Teller from Worthington Biochemical Corp. presented a paper at an ACS meeting in which he adopted Keston's enzymatic method for use with serum (5). This assay became the first commercial quantitative enzymatic glucose procedure for serum and was sold as the Worthington Glucostat™ test kit.

Huggett and Nixon in London adopted Keston's method for use with protein free filtrates of plasma. Their method was presented at a Society of Physiology meeting in March 1957 (6). In August 1957 they published the first full paper on the use of GO for the quantitative enzymatic assay of glucose in plasma and urine. That paper is presented here. They used o-dianisidine as oxygen acceptor with either protein free filtrates or direct assay with plasma. This paper went on to become the second most highly cited paper of all the papers published in *The Lancet* between 1961 and 1983 (7).

White and Secor in 1957 used o-tolidine coupled to a GO–HRP reaction mixture in a spray reagent to produce a blue chromophore to detect glucose in paper chromatography strips (8). Middleton and Griffiths in that same year adopted the Huggett and Nixon procedure for use with the oxygen acceptor o-tolidine (9). Trinder in 1969 replaced the potentially carcinogenic oxygen acceptors being used with p-aminophenazone (10–11). Kadish, Little and Sternberg used GO with a polarographic oxygen electrode to measure plasma and urine glucose in 20 seconds by monitoring the rate of oxygen consumption. This procedure developed into the first dedicated glucose analyzer for the clinical laboratory and was sold as the Beckman Model 777 Glucose Analyzer (12–13).

References

(1) Keilin, D. and Hartree, E.F. (1948) The use of glucose oxidase (notatin) for the determination of glucose in biological material and for the study of glucose-producing systems by manometric methods. Biochemical Journal. 42(2):230–238.

(2) Froesch, E.R. and Renold, A.E. (1956) Specific enzymatic determination of glucose in blood and urine using glucose oxidase. Diabetes. 5(1):1–6.

(3) Nelson, N. (1944) A photometric adaptation of the Somogyi method for the determination of glucose. Journal of Biological Chemistry. 153(2):375–380.

(4) Keston, A.S. (1956) Specific colorimetric enzymatic analytical reagents for glucose. *Abstracts of Papers*, 129th Meeting of the American Chemical Society: Dallas, Texas, April 8 to 13, 1956, American Chemical Society, Washington, DC, pgs 31C–32C.

(5) Teller, J.D. (1956) Direct, quantitative, colorimetric determination of serum or plasma glucose. *Abstracts of Papers*, 130th Meeting of the American Chemical Society: Atlantic City, N.J., September 16 to 21, 1956, American Chemical Society, Washington, DC, pg 69c.

(6) Huggett, A.St.G. and Nixon, D.A. (1957) The determination of blood glucose using mixed enzyme–oxygen acceptor reagent. Journal of Physiology. 137(1):3P.

(7) Garfield, E. (1984) 100 Classics from *The Lancet*. Current Comments. 39:295–305.

(8) White, L.M. and Secor, G.E. (1957) Glucose oxidase with iodide–iodate–starch or *o*-tolidine as a specific spray for glucose. Science. 125(3246):495–496.

(9) Middleton, J.E. and Griffiths, W.J. (1957) Rapid colorimetric micro-method for estimating glucose in blood and C.S.F. using glucose oxidase. British Medical Journal. 2(5060):1525–1527.

(10) Trinder, P. (1969) Determination of blood glucose using 4-aminophenazone as oxygen acceptor. Journal of Clinical Pathology. 22(2):246.

(11) Trinder, P. (1969) Determination of glucose in blood using glucose oxidase with an alternative oxygen acceptor. Annals of Clinical Biochemistry. 6(1–2):24–27.

(12) Kadish, A.H., Little, R.L., and Sternberg, J.C. (1968) A new and rapid method for the determination of glucose by measurement of rate of oxygen consumption. Clinical Chemistry. 14(2):116–131.

(13) Kadish, A.H. and Sternberg, J.C. (1969) Determination of urine glucose by measurement of rate of oxygen consumption. Diabetes. 18(7):467–470.

The Lancet. 1957; Vol. 270, no. 6991: 368–370
Reprinted with permission from Elsevier.

USE OF GLUCOSE OXIDASE, PEROXIDASE, AND O-DIANISIDINE IN DETERMINATION OF BLOOD AND URINARY GLUCOSE

A. St. G. Huggett

M.B., D.Sc., Ph.D. Lond.

PROFESSOR OF PHYSIOLOGY IN THE UNIVERSITY OF LONDON

D. A. Nixon

M.Sc. Lond.

LECTURER IN PHYSIOLOGY

From the Physiology Department, St. Mary's Hospital Medical School, London

It is well known that the estimation of blood-glucose by reducing methods lack specificity and give values which are too high. Enzymes, however, enable a greater selectivity of the reducing substances to be made.

In the presence of glucose oxidase, which is highly specific for glucose, glucose is oxidised to gluconic acid and hydrogen peroxide (Keilin and Hartree 1948). The hydrogen peroxide can, in the presence of peroxidase, oxidise a suitable oxygen acceptor to give chromogenic oxidation products (Keston 1956, Teller 1956), the intensity of whose colour is proportional to the amount of glucose initially present.

White and Secor (1957) have demonstrated glucose on paper chromatograms by adding glucose oxidase and detecting the resultant gluconic acid by an acid-iodide-iodate-starch reaction, and the hydrogen peroxide by using peroxidase and an oxygen acceptor.

The method described here is based upon that of Teller (1956) and has been found suitable for extensive determinations of blood-glucose, particularly in the presence of fructose. Since it seems to be of value for both clinical and research purposes, the factors determining its accuracy deserve publication. A preliminary account has been published (Huggett and Nixon 1957a and b).

Method

Reagent

A mixed enzyme-oxygen acceptor reagent is prepared as follows :

Glucose oxidase 125 mg., peroxidase 0·5 mg., and 0·5 ml. of 1% o-dianisidine (in 95% ethanol) are taken and made up to 100 ml. with 0·5 M $NaH_2PO_4,2H_2O$ (adjusted to pH 7·0 with NaOH) and filtered through paper. This preparation is called here the reagent.

The sources of the enzymes, when not prepared from *Penicillin notatum* (glucose oxidase) or horse-radish (peroxidase), is important.

Glucose-oxidase preparations are obtainable from the Sigma Chemical Company, St. Louis, Missouri, U.S.A.; G. F. Boehringer and Soehne, G.M.B.H., Mannheim, Germany; and Takamine Chemical Company, New Jersey, U.S.A. The peroxidase is obtainable from Boehringer and from Nutritional Biochemical Corporation, Cleveland, U.S.A. A mixture of the two enzymes and oxygen acceptor is marketed in the U.S.A. under the proprietary name of 'Glucostat' by the Worthington Corporation, New Jersey, U.S.A. The work in this paper was done with the Sigma glucose oxidase and the Boehringer peroxidase.

The reagent may be stored in a brown glass bottle at –5°C with little loss of activity for about a week. It goes brownish owing to auto-oxidation of the o-dianisidine, but this colour is allowed for in the blank.

Anticoagulation and Inhibition of Glycolysis

Anticoagulation may be achieved either with heparin or with a mixture containing sodium fluoride 10 mg. and potassium oxalate 30 mg. per 10 ml. of blood (Bayliss 1950). The latter method is used where the prevention of glycolysis in stored blood is desirable. A tenfold increase in the concentration of this mixture is still without any effect on the efficiency of the reagent.

Deproteinising

Concordant results can be achieved using zinc, barium, or cadmium hydroxides or ethanol as deproteinising reagents.

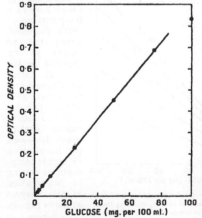

Fig. 1—The linear development of colour with increasing glucose concentrations in 0·1 ml. of glucose solutions and 2·5 ml. of reagent (read in 0·5-cm. cell after 60 minutes' incubation at 35°C).

Procedure

Whole blood.—It has been found convenient to use 0·2 ml. of a 1 : 10 deproteinised whole-blood filtrate. To this volume is added 2·5 ml. of the reagent. After 60 minutes' incubation in a water-bath thermostatically controlled at 35°–37°C the brownish-orange colour which develops is compared in a Spekker absorptiometer with 601 filters or in a Hilger spectrophotometer at 420 mμ with the colours produced by a set of glucose standards at 5, 10, and 15 mg. per 100 ml. treated in the same manner.

In all cases the volume of the unknown filtrates and standard solutions and/or the volume of the reagent may be varied to produce a suitable depth of colour commensurate with the concentration of glucose expected.

Plasma.—Glucose may be estimated directly in plasma without deproteinisation by this method, the inherent colour

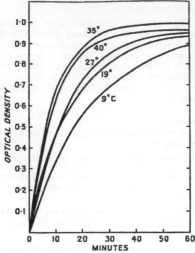

Fig. 2—Effects of different incubation temperatures on colour development in 0·1 ml. of 100 mg. per 100 ml. solution of glucose and 5 ml. of reagent (read in 1-cm. cell).

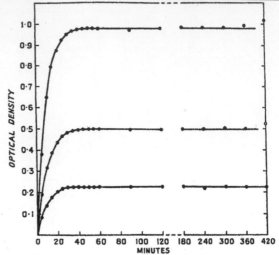

Fig. 4—Development and stability of colour of three different concentrations of glucose in 0·1-ml. glucose solutions with 5 ml. of reagent (read in 1-cm. cells after incubation at 35°C).

of the plasma being allowed for by its inclusion in a blank with phosphate buffer or heat-inactivated reagent.

Urinary glucose.—Direct determination of glucose in fresh urine, with an appropriate blank, leads to poor recoveries. However, these difficulties may be overcome by first treating the urine with charcoal to decolorise it ; a direct determination can then be made in the filtrate.

Results

Compliance with the Beer-Lambert Law.—The linear relationship of colour produced with increasing glucose concentrations is shown in fig. 1, the sensitivity being 1 µg. per 0·1 ml. By increasing the volume of the reagent to 10 ml. a straight-line relationship was obtained over the range 0–500 µg.

Variability.—The standard deviations of ten replicates with 10 and 25 µg. of glucose, expressed as a percentage of the mean, were 4·3% and 3·8% respectively.

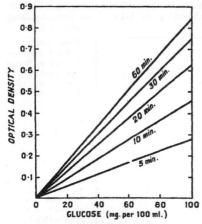

Fig. 3—Effects of different lengths of incubation at 30°C on development of colour, showing linear response in 0·1-ml. volumes of 25, 50, and 100 mg. per 100 ml. solutions of glucose with 5 ml. of reagent (read in 1-cm. cells).

Influence of incubation temperature.—With increased temperature there is an increase in the rate of colour development up to about 35°–37°C. The extent and rate of colour development decreased on incubation at higher temperatures owing to inactivation of the enzyme systems (fig. 2).

Influence of incubation time.—The linear relationship of developed colour with concentration is maintained at varying incubation times (fig. 3),

maximum development being attained after 45–60 minutes.

Stability of developed colour.—After 45–60 minutes' incubation the colour development is at a maximum and remains stable for about 5 hours (fig. 4). Accurate readings beyond this length of time cannot be made, because of progressive turbidity.

Influence of pH.—The system appears to have a pH optimum at 7·0, and a significant reduction in colour development is apparent when the pH value lies 0·5 of a unit either side of the optimum. The buffer contained in the reagent, however, has a high buffering capacity and can withstand the addition of acid or of alkali contained within the test solution under normal physiological conditions.

Optimum concentrations of glucose oxidase and peroxidase.—Increasing the concentration of glucose oxidase with a constant concentration of peroxidase increases the rate of colour development, but the degree of chromogenesis attained is reduced at high concentrations, possibly because of the presentation of the hydrogen peroxide to the peroxidase at a greater rate than it can utilise (fig. 5). In the presence of a constant concentration of glucose oxidase the optimum rate of colour development is given by 5·0 mg. of peroxidase (fig. 6).

Fig. 5—Influence of glucose-oxidase concentration on colour development in 0·1 ml. of 100 mg. per 100 ml. solution of glucose and 5 ml. of reagent containing variable amounts of glucose oxidase in the presence of 5 mg. of peroxidase per 100 ml. of reagent (read in 1-cm. cells after incubation at 35°C).

However, this colour is only 5% more than that given by a reagent containing only 0·5 mg. of peroxidase per 100 ml.

Recovery of added glucose.—Good recoveries are obtained by adding known weights of glucose to whole blood and urine. The results are shown in table I.

Parallel determination.—The comparison of the method with three other methods commonly used for determining blood-glucose showed good agreement (see table II).

Discussion

The use of enzymes in the determination of biological substances often enables a greater accuracy, due to high specificity, to be achieved. Though glucose is preferentially oxidised by glucose oxidase, galactose and mannose are also attacked. However, the rate of reaction of these two hexoses, as judged by the development of colour with the reagent, is considerably slower than that of glucose. With equivalent concentrations the degree of chromogenesis given by galactose and mannose is only about 0·06% and 0·80% respectively of that given by glucose. The phosphate esters glucose-1-phosphate and glucose-6-phosphate do not serve as substrates for the enzymic reaction.

Luntz (1957) notes that ascorbic acid may interfere with the development of colour when 'Clinistix' (commercial enzyme-impregnated paper for the rough quantitative assessment of urinary glucose) is used; but, when the present reagent was used with glucose solutions, concentrations of ascorbic acid equivalent to 5 mg. per 100 ml. of whole blood diluted 1 : 10 caused no reduction in chromogenesis.

The method is particularly suited for normal routine determinations of blood-glucose. The reagent is simple

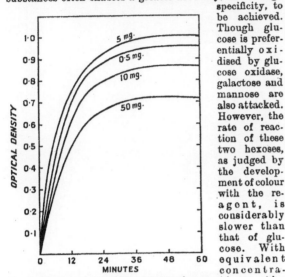

Fig. 6—Influence of peroxidase concentration on colour development in 0·1 ml. of 100 mg. per 100 ml. of solution of glucose and 5 ml. of reagent containing variable amounts of peroxidase in the presence of 125 mg. of glucose oxidase per 100 ml. of reagent (read in 1-cm. cells after incubation at 35°C).

TABLE II—PARALLEL DETERMINATION OF GLUCOSE IN WHOLE BLOOD BY REAGENT AND BY OTHER METHODS

Material = stored sheep blood.

	mg. per 100 ml.
Reagent	32
Hagedorn-Jensen	38
Somogyi colorimetric	33
Somogyi iodometric	35

The values cited represent the means of duplicate determinations of single samples of blood.

to prepare, and the method involves few operative steps. Since maximum chromogenesis takes place at 35°–37°C within 60 minutes and the colour is stable for several hours, accurate timing is unnecessary. When the assessment of the glucose concentration is urgent, a shorter incubation time can be used, but then accurate timing is essential. This is helped by adding 0·5 ml. of 0·05% sodium-cyanide solution, which inhibits further enzymic reaction.

Summary

A method applicable to the routine determination of blood-glucose is described which uses a mixed enzyme-oxygen acceptor reagent.

REFERENCES

Bayliss, R. I. S. (1950) Practical Procedures in Clinical Medicine. London : p. 17.
Huggett, A. St. G., Nixon, D. A. (1957a) *Biochem. J.* 66, 12P.
— — (1957b) *J. Physiol.* 137, 3P.
Keilin, D., Hartree, E. F. (1948) *Biochem. J.* 42, 230.
Keston, A. S. (1956) *Abstr. Amer. Chem. Soc.* 129th Meeting 31C.
Luntz, G. (1957) *Brit. med. J.* i, 499.
Teller, J. D. (1956) *Abstr. Amer. Chem. Soc.* 130th Meeting 69C.
White, L. M., Secor, G. E. (1957) *Science*, 125, 495.

TABLE I—RECOVERY OF ADDED GLUCOSE TO WHOLE BLOOD AND URINE

Whole blood (initial concentration 64 mg. per 100 ml.)		Urine (initial concentration 1 mg. per 100 ml.)	
Glucose added (mg.)	Glucose recovered (mg.)	Glucose added (mg.)	Glucose recovered (mg.)
3	4	1·2	1
3·5	5	7·4	5
10·6	13	17·1	17
20·8	20	33	34
38·4	35	82·9	80

page 279   **20. Wide, L. and Gemzell, C. A. (1960)**
An Immunological Pregnancy Test. Acta Endocrinologica (Copenhagen) 35:261–267.

COMMENTARY TO

I n the 1951 movie *People Will Talk*, Cary Grant played a physician who in an early scene has just told his patient, played by Jeanne Crain, that she is pregnant. She slumps into the office chair, her face buried in her hands. She asks if he is certain, the test was only done this morning she says. Grant assures her that the results are accurate and continues,

> Nowadays we find out about everything a lot more quickly than we used to, about life and even about death. They used to use a little pink rabbit for the pregnancy test. Now they use a frog, not as cute, but it's a lot faster. Only two hours and just as certain. The name of the frog, by the way, is *Rana pipiens*. Sounds like a movie star doesn't it? (1).

The use of animals for the laboratory detection of human pregnancy began with Aschheim and Zondek in 1927 and ended in 1960 with the introduction of an immunoassay for human chorionic gonadotropin (HCG) by Wide and Gemzell.

Aschheim and Zondek developed the first bioassay for pregnancy. It came to be called the A–Z test. Subcutaneous (sc) injections of HCG positive urine into female mice produced visible red hemorrhages in the animal's ovaries due to corpus luteum formation (2). The test took up to four days to complete. Improvements in the assay and the use of female rats reduced the test time to 6–24 hours (3–4). Friedman and Lapham injected urine into the ear vein of rabbits and were able to read the ovarian response in 12–24 hours (5). Galli-Mainini introduced the use of the male toad (*Bufo arenarum*) along with a new bio-marker for the detection of HCG (6–7). Urine was injected into the lateral lymphatic sacs of male toads. Their urine was collected over the next 1, 2 and 3 hours. HCG positive urines stimulated the release of spermatozoa into the toad's urine. Microscopic examination of the urine easily detected a positive response. Wiltberger and Miller applied this assay in 1948 to the male frog *Rana pipiens* (8).

The replacement of bioassay pregnancy tests began in July1960 with the preliminary report by Brody and Carlstrom on a complement fixation (CF) assay for serum and urine HCG (9). Antibody to HCG was produced in rabbits and used with sensitized sheep red cells in a 24-hour assay. McKean in September 1960 reported on the use of HCG antibody produced in rabbits in a precipitin ring (PR) test (10). The assay was run in 2×15 mm glass tubes. An antibody dilution was carefully overlaid onto a sample of the patient's urine or serum in the bottom of the tube. A positive HCG test was obtained with the appearance of a visible precipitin ring at the urine-antibody interface at 1, 2, 3 or 4 hours.

Wide and Gemzell were the first to report on a rapid immunoassay for HCG using hemagglutination inhibition (HI). Their paper was published in October 1960 and is presented here. For their assay they coated HCG onto formalin fixed and tannic acid treated sheep red cells. They prepared HCG antibody in rabbits. In the HI assay patient urine, sensitized red cells and HCG antibody were incubated in 12 mm diameter round bottom glass test tubes at room temperature. Assays were read visually at 1.5–2 hours. HCG if present in the urine preferentially bound to the antibody in solution leaving the red cells free to fall to the bottom of the tube and form a compact ring of cells. In the absence of HCG the antibody bound to the red cells formed cross bridges between the cells. The cells agglutinated and fell to the bottom of the tube in a clumped mat.

In 1962 Robbins, Hill, Carle, Carlquist and Marcus developed a similar agglutination assay using 0.8 μm latex beads (11). Over the next 2 years both HI tube tests and latex agglutination slide pregnancy tests became commercially available (12–14). In 1969, Cabrera evaluated 6 different commercial pregnancy test kits based on either latex or red cell agglutination inhibition (15). The kits were manufactured by Wampole, Organon, Ortho and Hyland, early pioneer companies in the development of commercial pregnancy test kits (16).

References

(1) *People Will Talk.* (1951) Staring Cary Grant, Jeanne Crain, Hume Cronyn and Walter Slezak. Written and Directed by Joseph L. Mankiewicz, Produced by Darryl F. Zanuck. Twentieth Century Fox, DVD Video™ 2003.

(2) Aschheim, S. (1930) The early diagnosis of pregnancy. Chorion-epithelioma and hydantidiform mole by the Aschheim-Zondek test. American Journal of Obstetrics and Gynecology. 19(3):335–342.

(3) Frank, R.T. and Berman, R.L. (1941) A twenty-four hour pregnancy test. American Journal of Obstetrics and Gynecology. 42(3):492–496.

(4) Zondek, B., Sulman, F., and Black, R. (1945) The hyperemia effect of gonadotropins on the ovary. Journal of the American Medical Association. 128(13):939–944.

(5) Friedman, M.H. and Lapham, M.E. (1931) A simple, rapid procedure for the laboratory diagnosis of early pregnancies. American Journal of Obstetrics and Gynecology. 21(3):405–410.

(6) Galli-Mainini, C. (1947) Pregnancy test using the male toad. Journal of Clinical Endocrinology. 7(9):653–658.

(7) Galli-Mainini, C. (1948) Pregnancy test using the male batrachia. Journal of the American Medical Association. 138(2):121–125.

(8) Wiltberger, P.B. and Miller, D.F. (1948) The male Frog, Rana pipiens, as a new test animal for early pregnancy. Science. 107(2773):198.

(9) Brody, S. and Carlstrom, G. (1960) Estimation of human chorionic gonadotrophin in biological fluids by complement fixation. Lancet. 276(7141):99.

(10) McKean, C.M. (1960) Preparation and use of antisera to human chorionic gonadotrophin. American Journal of Obstetrics and Gynecology. 80(3):596–600.

(11) Robbins, J.L., Hill, G.A., Carle, B.N., Carlquist, J.H., and Marcus, S. (1962) Latex agglutination reactions between human chorionic gonadotropin and rabbit antibody. Proceedings of the Society for Experimental Biology and Medicine. 109(2):321–325.

(12) Henry, J.B. and Little, W.A. (1962) Immunological test for pregnancy. Journal of the American Medical Association. 182(3):230–233.

(13) Goldin, M. (1962) The use of latex particles sensitized with human chorionic gonadotropin in a serologic test for pregnancy. American Journal of Clinical Pathology. 38(3):335–338.

(14) Noto, T.A. and Miale, J.B. (1964) New immunologic test for pregnancy. A two-minute slide test. American Journal of Clinical Pathology. 41(3):273–278.

(15) Cabrera, H.A. (1969) A comprehensive evaluation of pregnancy tests. American Journal of Obstetrics and Gynecology. 103(1):32–38.

(16) Hussa, R.O. (1987) The Clinical Marker HCG. Praeger, New York.

Acta Endocrinologicl. 1960; 35: 261–267
© 1960 Society for the European Journal of Endocrinology.
Reproduced by permission.

ACTA ENDOCRINOLOGICA
35, 261–267, 1960

Department of Obstetrics and Gynecology,
Karolinska Hospital
and King Gustaf V:s Research Institute, Stockholm

AN IMMUNOLOGICAL PREGNANCY TEST

By

Leif Wide and Carl A. Gemzell

ABSTRACT

An immunological method for the assay of chorionic gonadotrophin (HCG) in human urine has been described in detail. The method is useful as a simple and rapid pregnancy test and can be applied for quantitative determinations of HCG in urine.

Morning urine from 306 women were examined; 212 were found to be pregnant by the haemagglutination inhibition reaction and the pregnancy was confirmed by ordinary pregnancy tests; 94 women were not pregnant and the urine of these women gave in no case a haemagglutination inhibition reaction.

Quantitative determinations of HCG were performed in the urine of 103 women in early pregnancy. The urinary excretion of HCG increased following the missed menstrual period and reached in the 8th week of pregnancy a level of about 160.000 IU of HCG per liter of urine.

A method has been developed to assay urine for chorionic gonadotrophin (HCG) by means of a passive haemagglutination technique. This method can be used for a simple and rapid pregnancy test which if necessary can be applied for quantitative determinations of HCG in urine.

REAGENTS AND PREPARATIONS

Antigen. Ampuls containing 1500 IU of a commercial preparation (Pregnyl, Organon) of human chorionic gonadotrophin (HCG) was used.

Adjuvant. Ramon's adjuvant (*Ramon et al.* 1935) was prepared by mixing 48 ml of mineral oil, 24 ml of 0.9 per cent saline and 16 g of lanoline. The mixture was autoclaved and homogenized.

Antisera. Normal white adult male and female rabbits were used for immunization. Rabbit anti-HCG-sera was prepared by subcutaneous injection of 12 000 IU of HCG, suspended in 2 ml of Ramon's adjuvant at weekly intervals for five to six weeks. The sera were heated for 30 minutes at 56° C and stored frozen or with 1/10 000 Merthiolate at 4°C. The antisera could also be lyophilized. The mean antibody titre obtained from 4 rabbits was 1 : 1600 with a range of 1 : 800 to 1 : 6400.

Bloodcells. Sheep blood, diluted with Alserver's solution (*Bukantz et al.* 1946) was used. The blood was kept at 4° C while not in use. The blood was not used when it was older than 2 weeks.

Formalin. A 37 per cent formalin solution was brought to p_H 7–7.5 with 1 N NaOH and diluted to 3 per cent with saline.

Tannic acid. Baker's Analyzed Reagent Tannic Acid diluted 1/100 with saline was used. Immediately before use it was diluted to a final concentration of 1/40 000 with p_H 7.2-buffered saline.

Buffered saline. The p_H 7.2-buffered saline was prepared by mixing 100 ml of saline with 100 ml of a buffer containing 77 ml of 0.15 M Na_2HPO_4 and 23 ml of 0.15 M KH_2PO_4. The p_H 6.4-buffered saline was prepared by mixing 100 ml of saline and 100 ml of a buffer consisting of 35 ml of 0.15 M Na_2HPO_4 and 65 ml of 0.15 M KH_2PO_4. The p_H was carefully checked on a p_H-meter.

Preparation of formalized erythrocytes. The sheep red blood cells were treated with formalin by the method of *Weinback* (1958). One volume of a 3 per cent solution of formalin was added to 1 volume of a 8 per cent suspension of cells which previously had been washed three times with saline and was kept at 37° C for 18 to 20 hours. The formalin was removed by washing four times with distilled water. The cells were kept at 4° C in a 10 per cent suspension of saline with Merthiolate 1/10 000.

Preparation of tannic acid cells. The preparation of the tannic acid cells and the sensitization of the cells after formalinization was essentially the same as the technique of *Boyden* (1951). The time of treatment with tannic acid and antigen were extended to 30 minutes and one hour respectively as recommended by *McKenna* (1957) when using formalized cells. The formalinized cells were washed once with saline and kept in a 3.3 per cent suspension of saline. One volume of this suspension was incubated at 37° C for 30 minutes with one volume of 1/40 000 dilution of tannic acid. The cells were then washed two times with one volume of p_H 6.4-buffered saline.

Sensitization of tannic acid cells. One volume of a 3.3 per cent suspension of tannic acid treated formalinized cells in p_H 6.4-buffered saline was incubated for one hour at 37° C with one volume of p_H 6.4-buffered saline containing 200 IU of HCG per ml. The cells were washed two times with saline and finally stored in a 10 per cent solution of saline containing 1 per cent normal rabbit sera (NRS) and 1/10 000 Merthiolate. The suspension was either kept frozen or in a refrigerator at 4° C. A 10 per cent suspension of sensitized cells in saline containing 10 per cent saccarose and 4 per cent NRS was lyophilized (*Weinback* 1959). It was not necessary to wash away the saccarose which protect the cells before they were used in the haemagglutination reaction.

METHODS

Haemagglutination reaction. The haemagglutination reaction was performed in 13 × 35 mm glass-tubes with hemispherical bottom. Various dilutions of antisera were prepared with saline as a diluent. To each tube, containing 0.45 ml of diluted antisera, was added 0.05 ml of a 2.5 per cent suspension of sensitized cells in saline containing 1 per cent of NRS. The tubes were shaken and incubated at room temperature or at 4° C. The reaction was read after $1^{1}/_{2}$ to 2 hours. When incubated at 4° C the titre was somewhat higher and the reaction did not change during a period of 12 to 18 hours. The titre was expressed as the highest dilution of antisera which gave a definite reaction $(++)$.

The reaction was read and graded according to the pattern formed by the agglutinated cells on the bottom of the tubes. This grading was essentially the same as that described and illustrated by *Stavitsky* (1954) or by *Antoine & Ducrot* (1957). See Fig. 1.

$++$ = smooth mat on the bottom of the tube, the edge may be somewhat ragged or folded.

$+$ = narrow ring of cells around the edge of a smooth mat.

$-$ = the cells form a button in the center of the tube.

Haemagglutination inhibition reaction. For the estimation of the titre of HCG in urine, essentially the same technique was employed as the one described by *Read & Stone* (1958) for the determination of growth hormone in human plasma. The haemagglutination reaction was inhibited by the addition of HCG in sufficient amounts.

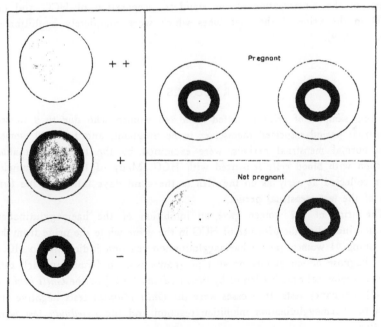

Fig. 1.
Appearance of patterns of haemagglutination.

263

Demonstration of HCG in urine (Pregnancy Test). The urine to be examined was filtered and diluted 1 : 5 with saline. Two glass-tubes with hemispherical bottom and 12 mm diameter were used. To each tube was added 0.25 ml of the diluted urine and 0.05 ml of the 2.5 per cent suspension of sensitized blood cells. To Tube 1 was added 0.20 ml of diluted antisera of appropriate titre and to Tube 2 0.20 ml of saline. The tubes were left in room temperature for 1¹/₂ hour and read according to Fig. 1. The reaction was »positive« (pregnant) when the two tubes showed a negative haemagglutination reaction. The reaction was »negative« (not pregnant) when Tube 1 showed a positive haemagglutination reaction.

Quantitative determination of HCG in urine. A serial dilution of filtered urine was performed in tubes containing 0.25 ml of saline. To the tubes were added 0.20 ml of diluted antisera of appropriate titre and 0.05 ml of the 2.5 per cent suspension of sensitized cells. The reaction obtained was compared with the inhibition reaction of a similar dilution of a standard HCG preparation in saline using the same titre of antisera and the same amount of sensitized cells. The lowest dilution of urine which gave a + + reaction was decided upon to be the endpoint of the reaction. By comparing the endpoints of the two serial dilutions, an approximative titre of HCG was found in the urine. The relative potency was obtained as the difference between the main logarithms of the concentration of the standard preparation of HCG and the HCG in the urine of the first tubes which were completely agglutinated (*Finney* 1952).

RESULTS

Morning samples of urine collected from 246 women who due to a missed menstrual period suspected themselves to be pregnant, and from 60 women with normal menstrual periods, were examined by the haemagglutination reaction with blood cells sensitized with HCG. Many of the urine samples were collected as early as on the 37th to the 42nd days following the first day of the last menstrual period.

The urine of 212 women gave an inhibition of the haemagglutination reaction indicating the presence of HCG in the urine while the urine from the remaining 94 women gave a haemagglutination reaction indicating no HCG. The diagnose of »pregnant« or »not pregnant« was confirmed in each case by gynaecological examination or by the *Friedman* (1947) or the *Galli Mainini* (1947) pregnancy tests. In 4 cases were the *Galli Mainini* tests negative but both the haemagglutination inhibition reaction and the *Friedman* test were positive.

A quantitative estimation of HCG was performed in the morning urine of

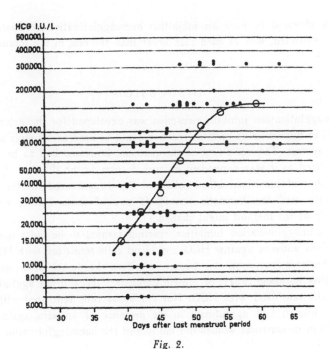

Fig. 2.

Increase in the urinary excretion of HCG during the first weeks of pregnancy. Each point represent a single determination. The open circles represent the mean logarithms of the determinations for a 3 day period.

103 women in early pregnancy. A serial dilution of urine was compared with a similar dilution of a standard preparation of HCG (Pregnyl, Organon). The urine samples were collected between the 38th and the 63rd days following the first day of the last menstrual period. Fig. 2 shows the individual levels of HCG and the average increase of HCG during the period of investigation. On the 60th day following the last menstrual period which approximately correspond to the 8th week of pregnancy a level of about 160 000 IU of HCG was found per liter of urine.

A young woman with an amenorrhea of 3 months duration and a diagnosed hydatidiform mole excreted about 3 millions IU of HCG per liter of urine. A quantitative assay of the urine of the same patient was performed 3 days later by the *Aschheim-Zondek* (1927) test and gave a concentration of about 2 millions IU of HCG per 24 hour urine.

Samples of morning urine from 6 women in early pregnancy were compared with samples of urine from the same women collected the same day in the afternoon. In each case the titre of HCG in the afternoon sample was sufficiently elevated to give an inhibition haemagglutination reaction but the titre, as expected, was in each case lower than the one in the morning.

DISCUSSION

A haemagglutination inhibition reaction was developed for the determination of HCG in human urine and the reaction could be used as a simple and rapid pregnancy test. The condition of the reaction, however, had to be carefully standardized as it was dependent on such factors as the dilution of antisera, the number of sensitized blood cells and the diluent used for the antisera, blood cells and the hormone (urine). If any of these factors varied the reliability of the method might change.

The haemagglutination inhibition reaction seemed to be quite specific for HCG when antisera against HCG and blood cells sensitized with HCG were used. Antisera prepared by human pituitary hormones such as follicle stimulating hormone (FSH) or growth hormone (GH) did not agglutinate the blood cells sensitized with HCG. Nor did blood cells sensitized with human FSH or human GH agglutinate in the presence of antisera against HCG. Urine from menopausal women did not inhibit the haemagglutination reaction for HCG.

A dilution of 200 to 300 IU of HCG per liter of urine gave a distinct inhibition of the haemagglutination reaction. A lower concentration of HCG in urine might be detected but the endpoints of the reaction, at least in some cases, became more difficult to recognize. However, the comparison with the ordinary biological pregnancy tests, such as the Friedman and the Galli Mainini tests the sensitivity of the haemagglutination inhibition reaction seemed to be of the same order of magnitude. In both cases a positive response was obtained in the urine of pregnant women as early as 40 to 41 days following the first day of the last menstrual period.

The haemagglutination inhibition reaction has a great advantage over the biological pregnancy tests as the results were obtained already after $1^1/_2$ hour. At room temperature the reaction changed somewhat and had to be read within 6 hours. As a temperature of 4° C the reaction was somewhat slower and should not be read before a period of two hours. On the other hand, at the lower temperature the reaction was more stable and stayed unchanged for at least 18 to 20 hours.

The levels of HCG in the urine of women in early pregnancy obtained by the haemagglutination inhibition reaction seemed to be slightly higher than the levels obtained after extraction of the urine and biological assay of the extract (*Loraine* 1950). Such differences might be expected as the extraction procedure probably caused some loss of biological activity. On the other hand, different standard preparations of HCG were used and thus, a comparison between levels might not be valid.

ACKNOWLEDGMENTS

N. V. Organon, Holland will further develop this method in such a way that it will be suitable as a routine pregnancy test.

REFERENCES

Antoine B. & Ducrot H.: Rev. Franc. d'Etudes Clinique et Biol. *2* (1957) 81.
Aschheim S. & Zondek B.: Klin. Wschr. *6* (1927) 1322.
Boyden S. J.: J. exp. Med. *93* (1951) 107.
Bukantz S. C., Rein C. R. & Kent J. F.: J. Lab. clin. Med. *31* (1946) 394.
Finney D. J.: Statistical Methods in Biol. Assay, Hafman Publ. Co., N. Y. (1952) 21.
Friedman M. H.: Amer. J. Physiol. *90* (1947) 653.
Galli Mainini C.: J. clin. Endocr. *7* (1947) 653.
Loraine J. A.: J. Endocr. *6* (1950) 319.
McKenna J. M.: Proc. Soc. exp. Biol. (N. Y.) *95* (1957) 591.
Ramon G., Lemetayer E. & Richou R.: Rev. Immunologie *1* (1935) 199.
Read C. H. & Stone D. B.: Amer. J. Dis. Child. *96* (1958) 538.
Stavitsky A. B.: J. Immunolog. *72* (1954) 860.
Weinback R.: Schweiz. Z. Path. Bakt. *21* (1958) 1043.
Weinback R.: Schweiz. Z. Path. Bakt. *22* (1959) 1.

COMMENTARY TO

21. Rahbar, S. **(1968)**
An Abnormal Hemoglobin in Red Cells of Diabetics. Clinica Chimica Acta 22(2): 296–298.

The second most important test used in the management of diabetes mellitus besides blood glucose itself is hemoglobin (Hb) A1c (also called glycohemoglobin, GHb). The discovery of elevated levels of Hb A1c in diabetics was made by Samuel Rahbar at the University of Tehran in 1968. Rahbar normally studied Hb variants using electrophoresis (1,2). On cellulose acetate he noticed that two patients showed a fraction that migrated with and broadened the Hb A band. On starch gel electrophoresis the band moved with Hb F. This abnormal Hb band was found only in patients with diabetes. Rahbar confirmed these findings in 47 more patients with diabetes. This short but insightful paper is the first demonstration of an elevated Hb fraction found associated with diabetes.

In 1969 Rahbar, Blumenfeld and Ranney reported that the best separation of the diabetic-associated Hb occurred with agar gel electrophoresis where it migrated between the Hb A and Hb F fractions (3). The diabetic-associated Hb on ion-exchange columns was found to be identical to the Hb A1c separated on electrophoresis. They reported that the normal reference ranges for Hb A1c were 4–6% of the total Hb. Hb A1c levels in diabetics ranged between 7.5 and 10.6%. Susceptibility to borohydride reduction and periodate oxidation suggested that the Hb A1c fraction in patients with diabetes contained an increased number of sugar moieties. In 1971 Trivelli and co-workers confirmed the two-fold increase in Hb A1c values compared to controls in 100 diabetics (4).

Hb A1c is created by the non-enzymatic condensation of glucose onto each of the beta-chains of Hb A. Formation of glycated HbA1c is irreversible and is directly proportional to the blood glucose concentration (5,6). Red blood cells have an average life span in the peripheral circulation of about 120 days. Any single HbA1c result will therefore represent the integrated value for the blood glucose values over the previous 6–8 weeks (7). The Diabetes Control and Complications Trial confirmed that the use of HbA1c levels as part of an intensive glycemic control program helped reduce the complications from Type 1 diabetes by up to 75% (8).

The most widely used methods for HbA1c include cation-exchange high performance liquid chromatography (HPLC) (9) and affinity chromatography in which glycated Hb is bound to boronate resin beads (10). Immunoassays for HbA1c have also been described (11). Roberts and co-workers in 2005 compared results of Hb A1c levels with 11 different commercial test kits (12). Rahbar and co-workers have gone on to study chemical inhibitors of glycation some of which they hope will result in drug therapies (13,14).

References

(1) Rahbar, S., Beale, D., Issacs, W.A., and Lehmann, H. (1967) Abnormal haemoglobins in Iran. Observation of a new variant-haemoglobin J Iran (α2 β2 77 His $\rightarrow$ Asp). British Medical Journal. 1(5541):674–677.

(2) Rahbar, S. (1968) Hemoglobin H disease in two Iranian families. Clinica Chimica Acta. 20(3):381–385.

(3) Rahbar, S., Blumenfeld, O., and Ranney, H.M. (1969) Studies of an unusual hemoglobin in patients with diabetes mellitus. Biochemical and Biophysical Research Communications. 36(5):838–843.

(4) Trivelli, L.A., Ranney, H.M., and Lai, H-T. (1971) Hemoglobin components in patients with diabetes mellitus. New England Journal of Medicine. 284(7):353–357.

(5) Rahbar, S. (1980–1981) Glycosylated hemoglobins. Texas Reports on Biology and Medicine. 40:373–385.

(6) Krishnamurti, U. and Steffes, M.W. (2001) Glycohemoglobin: a primary predictor of the development or reversal of complications of diabetes mellitus. Clinical Chemistry. 47(7):1157–1165.

(7) Sacks, D.B. (1999) Carbohydrates Chapter 24, in *Tietz Textbook of Clinical Chemistry*, 3rd Edition, Burtis, C.A. and Ashwood, E.R. (eds), W.B. Saunders Company, Philadelphia, pgs 790–797.

(8) The DCCT Research Group. (1993) The effect of intensive treatment of diabetes on the development and progression of long-term complications in insulin-dependent diabetes mellitus. New England Journal of Medicine. 329(14):977–986.

(9) The DCCT Research Group. (1987) Feasibility of centralized measurements of glycated hemoglobin in the diabetes control and complications trial: a multicenter study. Clinical Chemistry. 33(12):2267–2271.

(10) Wilson, D.H., Bogacz, J.P., Forsythe, C.M., Turk, P.J., Lane, T.L., Gates, R.C., and Brandt, D.R. (1993) Fully automated assay for glycohemoglobin with the Abbott IMx analyzer. Clinical Chemistry. 39(10):2090–2097.

(11) John, W.G., Gray, M.R., Bates, D.L., and Beachman, J.L. (1993) Enzyme immunoassay: a new technique for estimating hemoglobin A1c. Clinical Chemistry. 39(4):663–666.

288

(12) Roberts, W.L., Safar-Pour, S., De, B.K., Rohlfing, C.L., Weykamp, C.W., and Little, R.R. (2005) Effects of hemoglobin C and S traits on glyohemoglobin measurements by eleven methods. Clinical Chemistry. 51(4):776–778.

(13) Rahbar, S., Yerneni, K.K., Scott, S., Gonzales, N., and Lalezari, I. (2000) Novel inhibitors of advanced glycation endproducts (Part II). Molecular Cell Biology Research Communications. 3(6):360–366.

(14) Rahbar, S. and Figarola, J.L. (2003) Novel inhibitors of advanced glycation endproducts. Minireview. Archives of Biochemistry and Biophysics. 419(1):63–79.

Clinica Chim. Acta. 1968, 22(2): 296–298
Copyright 1968 Reprinted with permission from Elsevier.

An abnormal hemoglobin in red cells of diabetics

In a survey carried out on 1200 patients from Tehran University Hospitals, in addition to three rare hemoglobins which are under investigation both in our department here and at the University of Cambridge, two patients also showed an abnormal fast moving hemoglobin fraction: both were suffering from diabetes mellitus.

Studies were started to investigate the occurrence of this abnormal fraction in other diabetics, and in 47 cases examined in the last three months, including 11 children with severe diabetes mellitus, the additional fraction was detected. Routine hematological examination according to standard methods[2] gave normal results in the majority of cases.

Electrophoresis of hemoglobin was carried out on cellulose acetate according to Graham and Gruenbaum[3]; the abnormal fraction does not separate well by this method, but there is a broadening of the Hb A band. In starch gel electrophoresis with tris–EDTA–borate buffer pH 8.1 (ref. 1) the additional fraction moves a little faster than Hb A and slower than Hb J (Iran)[8] (Fig. 1).

Agar gel electrophoresis in citrate buffer pH 6.2 by the method of Robinson *et al.*[4] is the method of choice for the separation and demonstration of this fraction which moves in front of Hb A to the cathode in the same position as Hb F (Fig. 2).

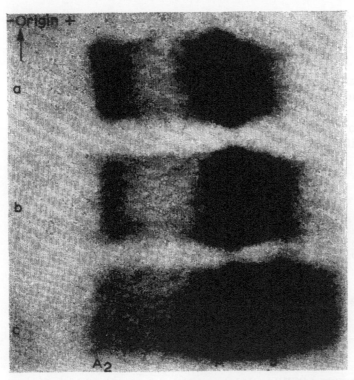

Fig. 1. Starch gel electrophoresis in tris–EDTA–borate buffer, pH 8.1. *o*-Dianizidine stain, ref. 7. a: normal; b: Hb A + Hb x; c: Hb A + Hb J (Iran).

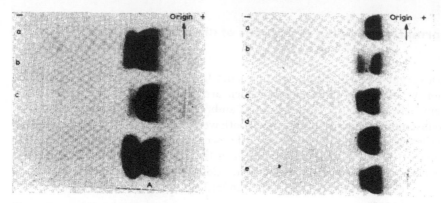

Fig. 2. Agar gel electrophoresis, pH 6.2. a: Hb A+x; b: normal; c: Hb A+Hb F.

Fig. 3. Agar gel electrophoresis, pH 6.2. a and d: normal; b: hemolysate (enriched) of abnormal fraction; c and e: diabetic hemoglobins.

Starch grain electrophoresis is performed with barbital buffer pH 8.6—0/05 in maize starch and the faster protion of the Hb A band which contains the abnormal fraction is removed, eluted from the starch, concentrated in vacuum and re-run on paper to enrich the abnormal fraction. The electrophoretic pattern of this enriched preparation can be seen in Fig. 3b.

For further investigation the hemolysate was reacted with p-chloromercuribenzoate (PCMB) by the original method of Bucci and Fronticelli[5] as modified by Rosemeyer and Huehns[6], and the peptide chains were separated by starch gel electrophoresis[1].

Fig. 4 shows that in diabetic hemolysate, in addition to normal αA^{PMB} and βA^{PMB} there exists an additional fast moving α chain. The separation of petide chains in urea gel according to Chernoff et al.[9] fails to separate the additional peptides.

No previous report has appeared in the literature on this subject and further work is needed to clarify the nature of the abnormality.

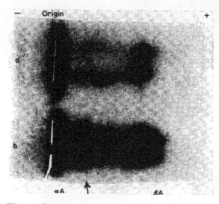

Fig. 4. Starch gel electrophoresis of PCMB-treated hemoglobins. a: normal; b: diabetic. Note the existence of an additional band faster than α A (↑).

Clin. Chim. Acta, 22 (1968) 296–298

298   BRIEF NOTES

Abnormal Hemoglobin Research   SAMUEL RAHBAR*
Laboratory, Department of
Immunology, Faculty of Medicine,
University of Tehran (Iran)

1 S. RAHBAR, *J. Med. Fac. Tehran*, 25, Suppl. 2 (1967) 10.
2 M. WINTROBE, *Clinical Hematology*, 5th ed., Lea & Febiger, Philadelphia, 1961.
3 J. L. GRAHAM AND W. GRUENBAUM, *Am. J. Clin. Pathol.*, 39 (1963) 567.
4 A. R. ROBINSON, M. ROBSON, A. R. HARRISON AND W. W. ZUELZER, *J. Lab. Clin. Med.*, 50 (1957) 745.
5 E. BUCCI AND C. FRONTICELLI, *J. Biol. Chem.*, 240 (1965) PC 551.
6 M. A. ROSEMEYER AND E. R. HUEHNS, *J. Mol. Biol.*, 25 (1967) 253.
7 H. LEHMANN, in *Man's Hemoglobin*, North Holland, Amsterdam, 1965.
8 S. RAHBAR, D. BEALE, W. A. ISAAC AND H. LEHMANN, *Brit. Med. J.*, i (1967) 674.
9 A. I. CHERNOFF AND N. M. PETTIT, *Blood*, 24 (1964) 750.

Received June 11, 1968

* Temporary address (till August 1969):
Albert Einstein College of Medicine, Yeshiva University, Eastchester Road and Morris Park Avenue, Bronx, N.Y. 10461.

Clin. Chim. Acta, 22 (1968) 296–298

COMMENTARY TO

22. Friedewald, W. T., Levy, R. I. and Fredrickson, D. S. (1972)
Estimation of the Concentration of Low-Density Lipoprotein Cholesterol in Plasma, Without Use of the Preparative Ultracentrifuge. Clinical Chemistry 18(6): 499–502.

The calculation for low-density lipoprotein (LDL) cholesterol presented here in this paper became known as the Friedewald equation. It allowed the concentration of the LDL cholesterol in plasma to be determined from the total cholesterol, the triglycerides and the high-density lipoprotein (HDL) cholesterol. The calculation when the lipid levels are expressed in mg/dL is:

$$\text{LDL Cholesterol} = [\text{Total Cholesterol}] - [\text{HDL Cholesterol}] - \frac{[\text{Triglycerides}]}{5}$$

Rapid and automated methods for total cholesterol, HDL cholesterol and triglycerides were readily available at the time of this publication. LDL cholesterol assays however required a 16-hour ultracentrifugation procedure (1). The Friedewald equation was compared to ultracentrifugation results in large numbers of patients and was soon validated (2–4).

Over the next 20 years measurement of LDL cholesterol took on major importance after it was determined that elevated levels were a major risk factor for cardiovascular disease (5). LDL cholesterol in the popular press became known as the "bad" cholesterol. Drugs called statins designed to lower LDL cholesterol became a billion dollar market. In the United States alone 13 million patients were on a statin drug by the end of the 1990s (5). The paper presented here by Friedewald, Levy and Fredrickson came at the right time and by 1998 it was the highest cited paper in the 50-year history of the journal *Clinical Chemistry* (6).

References

(1)   de Lalla, O.F. and Gofman, J.W. (1954) Ultracentrifugal analysis of serum lipoproteins, in *Methods of Biochemical Analysis*, Vol 1. Glick, D. (ed), Interscience Publishers, New York, pgs 459–478.

(2)   Niedbals, R.S., Schray, K.J., Foery, R., and Clement, G. (1985) Estimation of low-density lipoprotein by the Friedewald formula and by electrophoresis compared. Clinical Chemistry. 31(10):1762–1763.

(3)   DeLong, D.M., De Long, E.R., Wood, P.D., Lippel, K., and Rifkind, B.M. (1986) A comparison of methods for the estimation of plasma low- and very low-density lipoprotein cholesterol. The lipid research clinics prevalence study. Journal of the American Medical Association. 256(17):2372–2377.

(4)   Johnson, R., McNutt, P., MacMahon, S., and Robson, R. (1997) Use of the Friedewald formula to estimate LDL-cholesterol in patients with chronic renal failure on dialysis. Clinical Chemistry. 43(11):2183–2184.

(5)   Farmer, J.A. (2001) Learning from the cerivastatin experience. The Lancet. 358(9291):1383–1385.

(6)   Bruns, D.E. (1998) Citation classics in Clinical Chemistry. Clinical Chemistry. 44(3):698–699.

Clinical Chemistry 1972; 18(6): 499–502
© 1972 American Association for Clinical Chemistry.
Reproduced with permission.

Estimation of the Concentration of Low-Density Lipoprotein Cholesterol in Plasma, Without Use of the Preparative Ultracentrifuge

William T. Friedewald, Robert I. Levy, and Donald S. Fredrickson

A method for estimating the cholesterol content of the serum low-density lipoprotein fraction (S_f 0-20) is presented. The method involves measurements of fasting plasma total cholesterol, triglyceride, and high-density lipoprotein cholesterol concentrations, none of which requires the use of the preparative ultracentrifuge. Comparison of this suggested procedure with the more direct procedure, in which the ultracentrifuge is used, yielded correlation coefficients of .94 to .99, depending on the patient population compared.

Additional Keyphrases *hyperlipoproteinemia classification • determination of plasma total cholesterol, triglyceride, high-density lipoprotein cholesterol • beta lipoproteins*

An important requirement for classification of hyperlipidemia into the different types of hyperlipoproteinemia (*1, 2*) is the estimation of the concentration of plasma LDL[1] (S_f 0-20; the beta lipoproteins). This quantity is necessary for the assignment of the Type II pattern (*1–3*), which is defined as an increase in LDL concentration above some arbitrarily selected cut-off limit.

An indirect method is presented here for estimating the plasma LDL concentration in terms of the cholesterol contained in this lipoprotein (C_{LDL}). The method requires measurement of the concentrations of plasma total cholesterol, triglycerides, and C_{HDL}. This information can be obtained without ultracentrifugation and requires only routine lipid analyses in addition to a rapid precipitation of all plasma lipoproteins other than HDL. Two observations are used in the calculation. One is that the ratio of the mass of triglyceride to that of of cholesterol in VLDL is apparently relatively constant and about 5:1 in normal subjects (*1, 4*) and in patients with all types of hyperlipoproteinemia, except the rare Type III (*1, 2*). The other is that when chylomicrons are not detectable, most of the triglyceride in plasma is contained in the VLDL. Thus, in the vast majority of plasma samples in which chylomicrons are not present, the cholesterol in plasma attributable to VLDL can be approximated by dividing the plasma triglyceride concentration by five. The justification of this method for estimation of C_{LDL} is the subject of this paper.

Methods

Data were obtained from lipid and lipoprotein analyses performed by the Molecular Disease Branch of the National Heart and Lung Institute, on samples from patients with hyperlipidemia and from normal subjects. The results of the laboratory analyses are in the process of transfer to magnetic tapes for rapid retrieval and analysis. At the time of this study, complete lipoprotein analyses from 448 subjects classified as either normal, Type II, or Type IV primary hyperlipoproteinemia had been transferred to tapes and the data from all were used. The data reflect the research interests of the Branch, with most of the data coming from patients with familial hyperlipoproteinemia or their relatives, and as such do not represent an unbiased sample of the general population. The specimen from each patient chosen for analysis in this paper was the first sample on magnetic tape on which a complete lipoprotein analysis had been performed at the National Heart and Lung Institute. The subjects were receiving no dietary or drug treatment for hyperlipoproteinemia at the time of this sampling, except for two patients who were on caloric restriction but who were subsequently classified as Type IV.

The original plasma samples had been obtained 12 to 14 h after the last meal, mixed with EDTA (1 mg/ml), and immediately stored at 4 °C until analyzed. Total plasma cholesterol (*5*) and triglycerides (*6*) were measured, and data on C_{HDL}, C_{LDL}, and C_{VLDL} obtained by a combination of ultracentrifugation and precipitation procedures (*7*). In samples free of chylomicrons the C_{VLDL} was

From the Biometrics Research Branch (W.T.F.) and the Molecular Disease Branch (R.I.L. and D.S.F.), National Heart and Lung Institute, 9000 Rockville Pike, Bethesda, Md. 20014.

[1] Nonstandard abbreviations used: HDL, high-density lipoprotein; LDL, low-density lipoprotein; VLDL, very low-density lipoprotein; TG, plasma triglycerides; $C_{subscript}$, cholesterol concentration (mg/100 ml of plasma) in the fraction identified by the subscript; EDTA, ethylenediaminetetraacetic acid.

Received Feb. 29, 1972; accepted Mar. 13, 1972.

measured in two ways: (*a*) directly, by measuring the cholesterol content of the supernatant fraction after ultracentrifugation of plasma at D.1006 for 16 h at 100,000 × *g* in a Spinco 40.3 rotor, and (*b*) indirectly, by subtracting the cholesterol content of the infranatant fraction (the sum of the C_{HDL} and C_{LDL}) from the total plasma cholesterol. A methodologic error was presumed and the plasma reanalyzed if there was a large disparity between the results of the two methods.

Subjects were classified as normal or as having hyperlipoproteinemia Type I through V according to criteria previously described (*1, 2*).

Results

The results of lipid and ultracentrifuge lipoprotein determinations in 232 men and 216 women— 96 normal, 204 with Type II, and 148 with Type IV—were analyzed. Various statistics derived from these data are presented in Table 1.

C_{LDL} was also calculated for each person according to the following formula:

$$C_{LDL} = C_{plasma} - C_{HDL} - TG/5$$

Plots of each individual's C_{LDL} as calculated by this method vs. that obtained after preparative ultracentrifugation (*7*) are presented in Figures 1 to 3. For normal people and Type II patients the spread of points about the line of equality for the two methods does not appear excessive, as reflected in the high correlation coefficients, .98 and .99, respectively (*8*). However, in Type IV patients there are many outlying values and the correlation is somewhat lower, namely, .85. Closer scrutiny of these outliers revealed that most such patients had very high plasma triglyceride concentrations, and thus a plot was made of only those Type IV patients with plasma triglycerides less than 400 mg/100 ml (111 of the original 148 people). The spread of values is much smaller after these are

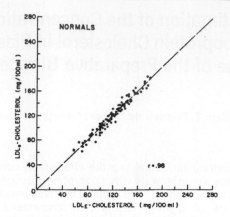

Fig. 1. Comparison of the plasma low-density lipoprotein cholesterol concentration in normal individuals as calculated by the estimation method (LDL_E) with that obtained by the ultracentrifuge method (LDL_U)

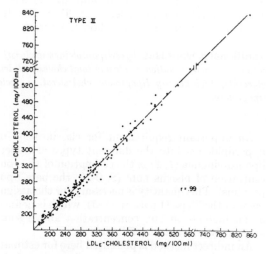

Fig. 2. Comparison of the plasma low-density lipoprotein cholesterol concentration in Type II patients as calculated by the estimation method (LDL_E) with that obtained by the ultracentrifuge method (LDL_U)

excluded (Figure 4), the correlation coefficient then being .94.

Further quantitative expressions for the disagreement between the C_{LDL} obtained by the two methods are presented in Table 2. Because each of the confidence intervals (*9*) contains zero, there is no compelling reason based on these data to believe that the estimate of C_{LDL} will be biased. The tolerance intervals (*10*) give the range of values that with probability .95 will include 95% of the differences between the two methods of measurement. The per cent error assumes that the C_{LDL} measured by the ultracentrifuge method is the standard with which the C_{LDL} estimate obtained by calculation is being compared. The distribution of values by each method was examined for each of the groups, by use of normal probability graph

Table 1. Mean, Standard Deviation, and Range of Plasma Lipids and Lipoproteins

	Normal (96)[a]	Type II (204)[a]	Type IV (148)[a]
		mg/100 ml	
Total plasma cholesterol	189 ± 33 (166–270)	359 ± 100 (217–888)	241 ± 57 (138–436)
Total plasma triglyceride	73 ± 7 (20–184)	126 ± 16 (25–656)	347 ± 61 (90–2502)
HDL cholesterol	53 ± 13 (29–77)	45 ± 13 (18–82)	38 ± 11 (15–74)
LDL cholesterol	122 ± 28 (62–185)	291 ± 99 (173–840)	135 ± 38 (28–231)
VLDL cholesterol	14 ± 9 (0–40)	24 ± 19 (0–78)	68 ± 55 (6–356)

[a] No. of patients.

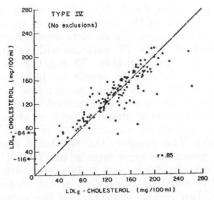

Fig. 3. Comparison of the plasma low-density lipoprotein cholesterol concentration in Type IV patients as calculated by the estimation method (LDL$_E$) with that obtained by the ultracentrifuge method (LDL$_U$); no exclusions

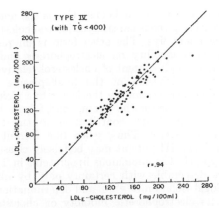

Fig. 4. Comparison of the plasma low-density lipoprotein cholesterol concentration in Type IV patients as calculated by the estimation method (LDL$_E$) with that obtained by the ultracentrifuge method (LDL$_U$), excluding individuals with serum triglycerides ≥ 400 mg/100 ml

paper. The values were reasonably normally distributed. Only in the Type IV patients was there evidence of skewness; this was only minimal, and in the negative direction.

We did not attempt to calculate the overall average error or to estimate the probability of misclassification by this estimation procedure because of the unusually large number of Type II and Type IV patients relative to normal people in the sample.

A subsample of 46 Type II patients was randomly chosen, a linear least-squares (11) fit of TG to C_{VLDL} was performed, and the equation so obtained was used in a separate subset of 55 Type II patients to estimate their C_{VLDL}. Analysis revealed that simple division of TG by five provided as accurate an estimate of C_{VLDL} as did this more complicated regression estimate.

Discussion

The method presented here for estimating plasma LDL concentrations provides a reasonable approximation that is useful for many purposes. There are, however, three important restrictions on its use. First, it is not applicable to plasma samples containing chylomicrons. However, such samples are characterized by a "cream" layer on top of plasma that has been stored at 4°C for 18 h or more. Chylomicrons are characteristic of lipoprotein patterns classified as Types I and V, in which the C_{LDL} concentration is not abnormally increased. Particles having similar appearance are also sometimes seen in Type III.

Second, the technique for estimating C_{LDL} gives erroneously high results in the rare patient with Type III hyperlipoproteinemia. In this disorder,

Table 2. Statistics on the Measurement of C$_{LDL}$ Utilizing the Ultracentrifuge vs. the Estimation Procedure

| | No. of values, (n) | (LDL$_U$–LDL$_E$)[b] | | | | |LDL$_U$–LDL$_E$|[c] | |
		Mean, (X̄)	Standard deviation, (SD)	95% confidence interval[d]	95% tolerance interval[e]	Mean	% error[f]
Normals	96	.3	5.9	[−.9, 1.4]	[−13.0, 13.5]	4.8	4%
Type II	204	1.3	11.9	[−.4, 2.9]	[−24.3, 26.8]	8.8	3%
Type IV	111[a]	.4	12.9	[−2.0, 2.9]	[−27.6, 28.6]	9.8	7%

[a] Only people with plasma triglycerides less than 400 mg/100 ml are included.
[b] LDL$_U$ = C$_{LDL}$ calculated with the preparative ultracentrifuge (see text).
   LDL$_E$ = C$_{LDL}$ calculated by the estimation procedure (see text).
[c] |LDL$_U$ − LDL$_E$| = the absolute value of (LDL$_U$–LDL$_E$).
[d] X̄ ± [t-value (.025, n − 1)] SD/n.
[e] X̄ ± [tolerance-value (.95, .95, n)] SD.
[f] % error = $\dfrac{\text{[mean of |LDL}_U\text{–LDL}_E\text{|]} \times 100}{\text{(mean of LDL}_U\text{)}}$

the VLDL are of two kinds (*12*). One is the normal variety having the usual triglyceride-to-cholesterol ratio of about five. The other form is unique in having beta mobility on electrophoresis and an abnormally high content of cholesterol relative to triglyceride. Division of the total plasma triglyceride concentration by the factor five yields a falsely low value for the "VLDL" and falsely high value for the "LDL" contribution to the total plasma cholesterol. Thus, when this formula is used, a Type III patient may be falsely classified as a Type II. The anomalous lipoproteins in Type III are detectable with certainty only by ultracentrifugal isolation of VLDL and determination of either its electrophoretic mobility or cholesterol and triglyceride content (*1–3*).

Third, C_{LDL} cannot always be accurately estimated when the plasma triglyceride concentration exceeds 400 mg/100 ml. It is noteworthy, however, that only two of the 204 Type II patients in this series had triglyceride concentrations of 400 mg/100 ml or greater. This suggests that few errors in classification would occur if patients with plasma triglycerides exceeding 400 mg/100 ml, in the absence of chylomicrons, were directly classified as Type IV. The frequency of this misclassification will no doubt depend in part on the cut-off limits used in defining an abnormal LDL concentration.

It is noteworthy that despite the good agreement between the estimation and actual measurement of C_{LDL}, simple division of the plasma triglyceride by five does not give a very accurate estimate of the VLDL cholesterol alone, even in normals or patients with Type II or Type IV. In normals and patients with Type II the average VLDL cholesterol concentration is low (see Table 1), and thus even small absolute errors yield large percentage errors. In Type IV the average VLDL cholesterol concentration is higher, but large percentage errors still result. However, when the estimate of C_{VLDL} is used to calculate C_{LDL} the percentage error does decrease to an acceptable level because the absolute error in C_{VLDL} estimation is small relative to the concentration of C_{LDL}.

Of some concern is the number of Type IV patients with relatively large values for $|LDL_U-$ $LDL_E|$ (see Table 2 and Figure 3), which of course greatly influences the per cent error of the whole group. To examine this problem more closely, we identified all Type IV patients with an $|LDL_U-LDL_E|$ value greater than 20 mg/100 ml and their entire laboratory profile on tape was reevaluated. Of the 16 people so examined, 13 had evidence of an undetected methodological error identified by a large disparity between the actual indirect and direct measurements of C_{VLDL} (see *Methods*). This suggests the possibility that the larger percentage error seen in the Type IV patients may be due in part to laboratory errors in the ultracentrifuge calculation of LDL cholesterol rather than greater inaccuracy of the estimation procedure.

References

1. Fredrickson, D. S., Levy, R. I., and Lees, R. S., Fat transport in lipoproteins—an integrated approach to mechanisms and disorders. *New Engl. J. Med.* **276**, 32, 94, 148, 215, 273 (1967).

2. Fredrickson, D. S., and Levy, R. I., Familial hyperlipoproteinemia. Chap. 28 in *The Metabolic Basis of Inherited Disease*, 3rd ed., McGraw-Hill, New York, N. Y. 1972, p 531.

3. Beaumont, J. L., Carlson, L. A., Cooper, G. R., Fejfar, Z., Fredrickson, D. S., and Strasser, T., Classification of hyperlipidaemias and hyperlypoproteinemias. *Bull. WHO* **43**, 891 (1970).

4. Hatch, F. T., and Lees, R. S., Practical methods for plasma lipoprotein analysis. *Advan. Lipid Res.* **6**, 1 (1968).

5. Total cholesterol procedure N-24b. Auto-Analyzer Manual, Technicon Instruments Corp., Tarrytown, N. Y., 1964.

6. Kessler, G., and Lederer, H., Fluorometric measurement of triglycerides. In *Automation in Analytical Chemistry, Technicon Symposia 1965*, L. T. Skeggs, Jr., et al., Eds. Mediad, New York, 1966, p 341.

7. Fredrickson, D. S., Levy, R. I., and Lindgren, F. T., A comparison of heritable abnormal lipoprotein patterns as defined by two different techniques. *J. Clin. Invest.* **47**, 2446 (1968).

8. Draper, N. R., and Smith, H., Applied Regression Analysis, John Wiley and Sons, Inc., New York, N. Y., 1966, p 33.

9. Dixon, W. J., and Massey, F. J., Jr., Introduction to Statistical Analysis, McGraw-Hill, New York, N. Y., 1957, p 127.

10. Dixon, W. J., and Massey, F. J., Jr., Introduction to Statistical Analysis, McGraw-Hill, New York, N. Y., 1957, p 130.

11. Draper, N. R., and Smith, H., Applied Regression Analysis, John Wiley and Sons, Inc., New York, N. Y., 1966, p 7.

12. Quarfordt, S., Levy, R. I., and Fredrickson, D. S., On the lipoprotein abnormality in Type III hyperlipoproteinemia. *J. Clin. Invest.* **50**, 754 (1971).

Instrumentation and Techniques

COMMENTARY TO

page 303

23. Durrum, E. L. (1950)
A Microelectrophoretic and Microionophoretic Technique. Journal of the American Chemical Society 72(7): 2943–2948.

Arne Tiselius received the Nobel Prize in Chemistry in 1948 for the development of moving boundary electrophoresis and adsorption chromatography. Moving boundary electrophoresis separated proteins within a buffer-filled glass U-tube. Bands of protein were detected by means of refractive index changes imaged onto photographic film. Separations required 1 mL or more of serum and a single analysis took up to 20 hours to complete (1). Companies in the United States and Europe soon manufactured moving boundary electrophoresis instruments based on Tiselius' work. The Perkin–Elmer Model 38 Tiselius Electrophoresis Apparatus weighed 45.5 kg (100 lbs) and measured 1.6 m (5.2 ft) long (2). Moving boundary electrophoresis made it possible for the first time to separate human serum proteins into the clinically significant bands, named by Tiselius, albumin, alpha, beta and gamma. Pauling and co-workers in 1949 demonstrated that hemoglobin from sickle cell patients differed in charge from normal hemoglobin based on moving boundary electrophoresis (3). They linked the charge difference between the two forms of hemoglobin with the gene coding for proteins. Their paper was the first time the term molecular disease had been used. Moving boundary electrophoresis despite its great promise remained a research tool until a simpler format was developed. This occurred with the introduction of zone electrophoresis.

Zone electrophoresis or electrophoresis on a porous support was first described using paper strips in 1937 by Paulo König in Sao Paulo, Brazil. Rosenfeld has reviewed the early development of zone electrophoresis in detail (4–6). König's first report in 1937 on the paper electrophoresis of proteins was published in Portuguese in a congress symposium (7). Two years latter Klobusitzky and König published a full report in a German journal (8). Both papers demonstrated the separation of proteins in snake venom. Unfortunately, neither paper received much attention. Twenty-one years after his first paper was published and 2 months before he died in 1958, König wrote the foreword for a book by Chales Wunderly, titled *Principles and Applications of Paper Electrophoresis with a Foreword by Dr. P. König, the Pioneer of Paper Electrophoresis*. König wrote that, "... the development, both qualitative and quantitative, which this method has experienced, were not anticipated by me" (9).

Emmett Durrum developed the first zone electrophoresis system for serum proteins while at the United States Army research facility in Fort Knox, Kentucky. He presented his technique at the American Chemical Society meeting in San Francisco on March 29, 1949. Tiselius was at the meeting and received a copy of the full report from Durrum (4). Twelve months latter in 1950 Durrum published the complete report that is presented here. A year latter Tiselius cited Durrum in his own 29-page paper on zone electrophoresis of serum proteins (10). To his credit Tiselius detailed the improvements made by paper electrophoresis over his own moving boundary technique.

Durrum's paper technique helped introduce electrophoresis into clinical chemistry. Serum samples of less than 10 µL were sufficient and the separated proteins could be stained and eluted easily from the paper for quantitative analysis. Paper electrophoresis in a plastic chamber the size of a kitchen toaster separated human serum proteins into five distinct bands comparable to the separations in the Tiselius apparatus (11). Up to 8 samples could be run at the same time. Durrum and Saul Gilford developed a paper strip densitometer mounted onto a Beckman photometer (12). Durrum verified the quantitative nature of protein staining using bromophenol blue (13) and reported on the protein serum profiles in a wide variety of disease conditions (14). His papers on lipoprotein staining and hemoglobin phenotyping by paper electrophoresis were among the earliest reports on these applications (15,16). In1958 Durrum co-authored a book on paper electrophoresis with Richard Block and Gunter Zweig. The book contained a bibliography with over 1800 articles on paper electrophoresis (17). In the next 11 years following Durrum's first report, alternative porous media were soon described including starch gel in 1955 (18), cellulose acetate in 1957 (19), acrylamide gel in 1959 (20) and agarose gel in 1961 (21).

References

(1) Tiselius, A. (1940) Electrophoretic analysis and the constitution of native fluids. Harvey Lecture, October 19, 1939. Bulletin of the New York Academy of Medicine. 16(12):751–780.

(2) [Advertisement]. (1953) The portable complete electrophoresis apparatus. The Perkin–Elmer corporation. Journal of Clinical Investigation. 32(6).

(3) Pauling, L., Itano, H.A., Singer, S.J., and Wells, I.C. (1949) Sickle cell anemia, a molecular disease. Science. 110(2865): 543–548.

(4) Rosenfeld, L. (1981) Origins of protein electrophoresis in paper [letter]. Clinical Chemistry. 27(11):1948–1949.

(5) Rosenfeld, L. (1982) Zone Electrophoreis on Paper in *Origins of Clinical Chemistry, The Evolution of Protein Analysis.* Academic Press, New York, Chapter 12, pgs 194–207.

(6) Rosenfeld, L. (1999) *Four Centuries of Clinical Chemistry.* Gordon and Breach Science Publishers, Amsterdam, pgs 427–429.

(7) König, P. (1937) Applicacao da Electrophorese nos Trabalhos Chimicos com Quantidades Pequenas. Actas e Trabalhos do Terceiro Congresso Sul-Americano de Chimica. Rio de Janeiro e Sao Paulo. 2:334–336.

(8) von Klobusitzky, D. and König, P. (1939) Biochemische Studien Uber die Gifte de Schlangengattung Bothrops. VI. Archiv fur Experimentell Pathologie und Pharmakologie. 192:271–275.

(9) Wunderly, C. (1961) *Principles and Applications of Paper Electrophoresis. With a Foreword by Dr. P. König, the Pioneer of Paper Electrophoresis.* Elsevier Publishing Company, Amsterdam, pgs vii–viii.

(10) Kunkel, H.G. and Tiselius, A. (1951) Electrophoresis of proteins on filter paper. Journal of General Physiology. 35(1):89–118.

(11) Koiw, E., Wallenius, G., and Gronwall, A. (1952) Paper electrophoresis in clinical chemistry; a comparison with Tiselius' original method. Scandinavian Journal of Clinical Laboratory Investigation. 4(1):47–54.

(12) Durrum, E.L. and Gilford, S.R. (1955) Recording integrating photoelectric and radioactive scanner for paper electrophoresis and chromatography. Review of Scientific Instruments. 26(1):51–56.

(13) Jencks, W.P., Jetton, M.R., and Durrum, E.L. (1955) Paper electrophoresis as a quantitative method. Serum proteins. Biochemical Journal. 60(2):205–215.

(14) Jencks, W.P., Smith, E.R.B., and Durrum, E.L. (1956) The clinical significance of the analysis of serum protein distribution by filter paper electrophoresis. American Journal of Medicine. 21(3):387–405.

(15) Durrum, E.L., Paul, M.H., and Smith, E.R.B. (1952) Lipid detection in paper electrophoresis. Science. 116(3016):428–430.

(16) Motulsky, A.G., Paul, M.H., and Durrum, E.L. (1954) Paper electrophoresis of abnormal hemoglobins and its clinical applications. A simple semiquantitative method for the study of the hereditary hemoglobinopathies. Blood. 9(9):897–910.

(17) Block, R.J., Durrum, E.L., and Zweig, G. (1958) *A Manual of Paper Chromatography and Paper Electrophoresis,* 2nd edition Academic Press, New York, revised and enlarged.

(18) Smithies, O. (1955) Zone electrophoresis in starch gels: group variations in the serum proteins of normal human adults. Biochemical Journal. 61(4):629–641.

(19) Kohn, J. (1957) A cellulose acetate supporting medium for zone electrophoresis. Clinica Chimica Acta. 2(4):297–303.

(20) Raymond, S. and Weintraub, L. (1959) Acrylamide gel as a supporting medium for zone electrophoresis. Science. 130(3377):711.

(21) Hjerten, S. (1961) Agarose as an anticonvection agent in zone electrophoresis. Biochimica et Biophysica Acta. 53(3):514–517.

[CONTRIBUTION FROM MEDICAL DEPARTMENT FIELD RESEARCH LABORATORY, U. S. ARMY]

A Microelectrophoretic and Microionophoretic Technique[1]

BY E. L. DURRUM

In performing electrophoretic and ionophoretic separations, several investigators have utilized an electrical potential applied across various packing materials intended to stabilize migrating boundaries by preventing convection currents in the electrolytes employed. Strain[1a] combined ionophoresis with chromatographic adsorption in the conventional Tswett adsorption column and mentioned utilizing columns filled with cotton for this purpose. Coolidge[2] was able to separate protein constituents in a column packed with ground glass wool across which a potential was applied. Consden, Gordon and Martin[3] described an ionophoretic technique suitable for the separation of certain amino acids which was carried out in silica jelly slabs made up with various buffers. These investigators utilized paper pulp to reinforce the mechanical strength of the silica jelly

slabs employed. They also reported an experiment in which their trough was filled with "paper powder saturated with liquids to be analyzed" but abandoned this variation of their method because current densities optimum for their purpose could not be employed. Butler and Stephen[4] have utilized asbestos fiber packed in a segmented polystyrene plastic tube and reported separating glycine from glycylglycine at pH 9.3 in this apparatus. None of the above processes was adapted to the separation of small quantities of material.

Recently, Haugaard and Kroner[5] applied electrical potentials across paper partition chromatographs during their development with phenol. They wove thin, flat, metallic electrodes into the edges of the paper which had first been treated with phosphate buffer solution and then dried prior to development with phenol. They reported that the degree of separation of basic and acidic amino acids attainable by paper partition

(1) Presented before the American Chemical Society, Division of Biological Chemistry, March 29, 1949, in San Francisco, California.

(1a) Strain, THIS JOURNAL, **61**, 1292 (1939).

(2) Coolidge, *J. Biol. Chem.*, **127**, 551 (1939).

(3) Consden, Gordon and Martin, *Biochem. J.*, **40**, 33 (1946).

(4) Butler and Stephen, *Nature*, **160**, 469 (1947).

(5) Haugaard and Kroner, THIS JOURNAL, **70**, 2135 (1948).

chromatography was enhanced by this expedient. Though their process is applicable to the separation of minute quantities of amino acids, it does not appear to be applicable to protein separations.

This paper is concerned with a microionophoretic or microelectrophoretic technique which has been found useful for the separation of both amino acids and protein constituents in which an electrical potential is applied across the ends of strips of filter paper saturated with buffer or other electrolyte solutions to which are applied, at narrowly circumscribed intermediate areas, mixtures of amino acids, peptides or proteins to be separated. The positions to which components have migrated are determined in the case of amino acids and peptides by spraying the dried strip with ninhydrin (Consden, Gordon and Martin[6]), and in the case of proteins, by "fixing" the protein *in situ* on the paper strips by heat or by coagulation with chemical agents followed by treating the paper strip with a dye selective for the coagulated protein constituents but easily washed from the filter paper in zones free of protein. A third method which has been employed either alone or in combination with the above methods in cases where radioactive constituents are concerned is that of making autoradiographs of the dried or "fixed" strips. The practical applicability of this method appears to be wide enough to make it desirable to report at this time, although its theoretical aspects remain to be investigated more thoroughly.

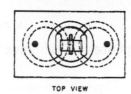

TOP VIEW

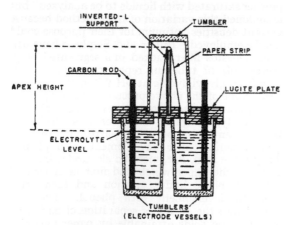

Fig. 1.—Diagram of apparatus.

(6) Consden, Gordon and Martin, *Biochem. J.*, **38**, 224 (1944).

Experimental

In preliminary experiments, narrow (1 cm.) strips of filter paper (about 0.16 mm. thick) were saturated with buffer solutions and the strips draped between two vessels containing the buffer solutions into which were inserted carbon rod electrodes. About the middle of the strips, a drop of serum or amino acid mixture was applied and then a potential of a few hundred volts applied across the carbon rods. These experiments served to show that separations could be practically effected in reasonable periods of time. There were, however, two disadvantages: (1) ill-defined zones of amino acids or proteins were obtained because of the syphoning of the buffer solutions to the low point of the paper with consequent "flooding" in this area; and (2) evaporation from the surface of the paper and temperature could not be controlled readily.

These difficulties were partially avoided by employing a glass bridge arrangement with a filter paper strip placed between somewhat wider plate glass strips resting on the electrode vessels. However, during many of the experiments, puddles of electrolyte were observed to collect irregularly and lateral to the edges of the paper with attendant uncertainties as to uniformity of field strength and as to diffusion of the amino acids into these areas.

This puddling of electrolytes is believed to be due to capillary action between the glass plates in the areas lateral to the paper strip. It was observed that this effect could be minimized by superposing at least three strips of filter paper which then separated the plates at the edges by at least about 0.5 mm. This is a promising method especially for multiple strips, and for single strips where a thicker filter paper is employed. This variation is being investigated further.

The apparatus was further modified to permit the use of single thickness strips of filter paper to which the electrolyte could be confined in a reproducible manner and all the experiments on which the present report is based were carried out in an apparatus of the type illustrated in Fig. 1. This apparatus is comprised of two 150-ml. glass tumblers carrying a lucite plate which seals their tops and supports an inverted L-shaped glass rod. The latter serves to support the apex of the filter paper strips which are draped over it. The ends of the strips pass into the electrolyte solution in the tumblers through slots in the lucite plate. Holes in the plate carry ordinary uncored arc carbon electrodes 8 mm. in diameter. The strips are isolated from the atmosphere by a third inverted 150-ml. tumbler. Annular grooves serve to improve the stability of this arrangement so that no external supports are required.

A larger version of the apparatus having electrode vessels of 500-ml. capacity and wide enough to support seven 1 cm. strips in parallel has proved to be quite convenient and useful when it is desired to compare known and unknown substances simultaneously under identical experimental conditions.

To adapt the apparatus shown in Fig. 1 for experiments of long duration the tumbler electrode vessels were replaced with U-tubes in order to separate the electrode reaction zones from the paper strip ends by a greater distance.

In all of the experiments described, filter paper strips cut from 32-cm. circles of Whatman No. 2 paper were used. Except where otherwise indicated strips 1 cm. wide were employed.

The use of this method is illustrated in the following experiments.

Experiment I.—Separation into five fractions of an equimolar amino acid mixture comprising 19 amino acids: arginine, lysine, histidine, glutamic acid, aspartic acid, glycine, alanine, valine, leucine, isoleucine, serine, threonine, cystine, methionine, tyrosine, tryptophane, phenylalanine, proline and hydroxyproline. Eighty ml. of buffer pH 5.9 prepared by mixing 50 ml. of 0.2 M potassium acid phthalate and 43 ml. of 0.2 M sodium hydroxide and diluting to 400 ml. were placed in each electrode vessel of the apparatus illustrated in Fig. 1. A pencil mark (x) was made across the middle of a 1 × 32 cm. strip of filter paper which was then draped across the glass support rod with the ends dipping about 1 cm. below the surface of the buffer solution in the electrode vessels. The apex of the paper strip was 14.5 cm. above the solution level. When in position, the paper strip was saturated with buffer solution applied to the apex with a medicine dropper. This served to wash from the paper strip any traces of amino acids picked up from the hands in the course of previous manipulations. The top tumbler was put in place and the apparatus allowed to stand for about thirty minutes to permit excess buffer to drain from the paper. About 20 micrograms of amino acid mixture in the form of a dry powder was then applied to the paper strip at the reference mark.

A potential of 600 volts direct current (supplied by a well filtered full wave rectifier) was applied across the carbon electrodes in series with a milliammeter and rheostat. The current was maintained at 1.0 milliampere by frequent adjustment of the rheostat for a period of one hundred and twenty minutes. Then the paper strip was transferred to a glass drying rack using forceps to avoid finger marks and taking care to maintain the apex upward during drying in order to prevent excess buffer at the ends of the paper from running back toward the apex and "smearing" the amino acid zones. The paper strip was dried in an oven at 90° for five minutes, then removed and sprayed with a 0.25% ninhydrin solution in water-saturated butanol (Williams and Kirby[7]) and replaced in the oven for five minutes. The strip showed the following pattern (with all measurements to the center of the spot concerned): (a) toward the anode, 75 mm. from the reference mark (x) a bluish spot corresponding to aspartic acid; 60 mm. from the reference mark, a lavender spot corresponding to glutamic acid; (b) toward the cathode, 4 mm. from the reference mark a dense mauve spot corresponding to the monoamino-monocarboxylic acid group; 23 mm. from the reference mark a grayish spot corresponding to histidine[8] and at 43 mm. a lavender-rose spot corresponding to arginine and lysine which were not completely separated in this experiment. A photograph of the significant portion of this strip is shown in Fig. 2. The initial pH was measured with the glass electrode and found to be 5.93. After the experiment was completed (one hundred and twenty minutes), the pH of the anode vessel was found to be 5.91 and the cathode 6.01.

To separate a solution rather than dry crystals, the following variation in technique is employed: The paper strip to be used is draped on a glass drying rack after the reference pencil mark is made and washed down by directing several milliliters of distilled water at the apex. The strip is air-dried and, being handled with forceps, inserted into the apparatus as described above. Next, about 0.01 ml. of hydrolysate or protein solution, such as blood serum, is applied to the reference mark (at the apex). Then, very carefully, buffer solution is applied with a medicine dropper *below* the apex of the strip at equal distances from the apex on either side, permitting the buffer to flow upward to the drop position by capillarity. In this manner, the solution is prevented from running down the filter paper as has been found to happen usually if even a minute drop of solution is applied to presaturated though drained paper with a resultant lack of sharpness in the patterns obtained. When this variation is used, it is not necessary to wait more than about ten minutes before applying the potential.

The above variation of technique is illustrated in the following example.

Experiment II.—Separation of human serum: 0.01 ml. of serum was applied from a micro-pipet to the reference mark of a 1-cm. paper strip as described above. Immediately, a 0.05 molar sodium diethylbarbiturate buffer solution (pH 8.6) was applied. A potential of about 350 volts was applied through a rheostat for one hundred and eighty minutes. The current was maintained at 0.5 milliampere by frequent adjustment of the rheostat. At the end of the run the strip was removed and dried for five minutes in an oven at 100°, then immersed for five minutes in a saturated solution of mercuric chloride in 95% alcohol to which had been added 0.1 g./100 ml. of brom phenol blue (tetrabromophenolsulfonphthalein). The strip was next removed and washed for ten minutes in running tap water. The strip was then dried. Four distinct blue zones were visible, all located toward the anodal side of the reference mark: the first, 35 mm. from the reference mark, corresponding to albumen; the second at 25 mm., probably corresponding to alpha₁-globulin; the third at 15 mm., corresponding to alpha₂-globulin and the fourth, 6 mm. from the reference mark, corresponding to beta-globulin. A fifth zone was located 13 mm. toward the cathode corresponding to gamma-globulin. The establishment of identity of these protein zones is discussed later in this paper.

Experiment III.—Separation of a mixture of alanine, valine, proline and tryptophane the apparatus of larger dimensions: in this apparatus, 500 ml. of 5 N acetic acid was placed in each electrode vessel. The apex height in this experiment was 11.5 cm. above the fluid level. Five paper strips were supported, washed down and saturated with electrolyte as described above. On one strip, a few micrograms of a mixture of these amino acids were placed at the reference mark (x) and on each of the other strips, one of the amino acids of the mixture was placed at the reference mark. A potential of 580 volts was applied across the carbon electrodes. The initial current was 1.5 milliamperes per 5 cm. (width). After one hundred and twenty minutes, the current had risen to 1.7 milliamperes per 5 cm. At this time, the strips were removed, dried and sprayed with ninhydrin. Portions of the resulting strips are shown in Fig. 3.

Experiment IV.—Separation of a mixture of glycine, isoleucine, phenylalanine and hydroxyproline. In an experiment exactly analogous to Experiment III, a mixture of the above amino acids was separated as illustrated in Fig. 4. (The faint zone on the hydroxyproline strip represents accidental contamination with isoleucine.)

Experiment V.—Separation of glycylglycine from glycyl-l-leucine in the apparatus of Experiments III and IV. Five strips were employed, to one of which a mixture of glycylglycine and glycyl-l-leucine was applied and to each of the remaining strips only one of these substances. Glycine and l-leucine were added to separate strips for comparison. The separation attained at the end of two hours is illustrated in Fig. 5.

Experiment VI.—Reproducibility of parallel runs. The reproducibility of this method on parallel runs is illustrated by Fig. 6 which shows sections of the paper strips obtained in a simultaneous run when a few micrograms of crystalline phenylalanine was applied to the origin of all strips. In this experiment, the electrolyte was 5 N acetic acid, the initial current 1.8 milliamperes per 6 cm. and the final current (one hundred and twenty minutes later) 1.7 milliamperes per 6 cm. (Ordinarily, the current has been observed to increase during the course of the runs. Rather marked line voltage fluctuations are sometimes noted which perhaps explain why the final current was recorded lower than the initial value.) The mean position of the phenylalanine was found to be 58.5 mm. from the origin with a standard deviation of ± 3.27 mm. It is evident that the reproducibility of parallel runs is of sufficient degree ordinarily to permit selection of "matching pairs of acids" as, for example, is illustrated in Figs. 3, 4 and 5.

Experiment VII: Rate of Migration of Phenylalanine.—Figure 7 illustrates the findings in an experiment in which the migration of phenylalanine toward the cathode was measured as a function of time. Six strips were set up in parallel with the apparatus previously described, the electrolyte being 5 N acetic acid. At thirty-minute intervals

---

(7) Williams and Kirby, *Science*, **107**, 481 (1948).

(8) It has been observed that pH 5.9 gives a good separation of the histidine from the arginine–lysine zone and the monoamino-monocarboxylic zone. As the pH is increased, the histidine tends to migrate at a velocity closer to the latter group, merging with it about pH 6.6. As the pH is decreased, the reverse has been observed with the histidine zone merging with the arginine–lysine zone at about pH 5.2.

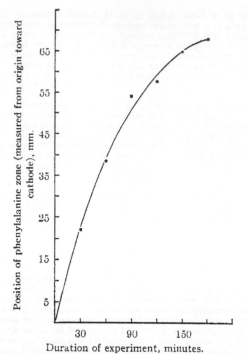

Fig. 7.—Migration of phenylalanine *vs.* time (experiment VII): 5 *N* acetic acid, Whatman No. 2 paper, 580-volt potential, 0.3 ma./cm. (width) average current, apex height 11.5 cm.

strips were removed. During the course of the experiment under a potential of 580 volts, the current per strip averaged 0.3 milliampere. It is evident that the migration of this amino acid down the paper under the conditions of these experiments is not a linear function of time.

Experiment VIII.—Comparison of zones derived from separated electrophoretic components with whole serum. Figure 8 illustrates an experiment in which electrophoretic components separated in a Tiselius apparatus are compared with the whole serum pattern from which these components were derived. In the electrophoretic apparatus of Moore and White,[9] a human serum sample was separated utilizing 0.1 *M* barbiturate buffer (*p*H 8.6, Longsworth[10]) into the following components: (a) albumin (from Zone IV ascending limb of Tiselius cell); (b) a mixture of albumin plus alpha₁-globulin (from Zone V); (c) pure gamma-globulin (from Zone I descending limb); (d) a mixture of gamma-globulin plus beta-globulin (from Zone II); (e) a mixture of gamma-, beta- and alpha₂-globulins (from Zone III).

Paper patterns of these fractions were prepared under the similar conditions enumerated in Fig. 8. The results of this experiment appear to establish the identity of all the components except alpha₁-globulin. It will be noted that in Fig. 8, paper patterns IV and V do not differ appreciably although alpha₁-globulin is presumably present in pattern V only. The probable explanation is that since this component is present in such low concentration it is scarcely visible on the whole serum pattern; it is then not surprising that it is not more evident in pattern V which material was diluted in the course of the prior Tiselius separation.

Experiment IX.—Comparison of zones derived from "immune globulin" and reference serum with electrophoretic patterns. A similar analysis of a sample of commercial

(9) Moore and White, *Rev. Sci. Inst.*, **19**, 700 (1948).
(10) Longsworth, *Chem. Revs.*, **30**, 323 (1942).

human "immune globulin" (Squibb) appears to be valid. Figure 9 illustrates an electrophoretic pattern prepared with 0.02 phosphate–0.15 *M* sodium chloride buffer (*p*H 7.4) in comparison with paper strips prepared with 0.05 *M* barbiturate buffer, *p*H 8.6 and 0.02 *M* phosphate buffer, (without added sodium chloride) *p*H 7.6.

The component migrating most rapidly appears to be albumin from comparison with the serum patterns. Superior resolution and correspondence seems in this case to be found with the barbiturate buffer although a greater migration has occurred in shorter time with the phosphate buffer, undoubtedly due to the greater field strength in this experiment.

Experiment X.—Comparison of human plasma and serum patterns. Figure 10 illustrates the patterns obtained in an experiment in which a serum and a heparin-plasma derived from the same sample of human blood were separated on paper with barbiturate buffer. Control experiments have established that heparin does not stain with brom phenol blue under the conditions used in these experiments. Therefore, the zone present near the origin of the plasma pattern but not evident in the serum pattern may be regarded as being derived from fibrinogen. In interpreting this pattern it is difficult to decide with certainty whether the position of the fibrinogen zone is due to its having been coagulated at the point of application (origin), as suggested by the circular configuration and size which has about the same dimensions as the circle resulting from the applied drop of plasma (about 0.01 ml.) at the beginning of the experiment, or is simply an expression of its low rate of diffusion and inherent electrical mobility, since in this pattern the point of origin falls coincidentally at a point intermediate between the gamma and beta globulins. It is of course well known that the fibrinogen boundary falls between these same constituents in conventional electrophoretic patterns obtained with sodium diethylbarbiturate buffer (Longsworth[10]).

Experiment XI.—Separation of radioactive inorganic iodide from protein-bound iodine. A 230-g. Wistar strain rat was injected intraperitoneally with 87 microcuries of I^{131}. The animal was sacrificed 210 minutes later. The thyroid was removed and all possible connective tissue carefully dissected from it. The resulting thyroid was ground in a Ten Broeck tissue grinder together with about 10 drops of 0.9% sodium chloride solution. The resulting material was centrifuged and the clear supernatant fluid applied to the reference marks of strips of filter paper which were separated in 0.05 *M* barbiturate buffer, *p*H 8.6, for various periods of time (from 0 to forty-five minutes) as illustrated in Fig. 11. The resulting strips were dried in an oven for five minutes and then autoradiographs were made of the strips with exposures of five hours and ninety-three hours as indicated. The following points may be noted. In the experiments, a distinct band of radioactivity is visible migrating rapidly toward the anode. This undoubtedly corresponds to inorganic iodide ion and it is seen that, in a comparatively short time, the paper in the zones retaining protein (identified by its property of being coagulated and dyed) is completely "cleared" of inorganic iodide, the residual activity being associated with protein and/or amino acid fractions. The lower photograph (five-hour exposure) is included to show detail in the protein–amino acid zones which is obscured by the longer exposure necessary to show the migration of the iodide ion.

Discussion

In the technique employed in all the experiments illustrated, it is believed that the paper strip plays merely a passive role as a carrier of the electrolyte. It probably may thus be regarded as analogous in a limited sense to the Tiselius cell.

In the course of several hundred experiments, no evidence of adsorptive phenomena has been noted under the experimental conditions employed, in either the case of protein[11] or amino acid separations. That is to say, (allowing for certain factors discussed below) the components seem to behave as they would be expected to in "free solution." Critical studies designed to answer this question have not been carried out and, therefore, the possible role of adsorption in the process must await elucidation. For this reason, it seems best for the time being to regard the separations described as ionophoretic or electrophoretic rather than "electrochromatographic" as are the separations described by Strain,[1a] or as "partition chromatography with applied voltage" as in the process described by Haugaard and Kroner.[5]

It will be realized that under the experimental conditions employed a relatively complicated equilibrium obtains which includes a number of simultaneously occurring processes which include at least the following: (a) migration of ions due to the electrical field; (b) diffusion; (c) electro-endosmotic flow; (d) evaporation of water from the paper strip due to heating of the strip incidental to the current flow; (e) hydrodynamic equilibrium on the paper strip between capillary forces and gravitational forces; (f) electrical resistance changes along the length of the paper strip principally due to concentration effects secondary to factors d and e.

Sufficient data for a criticial evaluation of these factors are not available. However, some of the more obvious relationships which appear to explain some of the experimental findings will be discussed briefly. We may consider that the field strength equation which is applied to the Tiselius cell is applicable as a first approximation at least to thin cross sections of the paper at any given level above the electrolyte level at any given instant. Then, migration velocity is proportional to field strength X and

$$X = I/qk_s$$

where I = current; q = cross sectional area of the paper; and k_s = conductivity of the solution on the paper at the cross section under consideration.

Limiting consideration to the case where the current is held constant, the cross section of the paper is constant and, therefore, must at any given level after equilibrium is established "contain" a given quantity of electrolyte. But, since the amount of electrolyte contained along the length of the paper varies due to the hydrodynamic and distillation equilibria mentioned, the "effective cross section" of the paper may be regarded as increasing as the electrolyte level is approached, and as decreasing as the apex is approached, reaching its minimum "effective cross section" at the apex. Therefore, the field strength may be expected to be highest at the apex and to decrease as the electrolyte level in the electrode vessels is approached.

The "drier" apex may be expected, therefore, to have more electrical resistance and, for a given current, would be expected to produce more heat than the "wetter" areas below. This factor would be expected to accentuate (or perhaps be principally responsible for) the "wetness gradient" down the paper. The above considerations appear to explain the lack of linearity of migration of ions with time as demonstrated for phenylalanine in Experiment VII (Fig. 7).

It is for the above reasons that the apex height has been recorded in experimental data, it having been observed that reproducibility of the method could not always be achieved unless this factor were carefully controlled, especially with protein separations where a certain optimal "degree of wetness" of the paper for a given current and buffer seems to be essential for satisfactory resolution.

Under the experimental conditions employed, due to the very large surface area–electrolyte volume ratio present in the paper strip, pronounced electroendosmotic currents toward the cathode would be anticipated. It is believed that this explains the apparent migration of the gamma-globulin toward the cathode as illustrated in the serum and plasma patterns (Figs. 8, 9 and 10). In conventional electrophoretic separations at pH 8.6, all the serum components are known to migrate to the anode. It is believed that this apparent migration of the gamma globulin toward the cathode can be explained by a displacement of the entire pattern toward the cathode due to this pronounced electroendosmotic current.

In view of the above considerations, a close correlation of the mobilities of protein constituents as measured in the Tiselius apparatus with these paper patterns is not to be expected.

It will be noted that the barbiturate buffer employed in these experiments is 0.05 M as compared with 0.1 M buffer often employed in conventional electrophoretic studies of human sera. It has been observed empirically that the more dilute buffer very much improves the degree of resolution attainable in human serum and plasma samples in the present technique. This may be due to the fact that the concentration of the buffer on the paper is increased above its value in the electrode vessels due to evaporation from the paper and, on the paper, thus approaches a concentration comparable with the optimum concentration observed in Tiselius separations.[12]

Acknowledgments.—It is a pleasure to acknowledge the helpful comments of Drs. L. G. Longsworth and D. H. Moore who have kindly read this manuscript.

(11) Professor A. Tiselius (personal communication), has evidence that the colored protein phycoerythrin from the alga *Ceramium rubrum* is adsorbed on paper under conditions similar to those described in this report.

(12) Since the present paper was submitted for publication, independent work by Wieland and Fischer, *Naturwissenschaften*, **35**, 29 (1948), has been brought to the author's attention. These workers report ionophoretic separations on filter paper saturated with acetate buffer and reported that at pH 5 they were able to separate a mixture of glutamic acid, alanine and histidine, at pH 7.5, a mixture of lysine and histidine, and at pH 3.7 glutamic acid and aspartic acid.

Summary

1. A micro-technique for the separation of amino acids, peptides and proteins has been developed.

2. The technique is carried out by applying an electrical potential across the ends of strips of filter paper saturated with electrolyte solution. At some intermediate position of these strips, the mixture to be separated is applied.

3. The course of separations is followed in the case of amino acids and peptides by ninhydrin treatment; in the case of protein separations by coagulation and selective dyeing *in situ* and in the case of radioactive components by autoradiography.

Fort Knox, Kentucky   Received July 28, 1949

COMMENTARY TO

24. Skeggs, L. T., Jr. (1957)
An Automated Method for Colorimetric Analysis. American Journal of Clinical Pathology 28(3): 311–322.

I n 1854 Louis Pasteur gave a speech at the University of Lille in Douai, France in which he said that chance favors the prepared mind (1). Articles by and by about Leonard T. Skeggs, Jr. help reveal how his mind was prepared to invent the first automated clinical chemistry analyzer (2–5).

Skeggs was head of the clinical chemistry laboratory at the Cleveland Veterans Administration Hospital. In 1950 all clinical chemistry tests were manual assays. The first step in a typical serum glucose or urea nitrogen assay was the preparation of a protein free filtrate with either centrifugation or filter paper filtration. This was followed by incubation with one or more reagents, color development and then optical density readings in a filter photometer one sample at a time. These assays were time consuming, contained numerous repetitive steps and were prone to human errors. Skeggs began to develop an automated instrument to perform these steps in an operator independent manner. He worked on the instrument in his spare time in his basement starting in 1950. The first proof of concept instrument was for urea nitrogen and was based on a continuous flowing stream (5). A peristaltic pump moved reagents through polyethylene tubing. Samples were introduced by holding the end of a tube in a bottle of urea standard for a fixed period of time. Protein free filtration was accomplished by means of a dialyzer. Skeggs was co-inventor in the late 1940s of an artificial kidney. In the Skeggs–Leonard kidney, blood passed between two sheets of cellophane clamped between a top and bottom rubber plate (6). Each plate had grooves cut into its under surface facing the cellophane through which saline was pumped. The saline removed urea and was sent to waste, the blood was returned to the patient. In the breadboard chemistry analyzer the dialyzed urea in saline was mixed with reagents and measured and the blood was sent to waste. Air bubbles that were introduced by accident into the flowing streams were soon introduced by design because Skeggs realized that they helped to prevent mixing of different patient samples. Skeggs spent almost 4 years trying to find a company to develop and commercialize his analyzer. After many rejections he met with the people at Technicon.

The Technicon Company was founded in 1939 in New York City. In 1954 Skeggs' prototype automated analyzer was capable of performing urea nitrogen or glucose assays at up to 30 samples/hour. This was the typical work volume of one technician in one day. When Skeggs demonstrated his analyzer to the people at Technicon many of the minds at that company were prepared for what they saw. People like Ray Roesh the field representative who heard about Skegg's analyzer and chased after him to see it; Edwin C. Whitehead the co-founder with his father of the company and engineers, like Andres Ferrari. Technicon was a manufacturer of histology laboratory equipment. Their main piece of equipment however was the Autotechnicon®, a robot-like device that moved baskets of tissue sections through 12 different fixation, dehydration, washing and staining solutions overnight completely unattended (7). Technicon produced other automated equipment like a 200-tube column chromatography fraction collector; an automatic pipette washer that held up to 200 pipettes and an air blower multi-slide dryer that produced dry slides in 7 minutes.

Skeggs signed an agreement with Technicon in 1954. The first report on the automated analysis of urea, glucose or calcium was presented by Skeggs at the September 1956 International Congress on Clinical Chemistry in New York City (8). In February 1957 the first full paper was published and that paper is presented here. Also in 1957, Technicon introduced the AutoAnalyzer® and sold 50 of them in the first year. In January 1958 the journal *Analytical Chemistry* ran a profile on the instrument (9). By 1969 over 18 000 had been sold (3). In 1964 Skeggs and Hochstrasser published a paper on an 8-channel analyzer with integrated sequential chemistries and a single chart report (10). Assays included the four electrolytes, albumin, total protein, urea and glucose. Their design led to the Sequential Multiple Analyzer (SMA®). These instruments ran up to 12 assays in 60 seconds (SMA 6/60 and 12/60) and latter the SMAC® (Sequential Multiple Analyzer with Computer) which ran 20 assays in 20 seconds (11).

Skeggs and Technicon did more than improve productivity and efficiency in clinical chemistry. The AutoAnalyzer stimulated a level of innovation that was not duplicated by any other automated instrument in clinical chemistry at the time. The Technicon AutoAnalyzer was an open system by design. Users were able to easily modify, improve and adapt new methods to automated analysis. Eric von Hippel in his book *The Sources of Innovation* included a study on the impact of the AutoAnalyzer on technical innovation in clinical chemistry (12). In 1977 in the United States 298 million chemistry tests in hospital laboratories were run on automated analyzers. Von Hippel reviewed the literature for the year 1977 and found that 46% of the hardware improvements on the Technicon AutoAnalyzer were developed first by the customers. In addition, 74% of the most frequently run tests on the AutoAnalyzer were first adapted onto the Technicon system by the customers. By contrast he studied a popular but proprietary closed system analyzer for the same period (the Dupont Automated Clinical Analyzer, aca®). He could find no tests or hardware improvements that had been developed by the customers.

References

(1) Peteson, H. (ed.) (1954) Louis Pasteur depicts the spirit of science [December 7, 1854], in *A Treasury of the World's Great Speeches*. Simon and Schuster, New York, pgs 469–474.
(2) Skeggs, L.T. (1966) New dimensions in medical diagnosis. Analytical Chemistry. 38(6):31A–44A.
(3) Henahan, J.F. (1970) Clinical chemistry's man on horseback. Leonard T. Skeggs, Jr. Chemical & Engineering News. 48(33): 54–58, August 10.
(4) Lewis, L.A. (1981) Leonard Tucker Skeggs — a multifaceted diamond. Clinical Chemistry. 27(8):1465–1468.
(5) Skeggs, L.T. Jr. (2000) Persistence... and prayer: from the artificial kidney to the AutoAnalyzer. Clinical Chemistry. 46(9):1425–1436.
(6) Skeggs, L.T. Jr., Leonards, J.R., and Heisler, C.R. (1949) Artificial kidney II. Construction and operation of an improved continuous dialyzer. Proceedings of the Society for Experimental Biology and Medicine. 72(3):539–543.
(7) [Advertisement] (1948) *Autotechnicon*. The Technicon Company. Science. 107(2784):5.
(8) Skeggs, L.T. Jr. (1956) An automatic method for colorimetric analysis. Clinical Chemistry. 2(4):241, Abstract 28.
(9) Muller, R.H. (1958) Automatic colorimetric analyzer eliminates need for many analytical procedures. Analytical Chemistry. 30(1):53A–57A.
(10) Skeggs, L.T. and Hochstrasser, H. (1964) Multiple automatic sequential analysis. Clinical Chemistry. 10(10):918–936.
(11) Schwartz, M.K., Bethune, V.G., Fleisher, M., Pennacchia, G., Menendez-Botet, C.J., and Lehman, D. (1974) Chemical and clinical evaluation of the continuous-flow analyzer "SMAC". Clinical Chemistry. 20(8):1062–1070.
(12) von Hippel, E. (1988) *The Sources of Innovation*. Oxford University Press, New York, pgs 93–101.

Am J Clin Pathol. 1957; 28: 311–322
© 1957 American Society of Clinical Pathologists.
Reprinted with permission.

TECHNICAL SECTION

AN AUTOMATIC METHOD FOR COLORIMETRIC ANALYSIS

LEONARD T. SKEGGS, JR., PH.D.

*Department of Pathology, Western Reserve University School of Medicine,
Cleveland, Ohio*

The staffs of laboratories of clinical chemistry are confronted with an ever-increasing number and variety of determinations. Many industrial laboratories, faced with similar increases in their work load, have derived benefit from the adoption of instrumental and semiautomatic methods of analysis. It seemed worthwhile, therefore, to explore the possibility of automatic methods for analyses in clinical laboratories.

Although one could possibly construct a robot that would be capable of performing all of the many steps required in a typical, conventional, colorimetric analysis, this approach seemed to be impractical. Instead of this, it was discovered that colorimetric analysis could be performed in continuously flowing streams, thus eliminating the need for stepwise measurement, addition, and processing of samples and reagents. One problem peculiar to the analysis of blood is the necessity for removal of blood proteins. This problem was solved by the use of a dialyzer that was designed to process continuously flowing streams of solutions.

In the method evolved, samples of whole blood or serum are successively pumped into a flowing stream of diluent. The diluted sample is passed through a dialyzer where diffusible constituents pass the membrane and are picked up in a flowing stream of reagent. This stream may then be mixed with other reagents and heated, or otherwise processed in a continuous manner, in order to produce a color that is specific for the desired constituent. The colored stream is finally passed through a flow-cell colorimeter that is equipped to record the results of the analysis.

EXPERIMENTAL

Several conventional chemical determinations have been successfully performed by the continuous method (Figs. 1 to 8). These include the determination of urea nitrogen and glucose in blood. Preliminary results have also been obtained in determinations of calcium, chloride, and acid and alkaline phosphatases in serum.

Only one determination will be described in this paper, with the intention of indicating the possibilities of the method and illustrating the technics and types of apparatus that are necessary for its performance. Subsequent publications will deal with this and other applications of the method, and with their reproducibility and accuracy under conditions of actual usage in hospital laboratories.

A schematic flow diagram illustrating the application of the Fearon reaction[1, 3, 4] to the determination of urea nitrogen in blood is reproduced in Figure 1.

Received, February 25, 1957; accepted for publication May 1.
Dr. Skeggs is Assistant Professor of Biochemistry.

It may be observed that several components* are required. These include a sample pick-up device (Fig. 3) that consists of a slowly rotating, replaceable disk of Plexiglas with 20 depressions in its periphery, into which samples (1.5 ml. or more are required) may be poured without measurement. Samples are aspirated in succession from the pick-up plate by means of a multiple-tubing type of pump (Fig. 4). The pump functions by rolling or squeezing out as many as 8 polyvinyl tubes that lie parallel to each other on a flat plate. By this means, reagents are drawn from bottles; combined with the sample, the reagents are then propelled through the flow circuit illustrated in Figure 1 (assembled with $\frac{1}{6}$-in., internal diameter, polyvinyl tubing) in constant proportion to one another. The dialyzer (Fig. 5) consists essentially of 2 tubular passages with 1 common wall of cellophane, across which dialysis may occur from one flowing stream to another. This is accomplished practically by means of pressing together 2 grooved plastic plates with a piece of cellophane between them. A heating bath is required as a means of accelerating the color reaction. This consists of a coil of glass tubing (Fig. 6) that is immersed in a thermostatically controlled (95 C. $\pm$ 0.2 C.) bath of ethylene glycol. The results are observed in a colorimeter (Figs. 7 and 8) that is especially adapted to the measurement of flowing solutions and is further equipped to record the results in terms of percentage of transmission.

In actual operation, samples of whole blood are successively aspirated by the pump (0.75 ml. per min. for 2-min. periods), from the pick-up plate, and they are then mixed with a flowing stream of physiologic solution of sodium chloride (2.1 ml. per min.), which is saturated with caprylic alcohol in order to reduce foaming. The solutions are repeatedly inverted and mixed by passing them through an 84-cm. length of 3-mm. (outside diameter) Pyrex tubing that is wound into a coil 2 cm. in diameter and lying on its horizontal axis. After mixing, air bubbles are added (0.75 ml. per min.), and the mixture is then passed through the dialyzer and finally discarded. While the diluted blood is passing through the dialyzer, urea crosses the membrane and is picked up in a flowing stream of 0.5 per cent diacetyl monoxime and 15 per cent solution of sodium chloride (2.5 ml. per min.) that contains bubbles of air (1.0 ml. per min.). In both instances, the air bubbles divide the liquid streams into small segments and mechanically prevent mixing of successive samples. After emerging from the dialyzer, the stream of diacetyl monoxime is converged and mixed in a second glass coil with 10 per cent arsenic acid in 50 per cent sulfuric acid (2.5 ml. per min.). The combined streams are then directed through a coil that is immersed in a heating bath maintained at 95 C., where color is developed in direct proportion to the concentration of urea. After development of the color, the solution is passed through the flow-cell of the colorimeter and a record of the density of color (480 mμ) is obtained.

A representative record is illustrated in Figure 2. In this instance, the colorimeter-recorder system was initially adjusted to 100 per cent transmission (base

---

* The responsibility for making commercially available apparatus for the continuous method has been assumed by the Technicon Co., Chauncey, New York.

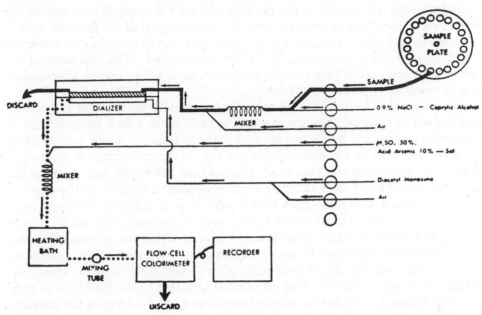

FIG. 1. Flow diagram of the continuous method for determining urea nitrogen.

line) while only the reagents were flowing through the system. Approximately 5 min. after a sample of blood was introduced, the recorder responded and indicated a value for the percentage transmission that represented the concentration of urea in the sample. Had the sample been pumped continuously, a long flat trace would have been recorded. However, 2 min. is more than adequate to provide for a maximal response. For this reason, samples were pumped for 2-min. periods, and, in order to provide a distinct separation of the results on the record, air was pumped between each of the samples and the next (for 1 min.), thereby permitting the recorder to return to its base line. Each sample is represented on the record, therefore, as an individual rise. The response to 9 different samples of blood may be observed at the left in Figure 2.

The response to a series of solutions of urea with from 10 to 100 mg. of urea N per 100 ml. is illustrated on the right in Figure 2. The peak values for transmission in such a record may be plotted on semilogarithmic paper in order to provide a calibration curve for the system. In the determination of urea, this curve is almost a straight line for values up to 60 mg. of urea N per 100 ml. When the amount is greater than 60 mg., the curve becomes somewhat nonlinear, although it is regularly reproducible. A similar lack of linearity occurs in the conventional method with diacetyl monoxime.[4] After obtaining a calibration curve, standard solutions of urea may be analyzed, as frequently as desired, in order to provide a constant check on the standardization.

Although 5 min. are required for the recorder to respond to the presence of a sample, analyses may be completed at the rate of 1 every 3 min. Two or more samples may be in different stages of processing at one time. It is necessary to introduce only the amount of sample that will yield a maximal response on the recording. The length of time required for the pumping of air between each 2 samples is that which produces an adequate cleft on the recording and thereby permits the various samples to be easily distinguished. With the method described for urea, it has been possible to observe satisfactory results when pumping as many as 40 samples per hour.

The factor that limits the maximal rate of analysis is the speed of response of the system as a whole. This is controlled, in turn, by at least 3 factors. First, it seems that some time is required for cellophane to establish a new rate of dialysis after a change in concentration on one side of the membrane. A method of accurately measuring this "lag time" has not been developed, but the time has been estimated to be from 15 to 30 seconds. Second, a solution in the flow-cell must be exchanged several times with a solution of a different density of color if a new and correct percentage of transmission is achieved. The cell illustrated in Figure 7, which will hold approximately 0.6 ml. of solution, represents a compromise between the desire to use a cell of small volume and, at the same time, have as long a path of light as feasible, in order to gain in colorimetric sensitivity. Finally, it is most important that air bubbles be added to the fluid lines as indicated in Figure 1. This has the effect of segmenting the fluid within the lines and mechanically preventing mixing. All fluid lines and "T" connections must be of such size, therefore, that air bubbles do not divide or coalesce after they are once formed.

Certain precautions are necessary in the design of the apparatus and in construction of the flow-circuit, in order to eliminate undue excursions of the recorder pen ("noise") that are not directly related to the concentration of urea in the sample. It is obvious that the pump should proportion with accuracy and that the colorimeter should have a stable base line. In order to prevent oscillations of the level of fluid within the flow-cell, a separate leveling tube is used (Fig. 7), and this also serves as a vent for the escape of air. It is important that the pattern of air bubbles established in the diacetyl monoxime reagent-line (Fig. 1) be regular, so that the proportioning is correct when it is converged with the arsenic reagent. Finally, the design and operation of the dialyzer is of some importance in this regard. Cellophane is an easily stretched and flexible membrane. Therefore, small differences in pressure exerted by the flowing solutions may distend the membrane and produce inequalities in the proportioning of diacetyl monoxime to arsenic acid reagents. This may be held within reasonable limits by using a narrow-grooved dialyzer, as well as by slightly increasing the pressure on one side of the membrane.

Although the method, as outlined, is adjusted to yield a satisfactory sensitivity and range for clinical analysis, these factors may be altered to yield (1) a high degree of sensitivity with a low range or (2) a low sensitivity with a wide range. This may be easily accomplished by changing the proportions of sample and physiologic saline solution that are pumped. Thus, 20 per cent transmission may be made to the equivalent of as little as 75, or as much as 150 mg. of urea N per 100 ml. Increasing or decreasing the combined flow-rate of arsenic acid and

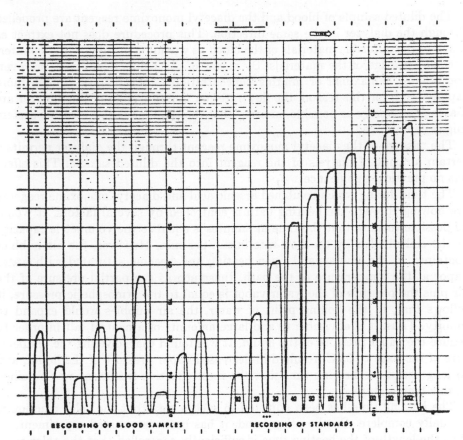

FIG. 2. Record of analyses for urea nitrogen, as performed by the continuous method.

diacetyl monoxime reagents also results in a change of sensitivity, although a certain rate of flow is required in order to obtain a sufficiently fast response or rate of change from one sample to another. The proportion of arsenic acid reagent to diacetyl monoxime has been adjusted to provide maximal colorimetric sensitivity.

Although the sensitivity of the method may be easily altered by a change in the proportion of sample to diluent, the sensitivity is also greatly affected by the effective area of membrane in the dialyzer, the length of the cell in the colorimeter, and the dimensions of the heating bath. In general, it is advisable to use a dialyzer with a large area. When analyzing for urea N, the concentration in the dialysate approaches that of the diluted blood and true equilibrium is almost obtained. With regard to glucose, the amount extracted is much less. In any case, the concentration of urea in the dialysate is in direct proportion to its concentration in the diluted sample of blood. It is of particular interest that cellophane membranes may be used for as long as a month, without detectable deterioration or clogging.

The diluted sample of blood and the diacetyl monoxime reagent are propelled through the dialyzer in the same direction. This procedure does not extract as much of the diffusible materials of the blood as would be possible with "counter-current" passage of the materials. By means of the 1-directional flow, however, one obtains a maximal or limiting concentration in the dialysate much more rapidly, and, for this reason, the flow-rates on opposite sides of the dialyzer are matched as closely as possible, in order that the time for response of the system might be improved.

The passage of the diacetyl monoxime reagent through the dialyzer permits a certain amount of dialysis of the reagent into the diluted sample. This effect has no practical significance.

The time during which combined arsenic acid and diacetyl monoxime reagents are flowing through the heating bath is controlled by several factors. It would be possible to state that the time in minutes is equal to the volumetric capacity of the coil divided by the flow-rate in ml. per min. However, the air bubbles contained in the line expand upon heating and become saturated with water vapor at the temperature of the bath. Inasmuch as the partial pressure of the air plus the pressure of the water vapor is equal to the atmospheric pressure, it follows that, at the boiling point of water, the air bubbles would expand to infinity. Thus, it is advisable to add large amounts of a neutral salt in order to reduce the vapor pressure of the liquid. It was found that lowering the temperature of the bath to 95 C. resulted in decreased vapor pressure of the liquid, and so lengthened the heating time that the most effective treatment was obtained at this temperature.

The heating time has been adjusted in such a way that the color reaction is virtually complete. On the other hand, it has been determined experimentally that the length of the heating time is sufficiently controlled that the color reaction need not proceed to completion as a requisite for obtaining accurate results.

The results obtained from the analysis of blood or serum may be easily verified by means of recovery experiments. These may be performed by continuously running a dilute standard solution of urea into the diluent tube, in place of physiologic saline solution (Fig. 1), while pumping the latter into the sample tube. A trace will then be described upon the recorder; the trace amount of urea may be read from a previously prepared calibration curve. A sample of blood with a known, low or normal concentration of urea is then added by way of the sample tube. The recorder responds by indicating a new and higher value for urea; this should represent the sum of the urea in the blood plus that in the standard solution. This procedure is similar to that of the usual recovery experiment, with the exception that the addition of the urea to the blood is accomplished by means of the pump.

The concentration of urea in a series of samples of blood was determined according to the continuous method and, also, the urease-aeration-titration method of Van Slyke and Cullen,[5, 6] as modified by Summerson.[2] The results are summarized in Table 1. The average difference for the 40 samples was only 0.85 mg. of N per 100 ml. The values obtained by the continuous method have been corrected by a factor of 0.95. The validity of this factor may be demon-

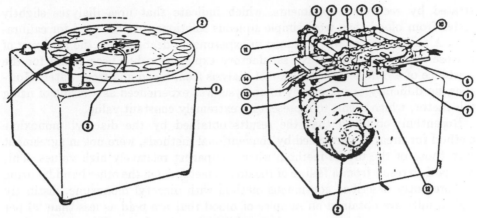

FIGS. 3 and 4

FIG. 3 (left). Device for pick-up of samples.

No. 1 indicates the box that provides a support for the mechanism.

No. 2 designates the plate of Plexiglas that is set on a shaft driven by a small timing motor, and clamped to the shaft either by friction or by detent. The speed of the motor and the diameter of the holes in the plate should be arranged to provide 2 min. of pick-up of sample, with an intermediate period (1 min.) for pick-up of air. The holes are 2 cm. in diameter and 1 cm. in depth.

No. 3 indicates the pick-up tube, which is hinged on the support column. A polyethylene tube (internal diameter, 0.034 in.) is inserted through the pick-up tube and into the holes in the pick-up plate. The plate rotates in a counter-clockwise direction, and the pick-up tube falls into and slides out of the holes in the plate.

FIG. 4 (right). Proportioning pump for use with the continuous method, with the following parts, as labeled:

1. Rigid box for support of the mechanism. It is desirable that this box be constructed of material that is resistant to chemicals.

2. Synchronous gear-head motor, approximately 1⁄50 horse power, with an output speed in the range of 10 r.p.m.

3. Stainless steel sprocket-chain (pitch, ½ in.), with rollers at each pin. This chain passes over 2 sprockets, *i.e.*, one on the motor, the other on the operating shaft. Any combination of speed of the motor and diameters of sprockets may be used in order to attain 13.3 r.p.m. at the operating shaft.

4. Stainless steel chain with rollers at each pin, and with roller attachment every 2 in. Total length of chain, 10 in.

5. Rollers constructed from stainless steel, with bushings to fit the pin attachment on the chain.

6. Platen prepared from reinforced plastic.

7 and 8. Springs of stainless steel with a low spring-rate. A pair of springs is used on both ends of the platen, each of the units of the pair displaced outwardly, toward the edge of the platen, as far as possible. The springs must be sufficiently strong to compress the tubing.

9. Six-tooth sprockets (pitch-diameter, 1 in.), secured to the operating shaft. The sprockets are constructed preferably of laminated plastic material.

10. A pair of side-plates of noncorrosive metal or reinforced plastic. These plates provide the track for the chain. The plates should be of a width that restricts lateral movement of the chain; the height should correspond to the root-diameter of the sprockets, and the length should be such that only minimal looseness of the chain is permitted.

11. Universal joint to permit upward rotation of the entire pumping-roller assembly, in order to provide access to the tubes. The center line of the universal joint should coincide with the center line of the hinge (not illustrated) at the far end of the assembly.

12. Latch to provide a means of holding the pumping-roller assembly firmly in position during operation. The tubing is unloaded when the pumping-roller assembly is in the nonoperating position.

13. Polyvinyl tubing with an internal diameter selected as desired. The optimal thickness of the wall depends on the inside diameter, but, in the range discussed, a thickness of 0.033 in. yields good results. Tubing with an inside diameter of 0.045 in. permits passage of 0.75 ml. per min., whereas an inside diameter of 0.083 in. permits passage of 2.5 ml. per min.

14. End blocks to restrain the tubing in position; the tubing is slightly stretched.

strated by recovery experiments, which indicate that urea dialyzes slightly faster from blood than from simple aqueous solutions that are used for calibration. A similar factor was obtained in experiments dealing with the recovery of glucose from whole blood. No satisfactory explanation has been found for the existence of a difference in the rate of dialysis from blood and from water. However, no difficulty or inaccuracy in analysis was experienced as a result of using this factor, which seems to represent an extremely constant value.

Rosenthal[4] observed that the results obtained by the diacetyl monoxime method for urea, as performed by conventional methods, were not in agreement with those of the urease method, when comparing relatively high values. This was said to result from a failure of the urease method. On the other hand, by using the presently described continuous method with diacetyl monoxime, distinctly high results are obtained on samples of blood that are read as less than 20 per cent transmission. This was easily established, not only by direct comparison with conventional methods with urease, but also by recovery experiments that were performed as previously described. Samples of serum tested in the same manner yielded completely reliable results throughout the entire range of possible values for urea. When samples of uremic blood were treated with urease and analyzed, it was found that all of the chromogens had been destroyed and no reading was obtained. Assuming that the preparation of urease used was completely specific, it follows that some unknown substance (or substances) may dialyze from whole blood, but not from serum, and that such substances result in an increase in the amount of color obtained from large amounts of urea. The correctness of this supposition was established by including in the arsenic acid reagent an amount of urea that was sufficient to elevate the response of the recorder to a level that corresponded to a highly uremic blood. When a trace was indicated on the recorder, suitable for establishing a control, a urease-treated sample of blood was brought into the system by way of the sample tube. Although the blood contained no urea, the tracing now indicated a much higher value. Urease-treated serum manifested no effect.

In the practical operation of the scheme illustrated in Figure 1, it has been necessary to re-analyze all of the samples that yield a reading less than 20 per cent transmission after dilution with an equal volume of physiologic saline solution. This is in agreement with good colorimetric practice, and it has not been a great disadvantage.

DISCUSSION

It is obvious that a colorimetric method, whether manual or automatic, is limited in its accuracy by the degree of specificity of the reactions involved; the method is no more precise than the inherent colorimetric sensitivity will permit. Apart from these considerations, it would seem that the accuracy and precision of the continuous method of analysis is limited only by the degree of reproducibility of the apparatus that is used. A similar situation exists in manual conventional analysis, which is ultimately limited by the accuracy of calibration of pipets, as well as by the quality of the colorimeter. It is believed that components for the continuous method may be easily made reproducible and accurate to a

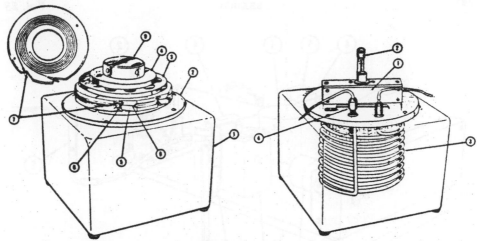

FIGS. 5 and 6

FIG. 5 (left). Diagram of the apparatus for dialyzing:

1. Box for support of the mechanism.

2. Tray of stainless steel, for the purpose of catching drippings.

3 and *4.* Clamping disks (stainless steel) designed to provide even pressure on the entire surface of the dialyzer plates. One set is required on each of the 2 sides of the plates (labeled *5* and *6*).

5 and *6.* Dialyzer plates (Plexiglas) that are mirror images of each other and pinned to provide registry within ±0.005 in. of the corresponding grooves cut into the faces. The grooves are semicircular, 0.060 ± 0.001 in. in diameter. Total length of the groove is 88 in. The grooves must be absolutely smooth, in order to prevent contamination, and the edges of the grooves should be sharp and clean. The thickness of the plates should be constant (within 0.001 in.) across the entire face, and the same requirement is applicable to items *3* and *4.* If this specification is not satisfied, adequate sealing does not occur. The fluids enter the dialyzing plates through nipples (*No. 7*).

7. Nipples prepared from Kel-F snugly pressed in place. The interior hole in the nipple is 0.042 in. in diameter and leads smoothly into the semicircular groove in the dialyzer plate.

8. Uncoated cellophane (Du Pont No. 300) is used as a membrane for dialysis. The material must be moistened with water and permitted to expand prior to its being placed smoothly in the dialyzer.

9. Clamping nut that operates on a central support shaft (not visible in the sketch). Antifriction qualities are desirable in order to permit the application of a sufficient amount of pressure.

FIG. 6 (right). Heating bath for the continuous method.

1. Thermostatic control, providing an accuracy of ±0.2 C. The box labeled "1" contains the various control elements, and the external knob permits adjustment of the operating temperature.

2. Thermometer for making observations of the temperature.

3. Glass coil, with an internal diameter of 1.6 mm. and a wall 0.6 mm. in thickness. Any configuration of coil is satisfactory if even amounts of heat are applied. The coil is 40 ft. long. Support is desirable in order to minimize the hazards of breakage. The coil is immersed in ethylene glycol that is contained in a beaker constructed of stainless steel (not illustrated in the sketch).

4. The leads for the coil are brought up through Kel-F chucks. Polyvinyl tubing is slipped over the extending portions of the glass coil.

value of 1 per cent or better, and that results obtained by this method should be equal in accuracy to those obtained by careful manual analysis. Experience indicates that this is generally true. Results of the continuous method have generally been superior to those obtained by methods used in routine clinical analyses.

More important, however, is the fact that the continuous method has a higher

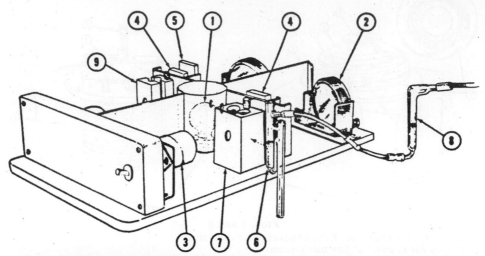

FIG. 7. Diagram of the colorimeter. All of the internal parts must be shielded from light and finished in dull black.

1. Pre-focused lamp, General Electric No. 1759, centrally located.
2. Front-face spherical mirrors for f3 system, with focal distance of 5.9 in.
3. One of the 2 selenium, barrier-type photo-cells, hermetically sealed.
4. Interference filter, 480 mμ, half-band width 18 mμ, 40 per cent transmission.
5. Aperture for the control of light.
6. Measuring cell; in the viewing region, this consists of tubing with an inside diameter of 6 mm. and an outside diameter of 8 mm. The exit from the viewing portion is dammed in order to control the height of the liquid at a level slightly above the viewing level. The unit is open to the atmosphere.
7. Holding block as a means of securely mounting the measuring cell.
8. Pre-mix cell, consisting of tubing with an inside diameter of 6 mm. The top of the cell should be vented.
9. Holding block for the standard cuvet, if this is used.

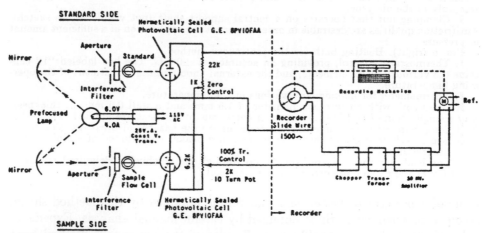

FIG. 8. Diagram of the optical and electrical features of the colorimetric-recorder system. The recorder is a Dynamaster, model TC-1PH1X560-51, zero to 100 mv., full scale, manufactured by the Bristol Company, Waterbury, Connecticut.

TABLE 1

COMPARISON OF DETERMINATIONS OF UREA PERFORMED BY THE CONTINUOUS METHOD
AND BY A CONVENTIONAL UREASE METHOD

Continuous Method	Urease Method	Difference
mg. of N per 100 ml.	*mg. of N per 100 ml.*	*mg.*
14.4	13.5	0.9
6.5	6.6	−0.1
19.3	17.9	1.4
20.5	22.0	−1.5
46.4	46.4	0.0
23.3	24.0	−0.7
28.3	29.0	−0.7
63.8	64.2	−0.4
41.8	40.8	1.0
66.8	66.0	0.8
74.8	74.1	0.7
14.6	14.9	−0.3
33.0	31.3	1.7
23.1	24.8	−1.7
27.1	26.5	0.6
16.3	15.6	0.7
17.6	16.1	1.5
15.9	16.0	−0.1
61.2	60.6	0.6
16.8	16.0	0.4
35.4	33.6	1.8
12.5	13.7	−1.2
22.8	24.0	−1.2
6.4	7.9	−1.3
15.3	15.4	−0.1
21.3	21.1	0.2
15.4	18.7	−3.3
15.3	15.5	−0.2
9.8	10.5	−0.7
10.1	11.2	−1.1
46.0	49.0	−3.0
14.7	14.7	0.0
12.8	12.2	0.6
16.8	16.1	0.7
13.6	13.5	0.1
11.4	11.1	0.3
13.1	12.0	1.1
11.9	11.8	0.1
11.2	11.4	−0.2
43.1	42.5	0.6
13.0	12.6	0.4

index of reliability than methods used routinely in clinical laboratories. In performing a number of analyses for urea in blood, a technician must perform many consecutive steps, each with care, in order to arrive at a correct result. The number of operations that must be performed provides ample opportunity for error. As is frequently the situation, the technician is required to perform several other kinds of analyses as well, and the opportunity for error becomes greatly magnified. By contrast, all of the operations in the continuous system, with unknowns and standards, are performed in an exactly similar manner and the results are recorded plainly and permanently for easy comparison.

Although it is probable that the continuous method would be most useful in the clinical laboratory, where it may be used for routine and emergency work, it is believed that the method may be of value in other fields as well. For example, it might be used in screening programs that are designed for the detection of diabetes or renal disease. During operation of the artificial kidney, multiple or continuous determinations of urea in the patient's blood would be possible, thereby providing a constant check on the progress of the treatment. With the adaptation of additional colorimetric methods, it might prove useful in various programs of research in which large numbers of determinations are sometimes required.

SUMMARY

An automatic, continuous colorimetric method of analysis is described. Samples are introduced in close succession by means of a pump and are propelled through a small dialyzer, together with a constant proportion of a suitable diluent. The substance to be determined is transferred, in the dialyzer, to a flowing stream of reagent, which is then continuously processed to produce a specific change in color. The degree of change in color and, therefore, the concentration in the original sample is measured by passing the flowing stream through a colorimeter that is equipped for recording.

SUMMARIO IN INTERLINGUA

Es describite un automatic e continue methodo colorimetric. Specimens es introducite in succession rapide per medio de un pumpa que propelle los a transverso un parve dialysator insimul con un proportion constante de un appropriate diluente. In le dialysator le substantia a determinar es transferite a in un fluxo currente del reagente, e isto es processate continuemente con le production resultante de alterationes specific de color. Le grado del alterationes de color—e per consequente le concentration in le specimenes original—es mesurate per conducer le fluxo currente a transverso un colorimetro que es equipate con un apparatura de registration.

REFERENCES

1. FEARON, W. R.: The carbamido diacetyl reaction, a test for citrulline. Biochem. J., **33:** 902–907, 1939.
2. HAWK, P. B., OSER, B. L., AND SUMMERSON, W. H.: Practical Physiological Chemistry. Ed. 12. Philadelphia: Blakiston, 1947.
3. ORMSBY, A. A.: Direct colorimetric method for determination of urea in blood and urine. J. Biol. Chem., **146:** 595–604, 1942.
4. ROSENTHAL, H. L.: Determination of urea in blood and urine with diacetyl monoxime. Anal. Chem., **27:** 1980–1982, 1955.
5. VAN SLYKE, D. D., AND CULLEN, G. E.: A permanent preparation of urease and its use in the determination of urea. J. Biol. Chem., **19:** 211–228, 1914.
6. VAN SLYKE, D. D., AND CULLEN, G. E.: The determination of urea by the urease method. J. Biol. Chem., **24:** 117–122, 1916.

page 327

25. Free, A. H., Adams, E. C., Kercher, M. L., Free, H. M. and Cook, M. H. (1957)
Simple Specific Test for Urine Glucose. Clinical Chemistry
3(3): 163–168.

D ry chemistry test papers or dipsticks for urine assays were developed by a number of researchers in the 19th century (1). For example, George Oliver in 1883 described dry-reagent dipsticks for urine glucose and protein. He was a London physician who needed a more convenient method for the bedside testing of patient urines. Oliver stated that he had "succeeded in all my reagents in abolishing the fluid state, and likewise the solid form, either powder, crystal, or pellet" (2). His glucose dipsticks contained dried Fehling's copper reduction reagent. Various test papers for protein were also described that contained dried picric acid or sodium tungstate reagents. He reported on the room temperature stability of the papers and compared the results to conventional wet chemistry assays. Oliver offered to make the papers available free of charge to any physician who requested them and asked that they report their experience with them to *The Lancet* (3).

Seventy-three years latter in 1956 Albert S. Keston from the New York University College of Medicine presented a paper at the 129th Meeting of the American Chemical Society (ACS) on an enzymatic urine glucose procedure. This was the first description of an enzymatic glucose assay in which the reagents were dried into filter paper (4). The reagent mixture utilized glucose oxidase (GO, EC 1.1.3.4) and horseradish peroxidase (HRP, EC 1.11.1.7) in the following two-step reaction.

(i) $\quad$ glucose $+ O_2 + H_2O \rightarrow$ D-glucuronic acid $+ H_2O_2$

(ii) $\quad H_2O_2 + o$-tolidine $\rightarrow H_2O +$ oxidized o-tolidine

The oxygen acceptor o-tolidine produced a blue color. Yellow filter paper was used to improve the visual reading of the results. Two years latter, Beach and Turner from the Biochemical Laboratory at the Metropolitan Life Insurance Company published a full report on the use and preparation of Keston's reagents for both liquid and dried reagent paper strip format for glucose in urine (5). The authors stated that Keston had provided his reagent formulations to them in a personal communication in March 1954, two years before Keston presented his own report at the ACS meeting.

At the same ACS meeting in 1956 where Keston presented his paper, J.P. Comer from Eli Lilly and Co. cited Keston and described a method for urine glucose in a dried paper strip format (6). A few months latter Comer published a full report on the paper strips that had already been released as the commercial product TesTape® (7). The dried paper came in a dispenser roll. A short length of the paper was torn off the roll, dipped into the urine and the blue color read in less than a minute.

Alfred H. Free and co-workers at the Ames Company Inc. also developed a dried paper format assay for glucose in urine based on Keston's work. Ames in 1956 was the producer of Clinitest®, the Benedict copper reduction reagent for urine sugars dried into a tablet format. The full description of the Ames dried paper strip method for enzymatic glucose was submitted in December 1956 and published in the June 1957 issue of *Clinical Chemistry*. That paper is presented here. In the Ames product the reagents of Keston utilizing o-tolidine were dried onto white filter paper strips that were attached to a plastic backing. Adams, Burkhart and Free published on the specificity of the dipsticks (8). In August 1958, a United States Patent was issued for what was sold by Ames as the Clinistix® urine dipstick (9). Studies soon appeared which compared and validated the performance of the commercial enzymatic urine test paper by Eli Lilly and the dipstick format by Ames (10–11).

In 1957 Joachim Kohn at Queen Mary's Hospital in London was the first to describe the use of the enzymatic urine glucose test strips with blood. He let fall a drop of blood from a fingertip or earlobe onto the Clinistix test strip. After 20 seconds, the blood was washed off with running tap water and the green-blue to dark-blue color was read visually against a color chart produced by him with commercial paints (12). A comparison in 545 patient samples was performed between the dipstick method and three alternate quantitative chemical methods. Kohn stated that the stiff format and white background of the dipstick facilitated the use of this product in bedside use for blood glucose. Lipscomb and co-workers in 1958 applied the Eli Lilly TesTape product to whole blood and published a similar comparison to results by conventional methods (13).

Over the next few years dried reagents in filter paper format were described for other assays in blood or serum. S.S. Kind in 1958 reported on a colorimetric dipstick for acid phosphatase (ACP, EC 3.1.3.2) in semen samples for forensic applications (14). The reagents consisted of alpha-naphthyl phosphate and diazo-*o*-dianisidine. The papers were stable for up to 1 year at room temperature when stored in an opaque glass bottle. The General Diagnostics Division of Warner-Chilcott in the early 1960s introduced an enzymatic dipstick for serum urea nitrogen. Capillary action caused serum to rise up the filter paper strip through a layer of dried urease. A plastic barrier prevented further rise of the liquid phase, however, the ammonia gas produced by the action of urease on urea continued past the semi-permeable plastic barrier into a dried bromocresol indicator dye. The indicator dye turned blue and its height provided a direct colorimetric reading of the urea nitrogen present in the sample (15). The product was sold as the Urograph® chromatography strip.

Dry reagent dipsticks for testing blood glucose improved greatly in 1964 when the group at Ames developed an enzymatic glucose assay specifically for use with blood and sold it as Dextrostix® (16). In these strips a clear semi-permeable barrier was attached over the filter paper that contained the dried GO−HRP−chromogen reaction mixture. Plasma in the drop of blood diffused into the reaction mixture and the color developed. Red cells and other cells remained on top of the barrier layer. This allowed cells to be easily washed off. This innovation prevented blood interference with the visual readings and opened up the possibility for an instrument-based reader. Ames introduced the first hand-held battery operated reflectance meter called the Ames Reflectance Meter (ARM™) in 1970 (17). A smaller unit was soon introduced called the Eyetone®. By the mid 1980s various companies offered up to 10 different enzymatic dipsticks for blood glucose testing along with 18 different models of reflectance meters (18).

Dry reagent chemistry systems based on filter paper elements for a wide range of analytes in blood or serum were commercialized over the next 10 years. Eastman Kodak Co. developed the multilayer Ektachem® systems (19,20); Ames produced a reflectance meter for blood assays in dipsticks called the Seralyzer® (21); Fuji Photo Film Co., Ltd. produced the Fuji Drichem® system and Boehringer Mannheim GmbH the Reflotron® system (22).

References

(1) Voswinckel, P. (1994) A marvel of colors and ingredients. The story of urine test strips. Kidney International. 46(Supplement 47):S3–S7.

(2) Oliver, G. (1883) On bedside urinary tests. The Lancet. 121(3100):139–140.

(3) Oliver, G. (1883) On bedside urinary tests concluded from pg 140. The Lancet. 121(3101):190–192.

(4) Keston, A.S. (1956) *Specific Colorimetric Enzymatic Analytical Reagents for Glucose*, Abstracts of Papers, 129th Meeting of the American Chemical Society: Dallas, Texas, April 8–13, American Chemical Society, Washington, DC. pgs 31C–32C.

(5) Beach, E.F. and Turner, J.J. (1958) An enzymatic method for glucose determination in body fluids. Clinical Chemistry. 4(6): 462–475.

(6) Comer, J.P. (1956) *A Semiquantitative Specific Test Paper for Glucose in Urine*, Abstracts of Papers, 129th Meeting of the American Chemical Society: Dallas, Texas, April 8–13, American Chemical Society, Washington, DC. pg 28B.

(7) Comer, J.P. (1956) Semiquantitative specific test paper for glucose in urine. Analytical Chemistry. 28(11):1748–1750.

(8) Adams, E.C., Burkhart, C.E., and Free, A.H. (1957) Specificity of a glucose oxidase test for urine glucose. Science. 125(3257):1082–1083.

(9) Free, A.H. (1958) Composition of matter. United States Patent No. 2,848,308. Issued August 19, 1958. Assigned to Miles Laboratories, Inc., Elkhart, IN. United States Patent Office.

(10) Luntz, G. (1957) A simple quick test for glucose in urine; report on use of clinistix. British Medical Journal. 51(5017): 499–500, March 2.

(11) Ackerman, R.F., Williams, E.F., Packer, H., and Ahler, J. (1958) Comparison of benedict's solution, clinitest, test-tape and clinistix. Diabetes. 7(5):398–402.

(12) Kohn, J. (1957) A rapid method of estimating blood-glucose ranges. The Lancet. 270(6986):119–121.

(13) Lipscomb, H.S., Bean, J., Dobson, H.L., and Green, J.A. (1958) The determination of blood sugar. A rapid screening method utilizing glucose oxidase paper. Diabetes. 7(6):486–489.

(14) Kind, S.S. (1958) Stable test-papers for seminal acid phosphatase. Nature. 182(4646):1372–1373.

(15) Erickson, M.M., Versun, R.M., and White, W.L. (1965) An evaluation of the urograph method for the determination of blood urea nitrogen. American Journal of Medical Technology. 31(5):335–338.

(16) Cohen, S.L., Legg, S., and Bird, R. (1964) A bedside method of blood-glucose estimation. The Lancet. 284(7365):883–884.

(17) Jarrett, R.J., Keen, H., and Hardwick, C. (1970) "Instant" blood sugar measurement using dextrostix and a reflectance meter. Diabetes. 19(10):724–726.

(18) McCall, A.L. and Mullin, C.J. (1986) Home monitoring of diabetes mellitus — A quiet revolution. Clinics in Laboratory Medicine. 6(2):215–239.

(19) Spayd, R.W., Bruschi, B., Burdick, B.A., Dappen, G.M., Eikenberry, J.N., Esders, T.W., Figueras, J., Goodhue, C.T., LaRossa, D.D., Nelson, R.W., Rand, R.N., and Wu, T.-W. (1978) Multilayer film elements for clinical analysis: applications to representative chemical determinations. Clinical Chemistry. 24(8):1343–1350.

(20) Curme, H. and Rand, R.N. (1997) Early history of Eastman Kodak Ektachem slides and instrumentation. Clinical Chemistry. 43(9):1647–1652.

(21) Walter, B. (1983) Dry reagent chemistries in clinical analysis. Analytical Chemistry. 55(4):498A–514A.

(22) Libeer, J.C. (1985) Solid phase chemistry in clinical laboratory tests: a literature review. Journal of Clinical Chemistry and Clinical Biochemistry. 23(10):645–655.

Clinical Chemistry 1957; 3(3): 163–168
© 1957 American Association for Clinical Chemistry.
Reproduced with permission.

Simple Specific Test for Urine Glucose

Alfred H. Free, Ernest C. Adams, Mary Lou Kercher, Helen M. Free, and Marion H. Cook

Procedures for the detection of sugar in urine are among the oldest tests known in clinical chemistry. Legends going back to earliest times describe the attraction of insects to urine containing sugar. In many instances it is difficult to separate legend from reality in tracing the history of tests for glycosuria.

Among such pioneer American clinical chemists as Stanley Benedict, Otto Folin, Victor Myers, and Donald Van Slyke, there was great interest in establishing and using tests for the detection of sugar in the urine. From the many tests described during the early part of the twentieth century, Benedict's copper reduction test emerged as the one of greatest popularity. During the past 10 years Clinitest,[1] which is a self-heating alkaline copper reduction test, has received wide acceptance in this country and in many other parts of the world. This test will subsequently be referred to as the tablet copper reduction test.

This report describes a new, simple test for the identification of glucose in the urine. The test depends on the action of an enzyme and thus bypasses the classical principle of alkaline reduction of metallic ions. The test is called Clinistix,[2] and will hereafter be referred to as the glucose oxidase test.

METHODS

The reagents involved in the glucose oxidase test are glucose oxidase, peroxidase, and orthotolidine. They are impregnated into a strip of stiff paper. In order to carry out a test, the strip is merely dipped into the

From the Miles-Ames Research Laboratory Elkhart, Indiana.

Presented at the International Congress on Clinical Chemistry, New York, September 9, 1956.

Received for publication December 22, 1956.

[1] Clinitest is a registered trademark of the Ames Company Inc., Elkhart, Ind.

[2] Clinistix is a registered trademark of the Ames Company Inc., Elkhart, Ind.

urine specimen. One minute later the strip is observed for the appearance of a blue color which indicates the presence of glucose.

The enzyme glucose oxidase was first identified by Muller (1) approximately 30 years ago. In the presence of oxygen the enzyme catalyzes the oxidation of glucose to gluconic acid and hydrogen peroxide. In the absence of oxygen the reaction of glucose and glucose oxidase may utilize certain organic compounds as hydrogen acceptors. In the glucose oxidase test the hydrogen peroxide formed reacts with orthotolidine, the reaction being catalyzed by a second enzyme—peroxidase. This reaction results in the development of a deep blue color.

EXPERIMENTAL

The glucose oxidase test is very sensitive for the detection of glucose. Less than 0.01% glucose in water will give a positive reaction. Somewhat greater amounts of glucose in urine are required for a positive reaction, with 0.01% usually being negative and 0.1% glucose in urine giving a positive reaction. As with most other urine qualitative tests, the precise sensitivity depends somewhat on the nature of the sample. Table 1 shows typical quantities of glucose which need to be added to urine to get positive reactions with the tablet copper reduction test, Benedict's qualitative test, and the glucose oxidase test. The sensitivity of the enzyme test is such that urines with a small quantity of glucose either added or naturally occurring may give a positive reaction with the glucose oxidase test, a negative reaction with Benedict's and a negative reaction with the tablet copper reduction test. Whether the great sensitivity of the new enzyme test will be an advantage or a disadvantage has not been established. Many laboratories including our own (2) have established that the high sensitivity of Benedict's test is a disadvantage over a less sensitive test. However, Benedict's test is a test for total reducing substances rather than for glucose.

Glucose oxidase has been studied for the specificity of its reaction with glucose. Keilin and Hartree (3) reported that the enzyme had a high specificity for glucose but would react slowly with mannose, xylose,

Table 1. TYPICAL CONCENTRATIONS OF GLUCOSE IN URINE REQUIRED FOR POSITIVE REACTION

Test	Glucose concentration (%)
Tablet copper reduction test	0.15
Benedict's copper reduction test	0.08
Glucose oxidase test	0.05

maltose, and galactose. The extreme sensitivity of the glucose oxidase test raises the question of whether it will react with concentrations of other sugars which may be encountered in the urine. It has been found that many commercial samples of sugars other than glucose give positive tests with the glucose oxidase test when tested in concentrated aqueous solutions. Purification of such sugars by fermentation with baker's yeast of any glucose present or by enzymatic removal of the glucose results in a nonreactive material. The sugars which have been established to be non-reactive with the glucose oxidase test at concentrations of 20% in urine are galactose, fructose, lactose, mannose, maltose, sucrose, xylose, D-ribose, L-arabinose, L-arabinose, and L-xylulose.

Glucose oxidase catalyzes the oxidation of β-D-glucose but has no effect on the oxidation of α-D-glucose. A freshly prepared solution of crystalline glucose contains only the alpha form and is essentially nonreactive with the glucose oxidase test. However, in solution an equilibrium mixture between the alpha and beta forms is readily established and within a few minutes reactivity with the glucose oxidase test appears, demonstrating the presence of β-D-glucose. In blood and urine containing glucose the equilibrium mixture of alpha and beta glucose is present so there is no problem of delayed reactivity, since β-D-glucose is available for immediate reaction with the enzyme.

In the treatment of diabetes, some clinicians regulate insulin dosage on the basis of the amount of sugar in the urine. For this reason quantitation of the amount of sugar in the urine is important. There are several factors which influence the amount of color in an enzyme test such as the glucose oxidase test. In addition to the effect produced by the sugar concentration the color is influenced by pH, temperature, and the concentration of antagonistic substances such as ascorbic acid. Accordingly, it has not been possible to quantitate the amount of glucose in urine. It is quite easy to demonstrate that with different amounts of sugar added to a given urine different colors are obtained. However, the variability of the urine and the inability of average operators to recognize small color differences makes it impossible to do little more than differentiate between small and large amounts of glucose in urine.

RESULTS

Table 2 shows the number of positive tests with the glucose oxidase test, the tablet copper reduction test, and Benedict's qualitative test obtained on a series of random urines from 352 healthy subjects. It will be seen that many positive reactions (trace) were obtained with Benedict's

Table 2. URINE SUGAR TESTS ON HEALTHY SUBJECTS

Test	Positive	Negative
Glucose oxidase test	2	350
Tablet copper reduction test	3	349
Benedict's qualitative	63	289

test but only 2 positive reactions were obtained with the glucose oxidase test and 3 positive reactions with the tablet copper reduction test. A series of tests were applied to the urines giving negative reactions with the glucose oxidase test and the tablet copper reduction test but trace Benedict's tests. These included paper chromatography, measurement of the amount of reducing substance in the urine with the Nelson-Somogyi procedure (4) before and after fermentation with baker's yeast, measurement of the amount of reducing substances with Nelson-Somogyi procedure in the urine before and after aeration with glucose oxidase, and attempts to form osazone crystals. The results of these tests indicated that the urines from healthy subjects which gave trace reactions with Benedict's test but negative reactions with the glucose oxidase test contained nonglucose reducing substances. Confidence in the results of such tests is achieved by reason of the fact that addition of glucose to negative urines in quantities of 0.1 per cent can readily be recognized by paper chromatography, by decrease in total Nelson-Somogyi reducing substances after yeast fermentation or after aeration with glucose oxidase, and by a positive glucose oxidase test reaction.

With some urines this quantity of glucose can be identified by forming osazone crystals, but with other urines osazone crystals have not been obtained with this small amount of glucose.

A large number of tests have been carried out over a period of more than a year on random hospital urine specimens using the glucose oxidase test and one or two of the standard copper reduction methods, the tablet copper reduction test, and Benedict's qualitative test. These urine samples were tested as received and also following the addition of glucose. Table 3 shows data on 2075 random urines from hospital patients tested with the glucose oxidase test and Benedict's qualitative test. It can be seen that 1423 urines were negative with both tests. Approximately one-fourth of all of the urines (571 out of 2075) gave trace reactions with Benedict's test but only 19 of these urines also gave a positive reaction with the glucose oxidase test. As with urines from healthy subjects the application of confirmatory tests to urines giving a trace reaction with Benedict's and a negative reaction with the glucose oxidase test showed that the reducing substance in these samples was not glucose. Fourteen

Table 3. COMPARISON OF TESTS ON RANDOM HOSPITAL URINES

	Benedict's qualitative		
	Negative	Trace	1+ or more
Glucose oxidase test			
Negative	1423	552	14
Glucose oxidase test			
Positive	36	19	31

Table 4. RESULTS OF BLIND TESTS ON RANDOM URINES WITH AND WITHOUT
ADDED GLUCOSE

Glucose Added	Glucose oxidase test		Tablet copper reduction test	
	Negative (%)	Positive (%)	Negative (%)	Positive (%)
0 (1925 urines)	98	2	99	1
0.1%–0.25% (845 urines)	3	97	18	82
0.35%–0.85% (776 urines)	1	99	0	100
1%–4% (964 urines)	0	100	0	100

of the urine samples were 1 + or more with Benedict's test but were negative with the glucose oxidase test. Two of these urines contained over 0.3% ascorbic acid which was a sufficient quantity to account for the positive test. In addition, among these urines were samples containing galactose, lactose, and fructose. There were 36 urine samples that gave positive reactions with the glucose oxidase test and negative reactions with Benedict's test. In these cases the amount of glucose present was quite small and not sufficient to give a positive Benedict's test. In these urines glucose was demonstrated by fermentation with baker's yeast with subsequent disappearance of the positive glucose oxidase test reaction and decrease in total Nelson-Somogyi reducing substance.

Evaluation

In order to assess the performance of the glucose oxidase test, samples of urine with or without added glucose were tested as unknowns by trained and untrained operators. Each day a few urines which gave negative reactions with the tablet copper reduction test and the glucose oxidase test were chosen. To each urine small amounts of concentrated aqueous glucose solutions were added to give several concentrations of glucose ranging from 0.1% to 4%. The samples were coded and given as unknowns to from one to five operators. Each operator tested each sample with the glucose oxidase test and recorded his result. Some rotation of operators occurred on different days so that every unknown was not tested by each operator. A separate trained operator tested each sample with the tablet copper reduction test. Results of the comparison of the copper reduction with the enzyme test are shown in Table 4. It is evident that the two tests have comparable accuracy with negative urines. The 2 per cent error with the glucose oxidase test and 1% error with the tablet copper reduction test on the negative urines was due to the operator's obtaining trace reactions on urines which had been classified as negative. With amounts of added glucose between 0.1% and 0.25% the tablet copper reduction test was positive 82 per cent of the time. This figure depends on the exact amount of glucose added as well as on the specific urine to which it is added. The glucose oxidase test was positive 97 per cent of the time with amounts of added glucose between 0.1% and 0.25%. This demonstrates the high sensitivity of the glucose oxidase test. With amounts of added glucose of from 0.35% to 0.85% the glucose oxidase test was positive in 99 per cent of the cases. The 1 per cent of samples giving negative results had extremely high concentrations of ascorbic acid.

DISCUSSION

From the results of these studies it is apparent that the glucose oxidase test is a simple, accurate, specific, sensitive test for urine glucose. It has potential usefulness in diabetes detection screening programs, in routine urinalyses in hospitals and physicians' offices, in the differential diagnosis of diabetic coma from insulin shock or other coma, in regular urine testing on known diabetics by both patients and physician, and in the differentiation of glucosuria from other meliturias.

SUMMARY

A simple, specific, sensitive, and speedy test for glucose in urine has been described. This enzymatic test based on the activity of glucose oxidase and peroxidase is called *Clinistix*. Data are presented to show that the test has a high accuracy with both positive and negative specimens.

REFERENCES

1. Muller, D., *Biochem. Z.* 199, 136 (1928).
2. Cook, M. H., Free, A. H., and Giordano, A. S., *Am. J. Med. Tech.* 19, 283 (1953).
3. Keilin, D., and Hartree, E. F., *Biochem. J.* 42, 221 (1948).
4. Hawk, P. B., Oser, B. L., and Summerson, W. H., *Practical Physiological Chemistry*, New York, Blakiston, 1954, p. 921.

page 335   **26. Severinghaus, J. W. and Bradley, A. F. (1958)**
Electrodes for Blood pO₂ and pCO₂ Determination. Journal of
Applied Physiology 13(3): 515–520.

T he development of the pCO_2 electrode has been well documented in numerous articles and book chapters (1–8). John W. Severinghaus who is credited with the development of the first practical pCO_2 electrode for blood levels has written the most complete historical reviews. Severinghaus perfected the pCO_2 electrode but Gesell, McGinty and Bean at the University of Michigan in 1926 described the first CO_2 electrode (9). These researchers covered a manganese dioxide pH electrode with a piece of wet peritoneal membrane from a dog. Diffusion of CO_2 across the membrane from expired air changed the measured pH at the electrode surface. Richard Stow and Barbara Randall in 1954 at an American Physiological Society Meeting described a pCO_2 electrode in which they covered a pH electrode with a rubber finger cot (10,11). A thin film of water was trapped between the semi-permeable rubber membrane and the pH electrode. CO_2 gas in a blood sample diffused across the semi-permeable rubber membrane and lowered the pH of the water at the electrode surface. The electrode proved to be unstable and Stow chose to not apply for a patent.

Severinghaus was present at the meeting in 1954 where Stow and Randall first presented their blood gas electrode. He listened to their paper and conceived of the idea of replacing the water between the rubber membrane and the pH electrode with a bicarbonate buffer. He discussed this idea with Stow at the meeting. Within a week Sveringhaus in his lab at the National Institutes of Health in Bethesda, Maryland had constructed the modified electrode. He replaced the water with a thin film of bicarbonate buffer between the membrane and the glass bulb of a Beckman pH electrode. The electrode was stable and provided a linear response as a function of the log pCO_2 from 5 to 700 mm of mercury (2).

Severinghaus' first report in 1958 on the new blood gas electrode for carbon dioxide tension is presented here. As early as 1957 Severinghaus and Bradley had a pH, oxygen and CO_2 electrode packaged into a single Plexiglas, temperature-controlled water bath. Their first complete blood gas analyzer is now housed at the Smithsonian Museum in Washington, D.C. (7). Numerous researchers including Severinghaus reported on the performance of this electrode (12–14). In December 1958 Yellow Springs Instrument Company in Ohio produced the first commercial blood gas analyzer which incorporated a Clark O_2 electrode and the Stow–Sveringhaus pCO_2 electrode. In 1960 Instrumentation Laboratory, Inc. introduced a blood gas instrument that measured pH, pCO_2 and pO_2 in a single sample of blood (15).

References

(1) Trubuhovich, R.V. (1970) History of pCO_2 electrodes. British Journal of Aneasthesia. 42(4):360.

(2) Astrup, P. and Severinghaus, J.W. (1986) AHA, in *The History of Blood Gases, Acids and Bases,* 1st Edition, Astrup, P. and Severinghaus, J.W. (eds), Munksgaard International Publishers, Copenhagen, pgs 264–295, Chapter XVIII.

(3) Severinghaus, J.W. and Astrup, P.B. (1986) History of blood gas analysis. III. Carbon dioxide tension. Journal of Clinical Monitoring. 2(1):60–73.

(4) Severinghaus, J.W. and Astrup, P.B. (1987) History of blood gas analysis. International Anesthesiology Clinics. 25(4): 78–95.

(5) Shapiro, B.A. (1995) The history of pH and blood gas analysis. Respiratory Care Clinics of North America. 1(1):1–5.

(6) Severinghaus, J.W., Astrup, P., and Murray, J.F. (1998) Blood gas analysis and critical care medicine. American Journal of Respiratory and Critical Care Medicine. 157(4 Pt 2):S114–S122.

(7) Hornbein, T.F. (2003) A tribute to John Wendell Severinghaus. Advances in Experimental Medicine and Biology. 543:1–6, Chapter 1.

(8) Severinghaus, J.W. (2004) First electrodes for blood pO_2 and pCO_2 determination. Journal of Applied Physiology. 97(5): 1599–1600.

(9) Gesell, R., McGinty, D.A., and Bean, J.W. (1926) The regulation of respiration. VI. Continuous electrometric methods of recording changes in expired carbon dioxide and oxygen. American Journal of Physiology. 79(1):72–90.

(10) Stow, R.W. and Randall, B.F. (1954) Electrical measurement of the pCO_2 of blood. American Journal of Physiology. 179(3): 678, Abstract.

(11) Stow, R.W., Baer, R.F., and Randall, B.F. (1957) Rapid measurement of the tension of carbon dioxide in blood. Archives of Physical Medicine & Rehabilitation. 38(10):646–650.

334

(12) Lunn, J.N. and Mapleson, W.W. (1963) The Severinghaus pCO_2 electrode: a theoretical and experimental assessment. British Journal of Anaesthesia. 35(11):666–678.

(13) Severinghaus, J.W. (1968) Measurement of blood gases: pO_2 and pCO_2. Annals of the New York Academy of Sciences. 148(Article 1):115–132.

(14) Smith, A.C. and Hahn, C.E. (1975) Studies with the "Severinghaus" pCO_2 electrode. I: electrode stability, memory and S plots. British Journal of Anaesthesia. 47(5):553–558.

(15) [Advertisement] (1960) The I.L. Meter, Instrumentation Laboratory, Inc. Journal of Clinical Investigation 39(2) February.

J. Appl. Physiol. 1958; 13(3) 'Electrodes for Blood pO$_2$ and pCO$_2$ Determination' by Severinghaus and Bradley
© 1958 American Physiological Society.
Reproduced with permission.

Electrodes for Blood pO$_2$ and pCO$_2$ Determination[1]

JOHN W. SEVERINGHAUS[2] AND A. FREEMAN BRADLEY[2]. *From the Anesthesia Research Laboratory, Clinical Center, and the Clinic of Surgery, National Heart Institute, National Institutes of Health, Bethesda, Maryland*

THE APPARATUS described herein has been developed to permit rapid and accurate analysis of oxygen and carbon dioxide tensions in gas, blood or any liquid mixture. There is no inherent connection between the two systems, except for their need of accurate temperature control, for which they have been mounted together in a miniature thermostated water bath.

The oxygen electrode, designed by Clark (1), is a polarographic cell, consisting of a silver reference anode and a platinum cathode charged to -0.5 v. The platinum surface is covered with an oxygen permeable membrane, e.g. polyethylene, on the other side of which is placed the unknown sample. Current passing through the cell is directly proportional to the oxygen tension outside the membrane.

The carbon dioxide electrode, first described by Stow (2), utilizes a CO$_2$ permeable membrane (Teflon) to allow CO$_2$ to diffuse from the unknown sample into an aqueous layer held in a cellophane film on the surface of a conventional pH glass electrode. The measured pH is altered in direct proportion to change in log pCO$_2$. The apparatus consists of the following: *a*) a small thermoregulated water bath containing pO$_2$ and pCO$_2$ electrodes and their cuvettes and a tonometer for calibration of the pO$_2$ electrode with blood (fig. 1). *b*) a transistor null balance meter for measuring the O$_2$ current. *c*) a Beckman model GS pH meter for measuring the pCO$_2$.

METHODS

Measurement of pO$_2$. The Clark electrode permits accurate determination of pO$_2$ provided the

following criteria are met: constant temperature, constant rapid stirring of the liquid at the electrode surface, constant pressure in the cuvette, no loss or gain of oxygen from the walls of the cuvette, or from openings or conducting tubing, and calibration with an equilibrated portion of the unknown sample. The 0.4-ml cuvette diagrammed in figure 2 is stainless steel and nylon. Oxygen is essentially insoluble in these materials. The Clark electrode (Yellow Springs Inst. Co., Yellow Springs, Ohio) has been shown in detail elsewhere (1, 3, 4). It is filled with saturated KCl made alkaline by the addition of 0.01 M KOH to decrease drift. It's 'O' ring, which holds the polyethylene membrane on, also seals the electrode into the cuvette.

The electrode stability is improved if the lips of the Lucite tip are ground back slightly on a hand stone, making the surface convex, which keeps the polyethylene film tightly applied to the surface. As supplied, the platinum is flush with, or slightly recessed below, the level of the Lucite. A 2-ml sample is sufficient to fill and wash through the cuvette to a stable reading. An antifoaming, detergent-washing solution is used. The antifoam agent prevents bubbles of gas from clinging to the cuvette walls when filling with sample. The detergent helps drain the membrane surface free of droplets for gas calibration, since tiny droplets will impede oxygen diffusion and reduce the reading. The following wash solution is satisfactory: General Electric Antifoam 60, 0.1 ml; Alconox, 0.1 gm; water, 1000 ml.

Response of O$_2$ Electrode. The current is proportional to pO$_2$, but is also dependent on the nature of the material containing oxygen. O$_2$ molecules must be supplied continually to the electrode, so solutions with high viscosity and low O$_2$ solubility have greater O$_2$ gradients in the vicinity of the membrane. This gradient depends on the permeability of the membrane, which controls the current. Use of a highly permeable membrane such as Teflon results in a high gradient in the solution and, therefore, a large difference between the reading obtained with gas and liquid of the same pO$_2$. Poorly permeable membranes such as Mylar (first suggested by Drs. George Polgar and Robert Forster, Univ. of Penna.) show very little

Received for publication May 13, 1958.
[1] Detailed construction drawings and response curves, as well as a sketch of the recent modification of the CO$_2$ electrode, have been deposited as Document number 5645 with the ADI Auxiliary Publications Project, Photoduplication Service, Library of Congress, Washington 25, D.C. A copy may be secured by citing the Document number and by remitting $1.25 for photoprints, or $1.25 for 35-mm microfilm. Advance payment is required. Make checks or money orders payable to: Chief, Photoduplication Service, Library of Congress.
[2] Present address: University of California Hospital, San Francisco 22, Cal.

Fig. 1. Assembled electrodes and water bath. Rear of bath contains a tonometer chamber, thermostat, heater and humidifier (vertical tube). Hood contains 2 motors.

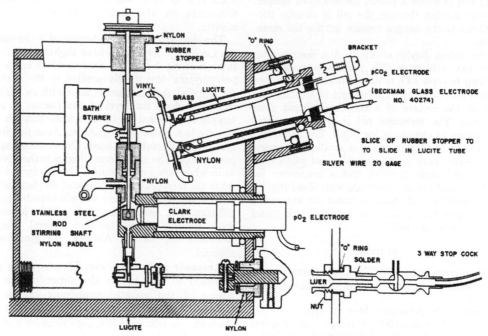

Fig. 2. Construction of cuvettes and CO_2 electrode.

difference between liquid and gas readings (table 1).

The background current in the absence of O_2 is relatively independent of the membrane. In our experience this current tends to increase gradually and erratically during use. It forms an insignificant portion of the total current with Teflon and polyethylene but a large share with Mylar.

The utilization of oxygen in the sample is, of course, directly proportional to the current. This becomes significant with Teflon, where 1% of the

O_2 dissolved in water in the cuvette (0.4 ml) may be used per minute. With polyethylene this use is about 0.2%/min. Drift was noted to be a serious problem with Mylar films, part of this drift being changes in background current. Mylar also had the slowest response time (table 2).

Of these three films, the most accurate readings of blood pO_2 can be made with polyethylene. Teflon's advantages of speed, high current and stability are outweighed by the rapid use of the oxygen in the sample and by the greater potential

TABLE 1. EFFECT OF VISCOSITY ON pO₂ ELECTRODE CURRENT

Substance	Viscosity Centipoise	Current Reading
Oxygen gas	0.02	100
Water	0.7	91
Aqueous glycerol 15.5%	1.0	88
38%	2.0	81
48%	3.0	75
54%	4.0	71
58%	5.0	70

37°C, Teflon 0.001″, platinum, diam. 2 mm, stirring 800 rpm, pO₂−713 mm Hg.

TABLE 2. COMPARISON OF VARIOUS MEMBRANES USED WITH OXYGEN ELECTRODE

Stirring Rate rpm	Teflon 0.001″	Polyethylene 0.001″	Mylar 0.00025″
Background current, μamp. pO₂ zero			
	.004−.02	.01−.02	.006−.01
Current with gas, μamp. pO₂ 713			
	32	7	.09
Response time to 99% (sec.)			
	20	40	120
Reading in liquid of pO₂ 713 mm Hg			
0			78−90
60			96−97
800	70	89	99−100
2400	81	93	99−100

pO₂—713 mm Hg, gas reading set to 100. Platinum, diam. 2 mm., 37°C, liquid test solution viscosity: 5 centipoise (glycerol 58% in water).

error of calibration of blood against gas which might result from changes in viscosity of the blood equilibration, protein layers on the membrane, or variations in stirring rate.

It is most convenient to calibrate the electrode with gas in the cuvette (e.g. air or oxygen) reading both the unknown and the equilibrated blood samples against gas. In this way many samples can be read against gas and the spare blood pooled for equilibration with the gas and calibration, which usually need be done only once for each subject and perhaps less often, depending on the scatter in calibration factor of differing bloods.

Oxygen Current Indicator. A null balance transistor amplifier and indicator were designed for use with the polarographic electrode (fig. 3). It is powered by penlight cells (standard or mercury), measures only 3″ x 5″ x 7″ and weighs 1.5 kg. It has sufficient sensitivity to detect changes of 0.5 mm Hg pO₂ using 0.001″ polyethylene membrane over a standard 2-mm platinum electrode. The O₂ tension is read from a 10-turn micrometer dial

(R3) having 1000 divisions. Sensitivity can be set (R4, R5) to make the dial read directly in mm Hg, or percentage of oxygen, or full scale with any O₂ concentration up to 100%. It has a provision for canceling out background (zero O₂) current (R6). The thermistor is used to reduce thermal drift. An external recorder may be used, in which case the micrometer dial (R3) may be used to offset zero to record a range such as 500−713 mm Hg.

A galvanometer may be used in place of this device. The null balance principle provides somewhat more accuracy, since it *1*) holds the voltage constant across the polarographic cell, *2*) can be read to 0.1%, *3*) does not depend on the linearity of an indicator meter and *4*) is vibration insensitive and lacks the long period and zero setting problems of a galvanometer.

CO₂ Electrode. The electrode is designed to allow pH measurement of a very thin film of bicarbonate solution which is in CO₂ equilibrium with the unknown sample. A Teflon membrane, permeable to CO₂ but impermeable to ions, is used to prevent the pH of the sample from affecting the pH at the glass electrode surface. The relationship between pCO₂ and pH in water and bicarbonate solutions is logarithmic, as shown in the following development.

THEORY

The reaction of CO₂ with water solutions of sodium bicarbonate is given by combining the equilibrium (Henderson-Hasselbalch) equations for the first and second hydrogen dissociations and the electro-neutrality equation:

$$\alpha p CO_2 = H_2CO_3 = \frac{A_H{}^2 + A_H A_{Na} - K_w}{K_1\left(1 + \dfrac{2K_2}{A_H}\right)} \quad (1)$$

In water without sodium bicarbonate, this approximates

$$\alpha p CO_2 = \frac{A_H{}^2}{K_1} \quad (2)$$

To compute the sensitivity, S, of the electrode, defined as

$$\frac{\Delta pH}{\Delta \log pCO_2}$$

(*equation 2*) is taken for two values of pCO₂ and H,' and ″, dividing and taking the negative logarithm:

$$-\log \frac{pCO_2{}'}{pCO_2{}''} = -\log \frac{A'_H{}^2}{A''_H{}^2}$$

or,

$$\log pCO_2'' - \log pCO_2' = 2(pH' - pH'')$$

$$\frac{\Delta pH}{\Delta \log pCO_2} = 0.5 = S \qquad (3)$$

When sodium bicarbonate is added, the second term in *equation 1* becomes dominant above 0.001M Na, and the equation approximates

$$\alpha pCO_2 = \frac{[A_H][A_{Na}]}{K_1} \qquad (4)$$

$(2K_2/A_H$ becomes significant at 0.1 M Na)

Taking two values, as in *2* and computing the sensitivity, S,

$$\frac{\Delta pH}{\Delta \log pCO_2} = 1.0 = S.$$

Thus, the inclusion of bicarbonate ion in the aqueous medium doubles the sensitivity. At higher concentrations, it falls off again as the carbonate ion appears. At 1.0 M NaHCO₃, $S = 0.88$ according to *equation 1*, but this is an approximation due to the changes in activity in high concentrations. This theoretical sensitivity curve has been plotted in figure 4.

To check this relationship, CO_2 of known tension was bubbled through water and bicarbonate solutions at 37°C in a system containing a glass and calomel electrode. The points obtained were in fairly good agreement in dilute solutions, but fell well below expected sensitivity in concentrated solutions (*triangles*, fig. 4). It was also noted that equilibration time was greatly prolonged in solutions above 0.02 M, requiring about 15 minutes at 0.1 M and more than 2 hours at 1 M, as compared with about 5 minutes for all more dilute solutions.

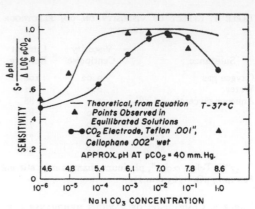

FIG. 4. Effect of bicarbonate concentration on sensitivity of CO_2 electrode.

The reasons for this delay and fall in sensitivity probably stem from the fact that more CO_2 must be exchanged as increasing amounts of carbonate occur. The pH values were taken from the water and bicarbonate solutions in equilibrium with 6.06% CO_2 and are in reasonable agreement with the calculated values.[3]

Design of the CO_2 Electrode. Stow's CO_2 electrode consisted of a very thin rubber membrane stretched tightly over a glass electrode, wet with distilled water between the rubber and glass. We have tried various modifications and find the most satisfactory system to be a Teflon membrane, a layer of cellophane (0.002″ when wet) to hold a better film of water between the glass and the Teflon and a solution of about 0.01 M NaHCO₃ and 0.1 M NaCl, in which the cellophane is soaked several hours. This system is more stable, twice as sensitive, faster in response and drifts much less, due to the stabilizing effect of the NaCl on the silver reference electrode, and the greater conductivity of the solution.

To assemble the electrode, the rubber stopper slice carrying the silver wire is slid on the glass electrode. A piece of wet cellophane film is stretched over the glass electrode and held in place with an 'O' ring. 0.001″ Teflon film (Dilectrix Corp., Farmingdale, L. I., N. Y.) is mounted on the end of the thin-walled Lucite tube with an 'O' ring, using silicone grease to seal the Lucite to the Teflon. This tube is filled with the electrolyte (NaHCO₃ - NaCl). A venting needle (20-gauge) is pushed through the rubber stopper slice. The glass electrode is inserted into the Lucite until its tip indents the Teflon membrane. The assembly is

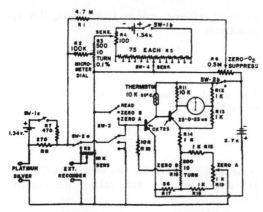

FIG. 3. Circuit of transistorized amplifier for O_2 measurement. 10K thermistor reduces thermal drift.

[3] At 37°C, emf = pH × 61.5 and on the microdial of the Beckman model GS pH meter R = (307.5 × pH) divisions.

TABLE 3. LINEARITY OF pCO₂ ELECTRODE

Change in % CO₂	Change in Reading, Divisions	Sensitivity S
1.38–3.16	77.5	0.922
3.16–6.06	81.5	0.937
6.06–11.37	102.5	0.927

NaHCO₃ conc., —0.005 M, Temp., 37°C.

slid into the chamber until the shoulder of the glass electrode seats on the nylon orifice of the cuvette. It is held in place with two screws and a bracket bearing on the outer end of the glass electrode cap. (The cellophane should not be mounted with the Teflon on the Lucite because of electrical and water leakage and trapping of air or liquid films between the two membranes.)

The silver reference electrode is 20-gauge sterling silver wire, given a chloride coat in 0.1 N HCl at 1 milliampere for 30 minutes. Drift may result from an insufficient chloride coat.

It was found necessary to have the water bath in direct contact with the Lucite electrode jacket for thermal equilibrium. Accordingly, the brass tube in which the electrode is mounted has several large holes on its sides. The Lucite tube is sealed to the brass tube at the outer end with an 'O' ring.

Response of the CO₂ Electrode. The relationship of pH to log pCO₂ is linear over the range tested, from 1.38% to 11.37% CO₂ (table 3). The response reaches equilibrium in about 2 minutes after a 4-fold rise in CO₂ and in about 4 minutes after a 4-fold fall in CO₂. The response time was not decreased by adding carbonic anhydrase to the electrolyte. Substitution of saturated KCl for the 0.1 M NaCl, as used by Dr. Fred Snell, did not substantially speed the response and resulted in serious drifting due to dissolution of the AgCl coat in the KCl solution (complex formation). Also, as suggested by Dr. Snell, we tried to replace the cellophane by a cellulose powder soaked in the electrolyte, to form a thinner film. We found sensitivity to be variable and lower and drifting to increase, but response was not accelerated. Attempts to omit the cellophane by using nonionic wetting agents failed for similar reasons. The rate or response can be increased by using thinner Teflon films, but these lack mechanical strength. The 99% response time with 0.00025″ Teflon was 60 seconds, as compared to 120 seconds for 0.001″ film. Polyethylene film (0.001″) required about 3 minutes. Silastic and rubber films were initially about the same, but became slower on standing overnight.

The sensitivity of the CO₂ electrode made with cellophane and Teflon membranes is plotted as a function of bicarbonate concentration in figure 4.

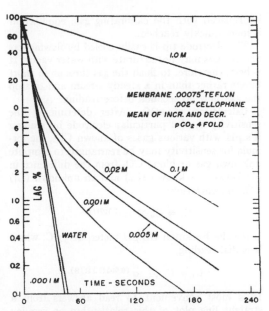

FIG. 5. Effect of bicarbonate concentration on response rate of CO₂ electrode.

In the range from 10⁻³ to 10⁻⁵ M the electrode sensitivity is distinctly less than that predicted from equations or measured in equilibrated water solutions. This loss of sensitivity may be a result of active buffer groups in the cellophane at these lower pH values. From 0.005 M to 0.02 M the sensitivity is such that 0.01 pH change is about equal to 2.5% change in pCO₂ at any pCO₂.

The time required for completion of response of the electrode after a change in pCO₂ is increased at high bicarbonate concentrations (fig. 5). In water, response is 99% complete after a 4-fold change in pCO₂ within 30 seconds. At 0.01 M, as we use it, about 2 minutes are required; this becomes 3–4 minutes at 0.1 M and 10–20 minutes at 1.0 M. The delay at higher concentrations is similar to that seen in the experiments in which gas was bubbled through solutions of bicarbonate. In all cases, the response appears to be faster by 50%–100% for a 4-fold rise than for a 4-fold fall in pCO₂. This is believed due to the logarithmic nature of the response, magnifying slight changes at low concentrations and making the tail more prominent.

Calibration. The pCO₂ reading was found to be identical in gas and various liquids of the same pCO₂, including water, blood, glycerine solutions, concentrated KCl, 0.1 N HCl, and phosphate buffer pH 7. This permits the electrode to be calibrated with gas in the cuvette. After a blood reading it is not necessary to rinse out the blood before the calibrating gas is read and if the blood is displaced with the calibrating gas, equilibrium is more quickly reached.

The electrode tip is easily cooled by flowing gas and the gas may not saturate with water vapor. It is best, therefore, to flush the gas through in short bursts rather than in a steady stream and to stop the gas flow 30 seconds before reading.

Calculation of pCO_2. After determining the sensitivity S, of a particular electrode by reading it's pH with various gases of known CO_2, the formula for sensitivity may be rearranged to compute unknown gas or blood CO_2 from the difference in pH between a known (C_o) and an unknown (C_i) CO_2 concentration:

$$C_i = C_o \times 10^{(\Delta pH/S)}.$$

Using the Beckman GS pH meter, at 37°C, where R is dial reading,

$$C_i = C_o \times 10^{(0.00325\Delta R/S)}.$$

It is most convenient to read answers from a straight line plot of this relationship on semilog paper. If the linear scale is calibrated in divisions change in the reading from a standard calibrating gas, such as 5% CO_2, the unknown may be read directly from the plot, in percentage of CO_2. Since compressed gases are used for calibration, it is better to prepare the graph in percentage of CO_2, which is unaffected by barometric pressure, rather than in pCO_2. The slope of the calibration curve, i.e. the sensitivity, is surprisingly constant, so that one calibration curve will suffice.

We wish to express appreciation for their suggestions and for their assistance in the development of the CO_2 electrode sensitivity equation to Drs. T. J. Kennedy and M. Eden and to Dr. Roger Bates, National Bureau of Standards, Dr. Leland C. Clark, Antioch College, Yellow Springs, Ohio, Dr. Fred Snell, Harvard University, Cambridge, Mass., and Dr. Julius Sendroy, Naval Medical Research Institute, Bethesda, Md.

ADDENDUM

The nylon orifice in the CO_2 electrode has been replaced by an 'O' ring (⁵⁄₁₆ in. i.d.) which adapts to variations in electrode diameter, forms the membrane better and has less tendency to tear it. The oxygen electrode cuvette, bath and amplifier will be commercially available (Yellow Springs Inst. Co.). The CO_2 electrode stability is improved by saturating the electrolyte with AgCl by adding a drop of dilute $AgNO_3$ solution.

REFERENCES

1. CLARK, L. C., JR. *Tr. Am. Soc. for Art. Int. Organs* 2: 41, 1956.
2. STOW, R. W., R. F. BAER AND B. F. RANDALL. *Arch. Phys. Med.* 38: 646, 1957.
3. KREUTZER, F., T. R. WATSON AND J. M. BALL. *J. Appl. Physiol.* 12: 65, 1958.
4. SPROULE, B. J., W. F. MILLER, I. E. CUSHING AND C. B. CHAPMAN. *J. Appl. Physiol.* 11: 365, 1957.

COMMENTARY TO

27. Friedman, S. M. and Nakashima, M. (1961)
Single Sample Analysis With the Sodium Electrode. Analytical
Biochemistry 2(6): 568–575.

E isenman, Rudin and Casby in 1957 described the chemical formula for the composition of cation sensitive glass that was selective for sodium or for other ions such as potassium with a change in the formulation (1). With one of the formulas of glass they constructed an ion selective electrode (ISE) for measuring sodium in aqueous solutions in the presence of potassium and other ions. Sydney M. Friedman and co-workers at the University of British Columbia obtained glass samples from Eisenman and constructed their own sodium electrode. The glass was blown into a cannula with a 0.5×2.0 cm^2 bulb that formed a flow through electrode (2). The electrode was used to measure Na$^+$ levels in-line in rabbit femoral arteries. They obtained potassium selective glass from Eisenman and built a two-electrode flow through system for Na$^+$ and K$^+$(3). The electrode was also used to measure Na$^+$ levels in-line in dog femoral arteries and to follow the effects of sodium ions on smooth muscle tone (4). J. A. M. Hinke in the laboratory at the Marine Biological Association in Plymouth, England obtained some of the Eisenman glass from Friedman and constructed electrodes with 70–90 μm tips and inserted them lengthwise into the giant axon of squids and measured intracellular sodium and potassium (5).

In the paper presented here, Friedman and Nakashima describe the performance of a static ISE for measurement of Na$^+$ in human blood. They constructed a 12-sample rotating valve that introduced 200 μL of blood sample into the electrode. Recovery, precision and correlation with flame photometry data are presented. Detailed construction diagrams and a parts list for sodium and potassium electrodes along with the description of the required electronic hardware were published by Friedman in 1962 (6). Circuit diagrams were published for construction of a computer interface that could monitor two electrodes in tandem (7). It was used to compare Na$^+$ and K$^+$ levels in blood by ISE to flame photometry in 1963 (8).

Commercial production of both sodium and potassium electrodes began as early as 1962. In 1981 the College of American Pathologists (CAP) proficiency survey report indicated that 29% of all participants used an ISE for sodium (9). In 1996 the percentage of labs using ISE's for sodium and potassium among 6 000 labs had risen to 99% (10).

References

(1) Eisenmann, G., Rudin, O., and Casby, J.U. (1957) Glass electrode for measuring sodium ion. Science. 126(3278):831–834.

(2) Friedman, S.M., Jamieson, J.D., Hinke, J.A.M., and Friedman, C.L. (1958) Use of glass electrode for measuring sodium in biological systems. Proceedings of the Society for Experimental Biology and Medicine. 99(3):727–730.

(3) Friedman, S.M., Jamieson, J.D., Nakashima, M., and Friedman, C.L. (1959) Sodium-and potassium-sensitive glass electrodes for biological use. Science. 130(3384):1252–1254.

(4) Friedman, S.M., Jamieson, J.D., Hinke, J.A.M., and Friedman, C.L. (1959) Drug-induced changes in blood pressure and in blood sodium as measured by glass electrode. American Journal of Physiology. 196(5):1049–1052.

(5) Hinke, J.A.M. (1961) The measurement of sodium and potassium activities in the squid axon by means of cation-selective glass micro-electrodes. Journal of Physiology. 156(2):314–335.

(6) Friedman, S.M. (1962) Measurment of sodium and potassium by glass electrodes. Methods of Biochemical Analysis. 10:71–106.

(7) Friedman, S.M. and Bowers, F.K. (1963) A computer solution of cation analysis with glass electrodes. Analytical Biochemistry. 5(June):471–478.

(8) Friedman, S.M., Wong, S.-L., and Walton, J.H. (1963) Glass electrode measurements of blood sodium and potassium in man. Journal of Applied Physiology. 18(September):950–954.

(9) Apple, F.S., Koch, D.G., Graves, S., and Ladenson, J.H. (1982) Relationship between direct-potentiometric and flame-photometric measurement of sodium in blood. Clinical Chemistry. 28(9):1931–1935.

(10) Scott, M.G., Heusel, J.W., LeGrys, V.A., and Siggaard-Anderson, O. (1999) Electrolytes and blood gases, in *Tietz Textbook of Clinical Chemistry*, Burtis, C.A. and Ashwood, E.R. (eds), 3rd Edition, W.B. Saunders Company, Philadelphia, PA, pg 1059, Chapter 31.

Anal. Biochem. 1961, 2(6): 568–575

ANALYTICAL BIOCHEMISTRY **2**, 568–575 (1961)

Single Sample Analysis with the Sodium Electrode[1]

SYDNEY M. FRIEDMAN AND MIYOSHI NAKASHIMA

*From the Department of Anatomy, The University of British Columbia,
Vancouver, British Columbia, Canada*

Received May 3, 1961

The determination of sodium by standard flame photometry is ordinarily limited to a precision of the order of $\pm 1\%$, to measurement of concentration rather than activity, and to the use of diluted samples. In 1957, Eisenman, Rudin, and Casby demonstrated that the specific affinity of glasses of the sodium aluminosilicate series for the alkalimetal cations is a systematic function of their composition (1). Their general conclusions also applied to glass systems in which lithium or potassium replaces sodium, boron replaces aluminum, or germanium replaces silicon. In turn, the affinity of these glasses for specific cations is a function of ionic activity rather than concentration.

In cooperation with these authors, we proceeded with the development of electrodes suitable for use in biological systems. In this work we were at first particularly concerned with the continuous monitoring of cationic activity in a moving stream (2, 3). The present report is concerned specifically with the determination of (Na^+) in single samples. We propose to show that the sample need not be diluted and that greater precision than that of flame photometry can be achieved.

METHODS

Sodium Electrode

A cannula-type electrode measuring approximately 2 mm internal diameter and 1.5 cm long was used throughout. The cannula was blown from NAS_{11-18} glass,[2] then silvered and copper plated at the central membrane area to make a metal-connected electrode. The method of fabrication has been described in detail elsewhere (4). The size is not critical. A calomel electrode in the line served as reference.

[1] This work was carried out with the aid of grants from the B. C. Heart Foundation and the National Research Council of Canada.

[2] Corning Glass Works, Corning, N. Y.

Electrometer

A Vibron electrometer was used.[3] A calibrated bucking voltage (calibrator) was interposed in the reference lead in order to back off the electrode potential to mid-scale on the electrometer. Thus the amplifier was essentially used as a null-point indicator, and readings were made on the calibrator.

Shielding

The electrode assembly was enclosed in a metal case which was solidly connected to the shield of the electrode cable. The shield was taken to earth, and the calibrator case was also taken to the same point. The apparatus is shown in Fig. 1.

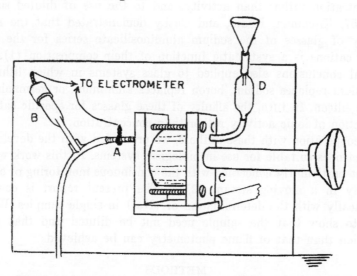

FIG. 1. Electrode assembly for measurement of sodium in single samples. The grounded enclosure contains the cannula electrode. (A) and calomel reference (B) in line. Samples can be loaded into the rotary sample chamber (C) or run in singly through an air gap (D).

Sample Delivery

Samples with a volume greater than 2 ml (diluted or undiluted) were fed into the electrode directly through a funnel. Samples with a volume less than this were first loaded into a sample changer devised for this work. The sample changer carried 12 samples of 0.2 ml each, and after loading was connected into the electrode path. The sample delivery components are shown in Fig. 2.

[3] Electronic Instruments Ltd., Richmond, Surrey, England.

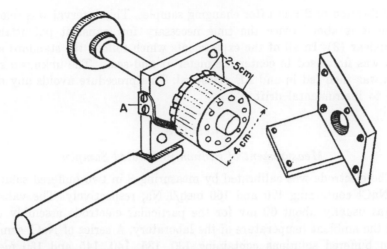

FIG. 2. Exploded view of rotary sample chamber. Capacity of each chamber is approximately 0.2 ml.

Analytical Measurements

NAS_{11-18}, the sodium-sensitive glass used, is insensitive at equilibrium to H^+ change above pH 5–6. All electrolyte solutions were thus prepared in a mixture of 0.1 N HCl and 0.2 N Tris [2-amino-2-(hydroxymethyl)-1,3,propanediol] buffered to pH 7. The plasma samples studied are, of course, biologically buffered.

Under these conditions,

$$E = E_0' + \frac{RT}{F} \ln [(Na^+) + k_{NaK}(K^+)]$$

where E is the measured electromotive force (E.M.F.), E_0' is a constant standard potential for the assembly, (Na^+) and (K^+) are the activities of sodium and potassium, and k_{NaK} is the empirically observed selectivity constant of the electrode for K^+ relative to Na^+. Since k_{NaK} for NAS_{11-18} is of the order of 0.004, it follows that in buffered solutions of a high Na^+ to K^+ ratio the electrode will be substantially unaffected by K^+ and will act as a pure Na^+ electrode at equilibrium.

Ideally, RT/F ln can be converted to 58 $\log_{10}$ in millivolts. In practice, we determine the actual value of this slope, S, by measuring E in two calibrating solutions of sodium:

$$\Delta E = S \log_{10} (Na_1 - Na_2)$$

The value of unknowns is then obtained directly from the measured value ΔE (relative to one standard) and the calibrated value of S.

All meter readings were taken after allowing a uniform period of equilibration of 2 min after changing samples. This interval was selected since it is about twice the time necessary for transient potentials to disappear (3). In all of the experiments which follow, the standard solution was first used to center the meter at mid-scale. The unknown solution was then fed in and ΔE measured. This procedure avoids any error due to instrumental drift.

RESULTS

Measurement of Sodium in Diluted Samples

The electrode was calibrated by measuring S in two buffered solutions of NaCl containing 120 and 160 meq/l Na, respectively. The value of S was usually about 60 mv for the particular electrode assembly used and the ambient temperature of the laboratory. A series of NaCl samples from buffered solutions containing 130, 135, 140, 145 and 150 meq/l, respectively, diluted 1:20 in distilled water, was then analyzed by the simple run-through method. Six separate runs were carried out. Exact values corresponding to the above known concentrations were obtained. In no case did the standard error of the mean exceed ±0.25 meq/l.

Measurement of Sodium in Undiluted Samples

The electrode was calibrated as before. A series of undiluted samples from buffered NaCl solutions 2 meq/l apart in Na^+ concentration in the range of 140–160 meq/l was analyzed. For purposes of method comparison, samples from the same solutions were analyzed by flame photometry in two ways. In one case the samples were diluted 1:20 before flame analysis; in the other the samples were undiluted so that their content of buffer was the same as that submitted to the electrode. Six separate trials, some on different days, were carried out. The results are shown in Fig. 3.

The precision of the electrode analysis measured over this range of concentrations was compared with the standard flame photometric results (Table 1). The electrode is evidently at least twice as precise.

Measurement of Sodium Added to Blood

Standard amounts of NaCl calculated to produce approximately 5 meq/l step rises in the plasma were added to aliquots taken from a pool of whole blood. Since this addition upsets the osmotic equilibrium between cells and plasma and since some sodium moves into cells, it is not possible to calculate the absolute values expected. Accordingly, plasma [Na] was analyzed by flame photometry and these results were used as a base for the electrode data. The electrode analyses were made

FRIEDMAN AND NAKASHIMA

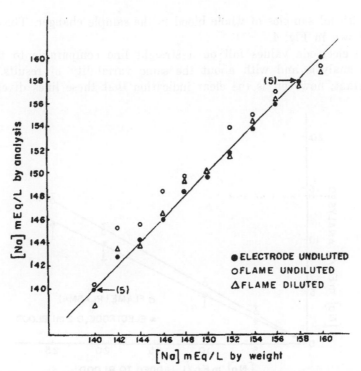

Fig. 3. Comparison of sodium analysis by flame photometry and glass electrode. Test NaCl solutions prepared by weight.

TABLE 1

COMPARISON OF ELECTRODE AND FLAME PHOTOMETRIC ANALYTICAL PRECISION[a]

	Sodium concentration			
	Electrode (0.2 ml)		Flame (diluted)	
Sample (meq/l)	Average	Standard error	Average	Standard error
140.0	140.0	±0.1	138.5	±1.20
142.0	142.8	±0.4	143.5	±0.72
144.0	143.3	±0.3	143.8	±1.35
146.0	146.0	±0.4	146.4	±0.84
148.0	148.4	±0.5	149.3	±1.56
150.0	149.6	±0.4	150.0	±0.01
152.0	151.6	±0.3	151.4	±1.20
154.0	153.7	±0.6	154.4	±1.34
156.0	155.9	±0.1	156.3	±1.26
158.0	157.9	±0.8	157.4	±0.90
160.0	160.0	±0.0	158.7	±0.81
Average % error	0.20	±0.06	0.43	±0.10

[a] Samples of NaCl at pH 7.

using 0.2-ml samples of whole blood in the sample changer. The results are shown in Fig. 4.

The electrode values fall on a straight line comparable to that of flame analysis and with about the same variability of results. Most important, however, is the clear indication that these lines diverge by

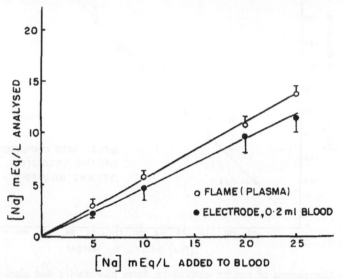

FIG. 4. Electrode analysis of sodium concentration in whole blood after addition of weighed amounts of NaCl. The actual plasma Na concentration achieved was measured by flame photometry.

a small constant. This strongly suggests that some proportion of the sodium in the plasma is inaccessible to the electrode; i.e., it is bound.

Measurement of Sodium Added to Plasma

The question of plasma binding of sodium is open to direct experimentation since, in contrast to the case with blood, it is possible to add NaCl to plasma in precisely known amounts. For this, NaCl was added to plasma aliquots from one original pool to form a series of samples of increasing [Na] in 5 meq/l steps. Samples from these were then analyzed by flame and electrode after 1:20 dilution. Any plasma binding should be reduced to vanishing by such dilution. A third set of samples was analyzed without dilution using the electrode sample changer. The results are shown in Fig. 5.

The results obtained on diluted samples analyzed either by flame or electrode evidently lie on the same line which, in turn, follows the anticipated 5 meq/l step. By contrast, the values for the undiluted electrode

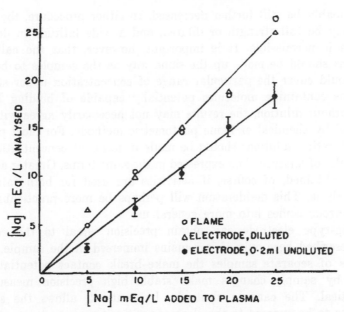

Fig. 5. Electrode analysis of sodium concentration in plasma after addition of weighed amounts of NaCl. The actual plasma Na concentration achieved was measured by flame photometry. Note that electrode analyses yield the same line as flame when 1:20 dilution is used but fall away when sample is not diluted.

samples lie on a line which diverges by a constant from that of diluted samples. This again suggests that some sodium in plasma is inaccessible to the electrode.

DISCUSSION

The sodium electrode of NAS_{11-18} glass is a practical instrument for the measurement of Na^+ in single samples of biological materials. It can be used alone provided the (H^+) of the solution is above pH 6. If precise measurements are to be obtained the (H^+) of the test solution should remain reasonably constant. The electrode does respond to K^+, but the affinity is very much lower than that to Na^+ and can be disregarded provided the test solution contains a high Na/K ratio. For practical purposes it is unresponsive to all other ions. The empirical formula of Eisenman *et al.* (1) given above permits one to assess beforehand whether K^+ is apt to present a problem in a particular application. For the routine analysis of Na^+ in blood and urine the electrode seems very useful.

The analysis of Na^+ can be carried out most simply using a run-through device requiring a sample of 2–3 ml. It can also be made on a sample as small as 0.2 ml using a sample changer. This latter amount can probably be still further decreased. In either procedure, the sample used may be full strength or diluted, and a wide latitude in degree of dilution is permissible. It is important, however, that the calibrating solutions should be made up the same way as the samples to be tested and should cover the particular range of concentration under study. If solutions containing molecules potentially capable of binding Na^+ are used without dilution, the results may not necessarily agree with those obtained by chemical or flame photometric methods. For most purposes the calibrating solutions should be made in terms of concentrations and the results of analysis then expressed in the same terms. Greater accuracy will be obtained, of course, if activities are used for both calibration and analysis. This desideratum will perhaps be more practicable when the electrode comes into more general use.

A dip-type electrode can obtain precision equal to the cannula-type described here only if it remains immersed in the sample. In the analysis of separate samples the make–break contact potentials occasioned by sample changes make stable high precision measurement impractical. The cannula electrode, by contrast, allows the separate solutions to be exposed to the electrode without such contact potentials.

Automation of the sample feed is quite practical and we are studying this. Direct readout of linearized sodium values as milliequivalents per liter (meq/l) rather than as millivolts on a logarithmic scale also seems eminently practical.

SUMMARY

The cannula-type metal-connected electrode is a practical device for the measurement of sodium in single aliquots of biological materials such as blood or urine. Measurements can be carried out with either diluted or undiluted samples using simple devices to present the solutions to the sensing membrane. Since the electrode responds to activity rather than concentration, ion binding has an important bearing on the analytical result. Evidence suggesting significant sodium binding in plasma is presented.

ACKNOWLEDGMENT

We are indebted to Mr. John Lees for his skill in blowing the glass electrodes.

REFERENCES

1. EISENMAN, G., RUDIN, D. O., AND CASBY, J. U., *Science* **126**, 831 (1957).
2. FRIEDMAN, S. M., JAMIESON, J. D., HINKE, J. A. M., AND FRIEDMAN, C. L., *Proc. Soc. Exptl. Biol. Med.* **99**, 727 (1958).
3. FRIEDMAN, S. M., JAMIESON, J. D., NAKASHIMA, M., AND FRIEDMAN, C. L., *Science* **130**, 1262 (1959).
4. FRIEDMAN, S. M., *Methods of Biochem. Anal.* **10**, in press (1961).

COMMENTARY TO

28. Anderson, N. G. (1969)
Analytical Technique for Cell Fractions XII. A Multiple-Cuvet Rotor for a New Microanalytical System. Analytical Biochemistry 28(1): 545–562.

The centrifugal analyzer was the first clinical chemistry analyzer designed for use with a microcomputer. The inventor, Norman G. Anderson sketched out the original concept for the centrifugal analyzer on the empty space in a full-page newspaper advertisement during a dinner in 1968 (1). A year latter a prototype had been built and tested. His first full report on the instrument was published in *Analytical Biochemistry* and that paper is presented here.

The design concept for the centrifugal analyzer was elegant and simple. Centrifugal forces moved reagents and samples into optical cuvetts positioned around the periphery of a spinning disk. Channels and risers along the linear axis or spokes of the spinning disk kept sample and reagents separated until angular momentum drove them to the periphery into the cuvette. Absorbance changes were monitored with a light source and photomultiplier arranged above and below the spinning disk of cuvettes. Reactions were monitored in real time on the fly and displayed on an oscilloscope. An attached microcomputer processed the stream of optical data. Anderson worked for the Oak Ridge national laboratory and the first centrifugal analyzer was called the GeMSAEC. This acronym stood for the sponsoring organizations, the National Institute of General Medical Sciences and the Atomic Energy Commission. Anderson published thorough reviews on the basic principals of centrifugal analyzers along with construction details (2–4). Rotors with up to 42 cuvettes were described (5). In one design iteration, the GeMSAEC analyzer spun a 17-place rotor at 600 rpm past a single beam photometer. Up to 10 200 absorbance readings per minute were fed into a Digital Equipment Corp. (DEC) PD8/I computer with 8k of fast memory and a magnetic tape back-up (6).

It is interesting that in the *Note* section to his 1969 paper presented here, Anderson reported that three different companies already offered prototype versions of the centrifugal analyzer. Two factors may have contributed to this rapid commercialization of a new technology. First, centrifugal analyzers with their high throughput seemed a viable alternative to continuous flow analysis, a technology that dominated clinical chemistry in the 1960s. A centrifugal analyzer, for example, with a 30-place rotor could run kinetic alkaline picrate creatinine assays at the rate of 30 samples in 80 seconds (7). Secondly, the technology was developed at Federal government research centers and was unencumbered by restrictive licensing issues.

The first commercial centrifugal analyzer was introduced in 1970 by Electro-Nucleonics, Inc. (ENI). It was called the GEMSAEC®. To avoid confusion with their original analyzer the Anderson group at Oak Ridge re-named their GeMSAEC analyzer the Oak Ridge National Laboratory or ORNL Centrifugal Fast Analyzer. By 1972 the Oak Ridge group had miniaturized their instrument (8). In 1974 they described a portable centrifugal analyzer that ran off batteries with an 8-position rotor housed in a case that measured $10 \times 10 \times 10$ cm ($3.9 \times 3.9 \times 3.9$ in.) (9).

Centrifugal analyzers performed both end point and kinetic assays. Kinetic enzymatic glucose assays with glucose oxidase required only 15 µL of serum and were completed in 70 seconds (10). They were widely used for the analysis of serum enzyme activities by kinetic analysis (11–12). Improvements in design and capability occurred throughout the 1970s. Tiffany *et al.* designed a fluorescence module for the centrifugal analyzer in 1973 (13–14). The enzyme multiplied immunoassay technique (EMIT®) was adapted to the centrifugal analyzer (15). Immunoassays were performed with kinetic light scattering nephelometric techniques in 30–200 seconds (16–17). Price and Spencer in 1980 edited a full textbook on centrifugal analyzers (18). By 1981 five different companies produced centrifugal analyzers; Union Carbide (CentrifiChem®); Electro-Nucleonics, Inc. (GEMSAEC® and GEMINI®); American Instruments (Rotochem®); Instrumentation Laboratories, Inc. (Multistat®). and Roche (Cobas-Bio) (19).

References

(1) Anderson, N.G. (1978) The birth and early childhood of centrifugal analyzers, in *Methods for the Centrifugal Analyzer*, The American Association for Clinical Chemistry, Savory, J. and Cross, R.E. (eds), Washington, DC, pgs xiii–xvi.

(2) Anderson, N.G. (1969) Computer interfaced fast analyzers. Science. 166(3903):317–324.

(3) Anderson, N.G. (1969) The development of automated systems for clinical and research use. Clinica Chimica Acta. 25(2):321–330.

(4) Anderson, N.G. (1970) Basic principles of fast analyzers. American Journal of Clinical Pathology. 53(5):778–785.

352

(5) Burtis, C.A., Johnson, W.F., Attrill, J.E., Scott, C.D., Cho, N., and Anderson, N.G. (1971) Increased rate of analysis by use of a 42-cuvet GeMSAEC fast analyzer. Clinical Chemistry. 17(8):686–695.

(6) Burtis, C.A., Mailen, J.C., Johnson, W.F., Scott, C.D., Tiffany, T.O., and Anderson, N.G. (1972) Development of a miniature fast analyzer. Clinical Chemistry. 18(8):753–761.

(7) Fabiny, D.L. and Ertingshausen, G. (1971) Automated reaction-rate method of serum creatinine with the centrifichem. Clinical Chemistry. 17(8):696–700.

(8) Scott, C.D. and Burtis, C.A. (1973) A miniature fast analyzer system. Analytical Chemistry. 45(3):327A–340A.

(9) Scott, C.D., Burtis, C.A., Johnson, W.F., Thacker, L.H., and Tiffany, T.O. (1974) A small portable centrifugal fast analyzer system. Clinical Chemistry. 20(8):1003–1008.

(10) Lutz, R.A. and Flucklger, J. (1975) Kinetic determination of glucose with the GEMSAEC (ENI) centrifugal analyzer by the glucose dehydrogenase reaction, and comparison with two commonly used procedures. Clinical Chemsitry. 21(10):1372–1377.

(11) Maclin, E. (1971) A systems analysis of GEMSAEC precision used as a kinetic enzyme analyzer. Clinical Chemistry. 17(8): 707–714.

(12) Tiffany, T.O., Jansen, J.M., Burtis, C.A., Overton, J.B., and Scott, C.D. (1972) Enzymatic kinetic rate and end-point analysis of substrate, by use of a GeMSAEC fast analyzer. Clinical Chemistry. 18(8):829–840.

(13) Tiffany, T.O., Burtis, C.A., Mailen, J.C., and Thacker, L.H. (1973) Dynamic multicuvette fluorometer–spectrophotometer based on the GeMSAEC fast analyzer. Analytical Chemistry. 45(9):1716–1723.

(14) Tiffany, T.O., Watsky, M.B., Burtis, C.A., and Thacker, L.H. (1973) Fluorometric fast analyzer: some applications to fluorescence measurements in clinical chemistry. Clinical Chemistry. 19(8):871–882.

(15) Mulligan, F.A., Fleetwood, J.A., and Smith, P.A. (1978) Enzyme multiplied immunoassay of phenytoin and carbamazepine using the centrifichem analyzer. Annals of Clinical Biochemistry. 15(6):335–337.

(16) Buffone, G.J., Savory, J., Cross, R.E., and Hammond, J.E. (1975) Evaluation of kinetic light scattering as an approach to the measurement of specific proteins with the centrifugal analyzer. I. Methodology. Clinical Chemistry. 21(12):1731–1734.

(17) Buffone, G.J., Savory, J., and Hermans, J. (1975) Evaluation of kinetic light scattering as an approach to the measurement of specific proteins with the centrifugal analyzer. II. Theoretical considerations. Clinical Chemistry. 21(12):1735–1746.

(18) Price, C.P. and Spencer, K. (eds) (1980) Centrifugal analysers in Clinical Chemistry, in *Praeger Special Studies*, Praeger Scientific, Eastbourne, UK.

(19) Sher, P.P. (1981) Centrifugal analyzers. Journal of Clinical Laboratory Automation. 1(1):60–61.

Anal. Biochem. 1969, 28(1): 545–562
Copyright 1969 Reprinted with permission from Elsevier.

ANALYTICAL BIOCHEMISTRY 28, 545–562 (1969)

Analytical Techniques for Cell Fractions

XII. A Multiple-Cuvet Rotor for a New Microanalytical System[1]

NORMAN G. ANDERSON

*Molecular Anatomy Program, Oak Ridge National Laboratory,[2]
Oak Ridge, Tennessee 37830*

Received September 17, 1968

Fast automatic analytical devices are required to handle the increasing number of different analyses desired on an ever-increasing number of samples in both biochemical research and clinical studies.

Mechanized analytical systems[3] have been divided into two major groups (1). Class I systems analyze a large number of samples for a single substance or activity, whereas those of class II analyze single samples for a number of different compounds or elements. Class I systems may be further subdivided into IA, in which many samples are analyzed simultaneously, and IB, in which they are done in sequence. A variety of hybrid systems is possible, some of which have been constructed. In this paper the problem of developing analyzers of class IA is considered, and the basic studies on the development of a new analytical system are presented.

BASIC REQUIREMENTS

1. *Simultaneity.* If a series of reactions done in parallel is carried to completion, it is not important that they all be started at exactly the same time. However, many colorimetric reactions are time-

---

[1] Research conducted under the joint NIH-AEC Molecular Anatomy (MAN) Program supported by the National Institute of General Medical Sciences, the National Cancer Institute, the National Institute of Allergy and Infectious Diseases, and the U. S. Atomic Energy Commission.

[2] Operated by Union Carbide Corporation Nuclear Division for the U. S. Atomic Energy Commission.

[3] Automation involves feedback control of a procss. Strictly speaking, nearly all analytical systems in present use are mechanized and use electronic control, sensing, and recording devices, but do not use feedback control. Hence, the words "automated" and "automation" do not apply to them despite previous use by the author and others.

dependent, e.g., with the color density increasing slowly, perhaps plateauing, and subsequently decreasing. In such instances, and where reaction rates are measured, as is the case with enzymatic reactions, it is of advantage to have all of a set of reactions start at as nearly the same time as possible, provided that all optical measurements can also be carried out over a short time span. If both the initiation of the reaction (mixing) and reading can be done in a few seconds or less, the reading may often be done very early in the course of a reaction. This is especially true when the set of samples is large enough to include a number of standards.

2. *Scaling*. The number of samples which may be analyzed with class IA analyzers should, in principle, be continuously variable. Some existing analyzers are adequate for larger loads but are not efficient for only a few (say, 1–10) samples. Small numbers of samples are usually done by manual methods. Hence, clinical and research analysts must often learn and use two different procedures for the same type of analysis, the choice depending on the volume of the analytical load. Obviously, methods that apply efficiently to both large and small sets of sample are desirable.

3. *Engagement*. If a sufficiently large number of analyses can be done in *one* apparatus and in a *short* time, the analyst can give full attention to the entire process. However, when data output is slow or a number of analyzers must be operated at the same time, the analyst's time is divided between them, with related tasks interleaved throughout the day. Because data output rate is slow, real-time computing is not economical, and data reduction is often postponed. Ideally, a set of analyses should be completed in the time required for the operator to determine whether a system of class IB is working properly. While truly automated variants which require no operator attention may ultimately be constructed, the potential for complete engagement (i.e., direct observation of reactions in real time) should still exist so that errors in dilution, reagent preparation, or procedure, or instrumental failure may be readily detected and proper action taken.

4. *Sample and reagent volumes*. Many analytical methods use enzymes and other expensive reagents. To minimize the reagent requirement, and to allow relatively small samples to be used, the techniques developed should be in the microliter range with sample and reagent volumes summing to between 0.2 and 3 ml. (It will be evident that the methods to be described are applicable to smaller or larger volumes, however.)

5. *Sample and reagent measurement*. Since discrete measurement is employed (as opposed to flow-ratio measurement and control),

the system developed should be adaptable to use with a variety of accurate volumetric devices.

6. *Sequencing of analytical elements.* If one considers each step in an analysis as an element, the following are the most common elements:

(*a*) Volume measurement.

(*b*) Measured volume transfer (two or more).

(*c*) Mixing.

(*d*) Incubation for a measured time at a known temperature.

(*e*) Precipitation (filtration or centrifugation).

(*f*) Extraction or washing of a precipitate.

(*g*) Extraction of an aqueous solution with an organic solvent.

(*h*) Absorbancy or other physical measurement, including radioactivity, pH, conductivity, etc.

(*i*) Data reduction.

The elements employed and their sequence may differ in different analyses. To be generally useful, an analytical system must be able to include most or all of these elements.

7. *Reaction monitoring.* The progress of all reactions being carried out simultaneously should be visible. A permanent chart record is not needed when reactions are carried out rapidly, if all results are monitored as they occur, and where data is reduced in real time.

8. *Data reduction.* Complete data reduction should be possible in a matter of seconds to minutes, and the computing system used should be programmed to handle both rate determinations and reactions carried to completion.

9. *Simplicity.* To be generally useful, the basic techniques must be simple, but should be applicable to a series of systems of varying complexity, ranging from a minimal-cost unit for teaching purposes to large units for handling the analytical load of a large hospital, research laboratory, or manufacturing plant.

In this and subsequent papers, the development of systems designed to fulfil these requirements is described. The present paper is concerned with demonstrating the feasibility of using a multiple-cuvet rotor to measure the absorbancies of a multiplicity of solutions simultaneously by using centrifugal force to load the cuvets, and variation in angular momentum to mix solutions.

BASIC DESIGN

The principle of double-beam spectrophotometry, in which the absorbancies of a reference solution and a sample are intercompared, either continuously or over time intervals short relative to the rate of change in intensity of the light source, is well known

To apply the same principle effectively to a larger number of samples requires that either they, or the light beam, move rapidly, one with respect to the other. We have chosen to move the samples rapidly past the light beam by using cuvet rotors. The centrifugal field inherent in this design has the advantage of providing the force for moving samples and reagents into the cuvets, for mixing them, for removing air bubbles, for sedimenting particulate matter, and for separating liquid phases. The electronic signal generated photoelectrically may be conveniently and continuously displayed on a cathode-ray tube and recorded photographically. In addition, the individual cuvet readings are made at a rate compatible with rapid electronic averaging of readings obtained over a short time period and with real-time data processing.

Cuvette Rotors. A flat 40 cuvet rotor (rotor G-I), with small, cylindrical test tubes as cuvets, was used in orienting studies. For more definitive studies, a 15 cuvet rotor (G-II) was designed. The G-II cuvets were formed by compressing a ring of $\sim \frac{1}{4}$ in. Teflon sheet between 2 discs of $\frac{1}{2}$ in. thick Pyrex. The Teflon section contains 15 round-bottom slots opening toward the center; the Teflon and upper Pyrex discs have a large hole in the center. Stainless-steel flanges above and below were connected with bolts to compress all parts together. A completed G-II rotor is shown in Figure 1, and a disassembled one in Figure 2. The circular apertures over the cuvets are $\frac{1}{4}$ in. in diameter, and the light beam is arranged to give a short, flat region on the end of the peaks as shown in Figure 3. When the rotor is spinning, 200 λ of solution is sufficient to fill the portion of the cuvet visible through the circular apertures in the stainless-steel end plates. The cuvet dimensions and volumes are chosen so that the rotor may be carefully brought to rest and re-accelerated without having the contents of the cuvets mix with each other when a total volume of 200–250 λ per cuvet is used. If larger volumes are to be decelerated to rest, longer cuvets must be employed.

Photoelectric sensors mounted next to the rotor edge provide synchronizing signals for the sweep circuits.

Sample and Reagent Addition. Samples and reagents are moved into the cuvets by centrifugal force. Since this may be done over a short period of time, all reactions start essentially together and may be followed continuously on the oscilloscope. Sample and reagent discs (Fig. 4) allow the sample and two reagents to be loaded into separate depressions designed not to allow fluids to mix at rest, but which all drain to the edge into the proper cuvets in a

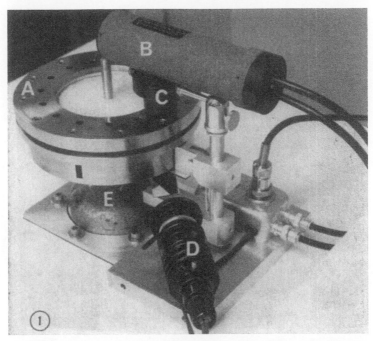

Fig. 1. G-II cuvet rotor: (A) cuvet rotor; (B) photomultiplier housing; (C) filter holder; (D) light source with diaphragm; (E) drive motor from IEC clinical centrifuge.

centrifugal field. Connections to the edge are through small capillary holes or past sloping surfaces which prevent mixing before spinning, but allow free horizontal drainage during rotation. The transfer discs may also be adapted to hold the transfer tubes previously described (1), or small, commercially available disposable microliter pipets. It is evident that these devices allow single or multiple addition reactions to be used, or reactions in which an incubation period occurs between two additions. It is also evident that precipitates formed during a reaction may be moved out of the optical path by centrifugal force, allowing the absorbancies of a clear supernatant to be measured.

Mixing. In many instances, especially where zonal centrifuge fractions are being analyzed, the reagents and the sample may differ considerably in density and viscosity. An effective means for achieving rapid mixing is therefore required.

In the cuvet rotor described, the radii to the top and the bottom of a 250 λ reaction volume differ by approximately 7 mm, giving a ratio of tangential velocities of 1.05. If the rotor is rapidly accel-

550 NORMAN G. ANDERSON

FIG. 2. Disassembled G-II rotor: (A) lower rotor housing; (B) lower gasket; (C) lower Pyrex plate; (D) Teflon cuvet spacer; (E) upper Pyrex window ring; (F) upper Teflon gasket; (G) upper stainless-steel end plate.

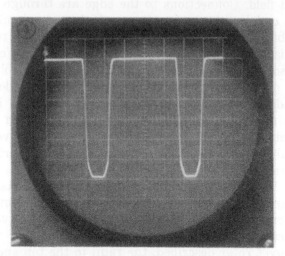

FIG. 3. Oscilloscope tracing of two peaks indicating flat tip. Baseline at top indicates 0% transmission (infinite optical density) line.

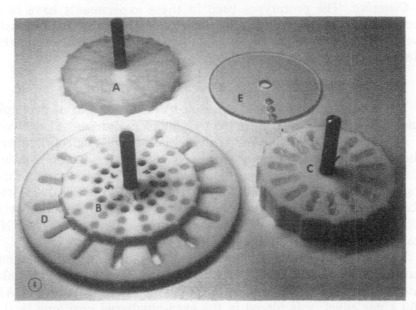

FIG. 4. Sample and reagent transfer and loading discs: (A, B) discs with cavities with top tilting toward the axis but connected by capillaries draining the bottom of the cavities to the edge; (C) disc having two concentric sets of fluid-holding cavities tilting outward so that fluid rises along cavity wall to drain through peripheral holes; (D) Teflon center piece of G-II rotor showing how loading disc fits cuvets in rotor; (E) cover plate used to cover disc and expose only cavities being filled.

erated and decelerated during fluid addition, effective circular flow is established in the cuvet because of the velocity differences between the centripetal and centrifugal surfaces of the fluid. In practice, therefore, the rotor is accelerated rapidly to transfer fluid, is rapidly decelerated (but not brought to rest), and then reaccelerated to the speed used for observation. The experiments described in subsequent sections demonstrate that adequate mixing may be obtained by this method within 10–15 sec after the rotor is started using protein samples in distilled water and a biuret reagent. With concentrated sucrose solutions, more rapid change in speed than could be effected with the present prototype will be required.[4]

[4] In subsequent studies the cuvets have been designed as syphons with a restricted entrance. Mixing may be achieved by bubbling air back through the cuvet syphon during rotation at slow speed. Detailed studies on mixing rates are included in a subsequent paper.

Cleaning. The prototypes used in these experiments have been hand-cleaned by using a fine stream of distilled water continuously removed by suction, followed by high-pressure air. An automatic cleaning device is under development.

EXPERIMENTAL

The studies recorded here are concerned with proof-of-principle and with more precise definition of problems remaining to be solved.

Leakage. Leakage from the cuvets to the edge during rotation was checked by filling the cuvets completely with water during rotation and observing the position of the fluid surface stroboscopically during rotation at 2000 rpm over an extended period. No leakage was observed. To determine whether leakage occurred between cuvets, alternate cuvets were filled with water and similarly observed. No leakage between cuvets was seen.

The possibility exists that fluid from the loading disc does not pass directly and quantitatively into the proper cuvet, but leaks laterally. A heavily stained solution of bovine serum albumin (BSA), 250 λ per cuvet, was placed in the even-numbered positions in the loading disc and moved by centrifugal force into the cuvets. The results, compared with a pattern obtained with all the cuvets full of water, are shown in Figure 5. It is seen that almost no light passed through the evennumbered cuvets, but full transmission was observed through the odd-numbered ones. The white Teflon-bottom inner surfaces of the latter cuvets were carefully examined by using stroboscopic illumination to verify that no leakage had occurred. (Note that in the oscilloscope patterns the lines connecting the peaks indicate 0% transmission or infinite absorbancy. The peaks therefore appear to be inverted when compared with those obtained with other analytical systems; however, the peak tips are all at the expected positions.)

Drainage from Loading Disc. Drainage from the loading disc was studied by using heavily stained BSA solutions. After spinning at 2000 rpm, only a very few traces of blue color could be observed in the disc. This was thought to be due to very small irregularities in the Teflon due to machining. The volumes involved appeared to be less than 1 λ. Further quantitative studies on solution transfer as a function of angular velocity, radius, disc design, and solution density, viscosity, and surface tension are indicated.

Calibration. The path length of the cuvets was measured directly by using an electronically indicating micrometer. The rotor cuvet end plates are not exactly parallel, but vary in a sine wave pattern as shown in Figure 6. This was confirmed by using identical dyed

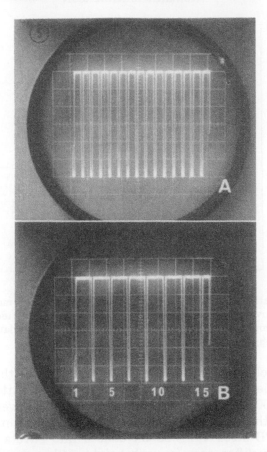

Fig. 5. Test for leakage between cuvets. (A) oscilloscope pattern with cuvets containing water. The ordinate for the tracing ranges from 0% transmission at the top to 100% transmission at the bottom of the trace. Cuvet numbers are in order from left to right. (B) Tracing seen when a solution of dyed BSA was added to even-numbered cuvets. A 550 nm interference filter was used.

protein solutions in each cuvet except the first, which contained water (Fig. 7B).

It is not desirable to depend on precision construction of cuvets to define path length or to assume that the blank (water) absorbancies of all cuvets are equal or constant. Instead blank absorbancies and absorbancies with standard solutions to determine path length should be redetermined at intervals when high precision is required. The values obtained are incorporated in the final calculations.

To see whether reporducible curves could be obtained with standard solutions, a solution containing 1.5 gm of crystalline BSA and 15 mg of bromphenol blue (BPB) in 100 ml was diluted to give

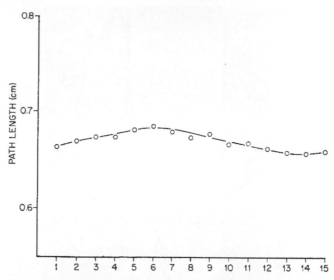

Fig. 6. Variation in cuvet path length of G-II rotor as measured with a micrometer (cuvet number on abscessa). This pattern is identical to that observed spectrophotometrically in Figure 7B. Maximum variation is ±2% of average path length.

a series of solutions containing 10% increments of the stock solution in distilled water. Figure 7A illustrates the pattern observed by using a 660 mµ filter and distilled water in all cuvets. In Figure 7B is shown the pattern obtained where a solution containing water and BSA-BPB stock 1:1 was introduced into cuvets 2–15 during rotation. The differences in peak height, although small, confirm those observed by direct measurement, as shown in Figure 6. In Figure 7C a complete series of incremented standards is shown in cuvets 3–12, with a duplicate of the stock solution used in 12 also in cuvet 14. The four remaining cuvets contained distilled water. Measurements were made on 8½ × 11 enlargements, and all peaks converted to $1/\%T$ by dividing the first blank by each subsequent reading in turn. The log of $1/T$ is the absorbance, which after blank subtraction, is then multiplied by a cuvet factor to give absorbancies for a 1 cm path length. The data obtained from Figure 7C in this manner are plotted in Figure 8.

Biuret Reaction for Protein. The experiments described thus far show that the cuvet rotor and associated electronics can be used to measure absorbancies of standard solutions, but do not demonstrate that the system can be used to follow reactions occurring in the cuvets. The biuret reaction for protein is a simple one-reagent analysis which is of general interest and is suitable for evaluation

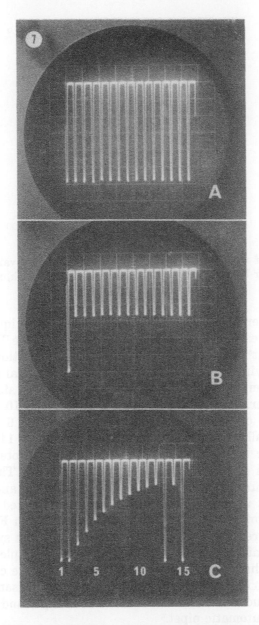

FIG. 7. Calibration curves and curves obtained with various dilutions of a dyed protein solution using 660 nm interfenrence filter: (A) all cuvets filled with distilled water; (B) cuvets 2–15 filled with a 1:1 dilution of the dyed protein stock solution; (C) dilutions of stock protein in 10% increments. Cuvets loaded as follows: water in cuvets 1, 2, 13, and 15; dilutions of stock solution in 10% increments in cuvets 3–12, with undiluted stock solutions in cuvets 12 and 14.

NORMAN G. ANDERSON

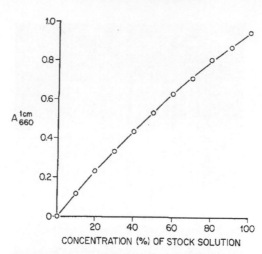

FIG. 8. Plot of results obtained in Figure 7, corrected for variation in absorbancies of water blanks and for variation in cell path length, and corrected to 1 cm light path.

of the transfer discs, of mixing, and of the technique of reading absorbancies early in the course of a reaction. The Weichselbaum biuret reagent[5] (2) may be used with protein solutions in a range of ratios varying from 0 to 50% reagent in the final mixture, provided that identical dilutions are used to obtain the standard curve.

In the experiments illustrated in Figures 9 and 10, 2 ml of various dilutions of BSA standard was mixed with 6 ml of biuret reagent and allowed to stand at room temperature 14 hr to ensure a stable color before reading. The dilutions of protein used were in increments of 0.2% protein from 0.2 to 2.0%. The results are shown in Figure 9B with water in cuvets 1, 3, 14, and 15, and the reagent blank in cuvet 2. The absorbancies, corrected to 1 cm path length, are plotted against protein concentration in Figure 10 and illustrate the results that may be expected with this system.

In the remaining studies the ratio between sample and reagent volume was changed to 1:1, and the reactions were carried out in the cuvets during rotation. Both the reagent and sample volumes were 200 λ and were introduced into the sample and reagent disc by using an automatic pipet.[6]

[5] The biurent reagent stock solution contained 45 gm sodium potassium tartrate, 15 gm copper sulfate, and 5 gm potassium iodide made up to 1 liter with 0.2 N sodium hydroxide. The reagent used was prepared by diluting 200 ml of this stock to 1 liter with 0.2 N NaOH containing 0.5% KI.

[6] Available from Baltimore Biological Laboratory, catalog No. 05-719, Baltimore, Maryland 21204.

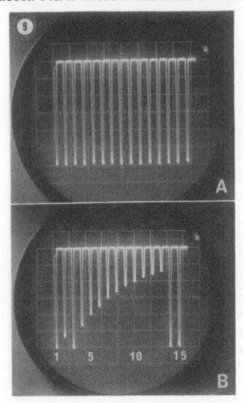

FIG. 9. Stable biuret reagent–protein standard mixtures loaded into G-II rotor with loading discs: (A) water blanks; (B) mixtures, etc. Mixtures prepared by mixing 2 ml of protein solution and 6 ml of Weichselbaum biuret reagent. Water blanks in cuvets 1, 5, 14, and 15; reagent blank in curvet 2. Series in cuvets 4–13 are prepared by using 0.2% increments of protein up to final concentration of 2.0%. All readings at 550 nm.

The reproducibility of measurements and the effect of reaction time are indicated in the two experiments shown in Figures 11 and 12. Three protein concentrations, 0.2, 0.4, and 0.6%, were used in the first experiment, the first two concentrations being run in quadruplicate, the third in pentuplicate. The pattern shown in Figure 11A was photographed 30 sec after the rotor was started, while that in Figure 11B was made after 25 min. Although a considerable increase in absorbancy is evident after 25 min, little difference in reproducibility is seen.

Figure 12 shows all the calibration data required for 30 sec protein determinations using a total of 15 blanks plus standards or samples. Water blanks are shown in Figure 12A and cuvet path comparisons obtained by using 0.6% protein standards and

558   NORMAN G. ANDERSON

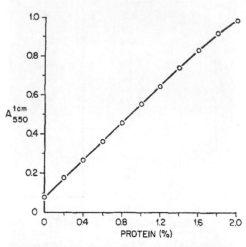

FIG. 10. Plot of data from Figure 9B corrected to 1 cm path.

one water blank are shown in Figure 12B. The series in Figure 12C was photographed 30 sec after starting the rotor (and therefore the reaction) with the sample and reagent disc loaded with water and reagent blanks, and duplicate standards having 0.2, 0.4, 0.6, 0.8, and 1.0% protein. The results are plotted in Figure 13.

DISCUSSION

The G-II cuvet rotor allows 15 reactions to be initiated simultaneously and the absorbancies to be observed and measured after very short time intervals. The rotor is an extension of the double-beam spectrophotometer concept where two absorbancies are repeatedly measured over intervals of time which are short relative to the variations in intensity that may occur in the light source.

Unlike sequential analyzers of class II, no carryover between samples is observed, and the tracing returns to 0% transmission (infinite absorbancy) between each reading. Since one or more water blanks are included in each series, readings for samples, 0, and 100% transmission are made during each revolution; thus at 1,200 rpm, 20 revolutions and 20 sets of measurements are made each second. If 1 sec exposure times are used, then the result represents the average of 20 readings. The time between peaks is ample to allow computer averaging of digitalized peak height by using the techniques described by Spragg and Goodman (3).

Errors may be due to pipetting, incomplete fluid transfer, incomplete mixing, nonlinearity in the electronics, lack of precision in measuring the photographs, and changes in the optical properties

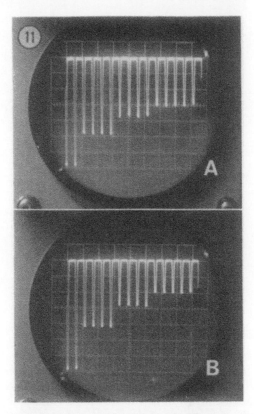

Fig. 11. Variation observed when multiple identical standards are analyzed 200 λ of protein standard or water and 200 λ of biuret solution placed on Teflon loading disc and moved into cuvets by centrifugal force. Cuvets loaded as follows: 1–2, water; 3–6, 0.2% protein; 7–10, 0.4% protein; 11–15, 0.6% protein. (A) Pattern observed 30 sec after starting rotor; (B) pattern 25 min after starting rotor.

of the cuvets during the experiment. Each of these can be examined in detail when fully automatic methods for digitalizing and printing-out results are completed, and statistical studies on the precision obtainable have been deferred until that time. It is concluded that the present studies justify completion of the entire system. It is evident that the basic concept presented here can be applied to a variety of analytical problems, especially those dependent on determination of reaction rate.

If small fluid volumes are added to the rotor initially, the rotor may be brought to a complete stop and a new sample-reagent disc inserted. In this way, reactions depending on sequential timed additions may be performed. The centrifugal capabilities of the rotor

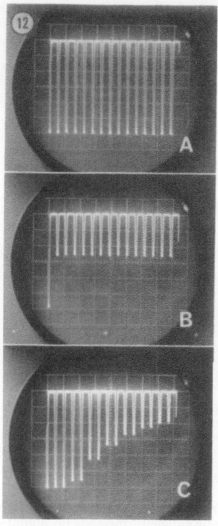

Fig. 12. Complete series for 30 sec biuret protein determination: (A) water
blanks; (B) path length standardization with 1:1 mixture of biuret reagent
and 0.4% BSA (read 10 min after mixing) in cuvets 2–15, and water in cuvet
1; (C) blanks and standard series read 30 sec after starting rotor. 1–3, water;
4–5, reagent blanks; 6–7, 0.2% protein; 8–9, 0.4%; 10–11, 0.6%; 12–13, 0.8%;
and 14–15, 1.0%.

may also be employed to sediment particulate matter and to ensure
that the solutions read are not turbid when read. In addition, the
decrease in turbidity may be observed and if necessary read during
centrifugation.

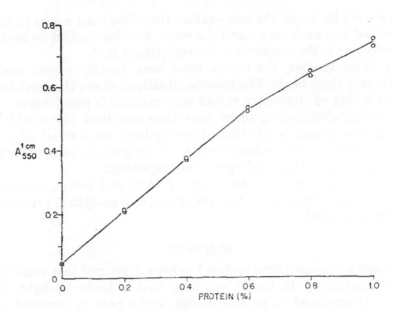

FIG. 13. Plot of data from Figure 12C corrected to 1 cm light path.

Other ways in which the rotor may be constructed include the use of removable or disposable cuvets of glass or plastic or the use of precision individual cuvets. A wide range of rotor sizes and numbers of cuvets is possible, and a number of different methods may be employed to load the sample and reagent discs. These will be explored in subsequent studies, which will also be concerned with other analytical procedures.

For the biuret reaction described to be useful for the analysis of zonal centrifuge fractions when sucrose is used as the gradient material, corrections for the color diminution produced by sucrose (4) must be made. In addition, more effective and more rapid rotor speed changes may be required to produce complete mixing with very viscous samples. Definitive studies on the biuret reaction in the G-II rotor are in progress.

More than one reaction may be carried out in one rotor at the same time. With NAD-linked assays, for example, a variety of different enzyme determinations may be done simultaneously. Decisions on how long data collection should be continued can be based on the readings obtained and the run terminated by a small computer as soon as sufficient information is obtained.

When analytical methods are employed which require incubation at elevated temperatures, the entire rotor may be heated. Standards

NORMAN G. ANDERSON

are always included; the temperature therefore need not be critical, provided it is uniform around the rotor. Bubbles formed by heating rapidly rise to the meniscus in the centrifugal field.

In these studies, the cuvets have been rapidly moved past a stationary light beam. The reverse situation, where the light beam scans a ring of stationary cuvets very rapidly, is also feasible.

Gradient-distributing rotors have been described (5) in which a continuous stream is distributed evenly between a series of tubes during rotation. This concept may allow reagents to be apportioned between cuvets, thus avoiding multiple pipettings.

In subsequent papers, the details of additional rotors, associated electronics, and data reduction programs, and analytical procedures will be presented.

SUMMARY

A new microanalytical system has been developed that employs a rotor containing 15 cuvets spinning past a beam of light. The signal is displayed on an oscilloscope, and a peak is observed continuously for each cuvet. Standards, samples, and reagents are placed in a central loading disc, and all solutions moved out into the cuvets by centrifugal force. Minimum volume to fill the cuvets is 200 λ. By using the Weichselbaum biuret reagent for proteins, 15 analyses may be completed in as little as 30 sec after the rotor is started. The data are obtained photographically.

Note added in proof. The General Medical Sciences-Atomic Energy Commission (GeMSAEC) system described is available in prototype form from Electro-Nucleonics, Inc., Caldwell, N. J.,; Tennecomp, Inc., Oak Ridge, Tennessee (computerized); and Union Carbide Corporation, Development Department, Tarrytown, New York.

ACKNOWLEDGMENTS

The electronic components for the G-I were designed and built by Mr. D. D. Willis, and those for G-II were designed and built by Mr. Douglas Mashburn.

REFERENCES

1. ANDERSON, N. G., *Anal. Biochem.* **23**, 207 (1968).
2. WEICHSELBAUM, T. E., *Am. J. Clin. Path.* **7**, 40 (1946).
3. SPRAGG, S. P., AND GOODMAN, R. F., paper presented at conference on "Advances in Ultracentrifugation Analysis," Feb. 15–17, 1968, in New York; to be published by *Ann. N.Y. Acad. Sci.*, New York (1969).
4. GERHARDT, B., AND BEEVERS, H., *Anal. Biochem.* **24**, 337 (1968).
5. CANDLER, E. L., NUNLEY, C. E. AND ANDERSON, N. G., *Anal. Biochem.* **21**, 253 (1967).

29. Glad, C. and Grubb, A. O. (1978)
Immunocapillarymigration—A New Method for Immunochemical Quantitation. Analytical Biochemistry 85(1): 180–187.

The first solid phase immunoassay was described by Catt in 1967 for use in radioimmunoassay (RIA) (1,2). He coated antibody onto polypropylene or polystyrene test tubes followed by incubation with radio labeled tracer and patient serum. At the end of the incubation the tubes were washed and the bound tracer on the walls of the test tubes was counted. In another version, 12.7 mm (0.5 in.) discs punched from poly tetrafluoethylene (PTFE) plastic sheets were coated with antibody and used as the solid phase. Eva Engvall and co-workers in 1974 extended this format for use with the enzyme-linked immunosorbent assay (ELISA). She coated the antibody or antigen onto the walls of micro titer wells instead of plastic test tubes (3).

Aalberse in 1973 may have been the first to describe a solid phase immunoassay on a porous membrane (4). The assay was a sandwich immunoassay for human IgG. Five-millimeter diameter discs were punched from sheets of Whatman Number 20 cellulose filter paper. The paper discs were activated through a cyanogen bromide reaction followed by covalent coupling of an anti-human sheep IgG antibody to the paper. The same antibody was also conjugated with fluorescein isothiocyanate (FITC). Serum and FITC labeled antibody were incubated with the antibody coated paper discs in test tubes. At the end of the incubation the discs were washed, the bound FITC labeled antibody eluted with alkali and the eluate read in a fluorometer.

Christina Glad and Anders Grubb in a proceedings abstract in 1977 described a generic immunoassay performed entirely in a membrane format (5). A full report on this method was published in *Analytical Biochemistry* the next year and that paper is presented here. In this 1978 paper the authors coated anti-human transferrin antibody onto polyvinyl chloride (PVC) plastic sheets impregnated with a silica gel filler. After drying, the sheets were cut into 8 mm wide by 70 mm long strips. In the assay, an antibody-coated strip was mounted vertically in a small beaker containing a buffer dilution of human serum. Capillary forces pulled the serum up the length of the plastic sheet. When the liquid reached the top of the sheet it was rinsed under tap water and then incubated with a second anti-transferrin antibody conjugated with FITC. Following a second washing the sheets were held under an ultraviolet lamp and the height of the second fluorescent antibody bound to the strip was read visually. The height of the fluorescent-labeled second transferrin antibody on the strip was an inverse measure of the transferrin in the patient's serum.

In the Discussion section to this paper the authors propose other possible labels for the second antibody in this non-instrument based immunoassay. Among the possible labels mentioned were enzymes and colored latex particles. In 1981 Glad and Grub followed up on their original report. They developed a similar assay for human serum C-reactive protein (CRP) but with two major improvements. First, they used antibody coated cellulose acetate strips as the solid support and in addition, the second antibody was labeled with horse-radish peroxidase (HRP, EC 1.11.1.7) (6).

In 1985 a group of workers at Syva and Syntex, where the enzyme-multiplied immunoassay technique (EMIT®) was invented, described a modification of the Glad and Grubb assay for the measurement of drug levels in serum or whole blood. Whatman chromatography paper strips were coated with anti-theophylline antibody. Serum or a whole blood sample along with an HRP-theophylline conjugate migrated up the strip of paper through capillary action. The height of the bound conjugate after color development was inversely related to the concentration of theophylline in the sample (7). This capillary immunochromatograhy format was expanded to include other therapeutic drugs and was sold by Syva as the AccuLevel™ device (8).

Progress in the development of rapid membrane based immunoassays that did not require an instrument progressed rapidly through out the late 1970s and 80s. The most important changes that occurred were that the capture antibody was coated as a line and a variety of different visual labels were used. Hsu replaced enzymes as labels with red colloidal gold nanospheres (9). Colored latex beads were described as the label in numerous patents for solid phase immunoassays issued in the 1980s. The Campbell, Wagner and O'Connell United States patent issued in 1987 that used 0.5 μm colored latex particles and a nitrocellulose support is an example (10).

Birnbaum *et al.* in 1992 described a latex-based immunoafinity chromatography assay for human chorionic gonadotropin (HCG) that was typical of the formats at the time (11). The assay used a monoclonal antibody pair. The first antibody was applied as a 0.5 mm line on a 40 × 5 mm strip of Nylon membrane. The second antibody was conjugated to 0.3 μm blue latex beads. Urine sample and antibody-latex bead conjugate were mixed and allowed to migrate up the strip. When HCG was present in the sample it formed a sandwich on the 0.5 mm line and turned the line blue. This technology eventually evolved into today's widely used lateral flow membrane assays.

References

(1) Catt, K. and Tregear, G.W. (1967) Solid-phase radioimmunoassay in antibody-coated tubes. Science. 158(808):1570–1572.

(2) Catt, K., Niall, H.D., and Tregear, G.W. (1967) A solid phase disc radioimmunoassay for human growth hormone. Journal of Laboratory and Clinical Medicine. 70(5):820–830.

(3) Voller, A., Bidwell, D., Huldt, G., and Engvall, E. (1974) A microplate method of enzyme-linked immunosorbent assay and its application to malaria. Bulletin of the World Health Organization. 51(2):209–211.

(4) Aalberse, R.C. (1973) Quantitative fluoroimmunoassay. Clinica Chimica Acta. 48(1):109–111.

(5) Glad, C. and Grubb, A.O. (1977) A new method for immunochemical quantification. Biochemical Society Transactions. 5(3):712–714.

(6) Glad, C. and Grub, A.O. (1981) Immunocapillarymigration with enzyme-labeled antibodies: rapid quantification of C-reactive protein in human serum. Analytical Biochemistry. 116(2):335–340.

(7) Zuk, R.F., Ginsberg, V.K., Houts, T., Rabbie, J., Merrick, H., Ullman, E.F., Flscher, M.M., Sizto, C.C., Stiso, S.N., and Litman, D.J. (1985) Enzyme immunochromatography — a quantitative immunoassay requiring no instrumentation. Clinical Chemistry. 31(7):1144–1150.

(8) Chen, R., Li, T.M., Merrick, H., Parrish, R.F., Bruno, V., Kwong, A., Stiso, C., and Litman, D.J. (1987) An internal clock reaction used in a one-step enzyme immunochromatography assay of theophylline in whole blood. Clinical Chemistry. 33(9):1521–1525.

(9) Hsu, Y. (1984) Immunogold for detection of antigen on nitrocellulose paper. Analytical Biochemistry. 142(1):221–225.

(10) Campbell, R.L., Wagner, D.B., and O'Connell, J.P. (1987) Solid phase assay with visual readout. United States Patent No. 4,703,017. Issued October 27, 1987.

(11) Birnbaum, S., Uden, C., Magnusson, C.G.M., and Nilsson, S. (1992) Latex-based thin-layer immunoaffinity chromatography for quantitation of protein analytes. Analytical Biochemistry. 206(1):168–171.

Anal. Biochem. 1978, 85(1): 180–187

ANALYTICAL BIOCHEMISTRY **85,** 180–187 (1978)

Immunocapillarymigration—A New Method for Immunochemical Quantitation

Cristina Glad and Anders O. Grubb

Department of Clinical Chemistry, University of Lund, Malmö General Hospital,
S-214 01 Malmö, Sweden

Received December 17, 1976; accepted October 4, 1977

A new simple and rapid method for immunochemical quantitation called immunocapillarymigration is described. It is based upon the attachment of antibodies to a porous insoluble support and the subsequent capillarymigration of the antigen-containing solution in the porous support. The migration of the antigen solute is specifically delayed in comparison to the migration of the solvent and other solutes in the process and the relative delay decreases with increasing antigen concentration. When applied to the quantitation of transferrin in human plasma, immunocapillarymigration gave results which agreed with those obtained by single radial immunodiffusion.

Since the first quantitative immunochemical principles were established by Heidelberger and Kendall (1) some 50 years ago, a large number of methods for immunochemical quantitation have been developed. Some of these methods, e.g., single linear diffusion according to Oudin (2), single radial immunodiffusion according to Mancini *et al.* (3) or according to Fahey and McKelvey (4), double linear diffusion according to Preer (5) or Ouchterlony (6) or according to Wieme and Veys (7), Farr's ammonium sulfate technique (8), Laurell's electroimmunoassay (9), automated nephelometric analysis (10), and various types of radioimmunoassays (11,12) are presently used in most laboratories. Some of the abovementioned methods are distinguished by simplicity and rapidity while others are distinguished by extremely high sensitivity. All the methods employ either mixing, diffusion, or an electrophoretic process to bring about the necessary interaction between antibody and antigen. In this work we present a new simple and rapid quantitative immunochemical method in which capillary force is used to bring about the necessary antigen–antibody interaction.

MATERIALS

Polyvinylchloride sheets containing silica gel as a filler (microporous plastic sheets) were purchased from Amerace Erna Corporation, New York, N. Y., fluorescein isothiocyanate from BDH Chemicals, Pool,

0003-2697/78/0851-0180$02.00/0
180

England, and Sephadex G 25 from Pharmacia Fine Chemicals, Uppsala, Sweden. All other chemicals were of reagent grade. Pure human IgG, albumin, and transferrin were available at our laboratory.

Preparation of antisera. Goat and rabbit antisera against human IgG, albumin, and transferrin were produced by repeated subcutaneous injections of the purified antigens emulsified in Freund's complete adjuvant. Sera from several bleedings were pooled and the pools were rendered monospecific by appropriate absorptions. Immunoglobulin fractions of the antiserum pools were produced by ammonium sulfate precipitation. The precipitates were dissolved in and dialyzed against distilled water, pH 5–6, and then lyophilized.

Preparation of fluorescein-labeled antibodies. Parts of the immunoglobulin fractions of the antisera were labeled with fluorescein isothiocyanate (13). The fluorescein-labeled immunoglobulins were freed of small molecular weight contaminants by gel chromatography on Sephadex G 25 in 0.1 M Tris–HCl buffer, pH 8.0, containing 8 mmol/liter of sodium azide. Aliquots of the solutions of fluorescein-labeled immunoglobulins were stored at −20°C until used. The mean fluorescein:protein molar ratio of the conjugates was measured as described by Brandtzaeg (14) and was about 4.0 in the various preparations. The protein concentration of the solutions was determined spectrophotometrically (14) and was found to be about 1%.

METHODS

Single radial immunodiffusion. The procedure of Mancini *et al.* (3) was followed. A diffusion period of 48 hr was used, since no further increase of the precipitate areas occurred after this time.

Attachment of antibodies to capillary-containing sheets. Microporous plastic sheets were washed for 3 hr in three changes of 0.1 M Tris–HCl buffer, pH 8.0, containing 8 mmol/liter of sodium azide and excess buffer was expressed from the material using filter paper. The porous material was then immediately incubated for 18 hr at 4°C in the above-mentioned Tris buffer containing 0.5 g/100 ml of goat immunoglobulins against human transferrin. After the incubation the sheets were washed by the procedure described above. The sheets were then dried between filter papers and left to equilibrate with the air humidity for a week in darkness at room temperature. Strips measuring 70 × 8 mm were then cut with a very sharp blade from the antibody-containing sheets and a mark was made at the height of 50 mm on each strip.

The immunocapillarymigration procedure. The samples to be analyzed were diluted with 0.1 M Tris–HCl buffer, pH 8.0, containing 8 mmol/liter of sodium azide and about 300 μl of the dilutions were poured into cylindrical cups with a diameter of about 9 mm and the cups were placed in a closed humidiated chamber. One end of an antibody-containing strip was then placed in each cup and the other end was allowed to lean against one of the walls of the chamber. By this procedure an equal portion of each strip was submerged in the sample solution. The solution started thereby to migrate up the strip by capillary force and when it had reached the mark on the strip, after about 45 min, the strip was removed from the cup and excess antigen solution was washed off the strip under running tap water for about 2 min. The antigen-covered area of each strip was then detected by incubating the strip for about 3 min in the above-mentioned Tris–HCl buffer containing fluorescein-labeled antibodies against the antigen of interest. Excess fluorescein-labeled antibodies were washed off the strip under running tap water as described above. The strip was then inspected in ultraviolet light (wave length, 254 nm) and the fluorescent area was marked by a pencil. The antigen concentration in each serum sample was then obtained by comparing the height of the fluorescent area produced by the serum sample with corresponding heights produced by serial dilutions of a standard serum.

RESULTS

If a serum sample is allowed to migrate by capillary force in a porous strip with adsorbed anti-transferrin antibodies and the strip is then washed and incubated for a short time in a solution of fluorescein-labeled antibodies against human transferrin the strip will show a bright fluorescent area (Fig. 1A). For strips containing the same amounts of adsorbed antibodies the height of this area was found to be proportional to the logarithm of the transferrin concentration in the sample (Figs. 1A and B).

Specificity. When rabbit or goat IgG from nonimmune sera was adsorbed to a porous strip and human serum was allowed to migrate in the strips no fluorescence could be demonstrated on the strips by means of fluorescein-labeled rabbit and goat antibodies against human transferrin, IgG, or albumin. Likewise no fluorescence could be detected when strips containing antibodies against human IgG were incubated in solutions of fluorescein-labeled anti-transferrin or anti-albumin antibodies, or when strips with adsorbed antibodies against transferrin were treated with fluorescein-labeled antibodies against IgG or albumin.

If human serum enriched with pure transferrin was allowed to migrate in a strip containing adsorbed antibodies against transferrin the fluorescent area produced by fluorescein-labeled antibodies against human transferrin increased in size compared to the area obtained when the native human serum was allowed to migrate in an identical strip. The recovery of the added transferrin was approximately 100%. Enrichment of the same human serum with human albumin or IgG did not result in any change in the size of the fluorescent area.

Sensitivity. When the antibody activity of the immunoglobulin fraction adsorbed to the strip was decreased the fluorescent area produced by

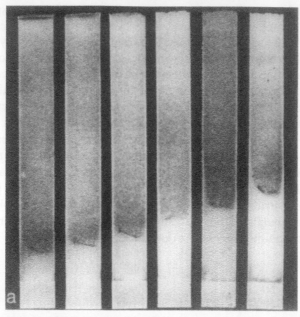

FIG. 1A. Strips with antibodies against human transferrin adsorbed. Serial dilutions of a standard serum were allowed to migrate in the strips and the transferrin-covered areas were determined by fluorescein-labeled antibodies against transferrin. The concentrations of transferrin in the different dilutions were 0.13, 0.16, 0.21, 0.31, 0.42, and 0.63 mg/ml.

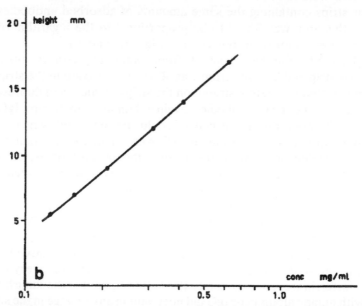

FIG. 1B. Standard curve relating the heights of the fluorescent areas of the strips in Fig. 1A and the antigen concentration of the standard serum dilutions. The heights of the strip portions submerged in the serum dilutions were subtracted from the total heights of the fluorescent areas.

184   GLAD AND GRUBB

migration of one and the same serum sample increased. But concomitantly with the increase in fluorescent area the intensity of the fluorescence decreased until it was impossible to recognize the boundaries of the fluorescent area. The dose–response curve was linear even for low antibody activities of the immunoglobulin fraction adsorbed to the strip when used for determination of low antigen concentrations. With the detection system described in this work the lowest transferrin and IgG concentrations that could be clearly differentiated from zero were about 40 μg/ml.

Quantitative comparison with single radial immunodiffusion. The transferrin concentration in 29 serum samples was determined by immunocapillarymigration and by single radial immunodiffusion. A correlation coefficient of 0.91 was obtained when the results of the methods were compared (Fig. 2). The standard curves in the procedures were calculated from dilutions of a human serum with known transferrin concentration.

Precision. As determined from duplicate determinations the standard deviation was found to be 11% of the mean for both high-range and low-range concentrations. When a sample series having various antigen concentrations was analyzed repeatedly on 5 successive days the precision was 14% for both high- and low-range concentrations.

Antiserum consumption. Calculated from the decrease in immunoglobulin concentration obtained on incubation of the porous material in the antibody solution as determined by absorption measurements at 280 nm, the immunoglobulin amount consumed in the preparation of one strip was found to be about 1 mg. The antibody concentration in the remaining solution was sufficient for preparation of additional strips.

Storage and stability. Newly prepared strips and strips stored dry in room temperature in a dark place for 1, 2, and 3 weeks and 1, 3, and 6 months gave results with a coefficient of variation of 14% when tested on six samples of various antigen concentrations stored at −24°C.

DISCUSSION

When an antigen solution is allowed to migrate by capillary force in a porous material to which antibodies are attached, the migration of the antigen solute will be specifically delayed in comparison to other molecules of the antigen solution and the relative delay will increase as the antigen concentration decreases. These phenomena form the basis for quantitation by immunocapillarymigration.

The specificity, sensitivity, and precision of the procedure will, aside from the properties of the antibodies, be determined by the properties of the porous material, the method used to attach antibodies to the material, and the system chosen to expose the area covered by antigen after migration of the antigen solution.

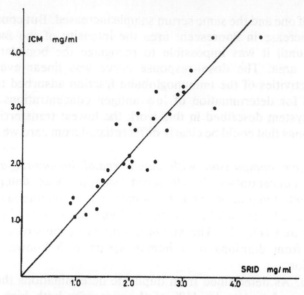

FIG. 2. Relation between the determinations of the transferrin concentration of 29 sera by single radial immunodiffusion (SRID) and by immunocapillarymigration (ICM). The equation $y = 1.08x$, where y is the estimation by ICM, gives the mathematical relation between the estimations as calculated by the method of least squares with zero point statistically weighted.

A porous material ideal for use in immunocapillarymigration should be easy to attach antibodies to and have a structure which promotes rapid and uniform capillarymigration of protein solutions. It should show no nonspecific binding of proteins and be tough enough to allow handling without special precautions. In this study we have used a material composed of polyvinylchloride and a filler, silica gel, which fulfills some of these criteria. Antibodies were attached to the material by simple adsorption and the treated material could be handled without special precautions, showed a rapid uniform capillarymigration of protein solutions and little nonspecific protein binding. Porous materials of cellulose and cellulose acetate were also tested for use in immunocapillarymigration but were found to have several drawbacks. Antibodies could not be adsorbed to these materials. Covalent linkage by comparatively sophisticated procedures involving reagents such as cyanogen bromide or glutardialdehyde was required. The treated materials also demonstrated considerable nonspecific protein binding and were very fragile.

Antibodies could easily be attached to the polyvinylchloride (PVC) material used in this work by a simple adsorption procedure. It is well known that hydrophobic materials like polyvinylchloride adsorb proteins from water solutions (15). The binding of antibodies to the PVC material seems to be comparatively strong since neither prolonged washing with neutral buffers containing 2 M urea nor extensive washing with acid or alkaline buffers could elute the antibodies. The antibodies were adsorbed

186   GLAD AND GRUBB

to the PVC material from a solution of immunoglobulins containing a high concentration of non-antibody protein molecules. The non-antibody protein serves the important function of saturating the protein binding structures of the PVC material so that the treated material will not show a nonspecific protein binding. Therefore when the amount of antibodies adsorbed to the PVC material is to be varied it is the antibody concentration, not the total concentration of protein, in the incubation medium that should be varied.

We performed some experiments to determine the relation between the amount of antibodies adsorbed to the PVC material and the size of the antigen-covered area produced by migration of an antigen solution in the material. There seemed to be an inverse proportionality between the adsorbed antibody amount and the size of the antigen-covered area, but further experiments are needed to establish this relation for a broad spectrum of antigen concentrations.

In this study a short incubation of the washed PVC material in a solution of fluorescein-labeled antibody was used to expose the antigen-covered area of the material. Several other ways of demonstrating this area are conceivable. The sensitivity of the quantitative procedure may perhaps be increased by the use of antibodies labeled with, e.g., radioactive isotopes or enzymes. If antibodies labeled with colored particles or cells are used, the inspection of the material can perhaps take place in visual light obviating the use of ultraviolet illumination. If trace amounts of antigen labeled in one of the above-mentioned ways are added to the antigen solutions to be analyzed, no incubation in an antibody solution is needed to visualize the antigen-covered area after the capillarymigration.

The antibody-containing material could be used for quantitative purposes without impairment after having been stored for 0.5 years at room temperature demonstrating the stability of antibodies when adsorbed to the porous PVC material. This is in agreement with the common observation that proteins coupled to insoluble-support media often are much more stable than proteins in solution.

Immunocapillarymigration performed as described in this work compares favorably with most other methods for immunochemical quantitation concerning rapidity, simplicity, and stability of reagents. It compares less favorably concerning sensitivity and precision at the present stage but readily allows quantitation of all major plasma proteins. It seems possible to develop immunocapillarymigration into a very rapid, simple, and sensitive quantitative method for general use.

ACKNOWLEDGMENTS

This work has been supported by grants from AB Kabi and AB Kabi Diagnostica, Stockholm, Sweden, and the Swedish Medical Research Council (Project Nos. B77-13X-581-13C and B78-13X-05196-01).

REFERENCES

1. Kabat, E. A. (1961) *in* Experimental Immunochemistry (Kabat, E. A., and Mayer, M. M., eds.), 2nd ed., pp. 22–96, and 361–383, Charles C Thomas, Springfield, Illinois.
2. Oudin, J. (1952) *in* Methods in Medical Research (Corcoran, A. C., ed.), Vol. 5, pp. 335–378, Year Book Publishers, Chicago.
3. Mancini, G., Carbonara, A. O., and Heremans, J. F. (1965) *Immunochemistry* **2**, 235–254.
4. Fahey, J. L., and McKelvey, E. M. (1965) *J. Immunol.* **94**, 84–90.
5. Preer, J. R. (1956) *J. Immunol.* **77**, 52–60.
6. Ouchterlony, Ö. (1967) *in* Handbook of Experimental Immunology (Weir, D. M., ed.), pp. 673–675, Blackwell Scientific Publishers, Oxford and Edinburgh.
7. Wieme, R. J., and Veys, E. M. (1970) *Clin. Chim. Acta* **27**, 77–86.
8. Farr, R. S. (1958) *J. Infect. Dis.* **103**, 239–262.
9. Laurell, C.-B. (1966) *Anal. Biochem.* **15**, 45–52.
10. Ritchie, R. F., Alper, C. A., Graves, J., Pearson, N., and Larson, C. (1973) *Amer. J. Clin. Pathol.* **59**, 151–159.
11. Ekins, R. P. (1960) *Clin. Chim. Acta* **5**, 453–459.
12. Yalow, R. S., and Berson, S. A. (1960) *J. Clin. Invest.* **39**, 1157–1175.
13. Holborow, E. J., and Johnsson, G. D. (1967) *in* Handbook of Experimental Immunology (Weir, D. M., ed.), pp. 580–581, Blackwell Scientific Publishers. Oxford and Edinburgh.
14. Brandtzaeg, P. (1973) *Scand. J. Immunol.* **2**, 273–290.
15. Herrmann, J., and Collins, M. (1976) *J. Immunol. Meth.* **10**, 363–366.

COMMENTARY TO
page 385

30. Towbin, H., Staehelin, T. and Gordon, J. (1979)
Electrophoretic Transfer of Proteins From Polyacrylamide Gels to Nitrocellulose Sheets: Procedure and Some Applications. Proceedings of the National Academy of Sciences of the United States of America 76(9): 4350–4354.

In 1975 E.M. Southern at the University of Edinburgh developed a method for the transfer of DNA from an agarose electrophoresis gel onto nitrocellulose membranes (1). Researchers soon began to call his transfer method Southern blotting (2). In 1977 Alwine, Kemp and Stark described an alternate technique for the transfer of RNA from agarose gels onto diazobenzyloxymethyl activated cellulose paper (3). This RNA transfer technique was called Northern blotting and an inside joke in the molecular biology community was begun. Towbin, Staehelin and Gordon in the 1979 paper presented here extended the transfer technique to the blotting of proteins from denaturing electrophoresis gels. Burnette in Seattle, Washington named the transfer of proteins, Western blotting and stated,

> With due respect to Southern, the established tradition of "geographic" naming of transfer techniques ("Southern," "Northern") is continued; the method described in this manuscript is referred to as "Western" blotting (4).

A year latter Reinhart and Malamud at the University of Pennsylvania described the transfer of proteins from native isoelectric focusing gels and wrote,

> While geographic and historical considerations suggested the use of Eastern blot for this technique, we have opted for the more descriptive name native blot (5).

Peferoen, Huybrechts and DeLoof in Leuven, Belgium modified Reinhart and Malamud's technique for native protein transfer and stated that, "Being Europeans, we find it hard to describe this method as Eastern blotting. Therefore we suggest to refer it as vacuum-blotting (6)."

The 1979 paper presented here by Towbin, Staehelin and Gordon for the transfer of proteins from electrophoresis gels, now called Western blotting, is number 7 among the top 50 papers in number of citations in the entire life sciences published in the last 40 years (7). This may be due to the fact that it is the most widely used of all the methods for protein transfer and not because it was the first report.

In 1946, Consden, Goron and Martin blotted amino acids and peptides onto filter paper sheets from silicate gels after electrophoresis (8). Estborn in 1959 blotted the isoenzymes of alkaline phosphatase (ALP EC 3.1.3.1) or acid phosphatase (ACP EC 3.1.3.2) onto cellulose filter paper after starch gel electrophoresis (9). Cellulose acetate membranes were used by Agostoni and co-workers in 1967 to blot proteins from Sephadex thin layer gels (10). Kohn in 1968 also blotted proteins from Sephadex thin layer chromatography gels onto cellulose acetate sheets and then ran immunoprecipitation reactions on the blotted proteins (11).

Renart and co-workers published a protein transfer method three months before the method of Towbin presented here (12). Their procedure required chemical modification of both the paper and the proteins in the gel before blotting. The transfer was through capillary action and took 24 hr. A similar method was published by Erlich and co-workers also in 1979 (13). Houvet and Clerc in 1979, blotted proteins onto filter paper after starch gel electrophoresis (14). Towbin's blot technique differed from these methods in that it used an electric field to drive proteins from the gel into unmodified nitrocellulose. The transfer was complete in less than 1 hr. Reinhart and Malmud (5) blotted non-denatured proteins onto unmodified nitrocellulose through capillary action without the use of an electric field. Peferoen and co-workers used vacuum to pull proteins from gels into unmodified nitrocellulose (6). Proteins blotted onto membranes were free of the gel matrix and easily probed with antibodies. An *in situ* sandwich enzyme linked immunosorbant assay (ELISA) could be performed on the transferred protein band. Western blotting followed by immunoassay of the transferred proteins found widespread use in clinical laboratory science especially in the areas of infectious diseases and autoimmunity (15). Western blotting soon became the confirmatory method for human immunodeficiency virus (HIV) antibody detection (16).

In 1982 a number of papers were published in which antigens or antibodies were spotted directly onto nitrocellulose membranes without the use of prior separation by electrophoresis. Immunoassay of the antigen or antibodies spotted on the membranes followed from the procedures developed in the original Towbin, Staehelin and Gordon 1979 paper. These assays in spot format on membranes came to be known as dot-immunobinding assays or dot-blots (17–19). Dot-blots offered all the technical features of conventional microwell ELISAs but with decreased reagent and sample volumes along with increased numbers of samples per assay. Spot applications of antigen or antibody were typically under 1 μL and the number of assay spots in a single 8×10 cm^2 membrane could be in the hundreds. Huang and co-workers, for example, described a sandwich immunoassay for 24 different cytokines in human serum (20). Up to 504 spots of capture antibody were formed in a 6×8 cm^2 area of membrane. The intensity of the dots in an entire membrane could be imaged and processed with charge coupled device (CCD) cameras using either fluorescence or chemiluminesence.

Rapid diagnostics were also developed in the dot-blot format in the 1980's. Hybritech Inc. produced the first of these devices with their ImmunoConcentration™ (ICON™) assay for human choriogonadotropin (HCG) (21). One of two monoclonal antibodies to HCG was bound as a dot on a white membrane. The second antibody with an enzyme label and the patient sample were poured onto the membrane. Absorbent wicks below the membrane pulled sample and conjugate through the membrane. After washing and color development, the sandwich ELISA spot was read with a reflectance meter.

References

(1) Southern, E.M. (1975) Detection of specific sequences among DNA fragments separated by gel electrophoresis. Journal of Molecular Biology. 98(3):503–517.

(2) Manning, R.F., Samois, D.R., and Gage, L.P. (1978) The genes for 18S, 5.8S and 28S ribosomal RNA of bombyx mori are organized into tandem repeats of uniform length. Gene. 4(2):153–166.

(3) Alwine, J.C., Kemp, D.J., and Stark, G.R. (1977) Method for detection of specific RNAs in agarose gels by transfer to diazobenzyloxymethyl-paper and hybridization with DNA probes. Proceedings of the National Academy of Sciences of the United States of America. 74(12):5350–5354.

(4) Burnette, W.N. (1981) "Western Blotting": electrophoretic transfer of proteins from sodium dodecyl sulfate-polyacrylamide gels to umodified nitrocellulose and radioiodinated protein A. Analytical Biochemistry. 112(2):195–203.

(5) Reinhart, M.P. and Malamud, D. (1982) Protein transfer from isoelectic focusing gels: the native blot. Analytical Biochemistry. 123(2):229–235.

(6) Peferoen, M., Huybrechts, R., and De Loof, A. (1982) Vacuum-blotting: a new simple and efficient transfer of proteins from sodium dodecyl sulfate-polyacrylamide gels to nitrocellulose. Federation of European Biochemical Societies (FEBS) Letters. 145(2):369–372.

(7) [web site] (2005) The most highly cited works in web of science® Thomson ISI, Science Citation Index. www.thomsonisi.com/wosbackfiles/mostcited.html [accessed January 5, 2005].

(8) Consden, R., Gordon, A.H., and Martin, J.P. (1946) Ionophoresis in silica jelly. a method for the separation of amino acids and peptides. Biochemical Journal. 40:33–41.

(9) Estborn, B. (1959) Visualization of acid and alkaline phosphatase after starch–gel electrophoresis of seminal plasma, serum and bile. Nature. 184(4699):1636–1637.

(10) Agostoni, A., Vergani, C., and Lomanto, B. (1967) Characterization of serum proteins by thin-layer gel filtration combined with immunodiffusion. Journal of Laboratory and Clinical Medicine. 69(3):522–529.

(11) Kohn, J. (1968) An immunochromatographic technique. Immunology. 15(6):863–865.

(12) Renart, J., Reiser, J., and Stark, G.R. (1979) Transfer of proteins from gels to diazobenzyloxymethyl-paper and detection with antisera: a method for studying antibody specificity and antigen structure. Proceedings of the National Academy of Sciences of the United States of America. 76(7):3116–3120.

(13) Erlich, H.A., Levinson, J.R., Cohen, S.N., and McDevitt, H.O. (1979) Filter affinity transfer. A new technique for the *in situ* identification of proteins in gels. Journal of Biological Chemistry. 254(23):2240–2247.

(14) Houvet, D. and Clerc, M. (1979) Reperage par empreinte des fractions electrophoretiques obtenues en gel d'amidon: application a l'alpha-1-antitrypsine. Medecine Tropicale: Revue Du Corps De Sante Colonial. 39(1):107–109.

(15) Towbin, H., Staehelin, T., and Gordon, J. (1989) Immunoblotting in the clinical laboratory. Journal of Clinical Chemistry and Clinical Biochemistry. 27(8):495–501.

(16) Carlson, J.R., Yee, J., Hinrichs, S.H., Bryant, M.L., Gardner, M.B., and Pedersen, N.C. (1987) Comparison of indirect immunofluorescence and western blot for detection of anti-human immunodeficiency virus antibodies. Journal of Clinical Microbiology. 25(3):494–497.

(17) Herbrink, P., Van Bussel, F.J., and Warnaar, S.O. (1982) The antigen spot test (AST): a highly sensitive assay for the detection of antibodies. Journal of Immunological Methods. 48(3):293–298.

(18) Hawkes, R., Niday, E., and Gordon, J. (1982) A dot-immunobinding assay for monoclonal and other antibodies. Analytical Biochemistry. 119(1):142–147.

(19) Huet, J., Sentenac, A., and Fromageot, P. (1982) Spot-immunodetection of conserved determinants in eukaryotic RNA polymerase. Journal of Biological Chemistry. 257(5):2613–2618.

(20) Huang, R-P., Huang, R., Fan, Y., and Lin, Y. (2001) Simultaneous detection of multiple cytokines from conditioned media and patient's sera by an antibody-based protein array system. Analytical Biochemistry. 294(1):55–62.

(21) Valkirs, G.E. and Barton, R. (1985) ImmunoConcentration™ — a new format for solid-phase immunoassays. Clinical Chemistry. 31(9):1427–1431.

Proceed. Nat. Acad. Sciences. 1979, 76: 4350–4354
Reprinted with permission from H. Towbin.

Proc. Natl. Acad. Sci. USA
Vol. 76, No. 9, pp. 4350–4354, September 1979
Biochemistry

Electrophoretic transfer of proteins from polyacrylamide gels to nitrocellulose sheets: Procedure and some applications

(ribosomal proteins/radioimmunoassay/fluorescent antibody assay/peroxidase-conjugated antibody/autoradiography)

HARRY TOWBIN*, THEOPHIL STAEHELIN†, AND JULIAN GORDON*‡

*Friedrich Miescher-Institut, P. O. Box 273, CH-4002 Basel, Switzerland; and †Pharmaceutical Research Department, Hoffman–La Roche, CH-4002 Basel, Switzerland

Communicated by V. Prelog, June 12, 1979

ABSTRACT   A method has been devised for the electrophoretic transfer of proteins from polyacrylamide gels to nitrocellulose sheets. The method results in quantitative transfer of ribosomal proteins from gels containing urea. For sodium dodecyl sulfate gels, the original band pattern was obtained with no loss of resolution, but the transfer was not quantitative. The method allows detection of proteins by autoradiography and is simpler than conventional procedures. The immobilized proteins were detectable by immunological procedures. All additional binding capacity on the nitrocellulose was blocked with excess protein; then a specific antibody was bound and, finally, a second antibody directed against the first antibody. The second antibody was either radioactively labeled or conjugated to fluorescein or to peroxidase. The specific protein was then detected by either autoradiography, under UV light, or by the peroxidase reaction product, respectively. In the latter case, as little as 100 pg of protein was clearly detectable. It is anticipated that the procedure will be applicable to analysis of a wide variety of proteins with specific reactions or ligands.

Polyacrylamide gel electrophoresis has become a standard tool in every laboratory in which proteins are analyzed and purified. Most frequently, the amount and location of the protein are of interest and staining is then sufficient. However, it may also be important to correlate an activity of a protein with a particular band on the gel. Enzymatic and binding activities can sometimes be detected *in situ* by letting substrates or ligands diffuse into the gel (1, 2). In immunoelectrophoresis, the antigen is allowed to diffuse (3) or electrophoretically move (4) against antibody. A precipitate is then formed where the antigen and antibody interact. Modifications have been described in which the antigen is precipitated by directly soaking the separation matrix in antiserum (5, 6). The range of gel electrophoretic separation systems is limited by the pore size of the gels and diffusion of the antibody. The systems are also dependent on concentration and type of antigen or antibody to give a physically immobile aggregate.

Analysis of cloned DNA has been revolutionized (7) by the ability to fractionate the DNA electrophoretically in polyacrylamide/agarose gels first and then to obtain a faithful replica of the original gel pattern by blotting the DNA onto a sheet of nitrocellulose on which it is immobilized. The immobilized DNA can then be analyzed by *in situ* hybridization. The power of immobilized two-dimensional arrays has been extended to the analysis of proteins by use of antibody-coated plastic sheets to pick up the corresponding antigen from colonies on agar plates (8). Sharon *et al.* (9) have used antigen-coated nitrocellulose sheets to pick up antibodies secreted by hybridoma clones growing in agar.

In this report we describe a procedure for the transfer of

proteins from a polyacrylamide gel to a sheet of nitrocellulose in such a way that a faithful replica of the original gel pattern is obtained. A wide variety of analytical procedures can be applied to the immobilized protein. Thus, the extreme versatility of nitrocellulose binding assays can be combined with high-resolution polyacrylamide gel electrophoresis. The procedure brings to the analysis of proteins the power that the Southern (7) technique has brought to the analysis of DNA.

MATERIALS AND METHODS

Immunogens and Immunization Procedures. *Escherichia coli* ribosomal proteins L7 and L12 were extracted (10) from 50S subunits and purified as described (11) by ion-exchange chromatography on carboxymethyl- and DEAE-cellulose. Antibodies were raised in a goat by injecting 250 μg of protein emulsified with complete Freund's adjuvant intracutaneously distributed over several sites. *Bacillus pertussis* vaccine (1.5 ml of Bordet–Gengou vaccine, Schweizerisches Serum- und Impfinstitut, Bern, Switzerland) was given subcutaneously with every antigen injection. Booster injections of the same formulation were given on days 38, 79, and 110. The animal was bled on day 117.

Subunits from chicken liver ribosomes (12) were combined in equimolar amounts, and 200-μg aliquots were emulsified with 125 μl of complete Freund's adjuvant injected at one intraperitoneal and four subcutaneous sites into BALB/c mice. Booster injections of 400 μg of ribosomes in saline were given intraperitoneally on days 33, 57, 58, and 59. The animals were bled on day 71.

Electrophoretic Blotting Procedures. Proteins were first subjected to electrophoresis in the presence of urea either in two dimensions (12) or in one-dimensional slab gels corresponding to the second dimension of the same two-dimensional system. The proteins were then transferred to nitrocellulose sheets as follows. The physical assembly used is shown diagrammatically in Fig. 1. A sheet of nitrocellulose (0.45 μm pore size in roll form, Millipore) was briefly wetted with water and laid on a scouring pad (Scotch-Brite) which was supported by a stiff plastic grid (disposable micropipette tray, Medical Laboratory Automation, Inc., New York). The gel to be blotted was put on the nitrocellulose sheet and care was taken to remove all air bubbles. A second pad and plastic grid were added and rubber bands were strung around all layers. The gel was thus firmly and evenly pressed against the nitrocellulose sheet. The assembly was put into an electrophoretic destaining chamber with the nitrocellulose sheet facing the cathode. The chamber contained 0.7% acetic acid. A voltage gradient of 6 V/cm was applied for 1 hr.

For polyacrylamide electrophoresis in the presence of sodium dodecyl sulfate (13) instead of urea, the procedure was as de-

‡ To whom reprint requests should be addressed.

386

Biochemistry: Towbin et al.

Proc. Natl. Acad. Sci. USA 76 (1979)   4351

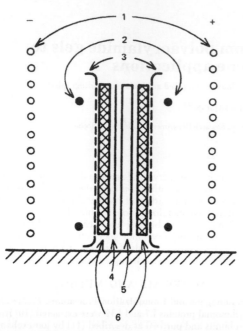

FIG. 1. Assembly for electrophoretic blotting procedure. 1, Electrodes of destainer; 2, elastic bands; 3, disposable pipette-tip tray; 4, nitrocellulose sheets; 5, polyacrylamide gel; 6, Scotch-Brite pads. Assembly parts are shown separated for visualization only.

scribed above except that the polarity of the electrodes was reversed and the electrode buffer was 25 mM Tris·192 mM glycine/20% (vol/vol) methanol at pH 8.3.

Staining for Protein. The blot may be stained with amido black (0.1% in 45% methanol/10% acetic acid) and destained with 90% methanol/2% acetic acid (see ref. 14).

Immunological Detection of Proteins on Nitrocellulose. The electrophoretic blots (usually not stained with amido black) were soaked in 3% bovine serum albumin in saline (0.9% NaCl/10 mM Tris·HCl, pH 7.4) for 1 hr at 40°C to saturate additional protein binding sites. They were rinsed in saline and incubated with antiserum appropriately diluted into 3% bovine serum albumin in saline also containing carrier serum with concentration and species as indicated in the legends. The sheets were washed in saline (about five changes during 30 min, total) and incubated with the second (indicator) antibody directed against the immunoglobulins of the first antiserum. As indicator antibodies we used ^{125}I-labeled sheep anti-mouse IgG. This had been purified with affinity chromatography on Sepharose-immobilized myeloma proteins and labeled by a modified version of the chloramine T method in 0.5 ml with 0.5 mg of IgG and 1 mCi of Na^{125}I (1 Ci = 3.7 × 10^{10} becquerels) for 60 sec at room temperature. The specific activity was approximately 1.5 μCi/μg of IgG. ^{125}I-Labeled IgG was diluted to 10^6 cpm/ml in saline containing 3% bovine serum albumin and 10% goat serum, and 3 ml of this solution was used for a nitrocellulose sheet of 100 cm^2. Incubation was in the presence of 0.01% NaN$_3$ for 6 hr at room temperature. The electrophoretic blots were washed in saline (five changes during 30 min, total) and thoroughly dried with a hair dryer. The blots were exposed to Kodak X-Omat R film for 6 days.

Fluorescein- and horseradish peroxidase-conjugated rabbit anti-goat IgG (Nordic Laboratories, Tilburg, Netherlands) were reconstituted before use according to the manufacturer's instructions. Fluorescein-conjugated antibodies were used at 1:50

dilution in saline containing 3% bovine serum albumin and 10% rabbit serum. After incubation for 30 min at room temperature, the blots were washed as above and inspected or photographed with a Polaroid camera under long-wave UV light through a yellow filter.

Horseradish peroxidase-conjugated IgG preparations were used at 1:2000 dilution in saline containing 3% bovine serum albumin and 10% rabbit serum. The blots were incubated for 2 hr at room temperature and washed as described above. For the color reaction (15), the blots were soaked in a solution of 25 μg of o-dianisidine per ml/0.01% H$_2$O$_2$/10 mM Tris·HCl, pH 7.4. This was prepared freshly from stock solutions of 1% o-dianisidine (Fluka) in methanol and 0.30% H$_2$O$_2$. The reaction was terminated after 20–30 min by washing with water. The blots were dried between filter paper. Drying considerably reduced the background staining. The blots were stored protected from light.

RESULTS

Electrophoretic Transfer of Ribosomal Proteins from Polyacrylamide Gels to Nitrocellulose Sheets. Most proteins or complexes containing protein adsorb readily to nitrocellulose filters (16), whereas salts, many small molecules, and RNA are usually not retained. These binding properties are widely used for binding assays with nitrocellulose filters. We found that proteins were retained on these filters equally well when carried towards the filter in an electric field. If the electric field was perpendicular to a slab gel containing separated proteins (see Fig. 1), we obtained a replica of the protein pattern on the nitrocellulose sheet. This is demonstrated with ribosomal proteins from E. coli; a conventionally stained gel (Fig. 2A) and a stained electrophoretic blot of an identical gel (Fig. 2B) are shown. All ribosomal proteins from chicken liver and E. coli ribosomes detectable on two-dimensional gels could be seen on the electrophoretic blots produced from them. An example of a blot from a two-dimensional gel is given in Fig. 3. When the original polyacrylamide gel was stained after blotting, no protein could be detected. Thus, the blotting procedure removed all protein from the gel.

To establish whether the proteins removed from the gels were quantitatively deposited on the nitrocellulose sheet, we separated ^{3}H-labeled proteins from chicken liver 60S ribosomal subunit by two-dimensional electrophoresis and compared the radioactivity that could be recovered from the blot with that recovered directly from the gel (Table 1). Single proteins or groups of poorly separated proteins were cut out and radioactivity was measured after combustion of the samples. The results were within the variability inherent to two-dimensional analyses. Variations could be accounted for by variable transfer of proteins into the second dimension gel and the acuity with which spots can be cut out.

At loads exceeding the capacity of nitrocellulose, losses of protein occurred. Titration with radioactive ribosomal proteins under blotting conditions showed that at concentrations below 0.15 μg/mm^2 all protein was adsorbed. Overloading became apparent when a second sheet of nitrocellulose directly underneath the first one took up protein or when protein became visible on the cathodal surface of amido black-stained blots.

The conservation of resolution together with the high recovery of ribosomal proteins simplifies the procedure for autoradiography. The common procedure involving drying of polyacrylamide gels under heat and reduced pressure (19), which is tedious and time consuming, may be eliminated. Because the proteins become concentrated on a very thin layer, autoradiography from ^{14}C- and ^{35}S-labeled proteins should be highly efficient even without 2,5-diphenyloxazole impregnation

387

4352   Biochemistry: Towbin *et al.*   *Proc. Natl. Acad. Sci. USA 76 (1979)*

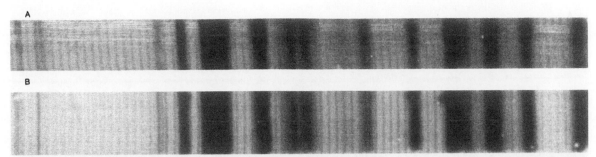

FIG. 2.   Electrophoretic blotting of ribosomal proteins from one-dimensional gels. Total ribosomal proteins from *E. coli* were separated on an 18% polyacrylamide slab gel containing 8 M urea. (*A*) A section of the gel was stained with Coomassie blue; (*B*) another section was electrophoretically blotted and the blot was stained as described in *Materials and Methods.* Electrophoresis was from left to right.

(19). We have successfully obtained such autoradiograms from gels of ^{35}S-labeled proteins (not shown). Further, preliminary experiments with tritiated proteins have shown that dried blots may be processed for fluorography by brief soaking in 10% diphenyloxazole in ether (20).

The above experiments were done with ribosomal proteins separated on polyacrylamide gels containing urea. We have electrophoretically blotted proteins from sodium dodecyl sulfate by the modified procedure also described in *Materials and Methods.* Again, there was no loss of resolution. However, differences of staining intensities between proteins on the gel

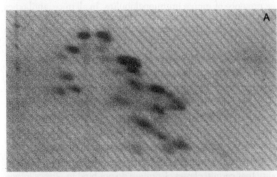

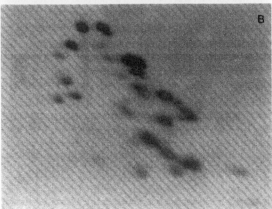

FIG. 3.   Electrophoretic blotting of ribosomal proteins from two-dimensional gels. Proteins (35 μg) extracted from the 60S ribosomal subunit of chicken liver (12) were separated by two-dimensional gel electrophoresis. (*A*) Gel stained with Coomassie blue; (*B*) blot of an identical gel. Electrophoresis: 1st dimension, from left to right (towards cathode); 2nd dimension, top to bottom.

and the blot were apparent. In spite of the apparently incomplete recovery, blots from polyacrylamide gels containing sodium dodecyl sulfate may be used for detection of antigen in the same way as described below for ribosomal proteins (unpublished experiments).

Detection of Antigen by Antibody Binding on Blots *In Situ.* We found that proteins transferred to nitrocellulose sheets remained there without being exchanged over several days. Because a blot could be saturated with bovine serum albumin to block the residual binding capacity of the sheet, it can be treated as a solid-phase immunoassay. In the following immunological applications, we used indirect techniques throughout. Thus, antibody bound by the immobilized antigen was detected by a second, labeled antibody directed against the first antibody, and in each case excess unbound antibody was washed out.

Table 1.   Efficiency of transfer of ribosomal proteins to nitrocellulose sheets

Protein or group of proteins analyzed	Recovery on blot, %
3	123
4, 4A	104
5	111
6	107
7, 8	86
9	80
10	112
11	79
12, 16	93
13	95
15, 15A, 18	125
17	115
19	139
21, 23	118
26	114
27	143
28, 29	69
31	117
33	131

Ribosomal large-subunit proteins from chicken liver were tritiated by reductive methylation (17) and separated by two-dimensional electrophoresis (12) in the presence of 35 μg of carrier protein. Two identical gels were run. One was stained; the other was electrophoretically blotted on a nitrocellulose sheet. Spots were identified according to our nomenclature for chicken ribosomes (12), which differs only in minor respects from that established for rat ribosomes (18). Corresponding spots or groups of spots were cut from the gel and the blot. Their radioactivity was determined after conversion to tritiated water in a sample oxidizer (Oxymat).

388

Biochemistry: Towbin *et al.*

Proc. Natl. Acad. Sci. USA 76 (1979)   4353

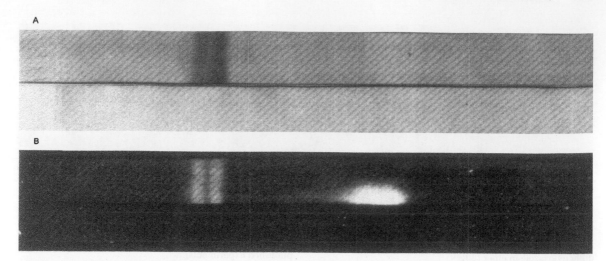

FIG. 4. Detection of *E. coli* ribosomal proteins L7 and L12 by (*A*) horseradish peroxidase- and (*B*) fluorescein-conjugated antibodies. Total ribosomal proteins from *E. coli* were separated and blotted as in Fig. 2. The anti-L7/L12 serum had a titer of 340 pmol of 70S ribosomes per ml of serum as determined by turbidity formation (20). Incubation was for 2 hr at room temperature in goat antiserum diluted 1:10 in saline containing 3% bovine serum albumin and 10% rabbit carrier serum and then with conjugated anti-goat IgG. In each case the lower strip is a control with preimmune antiserum. Electrophoresis was from left to right.

In Fig. 4 the detection of *E. coli* ribosomal proteins L7 and L12 with a goat serum specific for proteins L7 and L12 is shown. L7 is identical to L12, except for its N-acetylated NH_2-terminal amino acid (21). L7 and L12 fully crossreact immunologically (22) and are separated on acidic polyacrylamide gels (21). Both peroxidase- (Fig. 4*A*) and fluorescein-conjugated (Fig. 4*B*) antibodies were able to reveal immunoglobulin that was specifically retained by proteins L7 and L12. In each case, the lower gel is a control with preimmune serum. Peroxidase-conjugated antibodies were far more sensitive than fluorescein-conjugated ones. They could therefore be used at much higher dilution. This also permitted the detection of very small amounts of antigen. With a rabbit serum (23) we could detect 100 pg of L7 and L12 with serum and incubation conditions similar to those of the experiment described in Fig. 4 (not shown).

Because we can use the procedure to detect a specific antibody reacting with a specific protein after electrophoresis in polacrylamide, we should also be able to determine which proteins have elicited antibodies in a complex mixture of immunogens. In the experiment of Fig. 5, individual sera of five

mice immunized with chicken liver ribosomes were tested. We used ^{125}I-labeled sheep anti-mouse immunoglobulins to detect the presence of mouse immunoglobulins. In all mice, antibodies were preferentially produced against slowly moving proteins, presumably of high molecular weight. The procedure can thus characterize the antigen population against which specific antibodies have been raised in a mixture of immunogens.

DISCUSSION

The electrophoretic blotting technique described here produces replicas of proteins separated on polyacrylamide gels with high fidelity. We obtained quantitative transfer with proteins from gels containing urea. This was established here with ribosomal proteins. More generally, nitrocellulose membranes have been used to retain proteins from dilute solutions for their subsequent quantitative determination (16). Still, there remains the possibility that certain classes of protein do not bind to nitrocellulose. In this case absorbent sheets other than nitrocellulose or different blotting conditions may be helpful.

We have demonstrated that proteins immobilized on nitrocellulose sheets can be used to detect their respective antibodies.

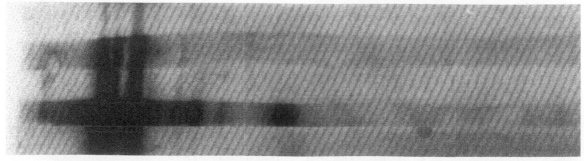

FIG. 5. Detection of immunoglobulin from individual mice directed against ribosomal proteins from chicken liver. Total protein from chicken liver ribosomes (12) was electrophoretically separated and blotted as in Fig. 2. Sera were obtained from five individual mice immunized against combined 40S and 60S subunits. The antisera were diluted 1:50 in saline containing 3% bovine serum albumin and 10% goat carrier serum. The blots were incubated in 250 μl of the diluted antiserum for 6 hr at room temperature. The blots were combined and treated with ^{125}I-labeled sheep anti-mouse IgG and autoradiographed. Electrophoresis was from left to right.

Proc. Natl. Acad. Sci. USA 76 (1979)

With radioactively labeled or peroxidase-conjugated antibodies the method is sensitive enough to detect small amounts of electrophoretically separated antigen, and this simple procedure can also be used to show the presence of small amounts of antibody in a serum of low titer. Because the antigen is immobilized on a sheet, the antibody is not required to form a precipitate with the antigen. The blotting technique therefore has the potential for immunoelectrophoretic analysis of proteins by using binding of Fab fragments or binding of antibodies against a single determinant, such as monoclonal antibodies produced by hybridomas (24). This could not be done by current immunoelectrophoretic techniques. If hybridoma clones are obtained from a mouse immunized with impure immunogen, it will be possible to use the technique to screen for clones making antibody directed against a desired antigen. Provided the desired antigen has a characteristic mobility in polyacrylamide gel electrophoresis, the appropriate clone can be selected without ever having pure antigen.

The procedure described here also has potential as a tool for screening pathological sera containing auto-antibodies—e.g., those against ribosomes (25–27). The precise identification of the immunogenic components may be a useful diagnostic tool for various pathological conditions.

A further advantage of immobilization of proteins on nitrocellulose is the ease of processing for autoradiography. Conventional staining, destaining, and drying of polyacrylamide gels takes many hours, and the exact drying conditions are extremely critical, especially for 18% gels as used in the second dimension for ribosomal proteins (12). When the proteins are transferred to a nitrocellulose support, as described here, the electrophoretic blotting takes 1 hr, staining and destaining less than 10 min, and drying an additional 5 min. This is thus both faster and simpler than conventional procedures, and it eliminates the tedious and hazardous procedure of soaking the gels in diphenyloxazole (19).

The technique has been developed to detect specific antisera against ribosomal proteins. However, it is applicable to any analytical procedure depending on formation of a protein–ligand complex. With the blotting technique, the usual procedure of forming a complex in solution and retaining it on a membrane would have to be reversed: the protein, already adsorbed to the membrane, would have to retain the ligand from a solution into which the membrane is immersed. Interactions that can possibly be analyzed in this way include hormone–receptor, cyclic AMP–receptor, and protein–nucleic acid interactions. The ligand may also be a protein. Enzymes separated on polyacrylamide gels could also be conveniently localized on blots by *in situ* assays. A critical requirement for these applications is that the protein is not damaged by the adsorption process and that binding sites remain accessible to ligands and substrates. In this respect, considerations similar to those in affinity chromatography and insoluble enzyme techniques pertain.

The method could also be adapted to the procedure of Cleveland *et al.* (28) for the analysis of proteins eluted from bands in polyacrylamide gels by one-dimensional fingerprints: one could label by iodination *in situ* on the nitrocellulose and then carry out the proteolytic digestion.

In preliminary experiments we have attempted to identify ribosomal RNA binding proteins by binding RNA to ribosomal proteins immobilized on nitrocellulose by the procedure of this paper, followed by staining for RNA (unpublished data), and have found a tendency for nonspecific binding. However, J. Steinberg, H. Weintraub, and U. K. Laemmli (personal communication) have independently developed a similar procedure for identifying DNA binding proteins.

We thank Drs. J. Schmidt and F. Dietrich for advice and help with immunization procedures and Mrs. M. Towbin for advice on setting up the peroxidase assay.

1. Gordon, A. H. (1971) in *Laboratory Techniques in Biochemistry and Molecular Biology*, eds. Work, T. S. & Work, E. (North-Holland, Amsterdam), p. 62.
2. Williamson, A. R. (1971) *Eur. J. Immunol.* 1, 390–394.
3. Grabar, P. & Williams, C. A. (1955) *Biochim. Biophys. Acta* 17, 67–74.
4. Laurell, C.-B. (1965) *Anal. Biochem.* 10, 358–361.
5. Zubke, W., Stadler, H., Ehrlich, R., Stöffler, G., Wittmann, H. G. & Apirion, D. (1977) *Mol. Gen. Genet.* 158, 129–139.
6. Showe, M. K., Isobe, E. & Onorato, L. (1970) *J. Mol. Biol.* 107, 55–69.
7. Southern, E. M. (1975) *J. Mol. Biol.* 98, 503–517.
8. Broome, S. & Gilbert, W. (1978) *Proc. Natl. Acad. Sci. USA* 75, 2746–2749.
9. Sharon, J., Morrison, S. L. & Kabat, E. A. (1979) *Proc. Natl. Acad. Sci. USA* 76, 1420–1424.
10. Hamel, E., Koka, M. & Nakamoto, T. (1972) *J. Biol. Chem.* 247, 805–814.
11. Möller, W., Groene, A., Terhorst, C. & Amons, R. (1972) *Eur. J. Biochem.* 25, 5–12.
12. Ramjoué, H.-P. R. & Gordon, J. (1977) *J. Biol. Chem.* 252, 9065–9070.
13. Laemmli, U. K. (1970) *Nature (London)* 227, 680–685.
14. Schaffner, W. & Weissmann, C. (1973) *Anal. Biochem.* 56, 502–514.
15. Avrameas, S. & Guilbert, B. (1971 *Eur. J. Immunol.* 1, 394–396.
16. Kuno, H. & Kihara, H. K. (1967) *Nature (London)* 215, 974–975.
17. Moore, G. & Crichton, R. R. (1974) *Biochem. J.* 143, 604–612.
18. McConkey, E. H., Bielka, H., Gordon, J., Lastick, S. M., Lin, A., Ogata, K., Reboud, J.-P., Traugh, J. A., Traut, R. R., Warner, J. R., Welfle, H. & Wool, I. G. (1979) *Mol. Gen. Genet.* 169, 1–6.
19. Bonner, W. M. & Laskey, R. L. (1974) *Eur. J. Biochem.* 46, 83–88.
20. Randerath, K. (1970) *Anal. Biochem.* 34, 188–205.
21. Terhorst, C., Wittmann-Liebold, B. & Möller, W. (1972) *Eur. J. Biochem.* 25, 13–19.
22. Stöffler, G. & Wittmann, H. G. (1971) *J. Mol. Biol.* 62, 407–409.
23. Howard, G., Smith, R. L. & Gordon, J. (1976) *J. Mol. Biol.* 106, 623–637.
24. Köhler, G. & Milstein, C. (1976) *Eur. J. Immunol.* 6, 511–519.
25. Schur, P. H., Moroz, L. A. & Kunkel, H. G. (1967) *Immunochemistry* 4, 447–453.
26. Miyachi, K. & Tan, E. M. (1979) *Arthritis Rheum.* 22, 87–93.
27. Gerber, M. A., Shapiro, J. M., Smith, H., Jr., Lebewohl, O. & Schaffner, F. (1979) *Gastroenterology* 76, 139–143.
28. Cleveland, D. W., Fischer, S. G., Kirschner, M. W. & Laemmli, U. K. (1977) *J. Biol. Chem.* 252, 1102–1106.

COMMENTARY TO

31. Jorgenson, J. W. and Lukacs, K. DeA. (1981)
Zone Electrophoresis in Open-Tubular Glass Capillaries. Analytical Chemistry 53(8): 1298–1302.

T he paper presented here by Jorgenson and Lukacs is the first full description of electrophoresis in a buffer filled glass capillary with an inside diameter (id) of less than 100 μm. Hjerten in 1970 used 3 mm id quartz capillaries (1) and Mikkers and co-workers in 1979 described electrophoretic separations in a Teflon™ capillary with an id of 200 μm (2). Jorgenson and Lukacs in this paper established the operational parameters for separations in small diameter open-tube capillaries. This form of electrophoresis came to be called capillary electrophoresis (CE). They demonstrated rapid separations of fluorescent-labeled amino acids and dipeptides in 75-μm id × 100 cm long glass capillaries. A similar report by these authors appeared in the journal *Clinical Chemistry* also in 1981 (3). Two years latter, Jorgenson and Lukacs separated human serum proteins by CE (4). The inner surface of the 75-μm id × 50 cm long fused silica capillary was coated with glycol-containing groups to reduce protein sticking. Separated proteins were recorded directly with an optical detector set at 230 nm and focused on a segment of the capillary that had been striped of its outer opaque coating.

Less than 17 years after this publication by Jorgenson and Lukacs over 6 000 articles and 30 books had been published on CE (10). Applications in clinical chemistry have been thoroughly reviewed in numerous publications (5–11). Chen and co-workers at Beckman Instruments in 1991 described 8-minute serum protein separations and 10-minute hemoglobin variant analysis in 75-μm id × 25 cm fused silica capillaries (12). An assay for hemoglobin A_{1c} was described that took less than 4 minutes and the results were compared to high performance liquid chromatography (HPLC) (13). The first commercial CE systems appeared in the late 1980s from Microphoretics, Applied Biosystems and Beckman Instruments. In 2000, workers at Molecular Dynamics reviewed the literature on a wide range of applications for CE in microfabricated devices or chip platforms (14). Up to 12 or more capillary channels less than 20 μm in depth in a chip 102 × 102 mm was described along with the potential clinical applications.

References

(1) Hjerten, S. (1970) Free zone electrophoresis. Theory, equipment, and applications. Methods of Biochemical Analysis. 18: 55–79.

(2) Mikkers, F.E.P., Everaerts, F.M., and Verheggen, T.P.E.M. (1979) High performance zone electrophoresis. Journal of Chromatography. 169(1):11–20.

(3) Jorgenson, J.W. and Kukacs, K.DeA. (1981) Free-zone electrophoresis in glass capillaries. Clinical Chemistry. 27(9): 1551–1553.

(4) Jorgenson, J.W. and Lukacs, K.DeA. (1983) Capillary zone electrophoresis. Science. 222(4621):266–272.

(5) Landers, J.P. (1995) Clinical capillary electrophoresis. Clinical Chemistry. 41(4):495–509.

(6) Lehmann, R., Liebich, H.M., and Voelter, W. (1996) Application of capillary electrophoresis in clinical chemistry — developments from preliminary trials to routine analysis. Journal of Capillary Electrophoresis. 3(2):89–110.

(7) Jenkins, M.A. and Guerin, M.D. (1996) Capillary electrophoresis as a clinical tool. Journal of Chromatography B. 682(1): 23–34.

(8) Lehmann, R., Voelter, W., and Liebich, W. (1997) Capillary electrophoresis in clinical chemistry. Journal of Chromatography B. 697(1–2):3–35.

(9) Jenkins, M.A. and Guerin, M.D. (1997) Capillary electrophoresis procedures for serum protein analysis: comparison with established techniques. Journal of Chromatography B. 699(1–2):257–268.

(10) Perrett, D. (1999) Capillary electrophoresis in clinical chemistry. Annals of Clinical Biochemistry. 36(Part 2):133–150.

(11) Shihabi, Z.K. (2002) Separation science in routine clinical analysis, in *A Century of Separation Science*. Issaq, H.J. (ed), Marcel Dekker, New York, pgs 601–610.

(12) Chen, F-T.A., Liu, C-M., Hsieh, Y-Z., and Sternberg, J.C. (1991) Capillary electrophoresis — a new clinical tool. Clinical Chemistry. 37(1):14–19.

(13) Doelman, C.J.A., Siebelder, C.W.M., Nijhof, W.A., Weykamp, C.W., Janssens, J., and Penders, T.J. (1997) Capillary electrophoresis system for hemoglobin A_{1c} determinations evaluated. Clinical Chemistry. 43(4):644–648.

(14) Dolnik, V., Liu, S., and Jovanovich, S. (2000) Capillary electrophoresis on microchip. Electrophoresis. 21(1):41–54.

1298   *Anal. Chem.* **1981,** *53,* 1298–1302

Zone Electrophoresis in Open-Tubular Glass Capillaries

James W. Jorgenson* and Krynn DeArman Lukacs

Department of Chemistry, University of North Carolina, Chapel Hill, North Carolina 27514

A system for performing zone electrophoresis in open-tubular glass capillaries of 75 μm inside diameter and with applied voltages up to 30 kV is described. The small inside diameter of these capillaries allows efficient dissipation of the heat generated by the application of such high voltages. However, the small inside diameter also necessitates the use of a sensitive on-column fluorescence detector to record the separation of solute zones. With this system, separation efficiency is proportional to the applied voltage, with efficiencies in excess of 400 000 theoretical plates demonstrated. Strong electroosmotic flow in the capillary allows both positive and negative ions of a variety of sizes to be analyzed in a single run with relatively short analysis times. High-efficiency separations of fluorescent derivatives of amino acids, dipeptides, and amines as well as separation of a human urine sample were obtained with analysis times of 10–30 min.

Several important causes of zone broadening may be identified when considering separation efficiency in zone electrophoresis. Molecular diffusion will certainly cause zone broadening, although its effects are generally negligible. More serious difficulties often arise from convection currents in the electrophoretic medium. These are usually minimized through the use of gels, paper, or other stabilizers. However, this approach may introduce additional zone-broadening problems such as adsorptive interactions between the solutes and stabilizer and "eddy migration" in the channels created by some stabilizers (1). Mikkers, Everaerts, and Verheggen (2) sought to solve these convection problems through the use of the "wall

effect" by performing zone electrophoresis in narrow-bore Teflon tubes. This approach appeared to solve the problem of convection in a simple way, avoiding the difficulties associated with stabilizers. They found that the concentration of sample ions must be kept well below the concentration of carrier electrolyte in order to achieve symmetric peak shapes. When the sample concentration is too high, the sample alters the conductivity of the medium in its own vicinity, resulting in a distorted electric field gradient and an asymmetric peak shape. If zone electrophoresis is performed in narrow-bore tubes using low concentrations of sample relative to carrier electrolyte, conditions arise where molecular diffusion, originally negligible, may become the predominant cause of zone broadening. The difficulty with this approach is in finding any suitable detection system capable of detecting minute quantities of solutes in small capillary tubes. In this study, zone electrophoresis was attempted in glass capillary tubes. Detection of solute zones was accomplished with an "on-column" fluorescence detector which detects fluorescent solutes while they are still in the glass capillary tube.

THEORY

Consider an electrophoresis system consisting of a tube filled with a buffering medium across which a voltage is applied. Charged species introduced at one end of the tube migrate under the influence of the electric field to the far end of the tube. If a suitable detection device is placed at the far end of the tube, the passage of each solute zone may be recorded, yielding an electropherogram.

The migration velocity of a particular species is given by

$$\nu = \mu E = \mu V / L \qquad (1)$$

0003-2700/81/0353-1298$01.25/0   © 1981 American Chemical Society

where ν is the velocity, μ the electrophoretic mobility, E the electric field gradient, V the total applied voltage, and L the length of the tube. The time, t, required for a zone to migrate the entire length of the tube is

$$t = L/\nu = L^2/\mu V \qquad (2)$$

If molecular diffusion alone is responsible for zone broadening, the spatial variance, σ_L^2, of the zone after a time, t, is given by the Einstein equation

$$\sigma_L^2 = 2Dt \qquad (3)$$

where D is the molecular diffusion coefficient of the solute in the zone. Substituting the expression for time from eq 2 into this expression yields

$$\sigma_L^2 = 2DL^2/\mu V \qquad (4)$$

The concept of separation efficiency expressed in terms of theoretical plates may be borrowed from chromatography as suggested by Giddings (3). The number of theoretical plates, N, is defined as

$$N = L^2/\sigma_L^2 \qquad (5)$$

Substituting eq 4 into this expression results in

$$N = \mu V/2D \qquad (6)$$

One may note several interesting aspects of this simple result. N is directly proportional to the applied voltage, which suggests the use of the highest voltages possible for high separation efficiency. Somewhat surprisingly, N is independent of tube length and analysis time. Finally, N is proportional to the ratio of the mobility to the diffusion coefficient, factors more or less intrinsic to the solute species and not easily manipulated to improve efficiency. Thus, the most direct route to improved separation efficiency seems to be increasing the voltage applied to the separation medium. Since eq 2 predicts that the analysis time is proportional to the square of the tube length and inversely proportional to the applied voltage, it appears that high voltages applied to short tubes would generate the greatest number of theoretical plates in the shortest length of time.

The principal difficulty with this approach lies in the limited ability to dissipate heat generated in the electrophoretic process. Heat is generated uniformly throughout the medium but is only removed at the inner surface and ends of the tube. Once thermal equilibrium is established, there will be a parabolic temperature gradient across the tube (1, 4, 5). Under extreme circumstances the temperature in the center of the tube will become high enough for the solvent to boil, leading to total breakdown of the electrophoretic process. However, before this effect is observed the undesirable consequences of a radial temperature gradient will be felt in the form of zone broadening. Electrophoretic mobility will increase as the temperature of the medium is increased, at a rate of approximately 2% °C⁻¹ (1). Solutes in the warmer center of the tube will migrate faster while those at the wall will migrate more slowly, resulting in zone broadening. The most effective way to minimize this effect is to reduce the tube radius. This approach should have two beneficial results. First, heat dissipation will be more efficient. According to Wieme (1), the temperature difference from the center to the wall of the tube is proportional to the square of the radius, so reducing the radius should reduce temperature differences markedly. There is a second beneficial effect to be expected from a reduction in tube radius. A temperature gradient is only undesirable to the extent that a solute molecule spends a larger than average fraction of its time in a particular portion of the radius of the tube. The radial position of individual solute molecules is constantly changed by diffusion. In a tube of reduced radius a solute molecule will diffuse back and forth

across the tube radius more often and thus be less likely to spend an abnormally large fraction of time in any one particular portion of the radius. By effectively randomizing or averaging the solute's radial occupancy of the tube the solute's migration velocity will also be averaged, and for a collection of solute molecules, deviations from the average will be small. Thus a reduction in tube radius not only should reduce radial temperature differences but should also diminish the impact of any temperature differences that remain. These two effects argue strongly for the use of small tube diameters.

EXPERIMENTAL SECTION

Apparatus. Straight lengths of glass tube (80–100 cm long; 75 μm i.d., 550 μm o.d.) were drawn from Corning Type 7740 Pyrex glass on a glass drawing machine (Shimadzu GDM-1B, Kyoto, Japan). These tube dimensions provided sufficient cooling efficiency to minimize difficulties associated with the aforementioned thermal effects. Larger diameter tubes inevitably led to problems with heat dissipation resulting in poor separation efficiency. The inside surface of these tubes was not modified in any way prior to filling with the electrophoresis buffer medium. A regulated high-voltage dc power supply (Megavolt Model RDC-30-10, Hackensack, NJ) delivering from 0 to +30 kV was used to drive the electrophoretic process. The operator was protected from accidental contact with the high voltage through an interlock system. The high voltage end of the system was enclosed in a Plexiglass box which automatically cut off the high voltage when opened. Detection was carried out by using a homemade "on-column" fluorescence detector. The detector used a high-pressure mercury arc lamp as the source of ultraviolet light, glass filters for isolation of excitation and emission wavelengths, and a photomultiplier tube detector.

Chemicals. Dansyl amino acids and fluorescamine were obtained from Sigma Chemical Co. (St. Louis, MO). Dipeptides were provided by C. Horvath of Yale University. Alkylamines were obtained from RFR Corp. (Hope, RI).

Procedure. Tubes were filled with 0.05 M pH 7 phosphate buffer. Filling was accomplished by dipping one end of the tube in the buffer solution and allowing capillary action to draw the buffer into the tube. The filling process could be accelerated by allowing the liquid to flow "downhill". In this way, complete filling of a tube required 15 min. After the tubes were filled, both ends of the tube were dipped in beakers containing the buffer medium. The end at which samples were introduced was connected via a graphite electrode to the positive high voltage supply. The detector end was connected via a graphite electrode to ground. Samples were applied to the tube by removing the beaker containing the buffer and replacing it with one containing the sample. High voltage was applied for a few seconds and then turned off. The beaker containing the buffer was replaced, the high voltage applied, and electrophoresis allowed to proceed.

RESULTS AND DISCUSSION

It is evident from the various electrophorograms shown that high separation efficiencies may be achieved within short analysis times by performing zone electrophoresis in glass capillaries. Figure 1 shows the separation of several amino acids as their fluorescent dansyl derivatives. The separation of basic, neutral, and acidic amino acids is completed in 25 min. Figure 2 shows the separation of several dipeptides as fluorescamine derivatives. Here, separation is based primarily on size rather than charge. Figure 3 shows the separation of normal propyl, butyl, and hexylamines as their fluorescamine derivatives; separation is based purely on size. The three amine derivatives are well resolved despite their very minor differences in overall size. Finally, Figure 4 is the separation of a human urine sample in which the compounds containing primary amine groups have been derivatized with fluorescamine. In this case, labeling was carried out by adding 10 mg of fluorescamine dissolved in 2 mL of tetrahydrofuran to 20 mL of diluted urine (diluted 10-fold with buffer). Electrophoresis reveals a complex mixture of amines of unknown identity and demonstrates the ability of this technique to deal

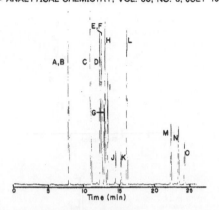

Figure 1. Zone electrophoretic separation of dansyl amino acids: A = unknown impurity, B = ε-labeled lysine, C = dilabeled lysine, D = asparagine, E = isoleucine, F = methionine, G = serine, H = alanine, I = glycine, J and K = unknown impurities, L = dilabeled cystine, M = glutamic acid, N = aspartic acid, O = cysteic acid. The concentration of each derivative is approximately 5 × 10⁻⁴ M, dissolved in operating buffer.

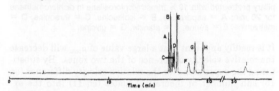

Figure 2. Zone electrophoretic separation of fluorescamine derivatives of dipeptides: A = phenylalanylleucine, B = phenylalanylvaline, C = valylleucine, D = glycyltyrosine, E = phenylalanylalanine, F = glycylproline, G = glycylalanine, H = glycylglycine, I = glycylaspartic acid. The concentration of each derivative is approximately 50 μg mL⁻¹, dissolved in operating buffer.

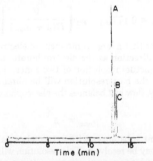

Figure 3. Zone electrophoretic separation of fluorescamine derivatives of amines: A = n-hexylamine, B = n-butylamine, C = n-propylamine.

with "real" samples. In all these separations the applied potential was 30 kV, the limit of the power supply. The current was approximately 0.1 mA.

It is important to note that at pH 7 all these substances bear a net negative charge and yet migrate toward the negative electrode. This apparent contradiction is the result of a strong electroosmotic flow occurring in the capillary. The magnitude of this flow is such that even small, triply charged anions are carried toward the negative electrode. Thus, the order of appearance of solutes in the electropherograms is cations, neutrals, and finally anions. The electroosmotic flow proves to be quite convenient for two reasons. First, without this flow only cations or anions could be analyzed in a single run, as sample ions of the "wrong" charge would not enter or migrate through the capillary. Second, without the flow, very large and/or weakly charged ions would require great lengths

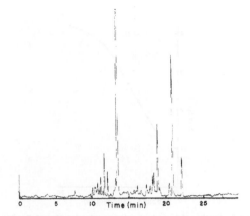

Figure 4. Zone electrophoretic separation of human urine derivatized with fluorescamine.

of time to travel the length of the tube. With the electroosmotic flow, ions of a variety of size and charge may be analyzed in a single run.

Electroosmotic flow will somewhat modify the equations describing the separation efficiency. Fortunately, the electroosmotic flow profile approximates a "plug" shape (6, 7) and thus the flow profile itself leads to minimal zone broadening. The velocity of electroosmotic flow may be given as

$$\nu_{osm} = \mu_{osm} E = \mu_{osm} \frac{V}{L} \tag{7}$$

where μ_{osm} is a coefficient similar to the electrophoretic mobility, relating the electroosmotic velocity to the electric field gradient. The net migration velocity of a substance is then given by

$$\nu = \mu \frac{V}{L} + \mu_{osm} \frac{V}{L} = (\mu + \mu_{osm}) \frac{V}{L} \tag{8}$$

The signs as well as the magnitudes of μ and μ_{osm} will be important, as the signs indicate the relative directions of the flow and the electrophoretic migration. The time it takes an ion to migrate the entire length of the tube is

$$t = \frac{L^2}{(\mu + \mu_{osm}) V} \tag{9}$$

and the resulting spatial variance is

$$\sigma_L^2 = \frac{2DL^2}{(\mu + \mu_{osm}) V} \tag{10}$$

The resulting separation efficiency is

$$N = \frac{(\mu + \mu_{osm}) V}{2D} \tag{11}$$

quite similar to the original expression for separation efficiency, eq 6. This new equation still predicts that the number of theoretical plates is proportional to the applied voltage. This prediction was tested by using the fluorescamine derivative of hexylamine as a solute. The number of theoretical plates was computed from peak profiles by using the formula

$$N = 5.54 \left(\frac{t}{w} \right)^2 \tag{12}$$

where w is the full peak width at the half-maximum points. In Figure 5 N is plotted vs. the applied voltage. The plot is essentially a straight line with two notable features. First, the data at low voltages extrapolate to a nonzero intercept, the meaning of which is not clear. The peaks themselves

396

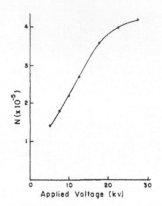

Figure 5. Number of theoretical plates as a functon of applied voltage.

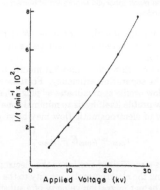

Figure 6. Relationship between analysis time and applied voltage.

appeared symmetric, suggesting that there was no serious overloading problem. However, overloading is a possible explanation of the nonzero intercept. Second, the plot shows a negative deviation from linearity at high voltages. This is interpreted as a result of the increased temperatures and temperature gradients that are expected at high voltages and currents.

In the presence of electroosmosis, eq 9 predicts an inverse relationship between applied voltage and analysis time. In Figure 6 reciprocal time is plotted vs. applied voltage giving the expected linear relationship. The positive deviation at high voltages is again indicative of higher temperatures, leading to increased mobilities and shorter analysis times.

Equation 11 suggests a misleading approach to improved separation efficiency. This is to promote very large values of μ_{osm}, electroosmotic flow, in the same direction as the electrophoretic mobility. Giddings (3) derived an expression for resolution in electrophoresis as

$$R_s = \frac{N^{1/2}}{4} \frac{\Delta \nu}{\bar{\nu}} \qquad (13)$$

where R_s is the resolution and $\Delta \nu / \bar{\nu}$ is the relative velocity difference of the two zones being separated. This ratio is equal to

$$\frac{\Delta \nu}{\bar{\nu}} = \frac{\mu_1 - \mu_2}{\bar{\mu}} \qquad (14)$$

where μ_1 and μ_2 are the mobilities of the two zones, and $\bar{\mu}$ is their average mobility. However, in the presence of electroosmosis this becomes

$$\frac{\Delta \nu}{\bar{\nu}} = \frac{\mu_1 - \mu_2}{\bar{\mu} + \mu_{osm}} \qquad (15)$$

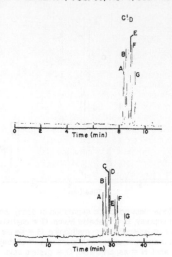

Figure 7. The effect of electroosmosis on resolution and analysis time of some dansyl amino acids: (upper) untreated capillary, (lower) capillary pretreated with 10% trimethylchlorosilane in dichloromethane for 20 min; A = asparagine, B = isoleucine, C = threonine, D = methionine, E = serine, F = alanine, G = glycine.

It is readily apparent that a large value of μ_{osm} will decrease the relative velocity difference of the two zones. By substituting the expressions for the relative velocity difference (eq 15) and number of theoretical plates (eq 11) into the expression for resolution (eq 13) we obtain

$$R_s = \frac{1}{4} \left[\frac{(\bar{\mu} + \mu_{osm})V}{2D} \right]^{1/2} \left[\frac{\mu_1 - \mu_2}{\bar{\mu} + \mu_{osm}} \right] \qquad (16)$$

and by rearranging

$$R_s = 0.177(\mu_1 - \mu_2) \left[\frac{V}{D(\bar{\mu} + \mu_{osm})} \right]^{1/2} \qquad (17)$$

Now it is clear that a large component of electroosmotic flow in the same direction as the electrophoretic migration will decrease the actual resolution of two zones. In fact, it may be seen that the best resolution will be obtained when the electroosmotic flow just balances the electrophoretic migration or

$$\mu_{osm} = -\bar{\mu} \qquad (18)$$

at which point substances with extremely small differences in mobility may be resolved. This resolution will be obtained, however, at a large expense in time, as may been seen by referring to eq 9 and imagining μ and μ_{osm} being nearly equal but opposite. In Figure 7 the effect of electroosmosis on resolution and analysis time is illustrated. The upper portion of the figure shows the separation of some neutral amino acids as dansyl derivatives. This separation was performed in a glass capillary as previously described. The lower portion of the figure shows the same separation, but this time in a capillary in which the inner surface was treated with trimethylchlorosilane in order to reduce electroosmotic flow. The improvement in resolution is obvious, especially in peaks C and D. However, this is at the expense of a large increase in analysis time.

The sample introduction technique described here was used because of the ease with which it could be carried out, as well as the fact that it introduces minimal zone broadening. This approach also eliminates any need for leak-free connections that would be required if introduction were accomplished with hydrostatic pressure. However, sample introduction with

1302   *Anal. Chem.* **1981**, *53*, 1302–1305

applied voltage will introduce substances based on their electrophoretic mobilities, which may complicate the matter of quantitative analysis. A detailed comparison of electric and hydrostatic sample introduction techniques with respect to zone broadening and quantitation is needed. If electric introduction proves viable with respect to the requirements of quantitation, the simplicity with which it could be automated would be an attractive advantage.

Extending the technique to a wide range of solutes will require development of alternative modes of detection. Conductometric, UV absorption, refractive index, and thermometric detectors (*2, 8*) have been described in conjunction with electrophoresis in larger bore tubes. Their application to detection of low concentrations of solutes in micron-sized tubes will be difficult, but any developments along these lines will be of great utility. Detectors of higher sensitivity will allow the use of even smaller diameter tubes, permitting the application of higher electric field gradients. This would open the way for the use of higher voltages and/or shorter tubes, with the result of even higher separation efficiencies and shorter analysis times.

Extension of the technique to separation of proteins and other macromolecules and particles is also of interest. Surfaces more inert and nonadsorptive than untreated glass will probably be necessary. Alternative detection modes will also be an advantage here. If these difficulties can be overcome, the possibilities for high-resolution separations of macromolecules will be quite promising.

The prospects for yet higher separation efficiencies hinge directly on the use of higher voltages in an apparently straightforward manner. Significant improvements in separation efficiency over the present work will require potentials in excess of 100 kV. These higher voltages may present certain practical difficulties in operation and safety, but these problems are probably resolvable. With higher voltages, the simple assumption that molecular diffusion is the dominant cause of zone broadening may also break down. Achievement of higher efficiencies will place stricter limitations on sample overloading, thermal gradients, adsorption, and the electroosmotic flow profile. Solving these difficulties in order to realize the benefits of higher voltages will require further refinement of the technique.

ACKNOWLEDGMENT

The authors gratefully acknowledge the kind gift of dipeptides from Csaba Horvath.

LITERATURE CITED

(1) Wieme, R. J. In "Chromatography: A Laboratory Handbook of Chromatographic and Electrophoretic Methods", 3rd ed., Heftmann, E., Ed.; Van Nostrand Reinhold: New York, 1975; Chapter 10.
(2) Mikkers, F. E. P.; Everaerts, F. M.; Verheggen, Th. P. E. M. *J. Chromatogr.* **1979**, *169*, 11–20.
(3) Giddings, J. C. *Sep. Sci.* **1969**, *4*, 181–189.
(4) Hinckley, J. O. N. *J. Chromatogr.* **1975**, *109*, 209–217.
(5) Brown, J. F.; Hinckley, J. O. N. *J. Chromatogr.* **1975**, *109*, 218–224.
(6) Pretorius, V.; Hopkins, B. J.; Schieke, J. D. *J. Chromatogr.* **1974**, *99*, 23–30.
(7) Rice, C. L.; Whitehead, R. *J. Phys. Chem.* **1965**, *69*, 4017–4024.
(8) Bier, M. In "An Introduction to Separation Science", Karger, B. L., Snyder, L. R., Horvath, C., Eds.; Wiley: New York, 1973; Chapter 17.

RECEIVED for review January 16, 1981. Accepted April 24, 1981. Support for this work was provided by the donors of Petroleum Research Fund, administered by the American Chemical Society, and the University Research Council of the University of North Carolina.

Chemometrics

32. Levey, S. and Jennings, E. R. (1950)
The Use of Control Charts in the Clinical Laboratory. American Journal of Clinical Pathology 20(11): 1059–1066.

Edward R. Tufte in his book *The Visual Display of Quantitative Information* wrote that, "Graphical excellence is that which gives to the viewer the greatest number of ideas in the shortest time with the least ink in the smallest space" (1). Walter A. Shewhart accomplished this when he designed the control chart in 1924. In 1931 he published a book on the use of his control charts in process quality control (2). Shewhart worked for Bell Laboratories in manufacturing quality control. His charts presented in a single graph the limits of acceptability and the performance over time of any selected observation made during the manufacturing process.

Stanley Levey and Elmer R. Jennings introduced the Shewhart or control chart into clinical laboratories for the first time in 1950. Their paper in the *American Journal of Clinical Pathology* is presented here. For many years, because of this paper, the Shewhart control chart was called the Levey–Jennings chart. Archibald in 1950 claimed that daily reference samples run with every batch of unknowns was a requirement in any laboratory quality control program (3). The Shewhart control chart was ideally suited to the graphical presentation of the results of these reference samples (pool serums) run every day in a clinical laboratory. Henry and Segalove in 1952 described the use of three different control charts for use in plotting daily quality control results one of which was the Shewhart chart (4). Freier and Rausch in 1958 utilized a pooled serum with every batch of unknowns. They established three standard deviations as a measure of precision and plotted the daily results on a Shewhart control chart (5). Richard J. Henry in 1959 credited the introduction of the control chart into clinical chemistry to Levey and Jennings and detailed the use of these charts in daily quality control monitoring (6).

Many modifications of the control chart were described over the years. Westgard and co-workers, for example introduced the decision limit cumulative sum quality-control chart in 1977 (7). In 1983 the Shewhart multi-rule chart was published as a selected method in clinical chemistry (8).

References

(1) Tufte, E.R. (1983) *The Visual Display of Quantitative Information.* Graphics Press, Cheshire, CT, pg 51, Third Printing.

(2) Shewhart, W.A. (1931) *Economic Control of Quality of Manufactured Product.* D. Van Nostrand, Princeton, Eight Printing.

(3) Archibald, R.M. (1950) Criteria of analytical methods for clinical chemistry. Analytical Chemistry. 22(5):639–642.

(4) Henry, R.J. and Segalove, M. (1952) The running of standards in clinical chemistry and the use of the control chart. Journal of Clinical Pathology. 5(4):305–311.

(5) Freier, E.F. and Rausch, V.L. (1958) Quality control in clinical chemistry. American Journal of Medical Technology. 24(4):195–207.

(6) Henry, R.J. (1959) Use of the control chart in clinical chemistry. Clinical Chemistry. 5(4):309–319.

(7) Westgard, J.O., Groth, T., Aronsson, T., and de Verdlier, C-H. (1977) Combined Shewhart–Cusum control chart for improved quality control in clinical chemistry. Clinical Chemistry. 23(10):1881–1887.

(8) Westgard, J.O., Barry, P.L., Hunt, M.R., and Groth, T. (1983) A multi-rule shewart chart for quality control in clinical chemistry, in *Selected Methods of Clinical Chemistry*, Vol 10. Cooper, G.R. (ed), American Association for Clinical Chemistry, Washington, pgs 29–37.

Am J Clin Pathol. 1950; 20: 1059–1066
© 1950 American Society of Clinical Pathologists.
Reprinted with permission.

TECHNICAL SECTION

THE USE OF CONTROL CHARTS IN THE CLINICAL LABORATORY*

STANLEY LEVEY, Ph.D., and E. R. JENNINGS, M.D.

From the Departments of Physiological Chemistry and Pathology, Wayne University College of Medicine, and Receiving Hospital, Detroit, Michigan

Constant supervision is necessary in order to obtain uniformly reliable values in a busy clinical laboratory. Normally the director of the laboratory does not have sufficient time to carry out a detailed supervision of the methods but, by the use of control charts, it is possible to determine at a glance whether the errors of analysis are beyond the statistical variation of the procedures employed.

Quality control by statistical methods is widely used in industry[7] to determine whether the variation observed among items produced by a single machine or an entire process is consistent with the hypothesis that a stable system of chance causes is operating. This hypothesis may be tested by taking a number of groups of observations at reasonable intervals of time, to see if the variations among the averages of each group are consistent with the variations observed within groups of observations made at the same time. The details of this technic are furnished by the theory of statistical quality control.[1]

In this laboratory the principle of the control chart provided a constant check on the reliability of the numerous determinations run each day. It made it possible to distinguish between what might be termed statistical fluctuations and actual error. In addition it offered a rational basis for action in correcting a defective procedure. While the chart will show when a method is out of control, *i.e.*, that the variation is greater than would be expected if chance alone were operating, it remains for the analyst to study the cause and prevention of the error. Since control-chart methods have not been widely used in the clinical laboratory, we undertook a study of their application in our laboratory.

METHOD

In setting up the type of control chart used here, it is necessary to have sufficient amounts of homogenous blood or plasma in which the concentration of the material to be analyzed is stable over a long period. Also, the concentration of the substance estimated should be approximately in the range of normal blood values, so that no special analytical steps would be required to check its level.

The pooled blood or plasma derived from this blood was obtained from lots discarded by the hospital blood bank. Whole blood was used to control the urea method, and plasma was used to study the total protein, albumin, globulin and chloride determinations. The blood or plasma was pooled and then distributed as 5-ml. samples in a large number of small test tubes. The latter were well corked, stored in a "deep-freeze" cabinet maintained at − 10 C. and kept frozen solid until ready for analysis. Sufficient samples were prepared at this

* Received for publication, May 15, 1950.

time to last approximately one year. On the morning of the analyses two samples each of plasma and whole blood were removed from the freezer and thawed at about 30 to 35 C.

For the control of the analyses of the carbon dioxide combining-power of plasma, the material obtained directly from the blood bank was not suitable since its carbon dioxide combining-power was approximately 12 volumes per 100 ml. plasma. This low value was partially due to the citric acid content of the preserving fluid. Sodium carbonate was added to the plasma in sufficient quantity to bring the carbon dioxide combining-power up to about 60 to 80 volumes per 100 ml. The plasma thus fortified was distributed in many small test tubes and stored in the frozen state.

Since the true value for the concentration of any of the control substances was not known, an estimate of the true value had to be obtained from the data. This estimate was obtained by averaging the individual values obtained from the first 20 pairs of samples analyzed. These values were obtained over a period of about a month so that any day-to-day factors influencing the analytical procedures were minimized. For the quality control study two samples for each determination studied were sent twice a week (usually Tuesday and Thursday) to the laboratory where they were numbered and treated as routine samples. It was most important in this work to make sure that the test samples were not given any preferential treatment. After the analyses were completed, the average value of the two samples and the value representing the difference in the two results (the range) were plotted on the control chart.

For the preparation of the control charts, the mean value of any one pair of determinations was plotted as the ordinate, with the date or order of analysis as the abscissa. In charting the variation between duplicate analyses, the range (the difference in values of duplicates) was plotted as the ordinate and the date or order of the analyses as the abscissa. The statistical limits for the mean and range were also placed on the control chart. Using two samples per determination, the limits for the mean and range are $\pm 1.88 \times \bar{R}$ and $3.27 \times \bar{R}$, respectively. The value of $\bar{R}$ was obtained by averaging the range values of the first 20 pairs of samples studied.[1] The control limits for the mean and range are approximately equivalent to 3 σ limits. This means that, if a stable system of chance causes is operating, only 3 out of 1000 observations should fall outside the 3 σ limits. If an observed average or range is beyond the limits given above, it is taken as evidence of lack of control since it has such a small probability of occurring under controlled conditions. Examples of the use of the control chart in the clinical laboratory are given below.

RESULTS

Urea Determination

The Karr Direct Nesslerization procedure was employed for the determination of urea in blood.[2] The control chart for this determination is given in Figure 1. It can be seen that on days 5, 6, 7 and 10 the values for the urea nitrogen were far out of control (exceeded the statistical limits assigned to the method).

Because of the large number of urea determinations (up to 80) run per day, two technicians usually worked as a team after the ammonia solutions were nesslerized, one technician matching the colors and the other recording the values on the reports sent to the wards. Since only one technician was working on the determinations during these days, it took considerably longer to complete a series of tests. Under these conditions the test samples which were placed near the end of the series developed turbidity at a more rapid rate than the standard. By reducing the size of a run or having two persons work on the determinations, this error was eliminated. On days 13, 19, 26 and 27 it was found that the average

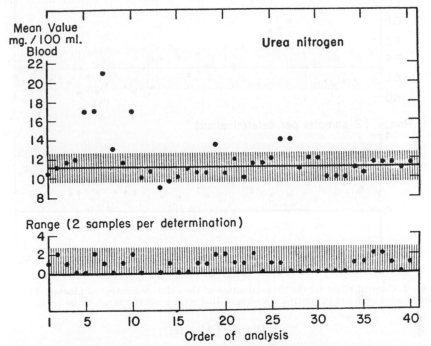

FIG. 1. Control chart for the urea nitrogen determination. The values in the shaded area are in control. The heavy line represents the mean value.

values for the urea determinations were again out of control. During the last two periods mentioned, an intensive study of the variables in this method was made and it was concluded that cooling the samples before nesslerization, a process used all summer, brought the values into control. Cooling of the samples before nesslerization had previously been done only during hot weather.

The allowable variations in the control of urea are relatively small from the clinical point of view, being only ± 1.6 mg. per 100 ml.* During the entire study

* The control limits used in this study are 3σ limits which were derived by calculations from the variations of the range values. In many of the charts, it is difficult to ascribe so many zero ranges to the operation of pure chance. The cause of this discrepancy appears to be the lack of sensitivity of the analytical methods for the mathematical procedures employed. There is probably considerable variation among samples listed as having a

1062   LEVEY AND JENNINGS

the range of the duplicate values was never beyond the limits of control. While the mean urea values might be far beyond the control limits (21 mg. per 100 ml.) the duplicates were only 1 mg. per 100 ml. apart. The range, therefore, was well in control. From this chart one notes that high values were obtained when the Karr urea method was out of control.

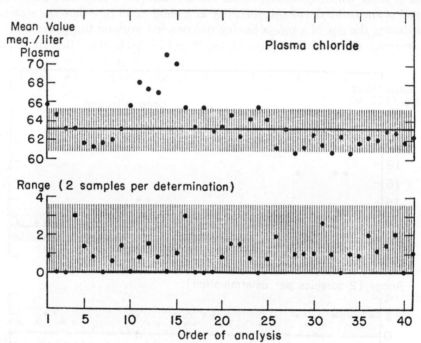

Fig. 2. Control chart for the determination of the chloride content of plasma. The values in the shaded area are in control. The heavy line represents the mean value.

Plasma Chlorides

The control chart concerned with the chloride concentration of the pooled plasma is shown in Figure 2. The method used for all the assays was the mercuric nitrate titration of Schales and Schales.[6] It should be noted that the chloride content of the test samples was markedly lower than the normal for blood. This is probably due to the dilution of the whole blood at the blood bank with a dextrose-citrate preserving solution. The determinations appeared to be in control up to day 9 (Fig. 2). On this day the technician who normally conducted the test left for her vacation. The new analyst, using the titration factor that the regular worker had determined earlier, reported values which showed the method to be out of control. On the fifteenth day the new technician determined

zero range value, but the method does not show it. This is similar to the variation which would occur in measurements of length of 0.01 mm., when the gauge employed was accurate to only 0.1 mm. Thus it should be recognized that the computed limits for the mean values are probably narrower than the true 3σ limits.

the titration factor herself and the values she reported immediately fell in line with the previously controlled data. The original technician returned on the

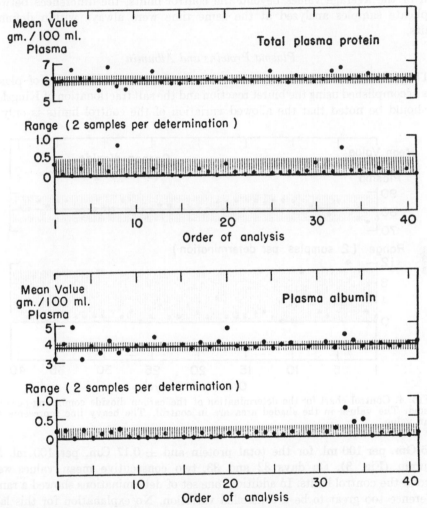

FIG. 3. Control charts for the determination of total plasma protein and albumin. The values in the shaded area are in control. The heavy line represents the mean value.

nineteenth day and, with the titration factor she had determined, all analyses were within the predicted limits until the twenty-eighth day, at which time it was suggested to the operator that the titration factor was low. When this was found to be true upon checking, a new factor was determined and used in all succeeding analyses.† The factor of the reagent is now rechecked weekly. Also,

† On examination of Figure 2 there appear to be two population groups present, one covering days 16 to 25 and the other consisting of all values beyond day 25. Using a mean of 64.2 mEq. per liter for the first group and σ of 1.11, one finds that the 3σ mean range

1064 LEVEY AND JENNINGS

because of individual variations in evaluating the end point of this titration, each analyst determines a titration factor herself. While there was some variation of the average values beyond the control limits, the differences between duplicate samples analyzed at the same time were always within definable limits.

Plasma Proteins and Albumin

The determination of the protein content of the pooled samples of plasma was accomplished using the biuret reaction and the salt fractionation of Kingsley.[3] It should be noted that the allowed variation of the control limits is only ±

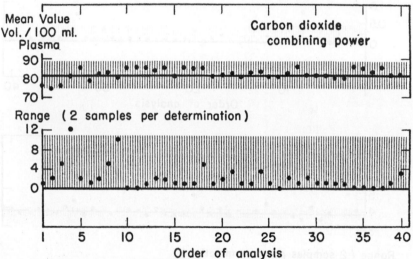

FIG. 4. Control chart for the determination of the carbon dioxide combining-power of plasma. The values in the shaded area are in control. The heavy line represents the mean value.

0.25 Gm. per 100 ml. for the total protein and ± 0.17 Gm. per 100 ml. for albumin (Fig. 3). On days 32 and 33, two consecutive mean values were beyond the control limits. In addition, one set of determinations showed a range·difference too great to be a controlled variation. No explanation for this lack of control could be found in spite of a rather thorough investigation of all the reagents and the use of new standards. On day 34 the analyses spontaneously came back into control and remained in control. The reason for this could not be ascertained. Since the determination of the protein content of serum or plasma is an analysis in which many errors can occur which are unknown to the analyst, quality control is a valuable tool for the over-all policing of this method.

derived from the data to be 60.9 to 67.5 mEq. per liter. The second group had a mean of 61.8 and a σ of 0.84; thus, the 3σ limits would be 59.3 to 64.3 mEq. per liter. The discrepancy between these mean limits and those derived from the range values given in Figure 2 is probably due to the relative insensitivity of the analytical procedure (too many zero range values).

The control chart for albumin showed scattered values which were out of control but there was no general trend that would indicate a constant error creeping into the determination. During the time the range values for the albumins were out of control, the total protein determination was fluctuating beyond the defined limits.

Carbon Dioxide Combining-Power of Plasma

This property of the test plasma was determined by the method of Van Slyke and Cullin,[5] employing the volumetric Van Slyke gasometric apparatus. The mean value for the test samples used was 81 volumes per 100 ml. plasma while the control limits ranged from 77.4 to 84.6 volumes per 100 ml. (Fig. 4). All the mean values were in control, although one of the range values appeared to be out of control. To be in control, the difference in duplicates could be as great as 11.7 volumes per 100 ml. plasma.

DISCUSSION

The control chart is an important aid in keeping the variation of any single chemical analysis of a clinical laboratory within set statistical limits. Some of the advantages of the control chart are as follows: The control chart offers a simple method of checking the resultant effect of all factors influencing the accuracy of a given test, *e.g.*, the reagents, standards, time factors and instruments used in the analysis. It offers a basis for action in initiating correction of a method that is not functioning properly. Also, it improves the general accuracy of the laboratory, since the technicians become control-conscious and readily detect and report a test that is out of control.

By using the control chart it is possible to tell at once whether or not a method is working well. If the method is out of control, the chart usually cannot give the reason and it is up to the analyst to determine the cause of the difficulty. Sometimes it is possible to note the deterioration of reagents or standards by observing a trend in the control curve. However, the possibility of deterioration of the substance in the frozen control standards should also be considered. In setting up control charts for the gamma globulin method of Kunkel,[4] it was found that the gamma globulin content of the normal plasma used as a standard had completely disappeared on storage in the frozen state after three months.

The data presented in this work were obtained by seven technicians and three student technicians; and yet with the exception of some of the chloride determinations, there were no constant analytical differences in the mean values which could be attributed to any one worker. But even with respect to chloride analysis, when the new worker determined a constant for the reagent, the analyses were again in control. It is possible by the use of the control chart to discover an inept analyst.

The use of control charts need not be limited to the determinations described here. Any of the chemical analyses routinely done in the laboratory, such as glucose, nonprotein nitrogen or hemoglobin, could be used for control studies.

1066   LEVEY AND JENNINGS

SUMMARY

The application of the principle of the control chart is made to the chemical division of the clinical laboratory. Examples of the use of these charts for the evaluation of the functioning of the determination of the urea content of the blood, and the chloride, protein and carbon dioxide combining-power of plasma are given. By using the control chart, it is possible to ascertain at a glance the over-all functioning of any determination thus studied. It also provides a basis of action in correcting a defective method.

Acknowledgments. The authors are indebted to Dr. Benjamin Epstein of the Department of Mathematics of Wayne University for aid in this study, and to Miss Jean Reed and Mrs. Violet Greenberg for technical assistance.

REFERENCES

1. American Society for Testing Materials: Manual on the Presentation of Data. Philadelphia: The Society, 1947, 47 pp.
2. KARR, W. G.: A method for determination of blood urea nitrogen. J. Lab. and Clin. Med, 9: 329–333, 1924.
3. KINGSLEY, G. R.: The direct biuret method for the determination of serum proteins as applied to photoelectric and visual colorimetry. J. Lab. and Clin. Med., 27: 840–845, 1942.
4. KUNKEL, H. G.: Estimation of alterations of serum gamma globulin by a turbidimetric technique. Proc. Soc. Exper. Biol. and Med., 66: 217–224, 1947.
5. PETERS, J. P., AND VAN SLYKE, D. D.: Quantitative Clinical Chemistry. Vol. 2. Baltimore: Williams & Wilkins Co., 1932, 957 pp.
6. SCHALES, O., AND SCHALES, S. S.: A simple and accurate method for the determination of chlorides in biological fluids. J. Biol. Chem., 140: 879–884, 1941.
7. WERNIMONT, G.: Use of control charts in the analytical laboratory. Ind. Eng. Chem. Anal. Ed., 18: 587–592, 1946.

COMMENTARY TO

33. Caraway, W. T. (1962)
Chemical and Diagnostic Specificity of Laboratory Tests. Effect of Hemolysis, Anticoagulants, Medications, Contaminants, and Other Variables. American Journal of Clinical Pathology 37(5): 445–464.

The effects of analytical and pre-analytical variables on the accuracy of test methods have been described in individual reports throughout the clinical chemistry literature. J. B. Ogden for example, in his book *"Clinical Examination of the Urine and Urinary Diagnosis"* published in 1900 cited morphine administration as a possible cause of an abnormal urine albumin test (1). Wendell Caraway 62 years latter was the first to review and pull together in one publication a wide range of analytical variables reported to interfere with laboratory procedures. Caraway's influential paper published in 1962 is presented here. An editorial in the journal *Clinical Chemistry* only nine years after its publication referred to his paper as a "classic" (2).

Caraway organized the factors effecting test accuracy into various categories including hemolysis, lipemia, and medications. In each category he cited specific articles in the literature that documented the effects. Wirth and Thompson three years latter compiled into tables of all of Caraway's literature citations along with added publications. They listed 86 different laboratory tests alphabetically followed by the condition or drug and their effects on the test (3). Cross and co-workers compiled a similar list for a number of urinary steroids and thyroid function tests (4). In 1968 Elking and Kabat, two pharmacy scientists, suggested that clinical pharmacists and clinical chemists should work together to ensure the better development of a database for monitoring drug effects on laboratory tests. They took Caraway's original data that Cross and co-workers had put into table form and prepared a drug list intended for ready reference by pharmacists (5). Lubran cited Caraway and prepared a list of tests organized by drug or drug class (6). Sunderman (7), a clinical pathologist and Christian (8), a clinical pharmacist each published tables of drug effects on laboratory tests in 1970. In 1972, Caraway updated his original list into an alphabetical tabulation of 44 different laboratory methods and the effects of analytical variables (9).

Donald S. Young and co-workers at the National Institutes of Health in Bethesda, Maryland in 1972 published the first computer based data file on drug effects called CLAUDE (computer list of anticipated and unintended drug effects). The database contained 9 000 entries on drug effects. The complete list was published as a special issue in the journal *Clinical Chemistry* in 1972 (10). In 1975, the list was expanded and again published as a special issue in *Clinical Chemistry* (11). The popularity and utility of Young's compilation produced three books by him and his co-workers. One covers the effects of pre-analytical variables (12), another is a two volume set on the effects of disease on laboratory tests (13) and finally, a two volume set on the effects of drugs. This latter book is in its fifth edition (14).

References

(1) Ogden, J.B. (1900) *Clinical Examination of the Urine and Urinary Diagnosis.* W.B. Saunders, Philadelphia, pg 120.

(2) Brody, B.B. (1971) The effect of drugs on clinical laboratory determinations. Clinical Chemistry. 17(5):355–357.

(3) Wirth, W.A. and Thompson, R.L. (1965) The effect of various conditions and substances on the results of laboratory procedures. American Journal of Clinical Pathology. 43(6):579–590.

(4) Cross, F.C., Canada, A.T., and Davis, N.M. (1966) The effect of certain drugs on the results of some diagnostic procedures. American Journal of Hospital Pharmacy. 23(5):235–239.

(5) Elking, M.P. and Kabat, H.F. (1968) Drug induced modifications of laboratory test values. American Journal of Hospital Pharmacy. 25(9):485–519.

(6) Lubran, M. (1969) The effects of drugs on laboratory values. Medical Clinics of North America. 53(1):211–222.

(7) Sunderman, F.W. (1970) Drug interference in clinical biochemistry. Chemical Rubber Company (CRC) Critical Reviews in Clinical Laboratory Sciences. 1(3):427–449.

(8) Christian, D.G. (1970) Drug interference with laboratory blood chemistry determinations. American Journal of Clinical Pathology. 54(1):118–142.

(9) Carawy, W. and Kammeyer, C.W. (1972) Chemical interference by drugs and other substances with clinical laboratory test procedures. Clinica Chimica Acta. 41:395–434, October.

(10) Young, D.S., Thomas, D.W., Friedman, R.B., and Pestaner, L.C. (1972) Drug interferences with clinical laboratory tests. Clinical Chemistry. 18(10):1041–1303.

(11) Young, D.S., Pestaner, L.C., and Gibberman, V. (1975) Effects of drugs on clinical laboratory tests. Clinical Chemistry. 21(5):1D–432D.

412

(12) Young, D.S. (1997) *Effects of Preanalytical Variables on Clinical Laboratory Tests*, 2nd Edition. American Association for Clinical Chemistry Press, Washington, DC.

(13) Young, D.S. and Friedman, R.B. (2001) in *Effects of Disease on Clinical Laboratory Tests*, Vols 1 and 2. 4th Edition. American Association for Clinical Chemistry Press, Washington, DC.

(14) Young, D.S. (2000) in *Effects of Drugs on Clinical Laboratory Tests*, Vols 1 and 2. 5th Edition. American Association for Clinical Chemistry, Washington, DC.

Am J Clin Pathol. 1962; 37: 445–464
© 1962 American Society of Clinical Pathologists.
Reprinted with permission.

THE AMERICAN JOURNAL OF CLINICAL PATHOLOGY
Vol. 37. No. 5, pp. 445–464   May, 1962
Copyright © 1961 by The Williams & Wilkins Co.
      Printed in U.S.A.

CHEMICAL AND DIAGNOSTIC SPECIFICITY
OF LABORATORY TESTS

EFFECT OF HEMOLYSIS, LIPEMIA, ANTICOAGULANTS, MEDICATIONS, CONTAMINANTS, AND OTHER VARIABLES

WENDELL T. CARAWAY, PH.D.

McLaren General Hospital, St. Joseph Hospital, and the Flint Medical Laboratory, Flint, Michigan

It is desirable that clinical laboratory procedures have a high degree of diagnostic specificity. Chemical analysis of biologic material is based upon the presence of functional groups in organic compounds or upon the separation and measurement of ions. Such analyses are inherently complicated and are becoming more so with the introduction of chemotherapeutic agents that are analogs of normal intermediary metabolites. The discussion in this paper deals with some of the variables encountered in clinical chemistry that affect the accuracy and also the interpretation of laboratory results. A special attempt has been made to consider potential sources of error in the individual specimen. The usual technics of quality control, such as concurrent analysis of standards or control serum, will not ordinarily detect or eliminate these individual variations.

In order to be assured of over-all accuracy of results, some degree of quality control must necessarily be practiced in the clinical chemistry laboratory. This implies adequate personnel and equipment, accurate standards, clean glassware, accurately calibrated pipets, and so on. As more instrumental analysis is used, an awareness of the limitations and potential sources of error in such equipment is imperative. Line voltage fluctuations, for example, have been demonstrated to produce errors in the coulometric titration of chloride ion.[186] Specific directions for performing methods should be available.[34, 140] Frequent analysis of control serum and interlaboratory exchange of specimens help to provide over-all checks on

Received, May 8, 1961; revision received October 2; accepted for publication February 7, 1962.

Dr. Caraway is Biochemist.

reagents and technics and to establish the precision or limits of error of a method under actual working conditions. The preparation and use of control serums and control charts are well described elsewhere and will not be considered in detail in this paper.[18, 70, 84, 156, 188] Inherent errors in all methods should be recognized in order to avoid attempts to read significance into slight changes in laboratory results from day to day. These inherent errors should not be regarded as "laboratory errors."[97]

CHEMICAL SPECIFICITY

Methods should be chosen that provide adequate specificity for the constituent to be determined. The classical example of non-specificity is the determination of "blood sugar" by means of the Folin and Wu procedure.[61] Numerous modifications designed to provide a measure of true blood glucose have been published.[170] Non-glucose-reducing substances such as glutathione are largely confined to the red cells. Somogyi[167] approached the problem by preparing protein-free filtrates in which these substances were eliminated. Benedict[17] modified the alkaline copper reagent in order to improve its specificity for glucose in the presence of other reducing substances. The use of serum or plasma obviates most of these differences, but variable amounts of saccharoids, such as glucuronic acid,[150] can also produce falsely elevated glucose values.[56, 57, 139, 176]

The icterus index, by means of visual comparison with dichromate standards, provides a rough estimate of bilirubinemia. When the visual method is adapted to photometers, a wave length of 460 rather than 420 mμ should be chosen, inasmuch as traces of hemoglobin absorb very strongly near 420 mμ.[88] Carotenemia and lipemia produce elevated values for the icterus index, not related to the concentration of bilirubin.

The determination of uric acid by means of reduction of phosphotungstic acid in alkaline medium is affected by other reducing substances, especially ascorbic acid and catechol derivatives.[194] The contribution of these reducing substances is relatively greater in methods that substitute carbonate or silicate for cyanide.[33, 106]

Urinary neutral 17-ketosteroids represent a number of individual chemical compounds separable into at least 7 major fractions by chromatography.[49]

In an attempt to decrease the 24-hr. incubation period for serum lipase determinations, tributyrin substrate was substituted for olive oil.[74] The tributyrinase activity was subsequently demonstrated to correlate with pseudocholinesterase activity and to provide no adequate measure of pancreatic lipase.[90] Similar observations have been made with Tween-derivative substrates.[7]

In the determination of cholesterol by means of the Liebermann-Burchardt reaction, bilirubin must be removed in order to prevent falsely elevated results. One milligram of bilirubin produces color equivalent to that produced by 5 mg. of cholesterol.[23] For this reason, procedures based on direct addition of acetic acid, acetic anhydride, and sulfuric acid to serum should not be used with jaundiced serum. At present, several reagent "kits" are commercially available that depend upon this reaction. For greatest accuracy the serum should be deproteinized and the cholesterol extracted into an organic solvent. A somewhat similar objection is applicable to the ferric chloride method for cholesterol when applied directly to serum.[196] Serum protein can react to produce appreciable color, and this interference seems to increase with the age of the serum.

In diabetes, high creatinine levels in serum are reported to be an artifact that is the result of interference by the higher levels of glucose and acetone present.[83] Interference is negligible, however, when optical density readings are made at the specified 10- or 15-min. interval after addition of alkaline picrate.

An evaluation of the benzidine test for occult blood in feces indicates that false-positive tests may be obtained from meat in the diet, bromides, iodides, and ferric ammonium citrate.[174] A mixture of benzidine, perborate, and biphthalate is reported to produce a false-positive test for blood in the presence of glucose.[152]

Determinations of protein in cerebrospinal fluid, based on the turbidity developed after the addition of sulfosalicylic acid, are affected by the albumin-globulin ratio, inasmuch as albumin produces nearly 2.5 times as much turbidity as an equivalent weight of globulin.[91] This differential is insignificant when trichloracetic acid is used as the protein precipitant.

Some analytic procedures have been based on the use of enzymes to increase the specificity of the reaction. Typical examples are the use of uricase to determine uric acid by ultraviolet spectrophotometry; urease to catalyze hydrolysis of urea; and glucose oxidase in the determination of glucose. Reluctance to adopt these methods stems in part from the possibility of inhibition of enzymes by drugs or serum proteins. Competition between naturally occurring substrates may also occur. The application of glucose oxidase methods to serum has been regarded as unacceptable on the basis of such inhibition,[29, 55, 147] although others have reported negligible inhibition with the use of higher dilutions of serum.[41, 109] Inhibitors of trypsin,[28] amylase,[121] and arginase[37, 172] have also been demonstrated in serum.

Conversely, depending on the type and specificity of the substrate, results can vary widely in the measurement of serum enzyme activity. Serum acid phosphatase may originate from the prostate gland, erythrocytes, blood platelets, and liver.[187] It has been demonstrated that alpha-naphthyl phosphate and beta-glycerophosphate have much greater specificity for prostatic acid phosphatase than for erythrocyte phosphatase.[8] Other substrates, such as phenyl phosphate, may be modified by the addition of tartaric acid to inhibit prostatic phosphatase or by the addition of cupric ion to inhibit erythrocyte phosphatase.[2]

An enzyme, depending on its source, may have variable properties. This is of considerable interest, inasmuch as it may be possible (with adequate technics) to demonstrate the tissue of origin. A number of such isoenzymes have been separated by starch gel electrophoresis. Serum lactic dehydrogenase has been separated into 4 or 5 fractions.[184, 191] That from heart muscle is the fastest moving component, whereas that from the liver is much slower. Heart muscle glutamic-oxalacetic transaminase has been separated into 2 fractions.[60] Serum alkaline phosphatase has been separated into 2 or 3 fractions corresponding to phosphatases from bone or liver.[105, 143] Amylase normally present in serum migrates predominantly with albumin, whereas that from salivary glands or pancreas associates with gamma globulin.[99, 121] Serum cholinesterase, aromatic esterase, and acid phosphatase have each been separated into various fractions by electrophoresis.[51] In a somewhat different approach, Kaplan and associates[104] have used analogs of diphosphopyridine nucleotide to demonstrate differences in lactic dehydrogenase from various tissues in a number of species. By means of suitable application of discriminating technics to various serum enzymes, it may be possible to perform a "chemical biopsy" of an organ by analysis of a small sample of blood.

ANTICOAGULANTS AND PRESERVATIVES

Chemical components of the blood are often present in different concentrations in the plasma and red cells. For this reason, values on whole blood may vary greatly with the hematocrit. Even for substances distributed uniformly in the body water, such as glucose or urea, the concentration per 100 ml. of plasma will be approximately 12 per cent greater than that of whole blood. This is explained by the fact that normal plasma contains 94 per cent water, whereas whole blood contains only 81 per cent.[4] Finally, changes in body tissues are reflected in the blood primarily by changes in the plasma component; hence, analysis of plasma or serum is more desirable.

Because of potential interference by the various anticoagulants, serum, rather than plasma, is preferred for most analyses. It is apparent that plasma containing oxalate, citrate, or ethylenediamine tetraacetic acid (EDTA) is unsuited for calcium determinations, and that plasma containing sodium or potassium salts of anticoagulants is unsuited for sodium or potassium determinations. Potassium oxalate causes shrinkage of red cells with resultant dilution of the plasma. Calculations indicate that 1 mg. of EDTA per ml. of blood should increase the non-protein nitrogen value by approximately 7 mg. per 100 ml., although Hadley and co-workers[79, 80] found experimentally that the error was somewhat less.

Heparin, as the sodium or ammonium salt, is an effective anticoagulant in such small quantities that there is no significant effect on electrolyte or urea nitrogen determinations. Preparations vary widely in their ammonium content, however, and should not be used to collect specimens for blood ammonia determinations.[44] Heparin is the most satisfactory anticoagulant for blood pH studies. In contrast, blood pH is decreased significantly by ammonium oxalate, sodium citrate, or disodium EDTA, and increased significantly by sodium or potassium oxalate.[72]

Anticoagulants may inhibit enzymatic activity. Oxalate has been demonstrated to inhibit lactic dehydrogenase[30] and acid phosphatase.[2] Fluoride ion prevents glycolysis and preserves blood glucose, but fluorided plasma is unsuitable for urea nitrogen determinations that are dependent on urease activity. Fluoride ion also inhibits acid phosphatase.[2] Amylase activity is activated by chloride and fluoride,[163, 168] but is inhibited by oxalate and citrate.[120] A study of errors in the use of anticoagulants for blood chemistry determinations has been reported by Clark.[42]

Plasma is generally unsuited for fractionation of proteins by electrophoresis because of interference by fibrinogen. Results of thymol turbidity tests on oxalated plasma were observed in some instances to be conspicuously less than on serum.[37] In addition, high enzymatic activity of leukocytes or platelets may obscure the true activity in the plasma. Carefully separated plasma reveals consistently less hemolysis than serum.[164] With reasonable care, however, serum can be obtained without visible hemolysis and used for electrolyte determinations.

Preservatives for 24-hr. specimens of urine depend upon the constituent to be measured. Glacial acetic acid, 5 ml., added to the container at the start of a 24-hr. collection, is satisfactory for most purposes. The final pH of the specimen is usually between 4 and 5; bacterial decomposition is prevented and the acetic acid tends to mask unpleasant odors. Specimens of urine for enzyme assay are preferably collected without preservative, but should be refrigerated during the collection period. Urine for quantitative urobilinogen assay should be collected in brown glass bottles containing 5 Gm. of sodium carbonate and 100 ml. of pertroleum ether.

STABILITY OF SPECIMENS AND REAGENTS

Indications of reagent stability should be included in descriptions of procedures. Many reagents are refrigerated unnecessarily and may be stored equally well at room temperature. Stability of final colors developed may be critical with respect to time, temperature, and light.

It is recognized that glucose will disappear from whole blood on standing. Human blood, collected with heparin or citrate anticoagulant and incubated at 37 C., reveals a glucose utilization of 1.5 to 2.0 μM per ml. of red cells per hr.[12] This rate is essentially independent of the concentration of glucose to a level of 10 mg. per 100 ml. From these data it may be estimated that the average loss in blood glucose amounts to approximately 15 mg. per 100 ml. per hr. at 37 C.

When blood is drawn, permitted to clot, and to stand at a warm room temperature, the serum glucose decreases approximately 7 per cent in 1 hr.[181] Glucose in separated serum seems to be relatively stable when stored at 4 C. At 25 C., variable stability, ranging from no loss to more than 50 per cent loss in 3 days, has been observed.[87]

Bilirubin in serum is sensitive to light. Losses of 50 per cent in 1 hr. can occur in serum exposed to direct sunlight.[127] Maximal sensitivity occurs with light of wave length 460 mμ.[45] There is an increase in oxidation-reduction potential and increase in absorbance at 600 mμ, presumably related to oxidation of bilirubin to biliverdin. Jaun-

diced babies, exposed to sunlight for 2 or 3 hr., had an average decrease in their serum bilirubin of 4 mg. per 100 ml. A slight loss of bilirubin occurs in serum when stored overnight at 4 C.[193]

The cephalin flocculation test is also light-sensitive, and tubes should be stored in the dark for the 24-hr. period of incubation, in order to avoid false-positive reactions.[13]

After overnight refrigeration of serum, thymol turbidity decreases slightly, phenol turbidity increases significantly,[193] and uric acid reportedly decreases slightly.[194] As a result of enzymatic hydrolysis of phosphate esters, inorganic phosphorus tends to increase on standing, especially in hemolyzed serum.[118]

Ascorbic acid is unstable at the pH of blood and undergoes oxidation by dissolved oxygen to dehydroascorbic acid. Above pH 7, dehydroascorbic acid also decomposes rapidly; the half-life at pH 7.25 and 38 C. is approximately 2 min.[10]

After blood is withdrawn, the content of ammonia rises approximately 0.3 μg. per 100 ml. per min. at 23 C.[158] It is important, therefore, that blood ammonia determinations be made on fresh blood.

The pH of blood increases rapidly, owing to loss of carbon dioxide, if the blood is exposed to air. When stored anaerobically at 37 C., the pH reveals a steady decline of 0.04 pH unit per hr. as a result of glycolysis.[169] Under anaerobic conditions, the pH of the blood remains constant for 3 hr. at 4 C.[72]

Salt fractionation of serum proteins may lead to denaturation of albumin, especially when sodium sulfite is used to precipitate the globulin.[148] The extent of denaturation is related to the rate of shaking and to the volume of air space above the mixture.

A number of enzymes assayed clinically are fairly stable, even at room temperature. This is an important factor when specimens are mailed or transported. The following enzymes seem to be stable for at least 1 week in serum stored at 4 C.: amylase,[87] lipase,[90] glutamic-oxalacetic and glutamic-pyruvic transaminases,[189] lactic dehydrogenase,[112, 190] and leucine aminopeptidase.[6, 73]

Activity of alkaline phosphatase has been

reported to increase as much as 10 per cent in serum refrigerated overnight.[22] This apparent increase has been explained on the basis of inadequate buffering and suboptimal pH in 1 method.[86] As the pH of serum rises on standing, the final pH of serum-substrate mixture more nearly approaches the optimum, and this results in a higher measured activity. Failure to provide adequate buffering could lead to negative errors of 35 per cent. On the other hand, Kaplan and Narahara,[103] using a method which seems to provide adequate buffering, have also reported an increase in serum alkaline phosphatase activity on standing.

Stability of acid phosphatase is related to the pH of the serum.[107] At pH 7.4 the enzyme is reasonably stable; when serum is separated from the clot, the pH rises and drastic loss in activity occurs. Serum should be separated from the clot in order to avoid contamination with red cell phosphatase, then analyzed immediately or refrigerated for not more than 2 or 3 hr. An alternate method is to add 0.01 ml. of 20 per cent acetic acid to each milliliter of serum. This decreases the pH to approximately 6.0 and stabilizes the enzyme.[98] Acid phosphatase is stable for at least 112 days in frozen serum.[48] In contrast, urinary acid phosphatase was found to be stable for 7 days at room temperature and for 15 days in the refrigerator, but was completely inactivated in 2 days in the frozen state.[155] Addition of protein to the urine protected the enzyme in the frozen state.

We have maintained pooled serum up to 3 years in the frozen state in our laboratories and have observed excellent stability for some 20 common constituents, including glucose and bilirubin. Intermittent freezing and thawing should be avoided. Serum that has thawed and refrozen without mixing will, of course, be concentrated at the bottom of the tube and nearly water-clear at the top as the floating ice melts and refreezes. It follows that frozen serum or urine should be completely thawed and well mixed prior to analysis.

HEMOLYSIS

Normal serum appears visibly hemolytic when the concentration of hemoglobin exceeds 0.02 Gm. per 100 ml.[15] Much higher levels could remain undetected in jaundiced serum.

A dilutional effect on serum components occurs on hemolysis if the concentration in the red cell is less than in plasma; thus, the concentration of sodium or chloride in whole blood is considerably less than in serum. The dilutional effect is negligible except in grossly hemolytic serum.

Hemoglobin may interfere directly in some determinations. A 50 per cent inhibition of activity of serum lipase was observed with concentrations of hemoglobin of 0.5 Gm. per 100 ml.[90] Diazotization procedures for serum bilirubin yield large negative errors in the presence of hemoglobin, possibly as a result of conversion of hemoglobin to methemoglobin by nitrous acid in the test but not in the control.[119, 122] Icterus index, determined photometrically, is falsely elevated by hemoglobin.[88] Color may also appear in protein-free filtrates. This results in falsely elevated values for bromide, urea nitrogen, iron, or other substances when absorbance is measured in the blue portion of the spectrum.[85] The absorbance of hemoglobin varies in acid or alkaline solutions and, thus, affects estimation of Bromsulphalein in some procedures.[159]

The red blood cells contain a high concentration of organic phosphate esters. The inorganic phosphate in hemolyzed serum increases rapidly as these esters are hydrolyzed by serum phosphatases.[118]

The pH of blood is decreased slightly by hemolysis. The pH of a concentrated suspension of red blood cells is approximately 0.1 unit less than that of plasma.[149]

Hemolysis increases consumption of prothrombin and causes interference in prothrombin consumption tests; however, hemolysis has negligible effect on the determination of prothrombin by means of the 1-stage technic.[133]

Some components of the red blood cell may diffuse into the serum in appreciable amounts in the absence of visible hemolysis. Potassium concentration increases in serum in contact with the clot,[180] and this increase is greater at 4 C. than at room temperature.[75] Acid phosphatase, and probably other enzymes, may diffuse from the red cells in unseparated serum. Acid phosphatase is also present in platelets and seems to be released to some extent into the serum during the clotting process.[197]

The problem of hemolysis is most important if the concentration of a measured constituent is much greater in the red cell than in the plasma. Depending upon the ratio of these concentrations, slight hemolysis can produce significant errors. Typical values are listed in Table 1. Ratios are an indication of maximal interference. This interference may be less than expected in some instances. For example, serum greatly inhibits red cell arginase activity.[172] The ratio of arginase in the red cell to that in the plasma is so great that some authors believe that activity of serum arginase represents a simple measure of hemolysis *in vivo*.[175]

Quantitative evaluation of error produced by hemolysis may be expressed in terms of the relative distribution of a substance in plasma and red cells. Serum hemoglobin is used as a measure of hemolysis. The average hemoglobin concentration in the red cell is

34 Gm. per 100 ml. of cells.[15] Addition of 0.03 ml. of hemolyzed red cells to 1.0 ml. of serum is equivalent to adding 1 Gm. of hemoglobin per 100 ml. of serum. Consequently, if: h = serum hemoglobin in Grams per 100 ml., then $0.03h$ = milliliters of hemolyzed red cells added per milliliter of serum.

Let s = concentration of a substance in nonhemolyzed serum or plasma; c = concentration of the substance in red cells; and x = concentration of the substance in hemolyzed serum, then:

$$(1)(s) + (0.03h)(c) = (1 + 0.03h)(x) \qquad \text{Eqn. 1.}$$

Whence:

$$x = \frac{s + 0.03hc}{1 + 0.03h} \qquad \text{Eqn. 2.}$$

Let p = per cent error = $\dfrac{x - s}{s} \times 100$.

Substituting the value of x from Equation 2

TABLE 1
CONCENTRATION OF SUBSTANCES IN ERYTHROCYTES AND PLASMA*

Substance	Erythrocytes	Plasma	Erythrocytes/ Plasma
Glucose, mg./100 ml.	74.0	90.0	0.82
Nonsugar-reducing substances, mg./100 ml.	40.0	8.0	5.00
Nonprotein N, mg./100 ml.	44.0	25.0	1.76
Urea N, mg./100 ml.	14.0	17.0	0.82
Creatinine (Jaffe), mg./100 ml.	1.8	1.1	1.63
Uric acid, mg./100 ml.[101]	2.5	.46	0.55
Total cholesterol, mg./100 ml.	139.0	194.0	0.72
Cholesterol esters, mg./100 ml.	0.0	129.0	0.00
Sodium, mEq./l.	16.0	140.0	0.11
Potassium, mEq./l.	100.0	4.4	22.70
Chloride, mEq./l.	52.0	104.0	0.50
Bicarbonate, mM/l.	19.0	26.0	0.73
Calcium, mEq./l.	0.5	5.0	0.10
Inorganic P, mg./100 ml.	2.5	3.2	0.78
Acid β-glycerophosphatase, units	3.0	0.25	12.00
Acid phenylphosphatase, units	200.0	3.0	67.00
Lactic dehydrogenase, units[106]	58,000.0	360.0	160.00
Transaminase, GO,† units[110]	31.5	0.8	40.00
Transaminase, GP,† units[110]	1.6	0.24	6.70
Arginase, units[87]			1000.00

* Average values are from the work of Behrendt,[15] unless otherwise indicated.

† GO = glutamic oxalacetic; GP = glutamic pyruvic.

and simplifying:

$$p = \frac{3h(c/s - 1)}{1 + 0.03h} \qquad \text{Eqn. 3.}$$

Inasmuch as h rarely exceeds 1 per cent, the term $0.03h$ is insignificant, whence:

$$p = 3h(c/s - 1) \qquad \text{Eqn. 4.}$$

Equation 4 is useful in calculating the percentage error when the serum hemoglobin and the ratio of concentrations in red cells and serum are known. For example, assume serum hemoglobin is 0.1 per cent and the ratio of potassium in red cells to serum is 20:1. The calculated error in serum potassium is: $p = (3) (0.1) (20 - 1) = + 5.7$ per cent. This calculated error agrees with that determined experimentally.[118]

LIPEMIA

Serum appears lipemic when the content of neutral fat is elevated. Although normally less than 200 mg. per 100 ml., the neutral fat may reach values of the order of 10,000 mg. per 100 ml. of serum. Colorimetric analytic methods, in which serum is still present in the final test mixture, but not in the blank, result in elevated optical densities in the presence of lipemia. Depending on the method used, falsely elevated values for total protein, thymol turbidity, amylase, transaminase, and so on, can be obtained. A suitable correction in most instances is to make a dilution of the serum in saline solution corresponding to the *final* dilution of serum in the cuvet. The optical density of this dilution is measured against distilled water, then subtracted from the optical density of the test specimen.

Results of determinations of blood hemoglobin may be falsely elevated as much as 2 or 3 Gm. per 100 ml. in the presence of gross lipemia. A suitable correction is made by means of diluting a second sample of blood with isotonic saline solution and centrifuging. The optical density or hemoglobin value measured on this control is subtracted from the test result.

Mild lipemia normally persists for several hours after eating and should cause no significant interference in most procedures, with the exception of thymol turbidity determination. Before obtaining blood for study of serum lipid or lipid fractions, a 12-hr. fast is recommended. In this connection it may be noted that after a standard breakfast a number of common chemistry determinations reveal no significant change in the blood.[5] These include: urea nitrogen, carbon dioxide, chloride, sodium, potassium, calcium, phosphorus, total protein, albumin, creatinine, uric acid, cholesterol, and cholesterol esters. Glucose, of course, increases greatly postprandially.

CONTROL OF TEMPERATURE

Control of temperature is obviously important for assays of enzyme. It should be noted that the rate of equilibration of temperature is much slower in an air incubator than in a water-bath; consequently, a water bath is recommended for prewarming substrates or for incubations of 1 hr. or less.

The pH of blood increases approximately 0.015 unit per degree decrease in temperature from 37 C.[169] If pH of blood is measured at other temperatures, appropriate corrections must be made. Buffers also change pH with temperature, but to a much lesser extent. As temperature increases, the pH of some buffers increases, whereas that of others decreases.[14]

Methods based on measurements of turbidity are often sensitive to ambient temperatures.[192] Values for thymol turbidity decrease significantly as the temperature increases from 15 to 37 C. Conversely, values for phenol turbidity increase greatly under the same conditions. Values for zinc sulfate turbidity reveal little change with temperature. A standard temperature of 25 C. is recommended for these measurements.

Specification of "room" temperature in a procedure is somewhat ambiguous. Laboratories in this country are usually maintained at between 20 and 30 C. Most glassware is calibrated for 20 C. Older physical constants are often reported for 15.5 C. In 1 report, a procedure for serum cholesterol was evaluated at 70 and 40 F., inasmuch as the latter was formerly regarded as being "room" temperature in continental China.[95]

Differences may be obtained in results, de-

452   CARAWAY   *Vol. 37*

pending upon whether a mixture is heated to boiling or simply heated in a boiling water bath. Acid hydrolysis of urinary 17-ketosteroids, for example, occurs at a higher temperature if the mixture is actively boiled, rather than heated in a boiling water bath.

Fixing of proteins on strips of paper after electrophoretic separation has been demonstrated to be critical with respect to temperature. As the temperature of drying was increased from 90 to 140 C., the apparent absolute concentrations of all globulins decreased, whereas the concentration of albumin increased. The apparent albumin-globulin ratio increased from 1.01 to 2.03. A drying temperature of 107 ± 3 C. was recommended.[89]

CONTROL OF pH

The pH of buffers may change on dilution. Thus, 2 M phosphate buffer, pH 6.6, increases to pH 7.0 on dilution to 0.01 M. On the other hand, 2 M acetate buffer reveals negligible change on dilution.[76]

Careful control of pH is obviously important for enzyme assays. A buffer should have adequate capacity to be relatively unaffected by the addition of the specimen. Serum and urine have significant buffer capacities and can change the final pH of a reaction mixture, often into nonoptimal ranges for enzyme assay. This has been described in procedures for amylase[87] and alkaline phosphatase.[86] Weak buffers above pH 7 can absorb carbon dioxide with resultant decrease in pH.

The buffer capacity of distilled water is extremely small. For example, addition of 0.01 ml. of 1 N hydrochloric acid to 1 l. of water theoretically should decrease the pH from 7.0 to 5.0. It would seem to be preferable in most instances to devise adequate buffer systems for a method rather than be concerned excessively over the pH of distilled water.

Some turbidity reactions are sensitive to pH changes. Globulin fractionation with ammonium sulfate is dependent on proper control of pH.[65] Thymol turbidity results vary somewhat with changes in pH between 7.55 and 7.80.[188]

CONTAMINATION

There are many sources of contamination in the clinical laboratory. Some of these may affect a random specimen and be overlooked or not detected, even with inclusion of standards and controls. Thorough cleaning of glassware is essential. Detergents frequently contain large quantities of sodium and inorganic phosphate, and these will interfere if not completely removed. A simple check to detect residual detergent consists of filling pipets or rinsing glassware with a dilute solution of phenolphthalein. No pink color should appear.

Other sources of contamination are more subtle. If cuvets are not rinsed adequately, phosphomolybdic acid from glucose determinations may remain to interfere with subsequent measurements of inorganic phosphate. Nessler's reagent can leave a film of mercury salts on glass that could inhibit subsequent assays of enzyme. Similarly, traces of iodine or mercury salts interfere in the determination of protein-bound iodine.

Calcium from tap water forms a film on glass that is difficult to remove by simple rinsing with distilled water. Subsequent use of such glassware for flame photometric calcium determinations involving the use of a sequestering agent (such as Sterox SE) can liberate calcium into the solution to produce falsely elevated results.[38] In other calcium procedures, serum protein alone tends to remove this film.

Evans blue dye also binds to glass surfaces. This could lead to errors in standardization for blood volume determinations unless the standard is further diluted immediately in plasma.[40]

Traces of cupric ion in distilled water interfere with the determination of oxyhemoglobin, presumably by conversion of oxyhemoglobin to methemoglobin.[53] Free chlorine may also be an unsuspected contaminant in distilled water. As such, it behaves as a strong oxidant and results in appreciable losses in the measurement of serum bilirubin and uric acid.[35]

Ordinary filter paper contains calcium, readily leached out by acid solution; hence, acid-washed filter paper should be used in determinations of calcium. On the other hand, acid-washed filter paper is not suitable for determinations of nitrogen, inasmuch as such papers readily absorb ammonia from the air.[157] Special low-fat filter papers, *e.g.*, Whatman No. 43, are available for use in estimations of lipid.

Some lots of reagent-grade absolute methanol contain impurities capable of reacting with Ehrlich's diazo reagent to produce a pink color. This results in falsely elevated bilirubin values by the Malloy-Evelyn technic in which the diazo reagent is omitted from the control. Methanol may be purified by shaking it with activated carbon and filtering.[36]

In determinations of serum amylase, great care should be taken in order to avoid contamination of mixtures with saliva. Results obtained by an amyloclastic procedure reveal that salivary amylase activity is nearly 700 times that of serum.[120] It is important, therefore, to deliver serum into the substrate from a pipet calibrated between 2 marks and to avoid use of pipets calibrated to contain or to "blow-out."

Strips of paper impregnated with glucose oxidase, peroxidase, and a chromogen are widely used for the detection of glucose in urine (Clinistix, Uristix, Combistix, the Ames Co.; Tes-Tape, Eli Lilly & Co.). The test for glucose depends on the following reactions:

$$\text{Glucose} + O_2 \xrightarrow{\text{glucose oxidase}} \text{gluconic acid} + H_2O_2$$

$$H_2O_2 + \text{chromogen} \xrightarrow{\text{peroxidase}} \text{color}$$

Numerous reports have emphasized the specificity of this preparation for glucose;[16, 43, 66, 68, 113, 115, 161] however, a number of strong false-positive reactions were encountered in our laboratories on urine specimens that were negative for glucose by copper reduction methods. Admitting the high specificity of glucose oxidase, it seemed probable that contamination had occurred either with hydrogen peroxide or with some strong oxidizing agent such as hypochlorite.

With Combistix test paper, a strong positive test for "glucose" was obtained with 0.003 per cent hydrogen peroxide or 0.005 per cent sodium hypochlorite. When hydrogen peroxide was added to random specimens of urine, a final concentration of 0.006 per cent usually resulted in a strong positive test. With sodium hypochlorite, a final concentration of 0.2 per cent was required to produce a strong positive test. Apparently, reducing substances normally present in urine destroy lesser quantities of hypochlorite. All such treated urines were negative for reducing sugar by copper reduction tests.

Several commercial bleaches contain approximately 5.5 per cent sodium hypochlorite, and various detergents contain larger quantities. If such detergents are used to clean urinals or specimen bottles, the containers must be rinsed thoroughly prior to collection of urine specimens for laboratory examinations. Our false-positives were traced in part to contamination with hydrogen peroxide. In 1 instance, all catheters during the cleaning process had been soaked routinely in 3 per cent hydrogen peroxide. In another instance, sterile bottles, used as containers for 3 per cent hydrogen peroxide for irrigating wounds, were subsequently used to collect urine specimens.

Addition of 2 drops of 3 per cent hydrogen peroxide to 50 ml. of urine is sufficient to produce a strong positive reaction for glucose by means of the glucose oxidase strip technic. It has also been observed that very dilute solutions of hydrogen peroxide (0.01 per cent) are not destroyed by autoclaving. A routine procedure at present is to eliminate glucose-negative urines with the strip test; those manifesting positive results are then graded by means of using a copper reduction tablet (Clinitest, Ames Co.). The presence of reducing sugars other than glucose may be detected by means of applying the copper reduction test to glucose-negative urines.

MEDICATIONS

Therapeutic agents may change the physiologic level of the substance measured. Many such changes are predictable or known from experience. Typical examples include: increased serum amylase activity from administration of morphine[24] and codeine;[77] increased serum glutamic-oxalacetic transaminase activity from opiates;[64] decreased serum uric acid after probenecid or salicylate therapy;[78] increased serum uric acid after chlorothiazide administration;[69] and decreased serum cholesterol when corn oil is substituted for usual dietary fat.[26]

Therapy with cortisone depresses excretion of urinary 17-ketosteroid, but, during administration, there will be an increase in urinary 17-hydroxycorticoids. Some high potency analogs of hydrocortisone, such as dexamethasone, are administered in such small doses that both 17-ketosteroid and 17-hydroxycorticoid excretion will be depressed.[114]

In mild renal insufficiency, serum potassium may be elevated by increasing the intake of fluids that contain appreciable concentrations of potassium.[47] Orange juice, for example, contains approximately 50 mEq. per l. of potassium.[21]

Medications may also interfere directly with the usual chemical analyses. In this event, diagnostic and therapeutic agents become contaminants, inasmuch as they affect the specificity of analytic methods. A typical illustration is the invalidation of protein-bound iodine determinations on patients after administration of x-ray contrast mediums. An extreme example has been reported for iophenoxic acid (Teridax) after its use in cholecystography. It has been estimated that a time interval of 33 years would be required for the concentration of Teridax in the blood to drop to the level of 1 μg. of iodine per 100 ml. of serum. More seriously, the material can apparently cross the placental membrane to such an extent that the newborn infant may have astronomical values for protein-bound iodine.[162]

Interference with determinations of serum calcium has been noted in patients being treated with EDTA.[39] Values for serum cholesterol, determined by means of ferric iron methods, are falsely elevated in the presence of bromide.[141] Bromsulphalein in serum can increase the apparent concentration of protein, as determined with alkaline biuret reagent. Oral administration of chloral hydrate and chlorobutanol has been reported to cause a rise in levels of blood urea nitrogen, as determined by Nesslerization technics.[133]

Many false-positive results associated with medications have been reported for tests performed with urine. Streptomycin can result in a false-positive Benedict test for reducing sugar.[126] Injection of radiographic contrast mediums, such as Hypaque, can produce within 15 to 30 min. an increase in specific gravity of urine, a false 4-plus reaction for albumin, and a heavy yellow precipitate with Ehrlich's aldehyde reagent.[96]

A metabolite of tolbutamide produces a false-positive test for protein in urine, when analyzed by means of turbidity procedures, such as heat and acetic acid, or sulfosalicylic acid. Tests based on the use of papers impregnated with indicators for protein are not affected.[67, 177]

Orally administered quinine and quinidine increase the apparent values for urinary catecholamine excretion. This effect is presumably caused by metabolic derivatives; direct addition of the drugs to urine was without effect on the determinations.[153]

Meprobamate is reported to interfere with the Zimmermann reaction for 17-ketosteroids.[151]

A test devised to detect the drug phenothiazine in urine was later demonstrated to produce a positive test with *p*-aminosalicylic acid. In addition, false-positive tests were obtained on urine specimens from patients with phenylketonuria or impaired liver function.[62, 63]

A test may have a reputation for being highly specific until personal experience or careful search of the literature reveals the possibility of false-positive reactions. A typical example is the determination of 5-hydroxyindoleacetic acid in urine as an aid in the diagnosis of malignant carcinoid. In 1 report, more than 1100 urine specimens were tested with no positive findings.[125] A second author reported on 1000 urines, of which 10 produced a light pink color in the test; however, none were intense enough to be regarded as positive.[181] Acetanilid interfered, but this drug is little used at present. These authors concluded that the specificity of the test was excellent.

Ingestion of bananas is reported to produce an elevation in the urinary excretion of 5-hydroxyindoleacetic acid (5HIAA).[46] After ingestion of 3 bananas, a normal excretion of 6.4 mg. per day rose to 12 mg.; after 6 bananas, to 23 mg. In a second experiment, after ingestion of 12 bananas, excretion rose from 5.9 mg. to 54 mg. per day. Certain muscle relaxants, such as mephenesin and methocarbamol, produce similar false-positive tests for 5HIAA. It is suggested that such medications be withheld for 48 hr. before repeating the test.[94] In our own experience a patient being treated with methocarbamol (Robaxin) was found to have an apparent excretion of 5HIAA of 140 mg. per 24 hr. (normal range is 3 to 10 mg.). The final colored extract and also pure 5HIAA exhibited maximal absorbance at 540 mu. Comparison of the entire absorption spectrum, however, revealed differences at higher wave lengths. The colors were visually different also. All medications were withheld for 3 days and the test became negative for 5HIAA.

With the continued introduction of many new drugs and new laboratory procedures it is most important to evaluate carefully any suspected interferences caused by medications. These interferences may produce false-negative, as well as false-positive results. For example, phenothiazine and its derivatives greatly *inhibit* the color development in the 5HIAA test.[144]

NORMAL PHYSIOLOGIC VARIATIONS

Normal values established for clinical purposes customarily include 95 per cent of the normal population studied. This necessarily implies that 5 per cent of laboratory results obtained on "normal" persons will be regarded as "abnormal." The significance of an abnormal result can be expressed only in terms of the probability that it could be excluded from the normal range. If we assume that the normal range of serum protein-bound iodine is 4 to 8 μg. per 100 ml. (95 per cent limits) or 3.5 to 8.5 (99 per cent limits), it follows that a result of 3.7 should be regarded as abnormal, with a confidence exceeding 95 per cent but not exceeding 99 per cent.

When normal limits are set too high, a test will lose discrimination, inasmuch as too many abnormal results will be included in the normal range. Even within the 95 per cent range, significant pathologic variations

may be obscured. Thus, a person with a protein-bound iodine of 7.8 μg. per 100 ml. would ordinarily be regarded as being within the normal range. It is conceivable, however, that this value could represent a hyperthyroid state if the true physiologic normal for this person happened to be 4.4 μg. per 100 ml. Similar considerations illustrate the difficulty and, indeed, the undesirability of drawing sharp demarcations between "normal" and "abnormal."

Williams[185] has emphasized the importance of considering biochemical individuality. Normally, interindividual differences range from 2- to 10-fold or more, but the pattern for a given person tends to remain fairly constant. Apparently, relatively little has been done to obtain repeated tests over long intervals on the *same* person in order to obtain a measure of diurnal, nocturnal, or seasonal variations. A few observations suggest that these variations can affect the clinical significance of laboratory results.

Diurnal variations in excretion of 17-hydroxycorticoid and catecholamine are well recognized. Plasma levels and urinary excretion of 17-hydroxycorticoids are approximately 3 times greater at 8 a.m. than at midnight, regardless of body activity.[124] Conversely, urinary excretion of epinephrine and norepinephrine is only approximately 1/3 as great during sleep, regardless of the time of day.[54]

Serum iron in normal persons manifests random variations as great as 50 per cent from the mean.[135] Diurnal variations between 15 and 100 μg. per 100 ml. occur with highest values in the early morning and lowest values in the afternoon.[136]

Levels of serum cholesterol seem to be relatively constant in some persons. In certain labile persons exposed to stress, the levels may fluctuate as much as 200 mg. per 100 ml. from day to day.[130]

Excretion of urinary 17-ketosteroid determined daily on a healthy male over a 6-year period manifested a mean value of 11.8 mg. per 24 hr. with a range of 9 to 16.4 mg. Great variations occurred from day to day which were independent of the 24-hr. volume of urine.[81]

Concentrations of uric acid in serum manifest a daily variation of approximately 0.5 mg. per 100 ml.[195]

Fecal excretion of urobilinogen varies greatly from day to day. A 4-day collection should be made for reliable interpretation.[178]

Josephson and Dahlberg[102] have studied the variation of some 24 parameters of the blood with respect to age, sex, and season. These studies emphasize the difficulty of assigning definite normal ranges to constituents of the blood.

Postural changes also affect certain determinations. In 1 study, healthy people were examined after lying in bed for 12 hr., and again at 15 min. and 60 min. after rising. The plasma volume decreased approximately 10 per cent in 15 min. when the posture changed from a lying to a standing position; plasma protein concentration increased 10 per cent and the hematocrit increased 7 to 8 per cent.[58] In another study, examinations were made before rising and again after 8 hr. of moderate activity. Under these conditions, the serum albumin revealed an increase averaging 0.4 Gm. per 100 ml., with no significant changes in hemoglobin, hematocrit value, or globulin.[182] Prolonged bed rest up to 3 weeks, however, resulted in a conspicuous *contraction* of plasma volume equivalent to a 9 per cent reduction in total blood volume.[178]

DIAGNOSTIC SPECIFICITY

A limited number of laboratory tests will be reviewed in this section, in order to examine their specificity in relation to a specific disease. Various tests have been proclaimed as being virtually pathognomonic, at least if a certain value is exceeded. Some tests have considerable diagnostic value; with others, early enthusiasm has been tempered with experience.

Many diagnostic tests for cancer, based on analysis of blood or urine, have been suggested and described, but apparently none has proved sufficiently reliable. In 1 test, a small drop of serum is floated on the surface of a mixture of phenol and glycerin adjusted to pH 8. Normal serum is said to float as a discrete drop, whereas serum from cancer patients spreads out and dissolves in approximately 2 min.[146] The chemical basis for this test is unknown. Claims for a high degree of specificity were not confirmed by others.[160]

Another suggested diagnostic test for cancer proposed the measurement of the optical density at 500 mμ of serum diluted with an equal volume of saline solution.[134] In general, serum from cancer patients exhibited lower optical densities than that from normal persons. Factors contributing to the optical density of serum at 500 mμ include bilirubin, hemoglobin, and iron-siderophyllin, but the cause for the differences in optical density was not postulated by the authors. A subsequent evaluation of this test was published, revealing a high incidence of false-positive and false-negative results.[137]

A proposed biochemical test for pregnancy, presumably based on the reaction of free estrone with *m*-dinitrobenzene or 2,4-dinitrophenylhydrazine to produce colored derivatives, was reported by the original author to have unusually high specificity.[142] These observations were essentially confirmed in a quantitative spectrophotometric modification of the original method.[123] Later reports have not confirmed these early findings.[59] In 1 study it was concluded, in fact, that the amount of estrone in pregnancy urine is too small to be detected by the original method.[50]

It is difficult to reconcile original findings on such tests, as reported above, with subsequent inability to reproduce the results. Various factors must be considered, such as unconscious bias (for or against), failure to duplicate experimental conditions exactly, and failure to include a sufficient range of pathologic material other than the specific condition under investigation.

Proposed diagnostic tests with no plausible scientific basis should be regarded with skepticism until a rational explanation is forthcoming. A more recent example is the *p*-toluenesulfonic acid test for acute systemic lupus erythematosus.[100] This test consisted of mixing 0.1 ml. of serum with 2 ml. of a 12 per cent solution of *p*-toluenesulfonic acid in glacial acetic acid and observing for a clot or precipitate after 20 min. Numerous subsequent studies have demonstrated that the test is nonspecific and will not distinguish between lupus and rheumatoid arthritis. The positive reactions seem to correlate with increased serum beta or gamma globulins and with other nonspecific compounds, such as C-reactive protein and serum mucoprotein.[11, 25, 52, 129, 171]

A specific enzyme, leucine aminopeptidase, has been described as being consistently elevated in the serum of patients with carcinoma of the pancreas.[73, 145] Subsequent studies have revealed that this is not invariably so, and that the enzymatic activity is elevated in patients with a wide spectrum of hepato-biliary-pancreatic diseases, including cirrhosis, viral hepatitis, infectious mononucleosis, common duct stone, metastases to the liver, acute cholecystitis, and acute pancreatitis.[6, 27, 93] Although nonspecific, the test still has limited value, in that a normal finding is presumptive evidence against the presence of carcinoma of the pancreas.

An elevated level of serum acid phosphatase is often regarded as pathognomonic for metastatic carcinoma of the prostate gland when the value exceeds 5 Bodansky units or 10 King-Armstrong units.[32] Ozar and co-workers[128] have evaluated various factors that might affect the reliability of this determination: hemolysis, endocrine therapy, prostatic manipulation, stability of substrates, and effect of anticoagulants. In 17 of a group of 20 men, the serum acid phosphatase rose to abnormal levels within 1 hr. after prostatic massage, but returned to normal within 24 hr.[92] Elevated values are encountered in the newborn[111] and in females receiving androgen therapy for metastatic carcinoma of the breast. Platelets may release the enzyme into the serum during the clotting process.[197] Some discrepancies may be related to the presence of several acid phosphatases in serum and to the lack of substrate specificity. The subject has been reviewed recently by Woodard.[187] A case has been reported with an elevation in both total and tartrate-inhibited serum acid phosphatase, although no demonstrable carcinoma of the prostate gland was found on autopsy.[165]

Estimation of serum amylase is a valuable aid in the diagnosis of acute pancreatitis.

Values of more than 1000 Somogyi units are believed by some authors to be pathognomonic.[9] In the absence of any pathologic process in the pancreas, however, values ranging from 1300 to 3200 have been reported in cases of peritonitis, mesenteric adenitis, and staphylococcal pneumonia.[20] Highly elevated values are also found in patients with biliary lithiasis.[19] In 1 series, 67 patients were explored surgically within 6 weeks of an attack of clinical acute pancreatitis. Fifty of these patients had values for serum amylase that exceeded 1000 Somogyi units, and, of these 50, there were 43 with biliary lithiasis, but with a grossly normal pancreas.[3] Serum amylase is also greatly elevated in mumps, although serum lipase is not.[31] It is recognized, of course, that the serum amylase is greatly elevated in acute pancreatitis,[1, 116] but this does not make the test specific for this disease.

The excretion of porphobilinogen in the urine is believed to occur rather specifically in acute idiopathic porphyria.[179] A routine survey of 1000 specimens of urine revealed no false-positive reactions.[82] In another study, however, false-positive reactions were observed to occur in approximately 50 per cent of patients with epilepsy.[117] The material manifesting the positive test was demonstrated not to be porphobilinogen.

In a recent study, the Congo red test was performed on 20 patients with a histologic diagnosis of amyloidosis.[71] Only 9 of these had more than 60 per cent disappearance of Congo red from the blood in 1 hr., and of these only 5 had more than 90 per cent disappearance. Again, it is recognized that a positive Congo red test is strong presumptive evidence for the presence of amyloidosis. Few false-positive tests occur, but false-negative tests may be expected in more than 50 per cent of patients with amyloidosis.

The estimation of serum protein-bound iodine is regarded as a measure of circulating thyroxine and is an excellent index of thyroid activity. Occasionally, normal values have been found in cretins.[154] Presumably the protein-bound iodine may be present as a biologically inactive iodinated protein. It is significant that these patients had very low serum levels of butanol-extractable iodine. The latter may be a somewhat more specific measure of circulating thyroxine in hypothyroidism. In the same report, 15 cases of hyperthyroidism are described with normal serum protein-bound iodine. It is suggested that this may reflect either random variation in the hyperthyroid group or a diminution in thyroxine-binding protein with concurrent elevation in metabolically active "free" thyroxine.

Bence Jones proteins in urine have the common property of precipitating when heated to 40 to 60 C., but redissolving at higher temperatures. Considerable variation in the structure of Bence Jones proteins has been revealed by a study of their N-terminal amino acids. Differences in molecular weights and electrophoretic mobilities have also been reported.[132] Apparently these chemical differences are not sufficient to invalidate clinical specificity. Assuming proper analytic technic, Snapper and Ores[166] have concluded that "the presence of Bence Jones proteinuria is pathognomonic for multiple myeloma."

It should be emphasized that many of the preceding tests do have considerable diagnostic value. Full appreciation of the limitations of certain tests further enhances their usefulness. A negative or normal result is often as significant as a positive finding. It may be anticipated that continued improvement of analytic technics in the laboratory will contribute substantially toward the goal of chemical and diagnostic specificity.

SUMMARY

This paper deals with a survey and discussion of variables and potential sources of error in the individual specimen that affect the chemical specificity of clinical laboratory tests. Special attention is directed to factors less subject to correction by means of the usual technics of quality control. These include consideration of chemical specificity, anticoagulants, stability of specimens, hemolysis, lipemia, control of temperature and pH, contamination, effect of medications, and normal physiologic variations in persons.

A number of so-called pathognomonic tests have also been examined with respect to their diagnostic specificity. These include (1) biochemical tests for cancer and pregnancy, (2) the p-toluenesulfonic acid test for lupus erythematosus, (3) leucine aminopeptidase for carcinoma of the pancreas, (4) amylase for acute pancreatitis, (5) acid phosphatase for carcinoma of the prostate gland, (6) porphobilinogen for acute porphyria, (7) Congo red absorption for amyloidosis, and (8) Bence Jones protein for multiple myeloma.

SUMMARIO IN INTERLINGUA

Iste communication es concernite con un revista e un discussion del variabiles e del causas potential de error in le specimen individual que affice le specificitate de tests al laboratorio clinic. Attention special es prestate al factores que es le minus apte a esser corrigite per le technicas usual del controlo de qualitate. Istos concerne factores como le specificitate chimic, le uso de anticoagulantes, le stabilitate del specimens, hemolyse, lipemia, regulation del temperatura e del pH, contamination, effecto de medicationes, e normal variationes physiologic in le subjecto individual.

Un numero de si-appellate tests pathognomonic esseva etiam examinate con respecto a lor specificitate diagnostic. Istos include (1) tests biochimic pro cancere e pregnantia, (2) le test a acido p-toluenosulfonic pro lupus erythematose, (3) aminopeptidase de leucina pro carcinoma del pancreas, (4) amylase pro pancreatitis acute, (5) phosphatase acidic pro carcinoma del glandula prostatic, (6) porphobilinogeno pro porphyria acute, (7) absorption de rubee Congo pro amyloidosis, e (8) proteina Bence Jones pro myeloma multiple.

REFERENCES

1. ABDERHALDEN, R.: Klinische Enzymologie. Die Fermente in der Pathogenese Diagnostic und Therapie. Stuttgart: Georg Thieme Verlag, 1958, p. 35.
2. ABUL-FADL, M. A. M., AND KING, E. J.: Properties of the acid phosphatases of erythrocytes and of the human prostate gland. Biochem. J., 45: 51–60, 1949.
3. ACKERMAN, L. V.: Surgical Pathology, Ed. 2. St. Louis: C. V. Mosby Company, 1959, p. 477.
4. ALTMAN, P. L., AND DITTMER, D. S.: Blood and Other Body Fluids. Washington, D. C.: Federation of American Societies for Experimental Biology, 1961, p. 19.

5. ANNINO, J. S., AND RELMAN, A. S.: The effect of eating on some of the clinically important chemical constituents of the blood. Am. J. Clin. Path., **31:** 155–159, 1959.

6. ARST, H. E., MANNING, R. T., AND DELP, M.: Serum leucine aminopeptidase activity: findings in carcinoma of the pancreas, pregnancy and other disorders. Am. J. M. Sc., **238:** 598–609, 1959.

7. BABSON, A. L.: Opportunities for application of the scientific method in the routine clinical laboratory. Am. J. M. Technol., **26:** 379–385, 1960.

8. BABSON, A. L., READ, P. A., AND PHILLIPS, G. E.: The importance of the substrate in assays of acid phosphatase in serum. Am. J. Clin. Path., **32:** 83–87, 1959.

9. BAILEY, H.: Emergency Surgery, Ed. 7. Baltimore: Williams & Wilkins Company, 1958, p. 334.

10. BALL, E. G.: Studies on oxidation-reduction. XXIII. Ascorbic acid. J. Biol. Chem., **118:** 219–239, 1937.

11. BARKOFF, J. R., AND SAWYER, F.: The *p*-toluene sulfonic acid test for systemic lupus erythematosus. A. M. A. Arch. Dermat., **80:** 445–446, 1959.

12. BARTLETT, G. R.: Influence of glucose and nucleoside metabolism on viability of erythrocytes during storage. A. M. A. Arch. Int. Med., **106:** 889–893, 1960.

13. BASSIR, O., AND HALL, J.: Photo-activation as a source of error in the cephalin cholesterol flocculation test. Scandinav. J. Clin. & Lab. Invest., **7:** 274–276, 1955.

14. BATES, R. G.: Electrometric pH Determinations: Theory and Practice. New York: John Wiley & Sons, Inc., 1954, p. 74.

15. BEHRENDT, H.: Chemistry of Erythrocytes. Springfield, Ill.: Charles C Thomas, Publisher, 1957.

16. BELL, W. N., AND JUMPER, E.: Evaluation of Tes-Tape as a quantitative indicator. J. A. M. A., **166:** 2145–2147, 1958.

17. BENEDICT, S. R.: The analysis of whole blood. II. The determination of sugar and of saccharoids (non-fermentable copper-reducing substances). J. Biol. Chem., **92:** 141–159, 1931.

18. BENENSON, A. S., THOMPSON, H. L., AND KLUGERMAN, M. R.: Application of laboratory controls in clinical chemistry. Am. J. Clin. Path., **25:** 575–584, 1955.

19. BERNARD, H. R., CRISCIONE, J. R., AND MOYER, C. A.: The pathologic significance of the serum amylase concentration. A. M. A. Arch. Surg., **79:** 311–316, 1959.

20. BERRYMAN, D. B., AND GEORGE, W. H. S.: The evaluation of serum amylase levels in nonpancreatic disease. A. M. A. Arch. Surg., **80:** 482–485, 1960.

21. BILLS, C. E., McDONALD, F. G., NIEDERMEIER, W., AND SCHWARTZ, M. C.: Sodium and potassium in foods and waters. Determination by the flame photometer. J. Am. Dietet. A., **25:** 304–314, 1949.

22. BODANSKY, A.: Phosphatase studies. II. Determination of serum phosphatase. Factors influencing the accuracy of the determination. J. Biol. Chem., **101:** 93–104, 1933.

23. VAN BOETZELAER, G. L., AND ZONDAG, H. A.: A rapid modification of the Pearson reaction for total serum cholesterol. Clin. Chim. Acta, **5:** 943–944, 1960.

24. BOGOCH, A., ROTH, J. L. A., AND BOCKUS, H. L.: The effects of morphine on serum amylase and lipase. Gastroenterology, **26:** 697–708, 1954.

25. BÖTTIGER, L. E.: Critical evaluation of the precipitation test for systemic lupus erythematosus. J. Lab. & Clin. Med., **52:** 909–911, 1958.

26. BOYER, P. A., LOWE, J. T., GARDIER, R. W., AND RALSTON, J. D.: Effect of a practical dietary regimen on serum cholesterol level. J. A. M. A., **170:** 257–261, 1959.

27. BRESSLAR, R., FORSYTH, B. R., AND KLATSKIN, G.: Serum leucine aminopeptidase activity in hepatobiliary and pancreatic disease. J. Lab. & Clin. Med., **56:** 417–430, 1960.

28. BUNDY, H. F., AND MEHL, J. W.: Trypsin inhibitors of human serum. I. Standardization, mechanism of reaction, and normal values. J. Clin. Invest., **37:** 947–955, 1958.

29. BUTLER, T. J.: The determination of blood glucose and urea nitrogen on a single microsample. Am. J. M. Technol., **27:** 205–213, 1961.

30. CABAUD, P. G., AND WRÓBLEWSKI, F.: Colorimetric measurement of lactic dehydrogenase activity of body fluids. Am. J. Clin. Path., **30:** 234–236, 1958.

31. CANDEL, S., AND WHEELOCK, M. C.: Serum amylase and serum lipase in mumps. Ann. Int. Med., **25:** 88–96, 1946.

32. CANTAROW, A., AND TRUMPER, M.: Clinical Biochemistry, Ed. 5. Philadelphia: W. B. Saunders Company, 1955, p. 253.

33. CARAWAY, W. T.: Determination of uric acid in serum by a carbonate method. Am. J. Clin. Path., **25:** 840–845, 1955.

34. CARAWAY, W. T.: Units of concentration. Am. J. Clin. Path., **29:** 493–495, 1958.

35. CARAWAY, W. T.: Chlorine in distilled water as a source of laboratory error. Clin. Chem., **4:** 513–518, 1958.

36. CARAWAY, W. T.: Microchemical Methods for Blood Analysis. Springfield, Ill.: Charles C Thomas, Publisher, 1960, p. 78.

37. CARAWAY, W. T.: Unpublished observations.

38. CARAWAY, W. T., AND FANGER, H.: Ultramicro procedures in clinical chemistry. Am. J. Clin. Path., **25:** 317–331, 1955.

39. CARR, M. H., AND FRANK, H. A.: Calcium analysis in patients being treated with EDTA. Clin. Chem., **3:** 20–21, 1957.

40. CASTER, W. O., SIMON, A. B., AND ARMSTRONG, W. D.: Analytical problems in the determination of Evans Blue caused by absorption on glass and protein surfaces. Anal. Chem., **26:** 713–715, 1954.

41. CAWLEY, L. P., SPEAR, F. E., AND KENDALL, R.: Ultramicro chemical analysis of blood glucose with glucose oxidase. Am. J. Clin. Path., **32:** 195–200, 1959.

42. CLARK, M. B.: Studies based on errors observed in the use of anticoagulants in blood chemistry determinations. Am. J. M. Technol., **17:** 190–197, 1951.

43. COMER, J. P.: Semiquantitative specific test paper for glucose in urine. Anal. Chem., **28:** 1748–1750, 1956.

44. Conn, H. O.: Effect of heparin on the blood ammonia determination. New England J. Med., **262:** 1103–1107, 1960.

45. Cremer, R. J., Perryman, P. W., and Richards, D. H.: Influence of light on the hyperbilirubinemia of infants. Lancet, **274:** 1094–1097, 1958.

46. Crout, J. R., and Sjoerdsma, A.: The clinical and laboratory significance of serotonin and catechol amines in bananas. New England J. Med., **261:** 23–26, 1959.

47. Darrow, D. C.: Body-fluid physiology: the role of potassium in clinical disturbances of body water and electrolyte. New England J. Med., **242:** 978–983; 1014–1018, 1950.

48. Davison, M. M.: Stability of acid phosphatase in frozen serum. Am. J. Clin. Path., **23:** 411, 1953.

49. Dingemanse, E., Huis in 't Veld, L. G., and Hartogh-Katz, S. L.: Clinical method for the chromatographic-colorimetric determination of urinary 17-ketosteroids. II. Normal adults. J. Clin. Endocrinol., **12:** 66–85, 1952.

50. Dobson, W. G., and Gornall, A. G.: Comments on the Richardson pregnancy test and the Rapp-Richardson saliva prenatal sex determination test. Am. J. Obst. & Gynec., **72:** 70–74, 1956.

51. Dubbs, C. A., Vivonia, C., and Hilburn, J. M.: Subfractionation of human serum enzymes. Science, **131:** 1529–1530, 1960.

52. DuBois, E. L., Rosenfeld, S., and Ohtomo, A.: Precipitation test and systemic lupus erythematosus. J. A. M. A., **168:** 813–814, 1958.

53. Ellerbrook, L. D., and Davis, J. H.: Effect of traces of copper on hemoglobin determinations. A potential source of error. Am. J. Clin. Path., **24:** 607–611, 1954.

54. von Euler, U. S., Hellner-Björkman, S., and Orwén, I.: Diurnal variations in the excretion of free and conjugated noradrenaline and adrenaline in urine from healthy subjects. Acta physiol. scandinav., **33** (Suppl. 118): 10–16, 1955.

55. Fales, F. W., Russell, J. A., and Fain, J. N.: Some applications and limitations of the enzymic, reducing (Somogyi), and anthrone methods for estimating sugars. Clin. Chem., **7:** 389–403, 1961.

56. Fashena, G. J.: On the nature of the saccharoid fraction of human blood. J. Biol. Chem., **100:** 357–363, 1933.

57. Fashena, G. J., and Stiff, H. A.: On the nature of the saccharoid fraction of human blood. II. Identification of glucuronic acid. J. Biol. Chem., **137:** 21–27, 1941.

58. Fawcett, J. K., and Wynn, V.: Effects of posture on plasma volume and some blood constituents. J. Clin. Path., **13:** 304–310, 1960.

59. Fischer, R. H., and McColgan, S. P.: The Merkel modification of the Richardson test for the diagnosis of pregnancy. Am. J. Obst. & Gynec., **65:** 628–632, 1953.

60. Fleisher, G. A., Potter, C. S., and Wakim, K. G.: Separation of 2 glutamic-oxaloacetic transaminases by paper electrophoresis. Proc. Soc. Exper. Biol. & Med., **103:** 229–231, 1960.

61. Folin, O., and Wu, H.: A system of blood analysis. A simplified and improved method for determination of sugar. J. Biol. Chem., **41:** 367–374, 1920.

62. Forrest, I. S., and Forrest, F. M.: Urine color test for the detection of phenothiazine compounds. Clin. Chem., **6:** 11–15, 1960.

63. Forrest, I. S., and Forrest, F. M.: Urine color test for the detection of phenothiazine compounds (supplement to). Clin. Chem., **6:** 362–363, 1960.

64. Foulk, W. T., and Fleisher, G. A.: The effect of opiates on the activity of serum transaminase. Proc. Staff Meet. Mayo Clin., **32:** 405–410, 1957.

65. Fredenburgh, E. J., and Hecht, B. P.: Sources of error in the turbidimetric determination of serum globulin with ammonium sulfate. Am. J. Clin. Path., **22:** 592–597, 1952.

66. Free, A. H., Adams, E. C., Kercher, M. L., Free, H. M., and Cook, M. H.: Simple specific test for urine glucose. Clin. Chem., **3:** 163–168, 1957.

67. Free, A. H., and Fancher, O. E.: Urine protein tests in presence of tolbutamide metabolite. Am. J. M. Technol., **24:** 64–65, 1958.

68. Free, H. M., Collins, G. F., and Free, A. H.: Triple-test strip for urinary glucose, protein and pH. Clin. Chem., **6:** 352–361, 1960.

69. Freeman, R. B., and Duncan, G. G.: Chlorothiazide-induced hyperuricemia; report of two cases. Metabolism, **9:** 1107–1110, 1960.

70. Freier, E. F., and Rausch, V. L.: Quality control in clinical chemistry. Am. J. M. Technol., **24:** 195–207, 1958.

71. Gafni, J., and Sohar, E.: Rectal biopsy for the diagnosis of amyloidosis. Am. J. M. Sc., **240:** 332–336, 1960.

72. Gambino, S. R.: Heparinized vacuum tubes for determination of plasma pH, plasma CO_2 content, and blood oxygen saturation. Am. J. Clin. Path., **32:** 285–293, 1959.

73. Goldbarg, J. A., and Rutenburg, A. M.: The colorimetric determination of leucine aminopeptidase in urine and serum of normal subjects and patients with cancer and other diseases. Cancer, **11:** 283–291, 1958.

74. Goldstein, N. P., Epstein, J. H., and Roe, J. H.: Studies of pancreatic function. IV. A simplified method for the determination of serum lipase, using aqueous tributyrin as substrate, with one hundred normal values by this method. J. Lab. & Clin. Med., **33:** 1047–1051, 1948.

75. Goodman, J. R., Vincent, J., and Rosen, I.: Serum potassium changes in blood clots. Am. J. Clin. Path., **24:** 111–113, 1954.

76. Green, A. A.: The preparation of acetate and phosphate buffer solutions of known pH and ionic strength. J. Am. Chem. Soc., **55:** 2331–2336, 1933.

77. Gross, J. B., Comfort, M. W., Mathieson, D. R., and Power, M. H.: Elevated value for serum amylase and lipase following the administration of opiates: a preliminary report. Proc. Staff Meet. Mayo Clin., **26:** 81–87, 1951.

78. Gutman, A. B., and Yü, T. F.: Current prin-

ciples of management in gout. Am. J. Med., **13:** 744–759, 1952.

79. HADLEY, G. G., AND LARSON, N. L.: Use of sequestrene as an anticoagulant. Am. J. Clin. Path., **23:** 613–618, 1953.

80. HADLEY, G. G., AND WEISS, S. P.: Further notes on use of salts of ethylenediamine tetraacetic acid (EDTA) as anticoagulants. Am. J. Clin. Path., **25:** 1090–1093, 1955.

81. HAMBURGER, C.: Six years' daily 17-ketosteroid determinations in one subject. Seasonal variations and independence of volume of urine. Acta endocrinol., **17:** 116–127, 1954.

82. HAMMOND, R. L., AND WELCKER, M. L.: Porphobilinogen tests on a thousand miscellaneous patients in a search for false positive reactions. J. Lab. & Clin. Med., **33:** 1254–1257, 1948.

83. HAUGEN, H. N.: Glucose and acetone as sources of error in plasma "creatinine" determinations. Scandinav. J. Clin. & Lab. Invest., **6:** 17–21, 1954.

84. HENRY, R. J.: Use of the control chart in clinical chemistry. Clin. Chem., **5:** 309–319, 1959.

85. HENRY, R. J., AND BERKMAN, S.: Absorbance of various protein-free filtrates of serum. Clin. Chem., **3:** 711–715, 1957.

86. HENRY, R. J., AND CHIAMORI, N.: Variation of pH of clinical samples as a source of error in enzyme determinations. Clin. Chem., **5:** 402–404, 1959.

87. HENRY, R. J., AND CHIAMORI, N.: Study of the saccharogenic method for the determination of serum and urine amylase. Clin. Chem., **6:** 434–452, 1960.

88. HENRY, R. J., GOLUB, O. J., BERKMAN, S., AND SEGALOVE, M.: Critique on the icterus index determination. Am. J. Clin. Path., **23:** 841–853, 1953.

89. HENRY, R. J., GOLUB, O. J., AND SOBEL, C.: Some of the variables involved in the fractionation of serum proteins by paper electrophoresis. Clin. Chem., **3:** 49–64, 1957.

90. HENRY, R. J., SOBEL, C., AND BERKMAN, S.: On the determination of "pancreatic lipase" in serum. Clin. Chem., **3:** 77–89, 1957.

91. HENRY, R. J., SOBEL, C., AND SEGALOVE, M.: Turbidimetric determination of proteins with sulfosalicylic and trichloroacetic acids. Proc. Soc. Exper. Biol. & Med., **92:** 748–751, 1956.

92. HOCK, E., AND TESSIER, R. N.: Elevation of serum acid phosphatase following prostatic massage. J. Urol., **62:** 488–491, 1949.

93. HOFFMAN, E., NACHLAS, M. M., GABY, S. D., ABRAMS, S. J., AND SELIGMAN, A. M.: Limitations in the diagnostic value of serum leucine aminopeptidase. New England J. Med., **263:** 541–544, 1960.

94. HONET, J. C., CASEY, T. V., AND RUNYAN, J. W., JR.: False-positive urinary test for 5-hydroxyindoleacetic acid due to methocarbamol and mephenesin carbamate. New England J. Med., **261:** 188–190, 1959.

95. HUANG, T. C., WEFLER, V., AND RAFTERY, A.: A rapid serum cholesterol determination by using a combination of the Liebermann-Burchard and Pearson, Stern, and Mc-

Gavack methods. Clin. Chem., **6:** 407, 1960.

96. HURT, R.: The effect of radiographic contrast media on urinalysis. Am. J. M. Technol., **26:** 122–124, 1960.

97. JACKLER, J. M.: Interpretations of discrepancies in laboratory reports. J. Maine M. A., **47:** 46–47, 1956.

98. JACOBSSON, K.: The determination of tartrate-inhibited phosphatase in serum. Scandinav. J. Clin. & Lab. Invest., **12:** 367–380, 1960.

99. JANOWITZ, H. D., AND DREILING, D. A.: The plasma amylase. Am. J. Med., **27:** 924–935, 1959.

100. JONES, K. K., AND THOMPSON, H. E.: Evaluation of simple precipitation test for systemic lupus erythematosus. J. A. M. A., **166:** 1424–1428, 1958.

101. JORGENSEN, S., AND THEIL NIELSON, A. A.: Uric acid in human blood corpuscles and plasma. Scandinav. J. Clin. & Lab. Invest., **8:** 108–112, 1956.

102. JOSEPHSON, B., AND DAHLBERG, G.: Variations in the cell-content and chemical composition of the human blood due to age, sex and season. Scandinav. J. Clin. & Lab. Invest., **4:** 216–236, 1952.

103. KAPLAN, A., AND NARAHARA, A.: The determination of serum alkaline phosphatase activity. J. Lab. & Clin. Med., **41:** 819–824, 1953.

104. KAPLAN, N. O., CIOTTI, M. M., HAMOLSKY, M., AND BIEBER, R. E.: Molecular heterogeneity and evolution of enzymes. Science, **131:** 392–397, 1960.

105. KEIDING, N. R.: Differentiation into three fractions of the serum alkaline phosphatase and the behavior of the fractions in diseases of bone and liver. Scandinav. J. Clin. & Lab. Invest., **11:** 106–112, 1959.

106. KERN, A., AND STRANSKY, E.: Beitrag zur kolorimetrischen Bestimmung der Harnsäure. Biochem. Ztschr., **290:** 419–427, 1937.

107. KING, E. J., AND JEGATHEESAN, K. A.: A method for the determination of tartrate-labile, prostatic acid phosphatase in serum. J. Clin. Path., **12:** 85–89, 1959.

108. KING, J.: A routine method for the estimation of lactic dehydrogenase activity. J. M. Lab. Technol., **16:** 265–272, 1959.

109. KINGSLEY, G. R., AND GETCHELL, G.: Direct ultramicro glucose oxidase method for determination of glucose in biologic fluids. Clin. Chem., **6:** 466–475, 1960.

110. KOJ, A., ZGLICZYNSKI, J. M., AND FRENDO, J.: Distribution of transaminases in human blood cells. Clin. Chim. Acta, **5:** 339–344, 1960.

111. LARON, Z., AND EPSTEIN-HALBERSTADT, B.: Activity of acid phosphatase in the serum of normal infants and children. Pediatrics, **26:** 281–284, 1960.

112. LAZARONI, J. A., JR., MAIER, E. C., AND GORCZYCA, L. R.: The stability of lactic dehydrogenase in serum. Clin. Chem., **4:** 379–381, 1958.

113. LEONARDS, J. R.: Evaluation of enzyme tests for urinary glucose. J. A. M. A., **163:** 260, 1957.

114. LIDDLE, G. W.: Tests of pituitary-adrenal

suppressibility in the diagnosis of Cushing's syndrome. J. Clin. Endocrinol., **20:** 1539–1560, 1960.

115. LONGFIELD, G. M., HOLLAND, D. E., LAKE, A. J., AND KNIGHTS, E. M., JR.: Comparison studies of simplified tests for glucosuria and proteinuria. Am. J. Clin. Path., **33:** 550–552, 1960.

116. MALINOWSKI, T. S.: Clinical value of serum amylase determination. J. A. M. A., **149:** 1380–1385, 1952.

117. MARKOVITZ, M.: Chromogens in the urine of normal individuals and of epileptic patients which react with Ehrlich's aldehyde reagent. J. Lab. & Clin. Med., **50:** 367–371, 1957.

118. MATHER, A., AND MACKIE, N. R.: Effects of hemolysis on serum electrolyte values. Clin. Chem., **6:** 223–227, 1960.

119. McGANN, C. J., AND CARTER, R. E.: The effect of hemolysis on the Van den Bergh reaction for serum bilirubin. J. Pediat., **57:** 199–203, 1960.

120. McGEACHIN, R. L., DAUGHERTY, H. K., HARGAN, L. A., AND POTTER, B. A.: The effect of blood anticoagulants on serum and plasma amylase activities. Clin. Chim. Acta, **2:** 75–77, 1957.

121. McGEACHIN, R. L., AND LEWIS, J. P.: Electrophoretic behavior of serum amylase. J. Biol. Chem., **234:** 795–798, 1959.

122. MEITES, S., AND HOGG, C. K.: Direct spectrophotometry of total serum bilirubin in the newborn. Clin. Chem., **6:** 421–428, 1960.

123. MERKEL, R. L.: A comparative study of chemical tests for the early diagnosis of pregnancy, including a new colorimetric determination. Am. J. Obst. & Gynec., **60:** 827–833, 1950.

124. MIGEON, C. J., TYLER, F. H., MAHONEY, J. P., FLORENTIN, A. A., CASTLE, H., BLISS, E. L., AND SAMUELS, L. T.: The diurnal variation of plasma levels, and urinary excretion of 17-hydroxycorticosteroids in normal subjects, night workers and blind subjects. J. Clin. Endocrinol., **16:** 622–633, 1956.

125. MOHLER, D. N.: Evaluation of the urine test for serotonin metabolites. J. A. M. A., **163:** 1138, 1957.

126. NEUBERG, H. W.: Streptomycin as a cause of false-positive Benedict reaction for glycosuria. Am. J. Clin. Path., **24:** 245–246, 1954.

127. O'HAGAN, J. E., HAMILTON, T., LE BRETON, E. G., AND SHAW, A. E.: Human serum bilirubin. An immediate method of determination and its application to the establishment of normal values. Clin. Chem., **3:** 609–623, 1957.

128. OZAR, M. B., ISAAC, C. A., AND VALK, W. L.: Methods for the elimination of errors in serum acid phosphatase determinations. J. Urol., **74:** 150–157, 1955.

129. PEARSON, C. M.: Analysis of the precipitation test for systemic lupus erythematosus. J. A. M. A., **169:** 30–33, 1959.

130. PETERSON, J. E., WILCOX, A. A., HALEY, M. I., AND KEITH, R. A.: Hourly variation in total serum cholesterol. Circulation, **22:** 247–253, 1960.

131. PIERCE, C.: Assay and importance of serotonin and its metabolites. Am. J. Clin. Path., **30:** 230–233, 1958.

132. PUTNAM, F. W., AND MIYAKE, A.: On the nonidentity of Bence-Jones protein. Science, **120:** 848–849, 1954.

133. QUICK, A. J., GEORGATSOS, J. G., AND HUSSEY, C. V.: The clotting activity of human erythrocytes: Theoretical and clinical implications. Am. J. M. Sc., **228:** 207–213, 1954.

134. QUINN, J. A., KATZ, S. A., AND RAPPOPORT, A. E.: Optical density of serum in cancer. Am. J. Clin. Path., **25:** 1128–1147, 1955.

135. RAMSAY, W. N. M.: The determination of iron in blood plasma or serum. Clin. Chim. Acta, **2:** 214–220, 1957.

136. RAMSAY, W. N. M.: Plasma iron. *In* SOBOTKA, H., AND STEWART, C. P.: Advances in Clinical Chemistry, Vol. I. New York: Academic Press, Inc., 1958, pp. 1–39.

137. RAY, H. E., HILL, J. H., AND PEACOCK, A. C.: The optical density of serum from cancerous and noncancerous patients. Am. J. Clin. Path., **29:** 25–28, 1958.

138. REINHOLD, J. G., AND YONAN, V. L.: The thymol test. A study of factors affecting its accuracy and description of a modified technic. Am. J. Clin. Path., **26:** 669–677, 1956.

139. REMP, D. G.: Influence of saccharoids on blood glucose values. Henry Ford Hosp. M. Bull., **1:** 24–26, 1953.

140. RICE, E. W.: Principles and Methods of Clinical Chemistry for Medical Technologists. Springfield, Ill.: Charles C Thomas, Publisher, 1960, p. 71.

141. RICE, E. W., AND LUKASIEWICZ, D. B.: Interference of bromide in the Zak ferric chloride-sulfuric acid cholesterol method, and means of eliminating this interference. Clin. Chem., **3:** 160–162, 1957.

142. RICHARDSON, G. C.: A new biochemical test for pregnancy. Am. J. Obst. & Gynec., **61:** 1317–1323, 1951.

143. ROSENBERG, I. N.: Zone electrophoretic studies of serum alkaline phosphatase. J. Clin. Invest., **38:** 630–644, 1959.

144. ROSS, G., WEINSTEIN, B., AND KABAKOW, B.: The influence of phenothiazine and some of its derivatives on the determination of 5-hydroxyindoleacetic acid in urine. Clin. Chem., **4:** 66–76, 1958.

145. RUTENBURG, A. M., GOLDBARG, J. A., AND PINEDA, E. P.: Leucine aminopeptidase activity. Observations in patients with cancer of the pancreas and other diseases. New England J. Med., **259:** 469–472, 1958.

146. SAGI, E. S., AND PLESS, J.: Differentiation of sera on phenol. Bull. New York Acad. Med., **30:** 693–710, 1954.

147. SAIFER, A., AND GERSTENFELD, S.: The photometric microdetermination of blood glucose with glucose oxidase. J. Lab. & Clin. Med., **51:** 448–460, 1958.

148. SAIFER, A., AND ZYMARIS, M. C.: Effect of shaking on the accuracy of salt fractionation methods for serum albumin. Clin. Chem., **1:** 180–189, 1955.

149. SALENIUS, P.: A study of the pH and buffer capacity of blood, plasma and red blood

cells. Scandinav. J. Clin. & Lab. Invest., **9:** 160–167, 1957.

150. SALTZMAN, A., CARAWAY, W. T., AND BECK, I. A.: Serum glucuronic acid levels in diabetes mellitus. Metabolism, **3:** 11–15, 1954.

151. SALVESEN, S., AND NISSEN-MEYER, R.: Influence of meprobamate therapy on the estimation of 17-ketosteroids and 17-ketogenic steroids. J. Clin. Endocrinol., **17:** 914–915, 1957.

152. SAWICKI, E.: Color test for sugar and blood in urine. Chem. Analyst, **45:** 45–46, 1956.

153. SAX, S. M., WAXMAN, H. E., AARONS, J. H., AND LYNCH, H. J.: Effect of orally administered quinine and quinidine on apparent values for urinary catecholamines. Clin. Chem., **6:** 168–175, 1960.

154. SCHNEEBERG, N. G.: Normal protein-bound iodine values in hyperthyroidism and cretinism. Am. J. M. Sc., **240:** 552–560, 1960.

155. SCHWARTZ, M. K., DANIEL, O., YING, S. H., AND BODANSKY, O.: Effect of storage in deep freeze upon activity of urinary acid phosphatase. Am. J. Clin. Path., **26:** 513–516, 1956.

156. SELIGSON, D.: Standardization of methods in clinical chemistry. Clin. Chem., **3:** 425–431, 1957.

157. SELIGSON, D., AND GENTZKOW, C. J.: General chemical technique. *In* SIMMONS, J. S., AND GENTZKOW, C. J.: Medical and Public Health Laboratory Methods. Philadelphia: Lea & Febiger, 1955, pp. 233–263.

158. SELIGSON, D., AND HIRAHARA, K.: The measurement of ammonia in whole blood, erythrocytes, and plasma. J. Lab. & Clin. Med., **49:** 962–974, 1957.

159. SELIGSON, D., MARINO, J., AND DODSON, E.: Determination of sulfobromophthalein in serum. Clin. Chem., **3:** 638–645, 1957.

160. SELLERS, T. F., JR.: An evaluation of the Sagi-Pless cancer detection test. J. Lab. & Clin. Med., **50:** 141–145, 1957.

161. SELTZER, H. S., AND LOVEALL, M. J.: Improved accuracy of Tes-Tape in estimating concentrations of urinary glucose. J. A. M. A., **167:** 1826–1830, 1958.

162. SHAPIRO, R.: The effect of maternal ingestion of iophenoxic acid on the serum protein-bound iodine of the progeny. New England J. Med., **264:** 378–381, 1961.

163. SHERMAN, H. C., CALDWELL, M. L., AND ADAMS, M.: A quantitative comparison of the influence of neutral salts on the activity of pancreatic amylase. J. Am. Chem. Soc., **50:** 2538–2543, 1928.

164. SHINOWARA, G. Y.: Spectrophotometric studies on blood serum and plasma. Am. J. Clin. Path., **24:** 696–710, 1954.

165. SIMON, H. B., AND NYGAARD, K. K.: Clinical interpretation of total serum and "prostatic" acid phosphatase level. J. A. M. A., **171:** 1933–1937, 1959.

166. SNAPPER, I., AND ORES, R. O.: Determination of Bence-Jones protein in the urine. J. A. M. A., **173:** 1137–1139, 1960.

167. SOMOGYI, M.: A method for the preparation of blood filtrates for the determination of sugar. J. Biol. Chem., **86:** 655–663, 1930.

168. SOMOGYI, M.: Modifications of two methods for the assay of amylase. Clin. Chem., **6:** 23–35, 1960.

169. STRAUMFJORD, J. V.: Determination of blood pH. *In* SELIGSON, D.: Standard Methods of Clinical Chemistry, Vol. II. New York: Academic Press, Inc., 1958, pp. 107–121.

170. SUNDERMAN, F. W., MacFATE, R. P., EVANS, G. T., AND FULLER, J. B.: Symposium on blood glucose. Am. J. Clin. Path., **21:** 901–934, 1951.

171. SYLVESTER, R. L., SLUKA, G. J., KANABROCKI, E. L., AND RUBNITZ, M. E.: Observations on the precipitation test for diagnosis of systemic lupus erythematosus. Am. J. Clin. Path., **32:** 45–47, 1959.

172. TAKEHARA, H.: On the action of arginase in the organism. III. Clinical investigation on the blood arginase. J. Biochem. (Japan), **28:** 451–462, 1938.

173. TAYLOR, H. L., ERICKSON, L., HENSCHEL, A., AND KEYS, A.: The effect of bed rest on the blood volume of normal young men. Am. J. Physiol., **144:** 227–232, 1945.

174. THORNTON, G. H. M., AND ILLINGWORTH, D. G.: An evaluation of the benzidine test for occult blood in the feces. Gastroenterology, **28:** 593–605, 1955.

175. VINCENT, D., AND SEGONZAC, G.: Le dosage de l'arginase sérique en pratique clinique. Rev. franç. études clin. et biol., **5:** 390–392, 1960.

176. VOLK, B. W., SAIFER, A., AND LAZARUS, S. S.: Blood saccharoids in diabetes mellitus. J. Lab. & Clin. Med., **57:** 367–376, 1961.

177. WACHTER, J. P., SMEBY, R. R., AND FREE, A. H.: Urinalysis and oral hypoglycemic agents. Am. J. M. Technol., **26:** 125–130, 1960.

178. WATSON, C. J.: Studies of urobilinogen. II. Urobilinogen in the urine and feces of subjects without evidence of disease of the liver or biliary tract. Arch. Int. Med., **59:** 196–205, 1937.

179. WATSON, C. J., AND SCHWARTZ, S.: A simple test for urinary porphobilinogen. Proc. Soc. Exper. Biol. & Med., **47:** 393–394, 1941.

180. WEBSTER, J. H., NEFF, J., SCHIAFFINO, S. S., AND RICHMOND, A. M.: Evaluation of serum potassium levels. Am. J. Clin. Path., **22:** 833–842, 1952.

181. WEISSMAN, M., AND KLEIN, B.: Evaluation of glucose determinations in untreated serum samples. Clin. Chem., **4:** 420–422, 1958.

182. WHITEHEAD, T. P., PRIOR, A. P., AND BARROWCLIFF, D. F.: Effect of rest and activity on the serum protein fractions. Am. J. Clin. Path., **24:** 1265–1268, 1954.

183. WIEG, H.: Elevation of blood urea N from the administration of chloral hypnotics. Am. J. M. Technol., **26:** 111–114, 1960.

184. WIEME, R. J.: Application diagnostique de l'enzymo-électrophorèse des déshydrogenases de l'acide lactique. Clin. Chim. Acta, **4:** 46–50, 1959.

185. WILLIAMS, R. J.: Biochemical Individuality. New York: John Wiley & Sons, pp. 1–214, 1956.

186. WISHINSKY, H., AND POOLE, E. L.: Fluctu-

ation of voltage as a source of error in coulometric titration. Am. J. Clin. Path., **34**: 195, 1960.

187. WOODARD, H. Q.: The clinical significance of serum acid phosphatase. Am. J. Med., **27**: 902–910, 1959.

188. WOOTTON, I. D. P.: Standardization in clinical chemistry. Clin. Chem., **3**: 401–405, 1957.

189. WRÓBLEWSKI, F.: The clinical significance of transaminase activities of serum. Am. J. Med., **27**: 911–923, 1959.

190. WRÓBLEWSKI, F., AND LaDUE, J. S.: Lactic dehydrogenase activity in blood. Proc. Soc. Exper. Biol. & Med., **90**: 210–213, 1955.

191. WRÓBLEWSKI, F., ROSS, C., AND GREGORY, K.: Isoenzymes and myocardial infarction. New England J. Med., **263**: 531–536, 1960.

192. YONAN, V. L., AND REINHOLD, J. G.: Effects of ambient temperature on thymol, phenol and zinc turbidity tests. Am. J. Clin. Path., **24**: 232–238, 1954.

193. YONAN, V. L., AND REINHOLD, J. G.: Effects of delayed examination on the results of certain hepatic tests. Clin. Chem., **3**: 685–690, 1957.

194. YÜ, T. F., AND GUTMAN, A. B.: Quantitative analysis of uric acid in blood and urine. Methods and interpretation. Bull. Rheumat. Dis., **7** (Suppl.): S17–S20, 1957.

195. ZAUCHAU-CHRISTIANSEN, B.: The variation in serum uric acid during 24 hours and from day to day. Scandinav. J. Clin. & Lab. Invest., **9**: 244–248, 1957.

196. ZLATKIS, A., ZAK, B., AND BOYLE, A. J.: A new method for the direct determination of serum cholesterol. J. Lab. & Clin. Med., **41**: 486–492, 1953.

197. ZUCKER, M. B., AND BORRELLI, J.: Platelets as a source of serum acid nitrophenyl-phosphatase. J. Clin. Invest., **38**: 148–154, 1959.

COMMENTARY TO

34. Vecchio, T. J. (1966)
Predictive Value of a Single Diagnostic Test in Unselected Populations.
New England Journal of Medicine 274(21): 1171–1173.

Tolbutamide (Orinase®) was the first successful oral hypoglycemic agent introduced into clinical practice in the 1950s (1). Tolbutamide, a sulfonylurea derivative, stimulates the beta cells in the pancreas to produce insulin. Unger and Madison in 1958 developed a tolbutamide challenge test to screen for diabetes (2). Patients with a glucose level of 80–84% of fasting levels after injection of a standard dose of tolbutamide were claimed to have a 50% probability of having diabetes (3). Thomas J. Vecchio and co-workers at Upjohn pharmaceutical, the manufacturer of tolbutamide, compared the tolbutamide test to the standard oral glucose tolerance test in 102 controls and 40 diabetics (4,5). The results were less than exciting. There was some crossover between the patients classified as diabetic by the glucose tolerance test and those termed diabetic by the new tolbutamide challenge test. Today, this challenge test is considered to have no value in the detection of diabetes (6).

These papers on tolbutaminde by Vecchio and co-workers and many others published during this period are of minor historical note. They demonstrate however how limited the statistical analysis of new tests were at the time. Vecchio in 1966 introduced the concept of the predictive value calculation and provided a statistical tool that accurately assessed the efficacy of a laboratory test. His 1966 paper presented here provided for the first time a single value on a % scale that directly stated the probability that a laboratory test would correctly predict a disease condition.

Vecchio first provides standard definitions for the sensitivity and specificity of a laboratory test. Sensitivity is the ability of a test to detect the disease. Specificity is the ability of a test to give a negative result in a patient who does not have the disease. The predictive value takes these analyses a step further. It calculates the probability that a positive or negative test result is correct. The positive predicative value of a test is obtained from,

$$\frac{\text{total positive test results}}{\text{true positive results} + \text{false positive results}} 100$$

The first application of Vecchio's predictive value model to published clinical laboratory data may have been by Robert S. Galen in 1974. In that same year a paper appeared in the literature that claimed that the presence of serum antibodies to either cow's milk protein or egg whites was a strong indicator of increased mortality following myocardial infarction (7). From the published data in the paper on the presence of antibodies to cow milk proteins Galen calculated the sensitivity of the test as 74.4%, the specificity as 54.0% and the predictive value as 26.6% (8). This means that almost three-fourths of all results will be false positives. Galen then applied the statistic to the results reported in another paper on a screening method for hepatocellular carcinoma using the alpha-fetoprotein test. He calculated a predictive value of 70% for this test which indicated a 30% false positive rate (9). In 1975 Galen, Reiffel and Gambino examined the efficiency of four cardiac marker enzymes and two isoenzyme markers in the diagnosis of acute myocardial infarction (10). In 100 patients studied only two indicators had a predictive value for myocardial infarction that was above 60%. The presence of an elevated CK-MB (Creatine kinase, EC 2.7.3.2) level or the presence of CK-MB on electrophoresis with a flipped LDH isoenzyme (Lactate dehydrogenase, EC 1.1.1.27) pattern both had a predictive value of 100%.

Galen and Gambino in 1975 published a book on the predictive value method (11). They applied the calculations to the data in numerous published articles and helped establish this parameter as one of the most important measures of a test's efficacy.

References

(1) Miller, W.L. and Dulin, W.E. (1956) Orinase, a new oral hypoglycemic compound. Science. 123(3197):584–585.

(2) Unger, R.H. and Madison, L.L. (1958) Comparison of response to intravenously administered sodium tolbutamide in mild diabetic and nondiabetic subjects. Journal of Clinical Investigation. 37(5):627–630.

(3) Carawy, W.T. (1970) Carbohydrates, in *Fundamentals of Clinical Chemistry*. Tietz, N.W. (ed), W.B. Saunders, Philadelphia, Chapter IV, pg 166.

(4) Vecchio, T.J., Smith, D.L., Oster, H.L., and Brill, R. (1964) Oral sodium tolbutamide in the diagnosis of diabetes mellitus. Diabetes. 13(1):30–36.

(5) Vecchio, T.J., Oster, H.L., and Smith, D.L. (1965) Oral sodium tolbutaminde and glucose tolerance tests. Archives of Internal Medicine. 115(2):161–166.

434

(6)   Sacks, D.B. (1999) Carbohydrates, in *Tietz Textbook of Clinical Chemistry*, 3rd Edition, Burtis, C.A. and Ashwood, E.R. (eds), W.B. Saunders, Philadelphia, Chapter XXIV, pg 776.

(7)   Davies, D.F., Johnson, A.P., Rees, B.W.G., Elwood, P.C., and Abernathy, M. (1974) Food antibodies and myocardial infarction. The Lancet. 303(7865):1012–1014.

(8)   Galen, R.S. (1974) Food antibodies and myocardial infarction. The Lancet. 304(7884):832.

(9)   Galen, R.S. (1974) False-positives. The Lancet. 304(7888):1081.

(10)   Galen, R.S., Reiffel, J.A., and Gambino, S.R. (1975) Diagnosis of acute myocardial infarction. Relative efficiency of serum enzyme and isoenzyme measurements. Journal of the American Medical Association. 232(2):145–147.

(11)   Galen, R.S. and Gambino, S.R. (1975) *Beyond Normality: The Predictive Value and Efficiency of Medical Diagnosis*. Wiley, New York.

New Engl. J. Med 1966; 1171–1173
Copyright ©1966 Massachusetts Medical Society. All rights reserved.
Reproduced with permission.

PREDICTIVE VALUE OF A SINGLE DIAGNOSTIC TEST IN UNSELECTED POPULATIONS

Thomas J. Vecchio, M.D.*

KALAMAZOO, MICHIGAN

WHEN a new test for a disease is being evaluated it is customary to perform the test in two selected groups of subjects: those with an indisputable diagnosis of the disease by other criteria; and those from the normal population who have no evidence of the disease and in whom all the factors known to result in a higher than normal risk of the disease can be excluded.[1-3] The test results may be expressed dichotomously as "positive" or "negative," or by some numerical units along a scale, usually with the bulk of values on either side of an arbitrary dividing point, but with some degree of overlap.

The reliability of such a test in distinguishing diseased from nondiseased persons is often defined by its *sensitivity* and *specificity*.[4] *Sensitivity* is the "ability of a test to give a positive finding when the person tested truly has the disease under study," and is calculated as follows:

$$\text{Sensitivity} = \frac{\text{diseased persons with positive test}}{\text{all diseased subjects tested}} \times 100.$$

Specificity is "the ability of the test to give a negative finding when the person tested is free of the disease under study." It is calculated as follows:

$$\text{Specificity} = \frac{\text{nondiseased persons negative to the test}}{\text{all nondiseased subjects tested}} \times 100.$$

*Manager of Medical Development, Upjohn Company.

Calculation of these indexes presupposes the ability to diagnose disease and nondisease more definitively by other means, which can be done during preliminary evaluation of the test.

Berkson[5] employs similar terms in his evaluation of test efficiency, using "utility" interchangeably with *sensitivity*, but employing the term "cost," which is the complement of specificity (100 per cent minus *specificity* in percentage).

When such a test is to be used in a large, unselected population it is important to know what its *predictive value* will be — that is, what is the likelihood that a subject yielding a positive test actually has the disease? Conversely, what is the likelihood that a subject with a negative test does not have the disease? This likelihood cannot be estimated directly from the *sensitivity* and *specificity* obtained in the preliminary test evaluation since it is related to the actual prevalence of the disease in the total population.[6]

The total population consists of a proportion (p) of subjects who actually have the disease, and the remainder of nondiseased subjects $(1 - p)$. The proportion of these subjects yielding a positive test will consist of pa, where a is the proportion of subjects with known disease yielding positive tests in the preliminary study *(sensitivity)*, plus $(1 - p)(1 - b)$, where b is the proportion of normal control subjects yielding a negative test in the preliminary study

1172   THE NEW ENGLAND JOURNAL OF MEDICINE   May 26, 1966

(specificity). It follows that the proportion of subjects in the total population yielding a negative test will consist of $p(1 - a)$ plus $(1 - p)b$. These relations are expressed in Table 1.

TABLE 1. *Proportions of Subjects in the Total Population Who Have or Do Not Have Actual Disease and the Proportions of Positive and Negative Tests in These Groups.**

CLASSIFICATION	TOTAL POPULATION	YIELDING POSITIVE TEST	YIELDING NEGATIVE TEST
Diseased	p	pa	p(1-a)
Nondiseased	(1-p)	(1-p)(1-b)	(1-p)b
Totals	p + (1-p)	pa + (1-p)(1-b)	p(1-a) + (1-p)b

* Symbols explained in text.

When test results are expressed dichotomously, the *predictive value* of a positive test may be defined as the percentage of times that a positive test will detect a diseased individual. This may be calculated as follows:

$$\text{Predictive value}^* = \frac{\text{number (or proportion) of diseased persons with positive test}}{\text{total number (or proportion) of persons with positive test}} \times 100.$$

From Table 1, this may be expressed as follows:

$$(1)\ \text{PV pos} = \frac{pa}{pa + (1 - p)(1 - b)} \times 100.$$

The *predictive value* of a negative test is the percentage of times that a negative test will detect a nondiseased person. This may be calculated as follows:

$$\text{Predictive value} = \frac{\text{number (or proportion of nondiseased persons with negative test}}{\text{total number (or proportion) of persons with negative test}} \times 100.$$

or, from Table 1:

$$(2)\ \text{PV neg} = \frac{(1 - p)b}{(1 - p)b + p(1 - a)} \times 100.$$

The differences of these equations from those for *sensitivity* and *specificity* (and "cost-utility") given above may be noted by comparison. The following example will illustrate this principle:

A test for diabetes mellitus is evaluated in 100 subjects with known diabetes and in 100 normal control subjects with no evidence of the disease nor of any factors known to result in increased risk of the disease. It is found that the diabetic group yields 95 per cent positive tests (*sensitivity*, 95 per cent), whereas the normal group has only 5 per cent positive results (*specificity*, 95 per cent). In comparison with other tests this test is considered highly accurate. What is the accuracy of a positive test in predicting diabetes in an unselected sample of 10,000 subjects, in whom the

*"Predictiveness" would be a better term but is, unfortunately, not recognized in *Webster's Unabridged Dictionary.*

actual prevalence of diabetes mellitus is 2 per cent?

By simple arithmetic one finds that there are 200 diabetic patients in the population, 190 of whom will have a positive test, and 10 a negative test. There are 9800 nondiabetic persons in this population, 9310 of whom will have a negative test, and 490 a positive test. Therefore, the *predictive value* of a positive test in detecting diabetes in the total population will be:

$$\frac{190}{190 + 490} = \frac{190}{680} = 0.279,\ \text{or } 27.9 \text{ per cent.}$$

Applying equation (1):

$$\text{PV pos} = \frac{(.02)(.95)}{(.02)(.95) + (.05)(.98)}$$

$$= \frac{.019}{.068} = 0.279,\ \text{or } 27.9 \text{ per cent.}$$

The *predictive value* of a negative test will be:

$$\frac{9310}{9310 + 10} = \frac{9310}{9320} = 0.999,\ \text{or } 99.9 \text{ per cent.}$$

Applying equation (2):

$$\text{PV neg} = \frac{(.98)(.95)}{(.98)(.95) + (.02)(.05)}$$

$$= \frac{0.931}{0.931 + 0.001} = 0.999\ \text{or } 99.9 \text{ per cent.}$$

RELATION OF PREDICTIVE VALUE TO PREVALENCE

Table 2 shows the *predictive values* of positive and negative tests at different actual disease prev-

TABLE 2. *Predictive Values of Positive and Negative Tests at Varying Disease Prevalences when Sensitivity and Specificity Each Equal 95 Per Cent.*

ACTUAL DISEASE PREVALENCE	PV POS	PV NEG
%	%	%
1	16.1	99.9
2	27.9	99.9
5	50.0	99.7
10	67.9	99.4
20	82.6	98.7
50	95.0	95.0
75	98.3	83.7
100	100.0	—

alences when *sensitivity* and *specificity* both are 95 per cent, as in the example given above. The predictive value of a positive test (PV pos) increases with increasing disease prevalence, and when prevalence is 50 per cent, PV pos equals *sensitivity* and PV neg equals *specificity*. Although the higher prevalences are unlikely in an unselected population, they may be obtained by preselection of the group to be tested, on the basis of historical or physical data, or some other test.

In the example given above, the actual prevalence of disease in the positive reactor group was shown to be 27.9 per cent under the conditions described. If a second test with similar, and not overlapping, sensitivity and specificity were to be performed in this group, its predictive value would be much increased, and would be somewhere over 82.6 per cent, as shown in Table 2, or actually 88.0 per cent, from equation (1):

$$PV\ pos = \frac{pa}{pa + (1 - p)(1 - b)} \times 100$$

$$= \frac{(.279)(.95)(100)}{(.279)(.95) + (.721)(.05)} = 88.0\ \text{per cent.}$$

The predictive value of a negative test (PV neg) is affected insignificantly except at very high prevalences.

RELATION OF PREDICTIVE VALUE TO SENSITIVITY AND SPECIFICITY

The value of PV pos when actual disease prevalence is 2 per cent is given for a range of sensitivities and specificities in Table 3. It is seen that the predictive value of a positive test is dependent

TABLE 3. *Predictive Value of a Positive Test (PV pos) over a Range of Sensitivities and Specificities when Actual Disease Prevalence is 2 Per Cent.*

SPECIFICITY	SENSITIVITY							
%	50%	60%	70%	80%	90%	95%	98%	99%
50	2.0	2.4	2.8	3.2	3.5	3.7	3.8	3.9
60	2.5	3.0	3.4	3.9	4.4	4.6	4.8	4.8
70	3.3	3.9	4.5	5.2	5.8	6.1	6.2	6.3
80	4.8	5.8	6.7	7.6	8.4	8.8	9.1	9.2
90	9.2	10.9	12.5	14.0	15.5	16.2	16.7	16.8
95	17.0	19.7	22.2	24.6	26.9	27.9	28.6	28.8
98	33.8	38.0	41.7	44.9	47.9	49.2	50.0	50.2
99	50.5	55.0	58.8	62.0	64.7	66.0	66.7	66.9

chiefly on the specificity of the test, but that at this disease prevalence PV pos has a maximal value of 66.9 per cent even at very high sensitivity (99 per cent) and specificity (99 per cent).

The values of PV neg at a 2 per cent disease prevalence are shown in Table 4. Sensitivity and specificity have relatively little effect on this parameter.

DISCUSSION

Despite apparent high efficiency of a diagnostic test in preliminary evaluation in groups of known diseased and nondiseased subjects, one must be careful when applying the results to unselected groups because of the magnification of false-positive errors by the relatively low prevalences of disease in the general population. The concepts of sensitivity and specificity are not in themselves adequate to predict test reliability under these circumstances. This may be done, however, if the new parameters PV pos and PV neg, the predictive values of positive

TABLE 4. *Predictive Value of a Negative Test (PV neg) over a Range of Sensitivities and Specificities when Actual Disease Prevalence is 2 Per Cent.*

SPECIFICITY	SENSITIVITY							
%	50%	60%	70%	80%	90%	95%	98%	99%
50	98.0	98.4	98.8	99.2	99.6	99.8	99.9	99.9
60	98.3	98.6	99.0	99.3	99.7	99.8	99.9	100.0
70	98.6	98.8	99.1	99.4	99.7	99.8	99.9	100.0
80	98.7	99.0	99.2	99.5	99.7	99.9	99.9	100.0
90	98.9	99.1	99.3	99.5	99.8	99.9	99.9	100.0
95	98.9	99.1	99.4	99.6	99.8	99.9	100.0	100.0
98	99.0	99.2	99.4	99.6	99.8	99.9	100.0	100.0
99	99.0	99.2	99.4	99.6	99.8	99.9	100.0	100.0

and negative tests, respectively, in unselected populations, are employed. These take into account the known or assumed actual prevalence of disease in the general population.

Errors may be minimized by an increase in the specificity of the test, even at the cost of decreased sensitivity, and by preselection of subjects to produce a higher disease prevalence in the population to be tested. Other tests should be performed in the positive group, when available, to attempt confirmation of the diagnosis and elimination of false-positive errors. The preselection of the population by the first test will increase greatly the predictive value of a second test.

SUMMARY

A technic is described for the estimation of the predictive value of diagnostic test results in the subject tested when the sensitivity and specificity of the test and the prevalence of the disease in the population are known.

I am indebted to O. S. Carpenter, J. A. Hagans and R. D. Remington for valuable criticism and suggestions.

REFERENCES

1. Mosenthal, H. O., and Barry, E. Criteria for and interpretation of normal glucose tolerance tests. *Ann. Int. Med.* **33**:1175-1194, 1950.
2. *Idem.* Evaluation of blood sugar tests: significance of non-glucose reducing substances and arterio-venous blood sugar difference. *Am. J. Digest. Dis.* **13**:160-170, 1946.
3. Unger, R. H., and Madison, L. L. New diagnostic procedure for mild diabetes mellitus: evaluation of intravenous tolbutamide response test. *Diabetes* **7**:455-461, 1958.
4. United States Public Health Service, Division of Chronic Diseases. Thorner, R. M., and Remein, Q. R. *Principles and Procedures in the Evaluation of Screening for Disease.* 24 pp. Washington, D.C.: Government Printing Office, 1961. (Monograph No. 67.)
5. Berkson, J. 'Cost utility' as measure of efficacy of test. *J. Am. Statist. A.* **42**:246-255, 1947.
6. Chiang, C. L., Hodges, J. L., Jr., and Yerushalmy, J. Statistical problems in medical diagnosis. In Berkeley Symposium on Mathematical Statistics and Probability (III). *Contributions to Biology and Problems of Health: Proceedings of Third Berkeley Symposium of Mathematical Statistics and Probability, v. 4:* Held at Statistical Laboratory, University of California, December, 1954, July and August 1955. 187 pp. Berkeley: Univ. of California Press, 1956.

I n 1944 E. J. King and co-workers determined that the standard used by most laboratories in the United Kingdom for the determination of blood hemoglobin had an assigned value that was about 7% too low (1). This, they claimed, finally explained why the normal reference range for hemoglobin in England was lower compared to that in the United States (US) or Germany. In the US in 1947 Belk and Sunderman reported on the results of a proficiency survey for seven different analytes (2). The results were disturbing. For the 3.15 mmol/L (12.6 mg/dL) calcium sample for example, the range of results reported back from the 59 participating laboratories was 1.75–7.15 mmol/L (7.0–26.6 mg/dL). The two most common reasons for this poor performance as reported by the laboratory directors who participated in the survey were poor training and inadequate numbers of technicians. R. J. Henry and M. Segalove in 1952 reviewed the Belk and Sunderman report and disagreed with these reasons. They claimed incorrect standards were a major cause of the inaccurate results (3). In 1972 Pragay and co-workers collected samples of the calcium standard used by 11 different laboratories. Each standard was assayed in a single laboratory by atomic absorption. The range of values found was 2.38–2.88 mmol/L (9.8–11.5 mg/dL) (4). In 2004 the National Institute of Standards and Technology (NIST) estimated that the health care costs due to inaccurate calibration of serum calcium assays by clinical laboratories was between 60 and 199 million US dollars per year (5).

The clinical chemistry literature in the 1960s used many different and often times confusing terms to describe a standard. These included serum standard, primary serum standard, reference serum, secondary calibrator and others. Nathan Radin in this 1967 paper presented here clarified these terms and helped to focus the field of clinical chemistry on the importance of standards. He defined a primary standard (a pure chemical added by weight to a defined solvent) and a secondary standard or calibrator (a chemical added to a protein or serum base and assigned a value by assay with a reference method). He emphasized the importance of pure chemicals and helped promote the use of Standard Reference Materials (SRMs) from the NIST as source material for the preparation of standards and calibrators. In fact in 1963 Radin and Gramza defined for the first time the purity requirements for a pure cholesterol preparation suitable for standardization of clinical assays (6). Schaffer, Bowers and Melville have reviewed the history of the development and the use of SRMs in clinical chemistry (7).

Radin also defined reference samples as the serum based, stabilized materials used for quality control monitoring. He emphasized the importance of the reference method in the standardization process of clinical chemistry methods. In 1999 Louis Rosenfeld edited a collection of biographies and essays on the history of clinical chemistry (8). The collection contained reprints of 79 articles, 34 of which were biographies and the remainder were articles on the history of clinical chemistry, except for one. Radin's original article on what is a standard was reprinted in its entirety.

References

(1)   King, E.J., Gilchrist, M., and Matheson, A. (1944) The haemoglobin equivalent of the B.S.I. Haldane standard. British Medical Journal. 1(4337):250–252.

(2)   Belk, W.F. and Sunderman, F.W. (1947) A survey of the accuracy of chemical analyses in clinical laboratories. American Journal of Clinical Pathology. 17:853–861.

(3)   Henry, R.J. and Segalove, M. (1952) The running of standards in clinical chemistry and the use of the control chart. Journal of Clinical Pathology. 5(4):305–311.

(4)   Pragay, D.A., Cross, C.L., and Chilcote, M.E. (1972) Inaccurate calcium standards in hospital laboratories. Clinical Chemistry. 18(9):1037.

(5)   Gallaher, M.P., Mobley, L.R., Klee, G.G., and Schryver, P. (2004) Planning report 04-1, in *The Impact of Calibration Error in Medical Decision Making*. National Institute of Standards and Technology, Gaithersburg, MD, April.

(6)   Radin, N. and Gramza, A.L. (1963) Standard of purity for cholesterol. Clinical Chemistry. 9(2):121–134.

(7)   Schaffer, R., Bowers, G.N., and Melville, R.S. (1995) History of NIST's contribution to development of standard reference materials and reference and definitive methods for clinical chemistry. Clinical Chemistry. 41(9):1306–1312.

(8)   Rosenfeld, L. (1999) *Biographies and other essays on the history of clinical chemistry*. American Association for Clinical Chemistry, Inc., Washington, pgs 160–181.

Reprinted from CLINICAL CHEMISTRY
Vol. 13, No. 1, January 1967
Copyright © by American Association of Clinical Chemists
HOEBER MEDICAL DIVISION of Harper & Row, *Publishers*
Printed in *U.S.A.*

What Is a Standard?

Nathan Radin*

This paper reviews the definition and the use of standards and the standards available for clinical chemists, and discusses the limitations of various so-called standards.

THE POOR QUALITY of reagent chemicals in the United States by 1916 led to the statement by Hillebrand (*1*) that "the situation is particularly deplorable just now." In 1919, to overcome the handicap of poor reagent chemicals, representatives of industry, together with representatives of reagent users, undertook jointly under the sponsorship of the American Chemical Society, the task of establishing standards of quality for reagent chemicals that would be defined by unambiguous and, as far as possible, quantitative methods of examination. Specifications for reagent chemicals are now issued in the book *Reagent Chemicals, ACS Specifications* (*2*).

The statement that "the situation is particularly deplorable now" can be used in looking at the standards situation in the field of clinical chemistry. Manufacturers are putting serum controls or standards on the market with some claims for their use that can be misleading to purchasers.

It is the purpose of this paper to review the definition and use of standards, the standards that are available for clinical chemists, and to discuss the limitations of various so-called standards.

From the Division of Biochemistry, Rochester General Hospital, Rochester, N. Y. 14621.

This paper had its origin in a desire on the part of the Committee on Standards and Controls to consider the present standards in clinical chemistry and the possible adoption of proposed definitions of standards and reference materials. It is not to be construed as an authoritative statement of the Committee or as an official view of the American Association of Clinical Chemists. The members of the Committee have not reached a consensus about the contents of the paper but do agree concerning the desirability of a publication on the subject. It is presented here as a review of the situation and an indication of the desirability of continuing improvement in analytical clinical chemistry.

The author gratefully expresses his appreciation to Dr. Alan Mather for his criticism and discussion of and contribution to some of the material that was used in this paper. Appreciation is also expressed to the Committee on Standards and Controls, and in particular to Dr. Joseph H. Boutwell, Jr., and Dr. Richard J. Henry, for their review and suggestions.

Received for publication Mar. 22, 1966; accepted for publication Oct. 11. 1966.

*Present address: Harrisburg Hospital Institute of Pathology and Research, Harrisburg, Pa. 17101.

Standards

Measurement can be interpreted as the estimate of the dimension, quantity, or capacity of an unknown in terms of the corresponding characteristics or properties of a known standard. Measurement can also be interpreted as the estimate of the ratio of the measure of a standard to that of an unknown. For example, the mass of an object is determined by balancing it with a set of weights that have been calibrated against a standard mass. The length of a line is measured with a ruler that has been calibrated against a standard length. Properties, such as absorbance, are often measured directly by setting up a standard series with respect to the property and assigning numbers, in arbitrary units, to each member of the series using appropriate operational procedures. Some member of this standard series can always be found that is equal in measure to that of an unknown; thus, the unknown is measured in terms of the standard series.

In setting up an analytical procedure an individual may define a standard for measuring a particular unknown. Others, using the same analytical procedure, must be able to reproduce the standard used by the originator. To be of value, any standard must be of demonstrated and widely accepted validity. An authority that establishes standards should be in a position to make sound recommendations regarding a particular standard. The authority itself should consist of both the originator and the user of each standard. If the originator and user of a standard disagree about the standard, a third party, perhaps a government agency, or an impartial group, should act as referee.

The National Bureau of Standards

It is probable that analytical chemists originally exchanged samples of a substance to check the accuracy of their analyses. The substance exchanged could be called the original reference standards. With the necessity for wider distribution of such samples a demand arose for a central source of such material. The National Bureau of Standards program on standard samples began soon after the organization of the bureau in 1901.

The primary purpose of the program is to provide a central basis for uniformity and accuracy of measurement by means of a continuing availability of Standard Materials from a common source for both science and industry. More than 600 different standards of metals, ores, ceramics, chemicals, and hydrocarbons are now available for distribution (3).

Nongovernmental Organizations

A partial list of professional organizations that publish official or standard methods and procedures include the American Pharmaceutical Association, the American Public Health Association, the American Society for Testing Materials, and the Association of Official Agricultural Chemists. Such organizations as the International Union of Pure and Applied Chemistry, the American Association of Clinical Chemists, the College of American Pathologists, the American Society of Clinical Pathologists, and the American Chemical Society have commissions and/or committees that are concerned with such matters as standards, quality control, and the standardization of nomenclature. These committees or commissions release periodic reports that contain recommendations pertaining to a particular study. In order for recommendations to be accepted as the standard specifications for a particular field, it is necessary that they be recognized as such by all concerned.

Specification Sources

The National Bureau of Standards (NBS) is the source of classifications and tolerances for laboratory weights (4). The NBS also lists specifications for volumetric glassware used in the laboratory (5). Further specifications for volumetric glassware were recommended by the Committee on Microchemical Apparatus of the American Chemical Society (6). Microchemical reagent or analyzed grade chemicals are manufactured, purified, and tested to ensure that the content of certain impurities is below specified maximal limits set by the Committee on Analytical Reagents of the American Chemical Society (2). The NBS prepares and distributes standards of certified properties or purity (3). For calculating purposes an atomic weight table is necessary. The present Table of Atomic Weights, based on carbon-12, was adopted in 1961 by the Commission on Atomic Weights of the International Union of Pure and Applied Chemistry (7) and revised in 1965 (8).

Spectrophotometry is a term that by universal usage refers to the relative measurement of radiant energy or flux as a function of wavelength (9). The significance of the word "relative" in this connection lies in the fact that the measurements are always made relative to some standard. What this standard may be in any case depends on the type of measurement. In transmission measurements it may be either the blank beam or a similar material, or a cell that is free of the absorbing constituent; but in all cases what is measured is the ratio of these qualities, wavelength by wavelength, throughout the spectral range of interest. The best procedure for checking the wavelength scale of a nonrecording spectrophotometer is by direct use of a source of radiant energy having spectral lines of suitable intensity adequately spaced throughout the range of interest. In the range from 220 to 1014 mμ (nanometers) the quartz mercury arc is by far the best single source. For nonrecording spectrophotometers, such as the Beckman Spectrophotometer, Model DU, the National Bureau of Standards lists sources and wavelengths suitable for the calibration of these spectrophotometers (9). NBS calibrated didymium glass standards are useful for the calibration of recording spectrophotometers. Wavelengths of maximum absorption bands are dependent on slit widths, and transmittances on the steep parts of the curve vary rapidly with slight wavelength changes. The use of didymium glass to check the wavelength calibration of a nonrecording spectrophotometer is considered much inferior from the standpoint of time, convenience, and reliability to the direct use of line sources as described above (9). Vandenbelt proposed that the spectrum of the oxide of the rare element holmium should be used for wavelength calibration of recording spectrophotometers in the ultraviolet and visible regions (10). Later reports (11, 12) also indicated such use of holmium oxide filters for wavelength calibration in the routine laboratory. General information regarding the spectral transmission characteristics of glass filters may be found in various NBS publications (13-16) and in circulars from industrial sources (17-23). Shortly after the advent of commercial photoelectric spectrophotometers the NBS instituted the service of issuing glass filters with accompanying certificates of calibration for use in checking the photometric scale of spectrophotometers (24-26). The spectral transmittances of aqueous solutions of copper sulfate, cobalt ammonium sulfate, and potassium chromate have been measured with sufficient care, and the solutions themselves are sufficiently stable, that they are recommended as suitable standards for checking the photometric scale of spectrophotometers (9, 27, 28).

In order to communicate within or between laboratories attempts have been made to standardize nomenclature. Language standardization in any field has for its primary objective a smoothing of channels of communication. Lists of definitions of terms commonly used by American chemists (29), abbreviations and symbols used in biochemistry (30), and suggested nomenclature in applied spectroscopy (31) have been published.

Standards in the Clinical Chemistry Laboratory

There has been no coordinated effort among various groups to bring about specifications and definitions for various standards and reference samples for the clinical chemistry laboratory. Suggested definitions and a review of the standards available to the clinical chemist will be presented in this discussion.

Primary Standard

In the reference frame of usual analytical laboratory practice the primary standard is defined as a pure chemical substance that is used for the purpose of assaying a volumetric solution of unknown strength or for the preparation of a solution of known concentration. Although it is not always possible to select a substance for a primary standard with all the desired characteristics, it is generally stated that the primary standard should be selected with the following criteria in mind (32, 33).

1. It must be a stable substance of definite composition.

2. It must be a substance that can be dried in the course of preparation, preferably at 105–110°, without change in composition.

3. It should have a high equivalent weight in order that weighing errors may have a relatively small effect.

4. It must be a substance that can be accurately analyzed.

5. Desired reactions should occur according to single, well-defined, rapid, and essentially complete processes.

6. The purity of a primary standard must be assured through well-defined qualitative tests of known sensitivity, or through preparation by a method that has been demonstrated to yield consistently a pure product, and by storage under conditions in which the product is entirely stable.

Primary Volumetric Standard

In volumetric analytical procedures the desired constituent is made to react with a measurable volume of a primary or a secondary volumetric standard by a titration technic. A primary volumetric standard is usually prepared by dissolving and diluting an accurately weighed primary standard to volume in a calibrated volumetric flask. Both mass and volume, which are covered by the NBS and ACS specifications (4–6), are measurable with a high degree of precision. Implicit in this description is the fact that the measured weight of the chemical substance used in the preparation of the standard solution must be that of the mass of the pure compound. This can be assumed only if the chemical substance is absolutely pure. Chemists have usually chosen for volumetric analysis primary standards that are capable of a high degree of purification.

Secondary Volumetric Standard

A reagent is dissolved and diluted to approximately the desired concentration. The concentration of the solution may be ascertained by titrating the solution (1) against a solution containing a known weight of a primary standard substance, or (2) against a measured volume of a primary standard solution. Alternatively, the solution to be standardized may be analyzed for the reagent substance by any analytical technic sufficiently accurate for this purpose.

Clinical Primary Standard

The classical characterization of a primary standard may not apply in the reference frame of the clinical chemistry laboratory. In the clinical chemistry laboratory, as in any other analytical laboratory, it is necessary to prepare standard solutions containing known amounts of the desired constituents. Many of the desired constituents found in the body fluids are not easily purified and many are not stable. In order to obtain a known amount of a desired constituent that is not absolutely pure by weight, the purity must be known. It is for this reason that a clinical primary standard is being defined as a body fluid chemical substance that can be prepared within purity limits that would be acceptable for measurement of the quantity of the molecular entity by weight.

The term "pure" must be used in an operational sense. It can be stated that a system of molecules is a pure compound if an exhaustive series of fractionations fails to produce fractions with different properties. What one calls a pure compound thus changes as new methods become available for separating material into fractions, or for measuring the properties of the fractions more accurately (34).

In establishing specifications for clinical primary standards the physical properties of fractions should be measured with the same chemical and instrumental measuring technics that will be used to measure the desired constituent. With this technic it should be possible to establish purity specifications within the sensitivity limits of the measurement method. It is also desirable to be able to check specifications for standards with equipment available in the user's laboratory. For example, spectrophotometric technics are much used in the clinical chemistry laboratory. Specifications for the standards used with these technics should include molar absorptivity values. Here, the criterion of purity for various fractions is usually based on the principle that the molar absorptivity increases with increasing sample purification. The molar absorptivity should be determined directly at a wavelength of an absorbance maximum for each purified fraction in a defined solvent system, or indirectly at a wavelength of an absorbance maximum that has been developed with each fraction by the same chemical reaction used for the desired constituent analysis. Exceptions to the foregoing may have to be made in some situations in which there is no absorbance maximum at the wavelength chosen for the measurement of the desired constituent.

The criterion of purity for a particular standard may be fallacious, as trace impurities not removed by the fractionation method or formed on standing may contribute significantly to the molar absorptivity. Hence, studies of standards should include fractionation by chromatographic technics so that the presence of structurally related (or non-related) compounds (impurities) may be shown. Further work to establish the contribution by the impurities to the physical property value, as the molar absorptivity, may be necessary so that limits of purity may be specified.

A complete description of the methods used to determine physical characterization values must be included with the specifications for each clinical primary standard. Physical characterization constants are dependent on the accuracy of the instrumental measuring system. Hence, standard procedures for instrument calibration are necessary. In a sense, an instrument itself might be thought of as a primary standard. For spectrophotometers, calibration procedures for the wavelength scale and photometric scale should be specified (see Specification Sources, above).

Studies of stability and storage conditions are necessary to establish the conditions for the proper handling of a standard. Conditions for drying each standard, if affected by moisture, should also be established.

Clinical Primary Standard Solution

A clinical primary standard solution is one in which an accurately known weight of a clinical primary standard substance is dissolved and diluted to an accurately known volume of solution. The solvent or solution for dissolving and diluting the standard must also be completely defined by a set of specifications.

Clinical Secondary Standard Solution

This is a solution that is prepared by dissolving and diluting a weighed amount of a chemical substance to a known volume with a defined solvent or solution. If the weighed material is not a primary standard, the concentration of the chemical substance is then determined by chemical analysis. The secondary clinical chemistry standard may be defined as a body fluid chemical substance that cannot be prepared within purity limits that would be acceptable for the measurement of the quantity of the molecular entity by weight. When this secondary standard is used for the measurement of the desired constituent it becomes necessary to analyze a solution of it in terms of a primary standard.

A prime example to illustrate a clinical secondary standard is seen in the method for determining serum proteins. It has been useful to prepare albumin solutions as standards. It is possible to obtain a crystalline albumin preparation with the purity estimated as better than 99% by the electrophoresis technic. If a crystalline albumin preparation with a purity of 99+% is acceptable as a clinical primary standard, then a clinical primary standard solution of known concentration can be prepared by dissolving and diluting a weighed amount of such an albumin preparation to a known volume with an appropriate solvent or solution. However, if an albumin solution be analyzed by the Kjeldahl technic with ammonium sulfate as a primary standard, then the solution is known as a clinical secondary standard solution for protein nitrogen. The conversion to protein concentration by the use of a factor is part of another problem for the clinical chemist. In this case, for example, the relationship between the analyzed nitrogen content of the albumin solution and the actual protein concentration involves assumptions that may not be valid.

Standards Available for the Clinical Chemistry Laboratory

A number of primary standards, which include benzoic acid, arsenious oxide, sodium oxalate, potassium dichromate, potassium hydrogen phthalate for the preparation of a pH 4 standard solution, and two phosphates, namely, potassium dihydrogen phosphate and disodium hydrogen phosphate for the preparation of a pH 6.8 standard solution, may be obtained from the National Bureau of Standards (3). Some of the foregoing, and others, such as ACS Analytical Reagents (2) may be obtained from chemical manufacturers for use as primary standards.

Blood pH Standard

The NBS has recommended a pH standard solution for use in the measurement of the pH of blood and other physiological fluids in the range of pH 7–8 (*35, 36*).

Cyanmethemoglobin

The Division of Medical Sciences of the National Academy of Sciences—National Research Council, in response to a request, organized an ad hoc Panel for the Establishment of a Hemoglobin Standard under its Subcommittee on Blood and Related Problems in 1954 (*37*). The Panel recommended that cyanmethemoglobin be adopted as a standard in clinical hemoglobinometry. Characterization constants and recommendations for the preparation of standard solutions were also included. The technical Subcommittee on Haemoglobinometry of the International Committee for Standardization in Haematology (ICSH), which met in Stockholm in September 1964, confirmed the NAS–NRC recommendation. The NAS–NRC was invited by the ICSH to coordinate the review of its proposal in the United States. That effort resulted in proposals (*38*) that were generally in agreement with the deliberations of the ICSH. Some of the proposals are as follows:

1. Clinical laboratories should use the cyanmethemoglobin method of hemoglobinometry exclusively, and the term "Gm. (of hemoglobin) per 100 ml. (of blood)" should be used to express the value of the measurement so determined.

2. Human hemoglobin is assumed to have a molecular weight of 64,458. The iron content of hemoglobin as computed from this molecular weight and the atomic weight of iron, assuming 4 atoms of iron per molecule of hemoglobin is 0.347% (w/w).

3. The millimolar extinction coefficient of cyanmethemoglobin is taken to be 44.0 at 540 mμ.

4. The purity is checked by inspecting the shape of the absorption curve between 450 and 700 mμ and by determining the quotient of absorption at 540 and 504 mμ. The quotient should be between 1.58 and 1.62.

Recommendations for the preparation of standard solutions were also included in these proposals (*38*). Formal adoption of the International Standard must await action by the Assembly of the ICSH during the International Congress of Hematology to be held in 1966 and the acceptance of a value for the molecular weight of hemoglobin by the Protein Commission of the International Union of Pure and Applied Chemistry.

Bilirubin

A committee composed of representatives from the American Association of Clinical Chemists, the American Academy of Pediatrics, the College of American Pathologists, and the National Institutes of Health recommended procedures for the establishment of a uniform bilirubin standard (*39*). In 1960, Henry *et al.* reported on their evaluation of the purity of a number of commercial bilirubin preparations, the preparation of pure bilirubin, and the purity of the purified product (*40*). A technic for the purification of bilirubin was presented. The physical properties of the commercial and purified bilirubin preparations studied included the melting point and the ultraviolet and infrared absorption curves. Since at that time there were no generally accepted criteria of the purity of bilirubin preparations, the Committee tentatively accepted the principle that increasing molar absorptivity at bilirubin's maximum in the visible region is an index of sample purity. The Committee examined six highly purified commercial and two privately purified crystalline preparations selected as giving the highest and most consistent molar absorptivity in chloroform at 453 mμ. The Committee recommended as acceptable a bilirubin giving a molar absorptivity in chloroform between 59,100 and 62,300 in a 1-cm. light path cuvet at 453 mμ at 25°.

The Committee also recommended that standard solutions for the assay of bilirubin in serum be made up in aqueous protein medium. The preparation of an acceptable standard bilirubin solution that was recommended is as follows: An accurately weighed quantity of an acceptable bilirubin is dissolved completely and as quickly as possible (in less than 5 min.) in subdued light at room temperature in 0.1 M sodium carbonate solution, the quantity of the latter being selected to constitute 2% of the final volume of the prepared standard. The clear red solution is immediately diluted with an acceptable serum diluent to a final volume selected to give a desired final concentration of not less than 5 mg. per 100 ml. An acceptable serum diluent used in the preparation of this standard and in any subsequent dilutions for calibration purposes, shall be defined tentatively as pooled serum having an absorbance of less than 0.100 at 414 mμ and 0.040 at 460 mμ at a dilution of 1:25 in 0.85% sodium chloride. Packaging, preservation, and calibration procedures for the preserved standards were also included in the Committee report.

Cholesterol

An investigation concerned with the evaluation of the purity of five commercial cholesterol preparations and one artificial mixture, the preparation of pure cholesterol, and a study of the properties of the purified cholesterols was reported by Radin and Gramza (*41*). Three recrystallization technics were used to fractionate the cholesterol preparations. Two technics involved the recrystallization of cholesterol from absolute ethyl alcohol and from glacial acetic acid respectively and the third involved the dibromide derivative method. Here, likewise, the criterion of purity of the cholesterol preparations was based on the principle that molar absorptivity increases with increasing sample purity. In order to check cholesterol preparations for use as standards, it was recommended that the molar absorptivity value of 1700 ± 30 (1 S.D.) for the defined Liebermann-Burchard procedure and 11,500 ± 100 (1 S.D.) for the defined sulfuric acid-iron procedure should serve as a guide to ascertain purity (*42*).

Proteins

At present, protein standards are based on Kjeldahl nitrogen analyses coupled with a conversion factor. Bovine serum albumin has been frequently used as a standard in protein analytical procedures, but its suitability has been questioned. In view of this and the fact that numerous forms of human and bovine albumin are now available, a study to find a product that would be a satisfactory standard for the analysis of total protein was reported by Pastewka and Ness (*43*). Human albumin Fraction V was recommended as the most suitable and practical choice for a standard to be used in total protein and albumin analyses of human body fluids.

Enzymes

In 1964 the International Union of Biochemistry recommended that enzyme activities be expressed in terms of micromoles of substrate transformed per unit, such activities to be termed International Units of the enzyme in question (*44*). Concentrations should be expressed in units per milliliter. The temperature should be stated and when practicable should be 30°. This study is another example of an attempt to standardize units that would make for easier communication among laboratories (*45*).

Biochemical Standards

While of limited application for the usual routine clinical chemistry laboratory, a book of interest to biochemists was prepared by the Committee on Biological Chemistry, Division of Chemistry and Chemical Technology, National Academy of Sciences-National Research Council with support by the National Institutes of Health (*46*). Compounds for which criteria and standards or specifications of purity for chemicals available for biochemical research are presented, and include amino acids, carbohydrates and related substances, coenzymes, lipids, purines, and pyrimidines.

Nomenclature

A system of units for clinical chemists was proposed by the Danish Society for Clinical Chemistry and Clinical Physiology. This system was transmitted to the IUPAC Commission on Clinical Chemistry in June 1963 (*47*). A subcommittee on Nomenclature and Usage of the Committee on Standards and Controls of the American Association of Clinical Chemists was assigned the task of studying the problem with a view to agreement upon a set of rational and internally consistent objectives for nomenclature reform. The recommendations of the subcommittee were in close agreement with those of the original Danish proposals as being desirable objectives.

Reference Samples

A number of evaluation studies of the accuracy of chemical tests in clinical laboratories in various countries have shown the need for improved performance (*48–53*). The necessity for improving the reliability of chemical tests led to the establishment of quality control measures. The first materials used for quality control programs were prepared by pooling serum samples. Simple statistical technics were used to evaluate the materials used as serum controls. The Council on Clinical Chemistry of the American Society of Clinical Pathologists has recommended a stepwise procedure for preparation of a control serum (*54*). Proper collections and preparation of pooled and frozen control serums in the manner suggested is not only time-consuming but requires considerable equipment and freezer space. Unfortunately, repeated freezing and thawing of pooled serum before use is known to alter the physiochemical properties of certain serum constituents. In addition, some constituents are not as stable in frozen serum as in lyophilized serum.

The use of ion-free serum* as a standard in clinical chemistry was discussed by Levy (*55*). Human serum pools are deionized with ion exchange resins. A cited advantage of an ion-free serum preparation is that it can be combined at one time with as many standard solutions as desired. Thus, serum standard values may be changed from day to day. Two commercial products are prepared by removing fibrinogen,

---

*Clinton Laboratories.

lipids, and cold water–insoluble globulins by a salt fractionation procedure, by inactivating enzymes in the remaining fraction with heat, by dialyzing to remove the salt used in the fractionation procedure, and by adding various constituents at various levels after analyzing the residual constituents. Aliquots of these preparations are sterile-packaged in individual bottles. Packaged samples are then analyzed by the producer laboratory staff members and by company-chosen reference laboratories to obtain the average value for each constituent of interest.

Various commercial normal and abnormal level lyophilized serum control samples are now flooding the market. Each sample aliquot is reconstituted by the user according to the producer's instructions, usually with a specified volume of distilled water. Charts with various constituent values are distributed with each sample. Many of the lyophilized serum controls* are prepared by pooling serum samples, processing the samples in various ways, and by adding or not adding various constituents, and are assayed by company-chosen reference laboratories. The mean constituent analytical values are then recorded for the user.

One producer† claims that the constituents, for which its preparations are standardized, are first removed from the serum, or reduced to fixed low levels by presently unspecified technics. It is claimed that each constituent of interest, in purified form, is accurately weighed-in to bring the serum to a standard known level. The values so obtained are given to the user with each lyophilized preparation. Klein and Weiss summarized the results of the analysis of the normal level product by a group of laboratories (56). The results of the analysis of refrigerated samples analyzed within 1 week of receipt and at designated intervals over a 6-month period were shown. It was reported for the lot examined that the manufacturer's value fell within one standard deviation of the mean value for each constituent examined by the evaluating groups. A later report by Klein and Weissman offered an evaluation of a product that contains constituents in the pathological or physiologically abnormal range (57). In this evaluation, with the exception of bilirubin, free cholesterol, and creatinine, the results came within the manufacturer's recommended 5% variation. In regard to two preparations that are meant to serve as serum enzyme controls, the manufacturer† states that "each enzyme in purified form is then

---

*Dade Reagents, Inc., Miami, Fla.; Clinton Laboratories, Los Angeles, Calif.; Hyland Laboratories, Los Angeles, Calif.; Michael Reese Research Foundation, Chicago, Ill.

†General Diagnostics Division, Warner-Chilcott Laboratories, Morris Plains, N. J.

precisely weighed-in to bring the serum to a predetermined level of enzyme activity." This could be done only if the enzymes were weighed-in in their crystalline form. This is not what is done. What is done is an analysis of a concentrated solution of the enzyme. Then weighed amounts of the enzyme are added. It is, of course, no more accurate to analyze before adding the concentrate than to do so after preparing the final dilution that is sold.

Discussion of Reference Samples

"Standard in serum," "serum standard and control," "a primary standard," "a primary serum standard," "clinical chemistry control serum," and "internal standard control," are among the descriptions found in the literature concerning the commercial preparations discussed above. It is proposed here that the term "reference sample" should be applied to a sample in which the chemical composition and the physical characteristics simulate the specimen analyzed. Lyophilized commercial preparations that on reconstitution fully simulate serum, laboratory, and commercially prepared serum pool aliquots all meet the definition of a reference sample or reference serum.

The values for various constituents in a reference sample should be known with a high degree of reliability (and probability). A minimal program for establishing the value of components in reference samples should require analysis by more than one laboratory and by more than one method. The analytical methods used should be acceptable to the users of the products. The statistical technics that are used to determine the mean values of the components and the limits of deviations about them should not only be specified but be acceptable to the users of the commercial reference samples. All reference samples should be accompanied by a certificate, as they are in most cases, with a statement containing the foregoing information.

Certified Reference Samples

At this time there is no generally acceptable impartial laboratory arrangement for certification of reference samples. If this does come about, the reference samples evaluated in the referee laboratory (impartial laboratory) should simply be called "certified reference samples."

Stability

One great difference between the standard samples certified by the NBS and the reference samples used in the clinical chemistry laboratory is in the stability of the preparations. Even though pooled serum samples are frozen, sterile-packed, or lyophilized, no such sample is absolutely stable. Even under ideal conditions it is possible for some of the constituents of the reference sample to deteriorate. From the time of certification by commercial companies to the time of arrival in a clinical chemistry laboratory, it is possible that conditions of transport and storage could contribute to the decomposition of the reference sample.

Uses for Reference Samples

Quality control   The reference samples are treated exactly as the specimen samples that are assayed concurrently. The proper use of the reference sample in the quality control program insures greater reliability of laboratory results.

In the author's laboratory various commercial reference samples are used in the quality control program. By using more than one lot of a particular commercial preparation and by using different commercial preparations it is possible to check a discrepancy in a component analytical value of a reference sample by the assumption that when any reference sample value is incorrect for any reason, it is improbable that the stated analysis of another lot or of another commercial reference sample would also be incorrect.

Accuracy   Reference samples with known constituent contents are of great value in establishing the accuracy of improved or new analytical methods.

The Use of the Reference Sample as a Standard

It has been stated or implied by some commercial producers that reference samples can be used as standards.

The component analytical values for reference samples are determined with reference to the usual primary or secondary standards. There is a greater uncertainty in the measured value of each component of the reference sample than in that of each standard solution so used. It should be recalled that a precise weighing technic (good to an estimate of tenths of milligrams) is used to measure clinical primary standards, and precise analytical technics are used in making up clinical primary standard solutions. There is a high degree of confidence in the calculated concentration of a properly prepared primary standard solution. The concentration values of clinical secondary standard solutions are not known as accurately as those of clinical primary standard solutions in as much as the values of the former are dependent on analytical methods with their inherent errors. The true value of each component in the commercial reference sample lies within a range indicated by the certificate of content. In discussing the use of commercial control serums Klugerman and Boutwell pointed out that over-all error is the summation of random and systematic errors (58). Random errors, closely associated with precision, are errors of chance, while systematic errors may be defined as constant errors or bias due to inaccurate standards or improperly prepared reagents. Since a reference sample is carried through every step of a procedure, it may be falsely assumed that the use of a single reference sample will compensate for both random and systematic error. Each reference standard component analytical value reflects systematic and random errors encountered in technics of measurement that are less precise than those used to prepare standard solutions. Actually, only the systematic error may be minimized or perhaps eliminated by this means. A large positive random error occurring in the assay for a particular component of the so-called standard, with a concomitant large negative error in the analysis of the unknown sample, could result in gross inflation of the final error.

In the case of enzymes in reference samples, users of such preparations may be tempted to extend the principle of compensation rather than of reliability to this field as well. Enzyme activity is determined, in most instances, under precisely defined conditions. In the use of control samples as enzyme secondary standards, the user may rely upon compensation to correct for deviations from established conditions under which the enzyme should be measured. This practice will tend to introduce bias because of the nonlinear effect of most of the variables.

In cases such as that of the Bilirubin Committee report, in which it was recommended that primary standards should be dissolved in serums of certain specifications, it is probable that manufactured lyophilized preparations of this kind would be more stable than home-made frozen preparations. Based on the definitions in this paper, a weighed clinical primary-standard grade bilirubin preparation in a medium composed of buffered albumin of acceptable purity can be called a clinical primary standard solution. If it is weighed into a pooled serum diluent, it is not a clinical primary standard solution. It is a reference sample. The reason for this is that at this time it is not possible to make specifications for the constituents in the serum. A recent paper showed that different spectral absorption curves are obtained for bilirubin in the presence of human versus bovine albumin (60).

The argument that reference samples called "weighed-in standards in serum" can be used as standards is based on the concept that these are defined and reproducible preparations. This argument is invalid as long as it is not shown that the dialyzed serum, to which weighed pure components are added, is completely defined chemically. Also, as long as the analyst uses procedures in which there is incomplete separation of the desired constituent from interfering substances, it is invalid to use a reference sample as a standard because the background spectral absorption is probably not the same as the background absorption of the unknown. The error due to varying background absorption becomes greater whenever absorbances as low as 0.1 are measured, as is true in some procedures.

In accordance with basic principles of analytical chemistry, in the isolation or partial isolation of a component in a measurable state, in many procedures, the reference sample and the unknown specimen are treated alike. The isolated, or partially isolated component, if that is all that is necessary, should be in the same state in which it is present in the standard solution. The isolated specimen solution and the primary standard solution should be in the same chemical and physical state for measurement. However, this would be an ideal situation only approached and not usually realized.

There is a tendency now to devise simple analytical procedures for the clinical chemistry laboratory. Commercial interests are promoting packaged reagents, standards, and procedures for constituents that are not simple to determine reliably. In the pursuit of simple procedures it is extremely important that the basic and fundamental aspects of the analytical process be carefully maintained.

Reagents

The value to the chemist of having his critical reagents supplied according to uniform specifications of quality can easily be appreciated, and indeed, in clinical chemistry today this is no longer merely a convenience; both the number and the sources of our reagents are proliferating, and we can scarcely spare the time to check all our materials and reagents in the laboratory. If we, then, transfer our proxy for this responsibility to the supplier, the need for closer control over and recognition of quality becomes increasingly important.

It should be pointed out that we are currently experiencing a not inconsiderable revolution in technology that is exerting increasing influence upon our ideas of reagent purity. Many commercially supplied organic materials that were considered to be highly "pure" a few years ago have been shown by gas chromatography to be contaminated to an unexpected extent. With almost instantaneous monitoring possible, these materials are beginning to be offered in unprecedented purity, sometimes in routine production, often as extremely expensive specialties. While prices of these products may or may not be justifiable, their costs are becoming so much of an item that it is absolutely essential that the laboratory customer get the quality for which he is paying.

This requires a readily available and extensive source of commonly accepted definitions of minimal criteria of quality to which the specification of "acceptable" material may be referred. Despite the enormity of compilation, and the vast difficulties of the working out of its broad acceptance and use, some such compendium is becoming increasingly necessary. What we demand of such a compendium, of course, is the specification of any reagent to such a standard of quality that its use in any desired reaction supplies only the effects of the pure reagent, that is, any residual contaminants cannot be detected by their effects and may safely be ignored. For any specific need this does not necessarily require a high state of purity in a reagent—a reagent need only be as pure as the application demands—but for general utility the standards of quality must be quite high to satisfy all potential requirements. It is quite possible that for a majortiy of reagents of importance in clinical chemistry, preparative or purification procedures can be found that yield products to which no one can find objection; if and when objection does arise, a new preparative procedure must be found and new specifications must be written. When several purification procedures appear to be incompatible it will probably be necessary to have separate definitions according to specific application.

For many purposes in clinical chemistry, the ACS requirements for analytical reagent quality probably coincide with our own. However, as pointed out above, definitions of purity are currently undergoing rather radical changes for many organic materials, and undoubtedly some ACS specifications will also undergo extensive revision within a few years. For example, ACS Reagent Acetic Acid is purportedly inadequate as a solvent for certain iron-sulfuric cholesterol reactions— actually this may not be true for certain other American commercial reagents. If it is adequately demonstrated that acetic acid bearing the ACS label probably requires supplemental oxidative distillation for this reaction, this might be employed as a new definition of reagent quality material. However, until acetic acid with new purity specifications is routinely supplied commercially, and until it can be demonstrated that for all reagent usage the potential oxidative products do not interfere with other reactions, separate definitions will be preferable.

"Official" Methods for the Analysis of Standards and References

For such procedures, we shall require official sanction and sponsorship for a single common "method-of-reference" for each constituent considered. When a technic is available that is above reproach, the solution will be easy; without this, the problem becomes one of selection and acceptance of the best available until a superior one can be demonstrated.

Definitive Methods

There would seem to be little argument that a "definitive assay" must be one in which the quantity measured accurately defines the concentration of the substance analyzed. For example, concentration of an acid is defined by the number of its titratable hydronium ions, and any analysis giving an accurate measure of titratable acidity is thus definitive. Concentrations of protein may be referred to protein nitrogen concentration; for this quantity we would probably all accept an accurately performed Kjeldahl determination of total nitrogen, suitably corrected for nonprotein nitrogen, as being definitive of this quantity although it does not define the mass concentration of protein in any biological sample. Of course, there are few definitive assays for the substances in which the clinical chemist is interested. Technically, the concentrations of many of the enzymes he measures are defined in terms of "activity equivalents" of products or reactants, but here the parameters governing activity are difficult to establish and reproduce.

Methods of Reference

For most substances we shall have to accept, somewhat arbitrarily in the best of circumstances, some common methods as reference methods. Possible methods for these assays need not, and probably would not be those most conveniently employed in routine analysis but would be chosen and accepted as giving results generally considered to be the most reliable indication of true concentration available to us. The choices will not be easy. While it may be tempting to provide more than one method of reference for convenience and flexibility, this would tend to be self-defeating. For example, certain instrumented methods of acknowledged superiority may have to be passed over in favor of more universally available "wet-chemical" technics.

"Acceptable" Methods; "Standard" Methods

For routine analysis, methodology is selected to fit the analytical situation best; the requirements for precision are balanced against reliability, speed, costs, equipment, personnel, and a variety of peripheral factors; methods in use range from those that may be considered highly reliable to those that may be of borderline merit. Any realistic approach toward our long-range objectives of interlaboratory standardization must balance practical considerations against uniform procedures. The Association's publication of *Standard Methods of Clinical Chemistry* (*59*) provides a continuing program of furnishing technics for the evaluation and use of arbitrarily selective methods while permitting relative freedom of choice to the analyst.

Conclusions

Despite extensive programs that have been carried on in the standardization of materials, there is actually very little in the way of official definitions of standards for the clinical chemist. The original objectives of the old formularies have been outgrown, and the pharmacopeias are aimed at problems differing from ours; the ACS definitions of analytical quality for chemical reagents may be only incidentally useful in the field of clinical chemical standards. Clearly, what is needed is a vigorous extension of the joint efforts of clinical chemists and others concerned with quantitative analysis, into the general area of standards.

References

1. Wichers, E., Standards for reagent chemicals. *Anal. Chem.* **23**, 1537 (1951).
2. American Chemical Society, *Reagent Chemicals: American Chemical Society Specifications, 1955*. Washington, D. C., 1955.
3. *Standard Materials*, National Bureau of Standards Miscellaneous Publication 241. Issued March 12, 1962, Washington, D. C.
4. Lashof, T. W., and Macurdy, C. B., *Precision Laboratory Standards of Mass and Laboratory Weights*, National Bureau of Standards Circular 547. Section I, Issued August 20, 1954, Washington, D. C.
5. Hughes, J. C., *Testing of Glass Volumetric Apparatus*, National Bureau of Standards Circular 602, Issued April 1, 1959, Washington, D. C.
6. Committee for the Standardization of Microchemical Apparatus, Division of Analytical Chemistry, American Chemical Society, Recommended specifications for microchemical apparatus. *Anal. Chem.* **23**, 523 (1951), *Anal. Chem.* **28**, 1993 (1956), *Anal. Chem.* **30**, 1702 (1958).
7. Cameron, A. E., and Wichers, E., Report of the International Commission on Atomic Weights (1961). *J. Am. Chem. Soc.* **84**, 4175 (1962).
8. Compt. Rend. Conf. Union Intern. Chem. Pure and Appl., Paris, 1965.
9. Gibson, K. S., *Spectrophotometry (200 to 1,000 Millimicrons)*, National Bureau of Standards Circular 484, Issued September 15, 1949, Washington, D. C.
10. Vandenbelt, J. M., Holmium filter for checking the wavelength scale of recording spectrophotometers. *J. Opt. Soc. Am.* **51**, 802 (1961).

11. McNeirney, J., and Slavin, W., A wavelength standard for ultraviolet-visible-near infrared spectrophotometry. *App. Opt.* **1**, 365 (1962).

12. MacDonald, R. P., Uses for a holmium oxide filter in spectrophotometry. *Clin. Chem.* **10**, 1117 (1964).

13. Coblentz, W. W., Emerson, W. B., and Lang, M. B., Spectroradiometric investigation of the transmission of various substances. *Bull. B. S.* **14**, 653 (1918), S325.

14. Gibson, K. S., Tyndall, E. P. T., and McNicholas, H. J., The ultraviolet and visible transmission of various colored glasses. National Bureau of Standards Techn. Pap. No. 148 (1920) T148.

15. Gibson, K. S., Spectral filters. *J. Opt. Soc. Am.* and *Rev. Sci. Instr.* **13**, 267 (1926).

16. Gibson, K. S., A filter for obtaining light at wavelength 560 millimicrons. *J. Res. Nat. Bur. St.* **14**, 545 (1935).

17. *Glass Color Filters*, Corning Glass Works, Corning, N. Y., 1962.

18. *Wratten Light Filters*, Eastman Kodak Co., Rochester, N. Y., 1962.

19. *Optical Interference Filters*, Baird-Atomic Inc., Cambridge, Mass., 1961.

20. *Multifilms*, Bausch & Lomb, Rochester, N. Y.

21. *Photo Filter Glass*, Tiffen Optical Co., Long Island, N. Y., 1961.

22. *Monochromator Transmission Type Interference Filters*, Photovolt Corp., New York, N. Y.

23. *Optical Filters for Visible and Near Infrared*, Barr & Stroud, Limited, Great Britain.

24. Gibson, K. S., Walker, G. K., and Brown, M. E., Filters for testing the reliability of spectrophotometers. *J. Opt. Soc. Am.* **24**, 58 (1934).

25. Gibson, K. S., and Balicon, M. M., Transmission measurements on the Beckman quartz spectrophotometer. *J. Res. Nat. Bur. St.* **38**, 601 (1947) RP1789.

26. Gibson, K. S., and Belknap, M. A., *Permanence of Glass Standards of Spectral Transmittance*, National Bureau of Standards Research Paper 2093, Washington, D. C., 1950.

27. Davis, R., and Gibson, K. S., *Filters for the Reproduction of Sunlight and Daylight and the Determination of Color Temperature*, National Bureau of Standards, Miscellaneous Publication M-114, Washington, D. C., 1931.

28. Hogness, T. R., Zscheile, F. P., Jr., and Sidwell, A. E., Jr., Photoelectric spectrophotometry, an apparatus for the ultraviolet and visible spectral regions: Its construction, calibration and application to chemical problems. *J. Phys. Chem.* **41**, 379 (1937).

29. Committee on Nomenclature, Division of Analytical Chemistry, American Chemical Society, A progress report of the Committee on Nomenclature, Division of Analytical Chemistry, June, 1952. *Anal. Chem.* **24**, 1348 (1952).

30. IUPAC-IUB Combined Commission on Biochemical Nomenclature. Abbreviations and Symbols for Chemical Names of Special Interest in Biological Chemistry. Revised Tentative Rules (1965). *J. Biol. Chem.* **241**, 527 (1966).

31. Joint Committee on Nomenclature in Applied Spectroscopy, Society for Applied Spectroscopy and American Society for Testing Materials, Suggested nomenclature in applied spectroscopy. *Anal. Chem.* **24**, 1349 (1952).

32. Farr, H. V., Butler, A. Q., and Tuthill, S. M., Commercial development of primary standards. *Anal. Chem.* **23**, 1534 (1951).

33. Stenger, V. A., Requirements for primary redox standards. *Anal. Chem.* **23**, 1540 (1951).

34. Eyring, H., Philosophy of the purity and identity of organic compounds. *Anal. Chem.* **20**, 98 (1948).

35. Bower, V. E., Paabo, M., and Bates, R. G., A standard for the measurement of the pH of blood and other physiologic media. *J. Res. Nat. Bur. St.* **65A**, 267 (1961)

36. Bower, V. E., Paabo, M., and Bates, R. G., A pH standard for blood and other physiologic media. *Clin. Chem.* **7**, 292 (1961).

37. Cannan, R. K., Proposal for a certified standard for use in hemoglobinometry. *Clin. Chem.* **4**, 246 (1958) and *Am. J. Clin. Pathol.* **30**, 211 (1958).

38. Cannan, R. K., Proposal for adoption of an international method and standard solution for hemoglobinometry, specifications for preparation of the standard solution, and notification of availability of a reference standard solution. *Am. J. Clin. Pathol.* **44**, 207 (1965).

39. Joint Bilirubin Committee, American Academy of Pediatrics, College of American Pathologists, American Association of Clinical Chemists, and National Institutes of Health, Recommendation on a uniform bilirubin standard. *Clin. Chem.* **8**, 405, (1962); *Am. J. Clin. Pathol.* **39**, 90 (1963); *Pediatrics* **31**, 878 (1963); *Standard Methods of Clinical Chemistry*, Vol. 5, S. Meites, Ed. Academic Press, N. Y., 1965, p. 75.

40. Henry, R. J., Jacobs, S. L., and Chiamori, N., Studies on the determination of bile pigments. 1. Standard for purity for bilirubin. *Clin. Chem.* **6**, 529 (1960).
41. Radin, N., and Gramza, A. L., Standard of purity for cholesterol. *Clin. Chem.* **9**, 121 (1963).
42. Radin, N., "Cholesterol (Primary Standard)." In *Standard Methods of Clinical Chemistry*, Vol. 5, S. Meites, Ed. Academic Press, N. Y., 1965, p. 91.
43. Pastewka, J. V., and Ness, A. T., The suitability of various serum albumin products as standards for the quantitative analysis of total protein and albumin in human body fluids. *Clin. Chim. Acta* **12**, 523 (1965).
44. Recommendations (1964) of the International Union of Biochemistry, *Enzyme Nomenclature*, Elsevier Publishing Company, Amsterdam, 1965.
45. Baron, D. N., *et al.*, Letter to the editor. *Clin. Chem.* **12**, 319 (1966).
46. Committee on Biological Chemistry, National Academy of Sciences, National Research Council, *Specifications and Criteria for Biochemical Compounds*, Publication 719, and Publication 719, Supplement, 1963, Washington, D. C., 1960.
47. Danish Society for Clinical Chemistry and Clinical Physiology, *Units in Clinical Chemistry*, Department of Clinical Chemistry, Rigshospitalet, Copenhagen Ø., Denmark, 1963.
48. Belk, W. P., and Sunderman, F. W., A survey of the accuracy of chemical analyses in clinical laboratories. *Am. J. Clin. Pathol.* **17**, 853 (1947).
49. Snavely, J. G., Golden, W. R. C., and Cooper, A. B., The accuracy of certain chemical determinations in Connecticut laboratories—the third survey. *Conn. State Med. J.* **16**, 894 (1952).
50. Maryland State Department of Health, Laboratory Bulletin for Medical and Public Health Laboratories, Ed. C. A. Perry, **9**, 1957, March.
51. Wootton, I. D. P., International biochemical trial 1954. *Clin. Chem.* **2**, 296 (1956).
52. Tonks, D. B., and Allen, P. H., Accuracy of glucose determinations in some Canadian hospital laboratories. *Can. Med. Ass. J.* **72**, 605 (1955).
53. Tonks, D. B., A study of the accuracy and precision of clinical chemistry determinations in 170 Canadian laboratories. *Clin. Chem.* **9**, 217 (1963).
54. Copeland, B. E., Ed., *Quality Control Manual. The American Society of Clinical Pathologists, Chicago, Ill., 1960.*
55. Levy, A. L., The use of ion-free serum as a standard in clinical chemistry. *Clin. Chem.* **8**, 174 (1962).
56. Klein, B., and Weissman, M., Study of dialyzed reconstituted dried serum as a clinical chemistry standard. *Clin. Chem.* **4**, 194 (1958).
57. Klein, B., and Weissman, M., Evaluation of a reconstituted dried serum as a clinical chemistry standard in the abnormal range. *Clin. Chem.* **7**, 149 (1961).
58. Klugerman, M. R., and Boutwell, J. H., Jr., Commercial control sera in the clinical chemistry laboratory. *Clin. Chem.* **7**, 185 (1961).
59. American Association of Clinical Chemists, *Standard Methods of Clinical Chemistry*, Vols. 1-5, Academic Press, New York, N. Y. 1958-1965.
60. Becker, S., Standards for microbilirubbin. *Clin. Chem.* **12**, 637, 1966.

COMMENTARY TO

36. Barnett, R. N., Cash, A. D. and Junghans, S. P. (1968)
Performance of "Kits" used for Clinical Chemical Analysis of Cholesterol.
New England Journal of Medicine 279(18): 974–979.

R oy N. Barnett was the first to develop a uniform testing and statistical analysis protocol for the comparison of clinical laboratory methods. The procedures he devised came from the need to evaluate the wide variety of commercial test kits in use in clinical chemistry at the time. In 1968, the year of the publication presented here, test kits had been in use in clinical chemistry for over 50 years. The LaMotte Chemical Products Company catalog in 1929, for example, offered complete test kits with pre-packaged reagents for serum urea, glucose (sugars), icterus index, bromides, calcium and chloride (1). Advertisements for complete kits for blood sulfonamide, bromide and other analytes routinely appeared in *The American Journal of Medical Technology* in 1943 (2). According to the Centers for Disease Control (CDC) there were 770 different test kits on the US market in 1970 (3). Twenty-seven of these kits were for cholesterol alone. By1975 48% of the 6 614 laboratories in the College of American Pathology (CAP) Proficiency Survey reported using a test kit for one or more of the reported results (4). The regulation and review of test kits however by the Food and Drug Administration (FDA) did not become law until late 1974 (5). In 1974 the total sales of clinical laboratory test kits in the US was $370 million and by 1978, the figure had increased to $955 million per year (6).

Barnett's first report on test kit performance was in 1963. He compared a newly released commercial hemaglutination inhibition antibody based pregnancy test kit with the standard male toad bioassay (7). Of the 111 patient urines tested by both methods, 16% gave a false positive with the test kit compared to 5% with the bioassay. The next year Barnett and co-workers compared serum cholesterol levels between a cholesterol test kit and results with the Abell–Kendall ether extraction method (8). The test kit results for paired samples varied from the Abell–Kendall method by as much as 3.70 mmol/L (143 mg/dL) higher to 1.79 mmol/L (69 mg/dL) lower (9). In 1968 the work on cholesterol was expanded to include the results with 12 different cholesterol kits. That report was published as a Special Article in *The New England Journal of Medicine* and it is presented here. Only two kits out of the 12 met the criteria used to assess test kit performance. Barnett and co-workers in the following years published additional reports on the performance of a wide range of different test kits. These included test kits for glucose (10); calcium (11); salicylates (12) and aspartate aminotransferase (EC 2.6.1.1) (13).

The method comparison scheme developed by Barnett was published as a complete paper in 1965 (14). This was the method used in the paper on cholesterol kits presented here. Barnett and Youdon expanded on the protocol in 1970 (15). In 1971, Barnett published a book on laboratory statistics that included the full protocol for test kit evaluation (16). Definitions and procedures were presented for the determination of precision; accuracy; selection of a reference method; recovery; and comparison of results with patient samples.

Barnett's work in the area of test kit comparisons helped promote interest in the field of quality laboratory assurance. In 1966, the American Association of Clinical Chemists (AACC) drafted a policy on test kits (17). Westgard (18) and Logan (19) separately produced a large body of literature that expanded on the use of method comparisons, statistical analysis and protocols. In 1979 the International Federation of Clinical Chemistry (IFCC) produced provisional guidelines for the evaluation of clinical chemistry kits (20). The Clinical and Laboratory Standards Institute (formerly The National Committee on Clinical Laboratory Standards, NCCLS) was founded in 1968. Many of their consensus guidelines were intended for use in method comparison studies (21).

Test kits and method comparison studies remain an integral part of clinical chemistry. Demacker *et al.* in 1983 (22) compared the accuracy of 20 different cholesterol test kits in a report reminiscent of what Barnett and co-workers began in 1964 (9). The January 2005 issue of the journal *Clinical Chemistry* contained twenty-four papers in the "Articles" section. Nineteen of the twenty-four papers used one or more commercial proprietary test kits to generate the bulk of the data presented in the published report (23).

References

(1) [Anonymous] (1929) *LaMotte Blood Chemistry Outfits for the General Practitioner's Routine Tests. (Product Catalog).* LaMotte Chemical Products Company, Baltimore, MD, pg 15.
(2) [Advertisement] (1943) The American Journal of Medical Technology. 9(4), July.

(3) Barnett, R.N. (1972) Reagent kits, in *Progress in Clinical Pathology*, Vol IV. Stefanini, M. (ed), Grune & Stratton, New York, pgs 181–198, Chapter 4.

(4) Brown, D.J. (1976) Clinical chemistry kits in the CAP survey program. American Journal of Clinical Pathology. 66(1): 223–233.

(5) Eavenson, E. (1974) In vitro diagnostic products and FDA regulation. Clinical Chemistry. 20(7):924–928.

(6) [Anonymous] (1980) The Frost and Sullivan report: the clinical diagnostic reagents and test kit markets. Surgical Business. 43(3):44–47.

(7) Barnett, R.N. (1963) Comparison of an immunologic and a toad test for pregnancy. American Journal of Clinical Pathology. 39(4):436–438.

(8) Abell, L.L., Levy, B.B., Brodie, B.B., and Kendall, F.E. (1952) A simplified method for the estimation of total cholesterol in serum and demonstration of its specificity. Journal of Biological Chemistry. 195(1):357–366.

(9) Hald, P.M., Weinberg, M.S., Barnett, R.N., Lavietes, P.H., and Seligson, D. (1964) Evaluation of a cholesterol kit. Clinical Chemistry. 10(3):241–245.

(10) Barnett, R.N. and Cash, A.D. (1969) Performance of "kits" used for clinical chemical analysis of glucose. American Journal of Clinical Pathology. 52(4):457–465.

(11) Barnett, R.N., Skodon, S.B., and Goldberg, M.H. (1973) Performance of "kits" used for the clinical chemical analysis of calcium in serum. American Journal of Clinical Pathology. 59(6):836–845.

(12) Kim, H., Skodon, S.B., and Barnett, R.N. (1974) Performance of "kits" used for the clinical chemical analysis of salicylates in serum. American Journal of Clinical Pathology. 61(6):936–942.

(13) Barnett, R.N., Ewing, N.S., and Skodon, S.B. (1976) Performance of "kits" used for the clinical chemical analysis of GOT (aspartate aminotransferase). Clinical Biochemistry. 9(2):78–84.

(14) Barnett, R.N. (1965) A scheme for the comparison of quantitative methods. American Journal of Clinical Pathology. 43(6): 562–569.

(15) Barnett, R.N. and Youdon, W.J. (1970) A revised scheme for the comparison of quantitative methods. American Journal of Clinical Pathology. 54(3 Part II):454–462.

(16) Barnett, R.S. (1971) Choice of methods, in *Clinical Laboratory Statistics*. Little, Brown and Company, Boston, Chapter 14, pgs 112–121.

(17) [Anonymous] (1966) AACC policy regarding reagent sets and kits. Clinical Chemistry. 12(1):43–44.

(18) Westgard, J.O. and Hunt, M.R. (1973) Use and interpretation of common statistical tests in method-comparison studies. Clinical Chemistry. 19(1):49–57.

(19) Logan, J.E. (1981) Criteria for kit selection in clinical chemistry, in *Clinical Biochemistry. Contemporary Theories and Techniques*, Vol 1. Spiegel, H.E. (ed), Academic Press, New York, pgs 43–85.

(20) Saris, N.-E. (1979) Provisional recommendation (1976) on evaluation of diagnostic kits. Part 2. Guidelines for the evaluation of clinical chemistry kits. (Stage 2 Draft 1). Clinical Chemistry. 25(8):1503–1505.

(21) NCCLS. (2002) Method comparison and bias estimation using patient samples: approved guideline-second edition, in *NCCLS document EP9-A2*. NCCLS, Wayne, PA.

(22) Demacker, P.N.M., Boerma, G.J.M., Baadenhuljaen, H., van Strik, R., Laijnse, B., and Jansen, A.P. (1983) Evaluation of accuracy of 20 different test kits for the enzymatic determination of cholesterol. Clinical Chemistry. 29(11):1916–1922.

(23) [Various Authors] (2005) Clinical Chemistry. 51(1):27–207, January.

New Engl. J. Med. 1968; 974–979

SPECIAL ARTICLE

PERFORMANCE OF "KITS" USED FOR CLINICAL CHEMICAL ANALYSIS OF CHOLESTEROL*

Roy N. Barnett, M.D., Ann D. Cash, M.T. (ASCP), and Siegfried P. Junghans, M.D.

S INCE about 1960 convenient "kits" for quantitative determination of blood chemical constituents have become increasingly popular, to the point
where several million are sold annually. They are
used predominantly in small laboratories that lack
the capacity to study product performance in a
scientific and systematic fashion. A voluntary pro-

*From the Division of Laboratories, Norwalk Hospital (address re-
print requests to Dr. Barnett at the Pathology Department, Norwalk
Hospital, Stevens St., Norwalk, Conn. 06852).

gram in which the manufacturer provides substan-
tial performance data, which are then checked independently, has been offered since 1964 by the
Standards Committee of the College of American
Pathologists, but this has been implemented for
only a few kits.

Because it seemed likely that unevaluated products might contain some incapable of producing valid results a study of diagnostic "kits" was initiated
in 1966. The present article is the first in a planned

series. Cholesterol was chosen because there are numerous widely used "kits" and because of our previous interest in the subject.[1]

MATERIAL AND METHODS

For the study, 12 of the currently commercially available cholesterol "kits" that we found advertised in medical or laboratory journals in June, 1966, were evaluated.* Attempts were made to buy these "kits" from area distributors. The "kits" not carried by distributors were bought directly from the manufacturer. Each "kit" was assigned a letter in the order in which it was received (Table 1). As a reference procedure, the Abell–Kendall method was chosen.[2]

TABLE 1. *Method Identification.*

CODE	NAME
AK	Abell – Kendall
A	Chole-Tech (Uni-Tech Chemical Manufacturing Co.)
B	Cholestex (Omni-Tech, Inc.)
C	Cholesterol Determination Set No. 650-15 (National Bio-Technical Laboratories)
D	Chole-stat (Scientific Chemical Co.)
E	Cholesterol Determination Set (Medi-Chem, Inc.)
F	Cholesterol Test Set #1 (Chem STAT)
G	Cholesterol Colorimetric Method (Boehringer Mannheim Corp.)
H	Cholesterol Determination (Hycel, Inc.)
I	Poly-Re-Sol (Hoppers Laboratories, Inc.)
J	Total Cholesterol Set Item No. 64149 (Hartman-Leddon Co., Inc.)
K	Uni-Test (Uni-Meter) (Bio-Dynamics, Inc.)
L	Hyland Cholesterol Test (Hyland Laboratories)

This is an alcohol extraction technic after alkaline hydrolysis using the Lieberman–Burchard color reaction and is generally considered a satisfactory reference method.[3-5] It is performed in duplicate, and the two values averaged to give a single result. Each of the other methods were performed as a single determination in accordance with the instructions of the manufacturer. No attempts were made to improve or alter any method, and the instructions of the producers were followed as closely as possible.

For all spectrophotometric readings, except Method K, the Coleman Jr. Spectrophotometer Model 6A was used. For Method K the distributor provided the Uni-Test colorimeter. Standard volumetric pipettes were used except for methods K and L, in which capillary pipettes supplied by the manufacturer were employed.

Data were collected, and later statistical analyses performed as described in an earlier article,[6] concerned with the details of comparing quantitative methods. This scheme outlined the performance of reproducibility studies, recovery studies and patient-comparison studies, as well as the statistical

*Some of these have since been given up, altered or replaced.

maneuvers used to evaluate the meaning of the data. The term "large discrepancy" used in the patient studies refers to a value obviously out of line with the reference method value after allowance is made both for a consistent difference between the two methods and for the inherent variability of both methods.

The reproducibility studies were done at three different levels. Three large serum sample pools were prepared at approximately 150 mg per 100 ml, for the low range, 250 mg per 100 ml for the medium range and 350 mg per 100 ml for the high range. The low-range sample was made from the blood of a single young person. Samples for the medium and high ranges were prepared by means of a "pool" made by freezing of all "normal"-appearing serum left over in the Chemistry Department each day until 1000 ml was collected, the pool then being thawed, mixed, centrifuged and frozen. Samples of the pool were analyzed once daily for 10 days, yielding a mean value of 215 mg per 100 ml. Appropriate amounts were then mixed with serum from a single patient whose serum cholesterol was 650 mg per 100 ml, yielding mixtures calculated to contain about 250 mg per 100 ml (medium range) and about 350 mg per 100 ml (high range). The three uniform mixtures were then divided into 7-ml aliquots sufficient for all the studies, and these were frozen until used. They remained stable throughout the experiment.

A pure cholesterol standard in alcohol prepared in the laboratory of Dr. Bradley Copeland, New England Deaconess Hospital, Boston, Massachusetts, by the dibromide method of Fieser[7] was used for the reference method. For all other methods, the standards recommended by the manufacturer were employed. All tests were done by one medical technologist (A.C.) who had several years of experience in hospital chemistry. Before the recorded studies were started, she was given time to familiarize herself with each method.

The nature of cholesterol makes recovery studies from serum unreliable. Ordinary crystalline cholesterol is not fully soluble in serum and cannot therefore be simply added by weight to a given serum sample. Although it is possible to dissolve cholesterol crystals in such solvents as alcohol, chloroform or acetic acid, use of added cholesterol so dissolved might not be compatible with each of the technics under study. For these reasons recovery studies were not attempted.

RESULTS

Reproducibility Studies

This series of analyses was designed to test the day-to-day reproducibility for each method. This procedure is based on the usual clinical practice in which patient samples are studied once a day or less often, particularly when cholesterol analyses are used to check response to treatment. We therefore analyzed each of the three large serum pools by each of the methods on each of 12 different days. If one result appeared out of line with the other results for that method we performed one more analysis on the thirteenth day.

An ampoule of each pool was completely thawed in a 37°C water bath. This provided enough serum to use in tests of all the methods for that day. The different methods were set up in order of convenience, according to technical details, such as time involved for the color reaction or heating or cooling of substances. Certain tests, which had about the same performance time, were grouped together. The reference method was set up as the first method each day, since it involved the longest time. To avoid certain environmental influences, such as temperature changes in the morning and afternoon and differences in the performance of the technologist after a certain number of working hours, the tests were performed each day in a different order. For example, those undertaken in the morning on one day were done in the afternoon on the next day. A total of 444 analyses were carried out. We did not find any evidence of progressive spoilage of the reagents or serums in the daily determinations. The photometric readings were recorded every day, but the calculations were postponed until all analyses were completed. This was done because of the large volume of tests performed and to avoid any possible operator bias.

Kits I and F were studied at a later date than the others. The reproducibility values for these kits were entirely comparable to those for the others.

The first 10 values for each method at each level were used for calculations unless there was reason to believe that a technical error had been made. If a value was considered to be in error it was discarded, and the eleventh value used in its place. In no analysis was more than one value in a series eliminated.

Figures 1-3 are bar graphs portraying the data. Each vertical bar represents the ±3 standard-deviation confidence limits of single determinations about a mean shown as the central line. The shaded band includes 99.7 per cent of the Abell–Kendall values, assumed to be the "correct" range in these experiments.

Table 2 indicates the coefficient of variation (standard deviation divided by the mean × 100) for each method at each level, arranged in ascending order.

These figures indicate vividly that certain "kits" provide values close to the reference values whereas others do not.

Patient Comparisons

The second part of the study was concerned with the performance of the "kits" on individual patient serum samples preserved by freezing. There were 40 consecutive patients on whom cholesterol determinations were ordered by an attending physician.

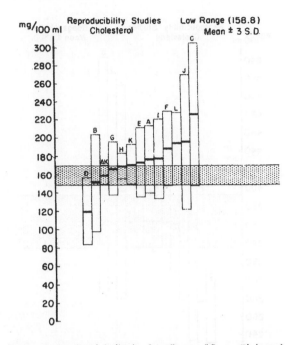

FIGURE 1. *Bar Graph Indicating Low Range of Serum Cholesterol. Bars indicate ±3 SD limits for each method; the central line is the mean. The shaded area is a projection of the reference method values. Bars are arranged from left to right in order of ascending means. Note the general correlation between accuracy and precision.*

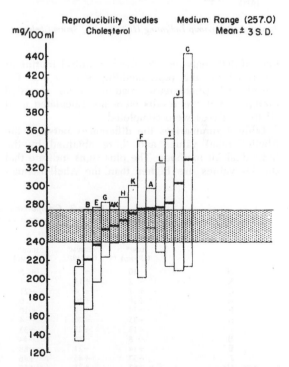

FIGURE 2. *Bar Graph Indicating Medium Range of Serum Cholesterol.*

Each day the frozen serums of five of these patients were thawed, and all 13 methods performed for a

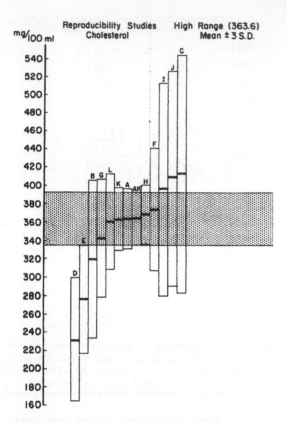

FIGURE 3. *Bar Group Indicating High Range of Serum Cholesterol.*

TABLE 2. *Reproducibility Studies.*

		COEFFICIENT OF VARIATION			
LOW RANGE (159 MG*)		MEDIUM RANGE (257 MG*)		HIGH RANGE (364 MG*)	
method	CV	method	CV	method	CV
AK	2.25	AK	2.29	AK	2.64
H	2.84	A	2.57	H	2.93
K	4.14	H	3.03	A	3.02
L	5.82	K	3.65	K	3.08
G	5.96	G	3.83	L	4.83
A	6.96	E	5.62	F	5.93
F	7.20	L	5.74	G	6.24
E	7.33	D	7.79	E	7.10
I	8.27	B	8.10	B	8.97
D	10.09	I	8.09	J	9.65
C	11.68	F	8.90	D	9.71
B	11.75	J	10.28	I	9.83
J	12.68	C	11.63	C	10.53

*Mean of AK values.

results, and the minus signs that the kit values are lower. The order in which the values are arranged is that of the total differences in the reproducibility studies, the first method having the smallest, and the last the greatest total divergence from the reference method. In the right half of the table the patients were divided into the specified ranges on the basis of the Abell–Kendall values.

Table 4 lists the means and the number of "large discrepancies" in the patient comparison study. "Large discrepancies"[6] are values lying outside the combined ±3 SD limits for both reference and test methods as calculated in the reproducibility studies, the mean patient values at each level for the kit being used as the midpoint. To calculate the combined standard deviations for two methods we squared each to obtain the variance, added the two variances and took the square root of the sum.

A number of additional statistical manipulations were carried out as previously described.[6] These included calculations of "runs," "male versus female," "odd vs even," "first 20 vs second 20" and "trend with time." One outlying value was identified in the reference method, and all results for this patient were eliminated from further calcu-

total of 520 analyses. The same technical routine as described for the reproducibility studies was followed. All photometric readings were recorded promptly, but the results were not calculated until all the analyses were completed.

Table 3 summarizes the differences between the Abell–Kendall values and those obtained by the individual kit methods. The plus signs indicate that the kit values are higher than the Abell–Kendall

TABLE 3. *Comparison with Abell–Kendall Values.*

METHOD	REPRODUCIBILITY STUDIES			PATIENT STUDIES		
	LOW* (159 MG)	MEDIUM* (257 MG)	HIGH* (364 MG)	LOW† (100-199 MG)	MEDIUM‡ (200-299 MG)	HIGH§ (300-399 MG)
H	+10	+ 7	+ 5	+20	+ 17	+ 14
K	+ 2	+14	− 1	+22	+ 4	− 2
G	+ 7	− 4	−21	− 1	− 40	− 49
A	+18	+19	0	+12	− 2	− 15
F	+30	+18	+10	+21	− 4	− 1
L	+35	+20	− 4	+52	+ 36	+ 17
I	+19	+25	+33	+24	+ 21	+ 4
B	− 8	−36	−44	−31	− 59	−158
E	+14	−21	−88	+38	0	− 33
J	+37	+46	+46	+42	+ 33	+ 44
C	+67	+71	+50	+59	+ 65	+ 52
D	−39	−84	+32	−54	−111	−146

*Based on 10 replicate analyses, 1/day, Abell–Kendall values.   ‡15 Abell–Kendall values, calculated from "unrounded" means.
†16 Abell–Kendall values, calculated from "unrounded" means.   §8 Abell–Kendall values, calculated from "unrounded" means.

TABLE 4. *Mean Values and Large Discrepancies in Patient Studies.*

METHOD	MEAN (MG/100 ML)	NO. OF LARGE DISCREPANCIES
D	139	3
B	159	7
G	202	2
AK	230	1
A	232	3
H	241	3
F	242	2
I	247	4
K	251	12
E	251	7
J	269	0
L	272	4
C	292	2

lations. Otherwise the data did not indicate the existence of any internal discrepancy in the experiment or any spoilage of reagents or serum samples.

DISCUSSION

Why is there so much variability in the performance of certain of these products? The accompanying instructions of some were vague or confusing and could be interpreted in more than one way. Some methods used standards that were well outside the usual clinical values, so that patient samples were not read in the most accurate portion of the spectrophotometer scale. Some reagents were cloudy and did not clear as the instructions stated they would. In others the color reactions were different from those stated in the instructions. Many other technical problems were encountered. Our studies were conducted in a clinical rather than a research laboratory setting, and our precision is not as great as could probably be achieved for any of the methods. For example, our coefficient of variation for the Abell–Kendall method ranged from 2.26 to 2.63 per cent, whereas the authors of the method[2] achieved a value of 1.28 per cent (calculated by us from their published data). For Kit H our coefficient of variation ranged from 2.84 to 3.03 per cent,

whereas the manufacturer's claims, confirmed by the Standards Committee of the College of American Pathologists,[3] ranged from 1.15 to 1.30 per cent. However, our comparative results are quite similar and should provide a reasonable guide to performance in clinical practice.

How accurate should cholesterol methods be for clinical use? We approached this from two points of view: the effect on population values; and the effect on values for individual patients. If the method standard deviation is one third that of the true population range it will spread the range by 5.4 per cent,[8] an acceptable figure. A frequently used normal range for cholesterol is from 130 to 250 mg per 100 ml,[9-11] a spread of 120 mg. This represents 95 per cent of the population, or ±2 standard deviations, so that the standard deviation for normal persons is 30 mg; one third of this is 10 mg.

To estimate what physicians particularly interested in cholesterol metabolism expected as the precision of methods, we made written inquiry of 10 consultants and obtained eight usable answers. They generally accepted 10 mg as an appropriate standard deviation for the method. They were concerned primarily with the medium and high ranges and expected less precision in the low range.

Combining the statistical, chemical and medical criteria, we arbitrarily established four criteria for evaluation of each method: the standard deviation in the middle range should not exceed 10, and that in the high range should not exceed 12.0 mg per 100 ml; the bias from the reference method should be small in both reproducibility studies and patient studies, so that comparison with published values is meaningful; the bias should be consistent in size and direction at the three levels (if it changes markedly from low to high values interpretation of results is impossible); and the number of "large discrepancies," indicative of random errors in the patient studies, should not exceed three.

Table 5 evaluates each method for each property:

TABLE 5. *Kit Evaluations.**

METHOD	PRECISION†	BIAS SMALL‡		BIAS CONSISTENT§		<3 LARGE¶ DISCREPANCIES
		REPRODUCIBILITY	PATIENT	REPRODUCIBILITY	PATIENT	
A	+	+	+	+	+	+
B	−	−	−	−	−	−
C	−	−	−	+	+	+
D	−	−	−	−	−	+
E	−	−	−	−	−	−
F	−	−	+	+	+	+
G	−	+	−	+	−	+
H	+	+	+	+	+	+
I	−	−	+	+	+	−
J	−	−	−	+	+	+
K	+	+	−	+	+	−
L	−	−	−	−	−	−

*+ = acceptable, − = acceptable with use of arbitrary criteria specified below.
†Not appreciably > 10 mg/100 ml in medium range & 12 mg/100 ml in high range.
‡Not >21 mg/100 ml different from reference method.
§No spread >28 mg/100 ml from low to high.
¶In patient comparisons.

+ indicates acceptable and − not acceptable according to these criteria.

Only two kits (A and H) of the 12 tested are acceptable according to all these criteria. In the 1963 National Cholesterol Survey[12] A and H were also found acceptable whereas of the others tested in both studies B, D and I were not acceptable in either. The manufacturer's claims for H, made in a fashion similar to those of this study, have been verified by the Standards Committee of the College of American Pathologists.[3]

The research on which this publication is based was performed pursuant to contract No. FDA 66-199 (NEG), with Food and Drug Administration, Department of Health, Education, and Welfare. This report reflects the opinion of the authors and not necessarily that of the Food and Drug Administration.

W. J. Youden, Ph.D., was the statistical consultant.

REFERENCES

1. Hald, P. M., Weinberg, M. S., Barnett, R. N., Lavietes, P. H., and Seligson, D. Evaluation of cholesterol "kit." *Clin. Chem.* **10**:241-245, 1964.

2. Abell, L. L., Levy, B. B., Brodie, B. B., and Kendall, F. E. Simplified method for estimation of total cholesterol in serum and demonstration of its specificity. *J. Biol. Chem.* **195**:357-366, 1952.

3. Muelling, R. J., and Copeland, B. E. Two meetings on selection of criteria for pure cholesterol preparation to be used for standardizing cholesterol measurements for medical diagnosis and therapy. *Am. J. Clin. Path.* **47**:654-657, 1967.

4. Straus, R., and Wurm, M. *Clinical Chemistry "Check Sample"* No. CC-15. Published by American Society of Clinical Pathologists, Chicago, Illinois, September, 1962.

5. Tonks, D. B. Estimation of cholesterol in serum: classification and critical review of methods. *Clin. Biochem.* **1**:12-29, 1967.

6. Barnett, R. N. Scheme for comparison of quantitative methods. *Am. J. Clin. Path.* **43**:562-569, 1965.

7. Fieser, L. F. Cholesterol and companions. VII. Steroid dibromides. *J. Am. Chem. Soc.* **75**:5421, 1953.

8. American Society of Clinical Pathologists, Council on Clinical Chemistry, Commission on Continuing Education. *Manual for Workshop on Statistical Methods in Clinical Laboratory.* Edited by R. N. Barnett. Chicago: The Society, 1965.

9. Cantarow, A., and Trumper, M. *Clinical Biochemistry.* Fourth edition. Philadelphia: Saunders, 1949. P. 150.

10. Henry, R. J. *Clinical Chemistry: Principles and technics.* New York: Harper, 1964. P. 862.

11. Todd, J. C., and Sanford, A. H. *Clinical Diagnosis by Laboratory Methods: 13th Edition, by Israel Davidsohn (and) Benjamin B. Wells.* Philadelphia: Saunders, 1962. P. 945.

12. Straus, R., and Wurm, M. *1963 National Cholesterol Survey.* Prepared by College of American Pathologists, Chicago, Illinois, 1963.

COMMENTARY TO

page 475

37. Pauling, L., Itano, H. A., Singer, S. J. and Wells, I. C. (1949)
Sickle Cell Anemia, a Molecular Disease. Science 110(2865): 543–548.

J ames Herrick in 1910 was the first to describe the peculiar shape of red cells in a patient suffering from a diverse set of symptoms including anemia (1). The patient was a 20-year old male from Granada who had come to Chicago to attend dental school. Herrick was unable to make a diagnosis but described the appearance of this patient's red cells under a microscope. He wrote, "…what especially attracted attention was the large number of thin, elongated, sickle-shaped and crescent-shaped forms" (1). He included four photomicrographs in his paper. Over the years, additional reports appeared in the literature confirming Herrick's findings in patients with anemia, all of whom were of African descent (2).

The term, sickle cell anemia to describe a disease condition was first used by Mason in 1922 (3). He described the sickle shape of the red cells in three African–American patients and proposed, based on case histories that the disease was hereditary in nature. Neel in 1949 used conventional family history analysis and speculated on the genetic basis of the disease. He claimed a homozygous inheritance for the sickle cell disease and heterozygous inheritance for the trait (4). Four months latter Linus Pauling and co-workers in the paper presented here determined that the hemoglobin molecule in sickle cell patients had a different electrical charge compared to normal patients. Electrophoresis demonstrated that sickle cell hemoglobin was altered in molecular structure. The disease was heritable and therefore the authors argued it was the genes that determined the molecular structure of proteins. It was the first time that the term molecular disease had been used in the title of a scientific paper and this paper helped launch the era of molecular medicine.

In 1949, the California Institute of Technology (Caltech) in Pasadena was one of the few institutions in the world that had a Tiselius moving boundary electrophoresis instrument. In moving boundary electrophoresis the separated proteins were detected by refractive index changes measured with schlieren optics recorded on photographic film. Swingle at Caltech improved the optical design and reduced the size of the optical bench of the Caltech instrument to 4.9 m (16 ft) (5). This was the instrument used by Pauling, Itano, Singer and Wells to separate the hemoglobin from sickle-cell and normal patient's blood. Electrophoresis demonstrated that hemoglobin from the two groups were separate proteins and that the difference was based on charge. Sickle-cell hemoglobin possessed a higher isoelectric point compared to normal hemoglobin. Sickle cell hemoglobin saturated with carbon monoxide moved at pH 7.0 like a positively charged protein whereas normal hemoglobin under the same conditions moved as a negatively charged protein.

The discussion section of their paper linked the charge differences in hemoglobin to an altered amino acid sequence in the molecule. Sickle-cell anemia had already been shown to be an inherited disease. Pauling and his co-workers then claimed that it was the gene that controlled the structure and hence the charge of the protein. For the first time a link was made between a protein, a gene and a specific disease. Seven years latter Ingram demonstrated that an amino acid substitution in the hemoglobin beta chain was the molecular basis of sickle cell disease (6).

In 1997, Richard Horton, a senior editor at *The Lancet* wrote an article in which he listed 27 works as constituting a core cannon in medical literature (7). The list included Hippocratic writings, Fleming's description of penicillin, Bernard's first paper on a successful heart transplant and Pauling's paper on sickle-cell hemoglobin.

References

(1) Herrick, J.B. (1910) Peculiar elongated and sickle-shaped red blood corpuscles in a case of severe anemia. Archives of Internal Medicine. 6:517–521.
(2) Savitt, T.L. and Goldberg, M.F. (1989) Herrick's 1910 case report of sickle cell anemia. The rest of the story. Journal of the American Medical Association. 261(2):266–271.
(3) Mason, V.R. (1922) Sickle cell anemia. Journal of the American Medical Association. 79(16):1318–1320.
(4) Neel, J.V. (1949) The inheritance of sickle cell anemia. Science. 110(2846):64–66.
(5) Swingle, S.M. (1947) An electrophoretic apparatus using parabolic mirrors. Review of Scientific Instruments. 18(2): 128–132.
(6) Ingram, V.M. (1956) A specific chemical difference between the globins of normal human and sickle-cell anaemia haemoglobin. Nature. 178(4537):792–794.
(7) Horton, R. (1997) A manifesto for reading medicine. The Lancet. 349(9055):872–874.

Sickle Cell Anemia, a Molecular Disease[1]

Linus Pauling, Harvey A. Itano,[2] S. J. Singer,[2] and Ibert C. Wells[3]

Gates and Crellin Laboratories of Chemistry,
California Institute of Technology, Pasadena, California[4]

THE ERYTHROCYTES of certain individuals possess the capacity to undergo reversible changes in shape in response to changes in the partial pressure of oxygen. When the oxygen pressure is lowered, these cells change their forms from the normal biconcave disk to crescent, holly wreath, and other forms. This process is known as sickling. About 8 percent of American Negroes possess this characteristic; usually they exhibit no pathological consequences ascribable to it. These people are said to have sicklemia, or sickle cell trait. However, about 1 in 40 (4) of these individuals whose cells are capable of sickling suffer from a severe chronic anemia resulting from excessive destruction of their erythrocytes; the term sickle cell anemia is applied to their condition.

The main observable difference between the erythrocytes of sickle cell trait and sickle cell anemia has been that a considerably greater reduction in the partial pressure of oxygen is required for a major fraction of the trait cells to sickle than for the anemia cells (11). Tests *in vivo* have demonstrated that between 30 and 60 percent of the erythrocytes in the venous circulation of sickle cell anemic individuals, but less than 1 percent of those in the venous circulation of sicklemic individuals, are normally sickled. Experiments *in vitro* indicate that under sufficiently low oxygen pressure, however, all the cells of both types assume the sickled form.

The evidence available at the time that our investigation was begun indicated that the process of sickling might be intimately associated with the state and the nature of the hemoglobin within the erythrocyte. Sickle cell erythrocytes in which the hemoglobin is combined with oxygen or carbon monoxide have the biconcave disk contour and are indistinguishable in that form from normal erythrocytes. In this condition they are termed promeniscocytes. The hemoglobin appears to be uniformly distributed and randomly oriented within normal cells and promeniscocytes, and no birefringence is observed. Both types of cells are very flexible. If the oxygen or carbon monoxide is removed, however, transforming the hemoglobin to the uncombined state, the promeniscocytes undergo sickling. The hemoglobin within the sickled cells appears to aggregate into one or more foci, and the cell membranes collapse. The cells become birefringent (11) and quite rigid. The addition of oxygen or carbon monoxide to these cells reverses these phenomena. Thus the physical effects just described depend on the state of combination of the hemoglobin, and only secondarily, if at all, on the cell membrane. This conclusion is supported by the observation that sickled cells when lysed with water produce discoidal, rather than sickle-shaped, ghosts (10).

It was decided, therefore, to examine the physical and chemical properties of the hemoglobins of individuals with sicklemia and sickle cell anemia, and to compare them with the hemoglobin of normal individuals to determine whether any significant differences might be observed.

EXPERIMENTAL METHODS

The experimental work reported in this paper deals largely with an electrophoretic study of these hemoglobins. In the first phase of the investigation, which concerned the comparison of normal and sickle cell anemia hemoglobins, three types of experiments were performed: 1) with carbonmonoxyhemoglobins; 2) with uncombined ferrohemoglobins in the presence of dithionite ion, to prevent oxidation to methemoglobins; and 3) with carbonmonoxyhemoglobins in the presence of dithionite ion. The experiments of type 3 were performed and compared with those of type 1 in order to ascertain whether the dithionite ion itself causes any specific electrophoretic effect.

Samples of blood were obtained from sickle cell anemic individuals who had not been transfused within three months prior to the time of sampling. Stroma-free concentrated solutions of human adult hemoglobin were prepared by the method used by Drabkin (3). These solutions were diluted just before use with the

[1] This research was carried out with the aid of a grant from the United States Public Health Service. The authors are grateful to Professor Ray D. Owen, of the Biology Division of this Institute, for his helpful suggestions. We are indebted to Dr. Edward R. Evans, of Pasadena, Dr. Travis Winsor, of Los Angeles, and Dr. G. E. Burch, of the Tulane University School of Medicine, New Orleans, for their aid in obtaining the blood used in these experiments.

[2] U. S. Public Health Service postdoctoral fellow of the National Institutes of Health.

[3] Postdoctoral fellow of the Division of Medical Sciences of the National Research Council.

[4] Contribution No. 1333.

appropriate buffer until the hemoglobin concentrations were close to 0.5 grams per 100 milliliters, and then were dialyzed against large volumes of these buffers for 12 to 24 hours at 4° C. The buffers for the experiments of types 2 and 3 were prepared by adding 300 ml of 0.1 ionic strength sodium dithionite solution to 3.5 liters of 0.1 ionic strength buffer. About 100 ml of 0.1 molar NaOH was then added to bring the pH of the buffer back to its original value. Ferrohemoglobin solutions were prepared by diluting the

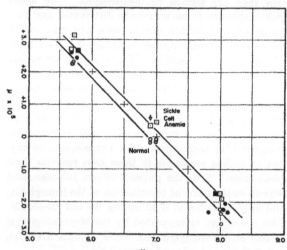

FIG. 1.　Mobility(μ)-pH curves for carbonmonoxyhemoglobins in phosphate buffers of 0.1 ionic strength. The black circles and black squares denote the data for experiments performed with buffers containing dithionite ion. The open square designated by the arrow represents an average value of 10 experiments on the hemoglobin of different individuals with sickle cell anemia. The mobilities recorded in this graph are averages of the mobilities in the ascending and descending limbs.

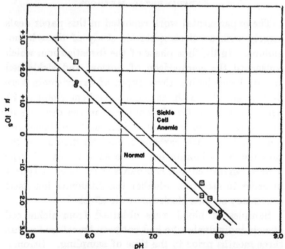

FIG. 2.　Mobility(μ)-pH curves for ferrohemoglobins in phosphate buffers of 0.1 ionic strength containing dithionite ion. The mobilities recorded in the graph are averages of the mobilities in the ascending and descending limbs.

concentrated solutions with this dithionite-containing buffer and dialyzing against it under a nitrogen atmosphere. The hemoglobin solutions for the experiments of type 3 were made up similarly, except that they were saturated with carbon monoxide after dilution and were dialyzed under a carbon monoxide atmosphere. The dialysis bags were kept in continuous motion in the buffers by means of a stirrer with a mercury seal to prevent the escape of the nitrogen and carbon monoxide gases.

The experiments were carried out in the modified Tiselius electrophoresis apparatus described by Swingle (14). Potential gradients of 4.8 to 8.4 volts per centimeter were employed, and the duration of the runs varied from 6 to 20 hours. The pH values of the buffers were measured after dialysis on samples which had come to room temperature.

RESULTS

The results indicate that a significant difference exists between the electrophoretic mobilities of hemoglobin derived from erythrocytes of normal individuals and from those of sickle cell anemic individuals. The two types of hemoglobin are particularly easily distinguished as the carbonmonoxy compounds at pH 6.9 in phosphate buffer of 0.1 ionic strength. In this buffer the sickle cell anemia carbonmonoxyhemoglobin moves as a positive ion, while the normal compound moves as a negative ion, and there is no detectable amount of one type present in the other.[4] The hemoglobin derived from erythrocytes of individuals with sicklemia, however, appears to be a mixture of the normal hemoglobin and sickle cell anemia hemoglobin in roughly equal proportions. Up to the present time the hemoglobins of 15 persons with sickle cell anemia, 8 persons with sicklemia, and 7 normal adults have been examined. The hemoglobins of normal adult white and negro individuals were found to be indistinguishable.

The mobility data obtained in phosphate buffers of 0.1 ionic strength and various values of pH are summarized in Figs. 1 and 2.[5]

[4] Occasionally small amounts (less than 5 percent of the total protein) of material with mobilities different from that of either kind of hemoglobin were observed in these uncrystallized hemoglobin preparations. According to the observations of Stern, Reiner, and Silber (12) a small amount of a component with a mobility smaller than that of oxyhemoglobin is present in human erythrocyte hemolyzates.

[5] The results obtained with carbonmonoxyhemoglobins with and without dithionite ion in the buffers indicate that the dithionite ion plays no significant role in the electrophoretic properties of the proteins. It is therefore of interest that ferrohemoglobin was found to have a lower isoelectric point in phosphate buffer than carbonmonoxyhemoglobin. Titration studies have indicated (5, 6) that oxyhemoglobin (similar in electrophoretic properties to the carbonmonoxy compound) has a lower isoelectric point than ferrohemoglobin in

The isoelectric points are listed in Table 1. These results prove that the electrophoretic difference between normal hemoglobin and sickle cell anemia hemoglobin

TABLE 1
ISOELECTRIC POINTS IN PHOSPHATE BUFFER, $\mu = 0.1$

Compound	Normal	Sickle cell anemia	Difference
Carbonmonoxyhemoglobin	6.87	7.09	0.22
Ferrohemoglobin	6.87	7.09	0.22

exists in both ferrohemoglobin and carbonmonoxyhemoglobin. We have also performed several experiments in a buffer of 0.1 ionic strength and pH 6.52 containing 0.08 M NaCl, 0.02 M sodium cacodylate, and 0.0083 M cacodylic acid. In this buffer the average mobility of sickle cell anemia carbonmonoxyhemoglobin is 2.63×10^{-5}, and that of normal carbonmonoxyhemoglobin is 2.23×10^{-5} cm/sec per volt/cm.[6]

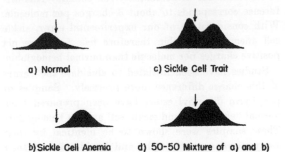

a) Normal   c) Sickle Cell Trait

b) Sickle Cell Anemia   d) 50-50 Mixture of a) and b)

FIG. 3. Longsworth scanning diagrams of carbonmonoxyhemoglobins in phosphate buffer of 0.1 ionic strength and pH 6.90 taken after 20 hours' electrophoresis at a potential gradient of 4.73 volts/cm.

These experiments with a buffer quite different from phosphate buffer demonstrate that the difference between the hemoglobins is essentially independent of the buffer ions.

Typical Longsworth scanning diagrams of experiments with normal, sickle cell anemia, and sicklemia carbonmonoxyhemoglobins, and with a mixture of the first two compounds, all in phosphate buffer of pH 6.90 and ionic strength 0.1, are reproduced in Fig. 3. It is apparent from this figure that the sicklemia material contains less than 50 percent of the anemia component. In order to determine this quantity accurately some experiments at a total protein concentra-

the absence of other ions. These results might be reconciled by assuming that the ferrous iron of ferrohemoglobin forms complexes with phosphate ions which cannot be formed when the iron is combined with oxygen or carbon monoxide. We propose to continue the study of this phenomenon.

[6] The mobility data show that in 0.1 ionic strength cacodylate buffers the isoelectric points of the hemoglobins are increased about 0.5 pH unit over their values in 0.1 ionic strength phosphate buffers. This effect is similar to that observed by Longsworth in his study of ovalbumin (7).

tion of 1 percent were performed with known mixtures of sickle cell anemia and normal carbonmonoxyhemoglobins in the cacodylate-sodium chloride buffer of 0.1 ionic strength and pH 6.52 described above. This buffer was chosen in order to minimize the anomalous electrophoretic effects observed in phosphate buffers (7). Since the two hemoglobins were incompletely resolved after 15 hours of electrophoresis under a potential gradient of 2.79 volts/cm, the method of Tiselius and Kabat (16) was employed to allocate the

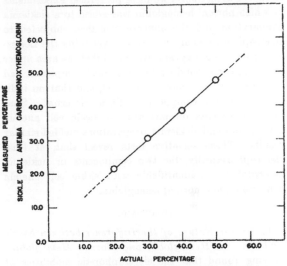

SICKLE CELL ANEMIA CARBONMONOXYHEMOGLOBIN

FIG. 4. The determination of the percent of sickle cell anemia carbonmonoxyhemoglobin in known mixtures of the protein with normal carbonmonoxyhemoglobin by means of electrophoretic analysis. The experiments were performed in a cacodylate sodium chloride buffer described in the text.

areas under the peaks in the electrophoresis diagrams to the two components. In Fig. 4 there is plotted the percent of the anemia component calculated from the areas so obtained against the percent of that component in the known mixtures. Similar experiments were performed with a solution in which the hemoglobins of 5 sicklemic individuals were pooled. The relative concentrations of the two hemoglobins were calculated from the electrophoresis diagrams, and the actual proportions were then determined from the plot of Fig. 4. A value of 39 percent for the amount of the sickle cell anemia component in the sicklemia hemoglobin was arrived at in this manner. From the experiments we have performed thus far it appears that this value does not vary greatly from one sicklemic individual to another, but a more extensive study of this point is required.

Up to this stage we have assumed that one of the two components of sicklemia hemoglobin is identical with sickle cell anemia hemoglobin and the other is identical with the normal compound. Aside from the genetic evidence which makes this assumption very probable (see the discussion section), electrophoresis experiments afford direct evidence that the assumption is valid. The experiments on the pooled sicklemia carbonmonoxyhemoglobin and the mixture containing 40 percent sickle cell anemia carbonmonoxyhemoglobin and 60 percent normal carbonmonoxyhemoglobin in the cacodylate-sodium chloride buffer described above were compared, and it was found that the mobilities of the respective components were essentially identical.[7] Furthermore, we have performed experiments in which normal hemoglobin was added to a sicklemia preparation and the mixture was then subjected to electrophoretic analysis. Upon examining the Longsworth scanning diagrams we found that the area under the peak corresponding to the normal component had increased by the amount expected, and that no indication of a new component could be discerned. Similar experiments on mixtures of sickle cell anemia hemoglobin and sicklemia preparations yielded similar results. These sensitive tests reveal that, at least electrophoretically, the two components in sicklemia hemoglobin are identifiable with sickle cell anemia hemoglobin and normal hemoglobin.

DISCUSSION

1) *On the Nature of the Difference between Sickle Cell Anemia Hemoglobin and Normal Hemoglobin*: Having found that the electrophoretic mobilities of sickle cell anemia hemoglobin and normal hemoglobin differ, we are left with the considerable problem of locating the cause of the difference. It is impossible to ascribe the difference to dissimilarities in the particle weights or shapes of the two hemoglobins in solution: a purely frictional effect would cause one species to move more slowly than the other throughout the entire pH range and would not produce a shift in the isoelectric point. Moreover, preliminary velocity ultracentrifuge[8] and free diffusion measurements indicate that the two hemoglobins have the same sedimentation and diffusion constants.

The most plausible hypothesis is that there is a difference in the number or kind of ionizable groups in the two hemoglobins. Let us assume that the only groups capable of forming ions which are present in carbonmonoxyhemoglobin are the carboxyl groups in the heme, and the carboxyl, imidazole, amino, phenolic hydroxyl, and guanidino groups in the globin. The number of ions nonspecifically adsorbed on the two proteins should be the same for the two hemoglobins

[7] The patterns were very slightly different in that the known mixture contained 1 percent more of the sickle cell anemia component than did the sickle cell trait material.

[8] We are indebted to Dr. M. Moskowitz, of the Chemistry Department, University of California at Berkeley, for performing the ultracentrifuge experiments for us.

under comparable conditions, and they may be neglected for our purposes. Our experiments indicate that the net number of positive charges (the total number of cationic groups minus the number of anionic groups) is greater for sickle cell anemia hemoglobin than for normal hemoglobin in the pH region near their isoelectric points.

According to titration data obtained by us, the acid-base titration curve of normal human carbonmonoxyhemoglobin is nearly linear in the neighborhood of the isoelectric point of the protein, and a change of one pH unit in the hemoglobin solution in this region is associated with a change in net charge on the hemoglobin molecule of about 13 charges per molecule. The same value was obtained by German and Wyman (5) with horse oxyhemoglobin. The difference in isoelectric points of the two hemoglobins under the conditions of our experiments is 0.23 for ferrohemoglobin and 0.22 for the carbonmonoxy compound. This difference corresponds to about 3 charges per molecule. With consideration of our experimental error, sickle cell anemia hemoglobin therefore has 2–4 more net positive charges per molecule than normal hemoglobin.

Studies have been initiated to elucidate the nature of this charge difference more precisely. Samples of porphyrin dimethyl esters have been prepared from normal hemoglobin and sickle cell anemia hemoglobin. These samples were shown to be identical by their x-ray powder photographs and by identity of their melting points and mixed melting point. A sample made from sicklemia hemoglobin was also found to have the same melting point. It is accordingly probable that normal and sickle cell anemia hemoglobin have different globins. Titration studies and amino acid analyses on the hemoglobins are also in progress.

2) *On the Nature of the Sickling Process*: In the introductory paragraphs we outlined the evidence which suggested that the hemoglobins in sickle cell anemia and sicklemia erythrocytes might be responsible for the sickling process. The fact that the hemoglobins in these cells have now been found to be different from that present in normal red blood cells makes it appear very probable that this is indeed so.

We can picture the mechanism of the sickling process in the following way. It is likely that it is the globins rather than the hemes of the two hemoglobins that are different. Let us propose that there is a surface region on the globin of the sickle cell anemia, hemoglobin molecule which is absent in the normal molecule and which has a configuration complementary to a different region of the surface of the hemoglobin molecule. This situation would be somewhat analogous to that which very probably exists in antigen-antibody reactions (9). The fact that sickling occurs only when the partial pressures of oxygen and carbon monoxide are low suggests that one of these sites is very near to the iron atom of one or more of the hemes, and that when the iron atom is combined with either one of these gases, the complementariness of the two structures is considerably diminished. Under the appropriate conditions, then, the sickle cell anemia hemoglobin molecules might be capable of interacting with one another at these sites sufficiently to cause at least a partial alignment of the molecules within the cell, resulting in the erythrocyte's becoming birefringent, and the cell membrane's being distorted to accommodate the now relatively rigid structures within its confines. The addition of oxygen or carbon monoxide to the cell might reverse these effects by disrupting some of the weak bonds between the hemoglobin molecules in favor of the bonds formed between gas molecules and iron atoms of the hemes.

Since all sicklemia erythrocytes behave more or less similarly, and all sickle at a sufficiently low oxygen pressure (11), it appears quite certain that normal hemoglobin and sickle cell anemia hemoglobin coexist within each sicklemia cell; otherwise there would be a mixture of normal and sickle cell anemia erythrocytes in sicklemia blood. We might expect that the normal hemoglobin molecules, lacking at least one type of complementary site present on the sickle cell anemia molecules, and so being incapable of entering into the chains or three-dimensional frameworks formed by the latter, would interfere with the alignment of these molecules within the sicklemia erythrocyte. Lower oxygen pressures, freeing more of the complementary sites near the hemes, might be required before sufficiently large aggregates of sickle cell anemia hemoglobin molecules could form to cause sickling of the erythrocytes.

This is in accord with the observations of Sherman (11), which were mentioned in the introduction, that a large proportion of erythrocytes in the venous circulation of persons with sickle cell anemia are sickled, but that very few have assumed the sickle forms in the venous circulation of individuals with sicklemia. Presumably, then, the sickled cells in the blood of persons with sickle cell anemia cause thromboses, and their increased fragility exposes them to the action of reticulo-endothelial cells which break them down, resulting in the anemia (1).

It appears, therefore, that while some of the details of this picture of the sickling process are as yet conjectural, the proposed mechanism is consistent with experimental observations at hand and offers a chemical and physical basis for many of them. Furthermore, if it is correct, it supplies a direct link between the existence of "defective" hemoglobin molecules and the pathological consequences of sickle cell disease.

3) *On the Genetics of Sickle Cell Disease*: A genetic basis for the capacity of erythrocytes to sickle was recognized early in the study of this disease (4). Taliaferro and Huck (15) suggested that a single dominant gene was involved, but the distinction between sicklemia and sickle cell anemia was not clearly understood at the time. The literature contains conflicting statements concerning the nature of the genetic mechanisms involved, but recently Neel (8) has reported an investigation which strongly indicates that the gene responsible for the sickling characteristic is in heterozygous condition in individuals with sicklemia, and homozygous in those with sickle cell anemia.

Our results had caused us to draw this inference before Neel's paper was published. The existence of normal hemoglobin and sickle cell anemia hemoglobin in roughly equal proportions in sicklemia hemoglobin preparations is obviously in complete accord with this hypothesis. In fact, if the mechanism proposed above to account for the sickling process is correct, we can identify the gene responsible for the sickling process with one of an alternative pair of alleles capable through some series of reactions of introducing the modification into the hemoglobin molecule that distinguishes sickle cell anemia hemoglobin from the normal protein.

The results of our investigation are compatible with a direct quantitative effect of this gene pair; in the chromosomes of a single nucleus of a normal adult somatic cell there is a complete absence of the sickle cell gene, while two doses of its allele are present; in the sicklemia somatic cell there exists one dose of each allele; and in the sickle cell anemia somatic cell there are two doses of the sickle cell gene, and a complete absence of its normal allele. Correspondingly, the erythrocytes of these individuals contain 100 percent normal hemoglobin, 40 percent sickle cell anemia hemoglobin and 60 percent normal hemoglobin, and 100 percent sickle cell anemia hemoglobin, respectively. This investigation reveals, therefore, a clear case of a change produced in a protein molecule by an allelic change in a single gene involved in synthesis.

The fact that sicklemia erythrocytes contain the two hemoglobins in the ratio 40:60 rather than 50:50 might be accounted for by a number of hypothetical schemes. For example, the two genes might compete for a common substrate in the synthesis of two different enzymes essential to the production of the two different hemoglobins. In this reaction, the sickle cell gene would be less efficient than its normal allele. Or, competition for a common substrate might occur at some later stage in the series of reactions leading to the synthesis of the two hemoglobins. Mechanisms of this sort are discussed in more elaborate detail by Stern (13).

The results obtained in the present study suggest that the erythrocytes of other hereditary hemolytic anemias be examined for the presence of abnormal hemoglobins. This we propose to do.

Based on a paper presented at the meeting of the National Academy of Sciences in Washington, D. C., in April, 1949, and at the meeting of the American Society of Biological Chemists in Detroit in April, 1949.

References

1. BOYD, W.　*Textbook of pathology.*　(3rd Ed.) Philadelphia: Lea and Febiger, 1938.　P. 864.
2. DIGGS, L. W., AHMANN, C. F., and BIBB, J.　*Ann. int. Med.*, 1933, 7, 769.
3. DRABKIN, D. L.　*J. biol. Chem.*, 1946, 164, 703.
4. EMMEL, V. E.　*Arch. int. Med.*, 1917, 20, 586.
5. GERMAN, B. and WYMAN, J., JR.　*J. biol. Chem.*, 1937, 117, 533.
6. HASTINGS, A. B. et al.　*J. biol. Chem.*, 1924, 60, 89.
7. LONGSWORTH, L. G.　*Ann. N. Y. Acad. Sci.*, 1941, 41, 267.
8. NEEL, J. V.　*Science*, 1949, 110, 64.
9. PAULING, L., PRESSMAN, D., and CAMPBELL, D. H.　*Physiol. Rev.*, 1943, 23, 203.
10. PONDER, E.　*Ann. N. Y. Acad. Sci.*, 1947, 48, 579.
11. SHERMAN, I. J.　*Bull. Johns Hopk. Hosp.*, 1940, 67, 309.
12. STERN, K. G., REINER, M. and SILBER, R. H.　*J. biol. Chem.*, 1945, 161, 731.
13. STERN, C.　*Science*, 1948, 108, 615.
14. SWINGLE, S. M.　*Rev. sci. Inst.*, 1947, 18, 128.
15. TALIAFERRO, W. H. and HUCK, J. G.　*Genetics*, 1923, 8, 594.
16. TISELIUS, A. and KABAT, E.　*J. exp. Med.*, 1939, 69, 119.

COMMENTARY TO

page 483

38. Hyypia, T., Jalava, A., Larsen, S. H., Terho, P. and Hukkanen, V. (1985)
Detection of Chlamydia Trachomatis in Clinical Specimens by Nucleic Acid Spot Hybridization. Journal of General Microbiology 131(Pt 4): 975–978.

The hybridization of complimentary stands of DNA in solution was first described in the early 1960's (1,2). Rapid advances took place over the next few years with the development of solid phase hybridization assays on nitrocellulose membranes (3,4). Automated synthesizers made probes of 18–100 nucleotides readily available and further stimulated the development of probe-based assays. Nucleic acid hybridization assays for a wide variety of clinically important infectious diseases were developed in numerous laboratories throughout the world during the early 1980's.

Moseley *et al.* in 1980 reported on a method to screen for enterotoxigenic *Escherichia coli* in stool specimens (5). They captured bacterial DNA on nitrocellulose membranes and used a radiolabled probe for detection. Hyypia *et al.* in Finland expanded on the membrane based assay to hybridize nucleic acids in *Chlamydia trachomatis* strains in a spot hybridization assay on nitrocellulose (6). In the paper presented here by Hyypia *et al.* a probe assay was presented for the detection of *C. trachomatis* in clinical specimens. DNA was extracted from male urethral and female cervical specimens collected onto swabs. Recombinant plasmid DNA was purified from pure colonies of the organism, radiolabeled and used as the probe. The extracted DNA from the clinical specimens was spotted onto the membrane filters and hybridization detected with autoradiography.

In 1985 Hyypia described a spot hybridization assay for the detection of adenovirus in nasopharyngeal specimens (7). In addition to the standard radiolabeled probe, Hyypia used a biotin-streptavidin-alkaline phosphatase (EC 3.1.3.1) conjugated probe. Enzyme-labeled probes had been described by Ward and co-workers in 1981 (8,9). The use of enzyme probes as an alternative to radioactive labels helped expand hybridization assays into routine clinical use. The first FDA cleared clinical diagnostic test however was for a radiolabeled DNA probe assay for culture confirmation of Legionnaires disease (10). This first DNA-based test was released in 1985 by the Gen-Probe Corporation. In 1987 Gen-Probe obtained FDA clearance for the first enzyme labeled probe test. The assay was for *C. trachomatis* in clinical specimens (11).

References

(1) Hall, B.D. and Spiegelamn, S. (1961) Sequence complementarity of T2-DNA and T2-specific RNA. Proceedings of the National Academy of Sciences USA. 47(2):137–146.

(2) Marmur, J. and Doty, P. (1962) Determination of the base composition of deoxyribonucleic acid from its thermal denaturation temperature. Journal of Molecular Biology. 5(1):109–118.

(3) Nygaard, A.P. and Hall, B.D. (1964) Formation and properties of RNA–DNA complexes. Journal of Molecular Biology. 9(1):125–142.

(4) Gillespie, D. and Spiegelman, S. (1965) A quantitative assay for DNA–RNA hybrids with DNA immobilized on a membrane. Journal of Molecular Biology. 12(3):829–842.

(5) Moseley, S.L., Huq, I., Alim, A.R.M.A., So, M., Samadpour-Motalebi, M., and Falkow, S. (1980) Detection of enterotoxigenic *Escherichia coli* by DNA colony hybridization. Journal of Infectious Diseases. 142(6):892–898.

(6) Hyypia, T., Larsen, S.H., Stahlberg, T., and Terho, P. (1984) Analysis and detection of chlamydial DNA. Journal of General Microbiology. 130(Part 12):3159–3164.

(7) Hyypia, T. (1985) Detection of adenovirus in nasopharyngeal specimens by radioactive and nonradioactive DNA probes. Journal of Clinical Microbiology. 21(5):730–733.

(8) Langer, P.R., Waldrop, A.A., and Ward, D.C. (1981) Enzymatic synthesis of biotin-labeled polynucleotides: novel nucleic acid affinity probes. Proceedings of the National Academy of Sciences USA. 78(11):6633–6637.

(9) Leary, J.J., Brigati, D.J., and Ward, D.C. (1983) Rapid and sensitive colorimetric method for visualizing biotin-labeled DNA probes hybridized to DNA or RNA immobilized on nitrocellulose: bio-blots. Proceedings of the National Academy of Sciences USA. 80(13):4045–4049.

(10) Wilkinson, H.W., Sampson, J.S., and Plikaytis, B.B. (1986) Evaluation of a commercial gene probe for identification of *Legionella* cultures. Journal of Clinical Microbiology. 23(2):217–220.

(11) Peterson, E.M., Oda, R., Alexander, R., Greenwood, J.R., and De La Maza, L.M. (1989) Molecular techniques for the detection of *Chlamydia trachomatis*. Journal of Clinical Microbiology. 27(10):2359–2363.

J. Gen. Microbiol. 1985; 131 (Pt 4): 975–978.
Copyright ©1985 Society for General Microbiology.
Reproduced with permission.

Journal of General Microbiology (1985), **131**, 975–978. *Printed in Great Britain*

975

SHORT COMMUNICATION

Detection of *Chlamydia trachomatis* in Clinical Specimens by Nucleic Acid Spot Hybridization

By TIMO HYYPIÄ,[1]* ANNIKA JALAVA,[1] STEVEN H. LARSEN,[2]
PERTTI TERHO[1,3] AND VEIJO HUKKANEN[1]

[1] *Department of Virology, University of Turku, Kiinamyllynkatu 13, SF-20520 Turku, Finland*
[2] *Department of Microbiology and Immunology and* [3] *Department of Medicine,
Indiana University School of Medicine, Indianapolis, Indiana 46223, USA*

(*Received 22 November 1984*)

A nucleic acid spot hybridization assay was used to detect *Chlamydia trachomatis* DNA. The hybridization probes included DNA isolated from elementary bodies of lymphogranuloma venereum (LGV) strains and cloned fragments of both chromosomal and plasmid DNA. The sensitivity of the test was in the range 10 to 100 pg homologous DNA and 10 *in vitro* infected cells. Cross-reactivity with bacterial DNA was avoided when purified chlamydia-specific DNA fragments were used as probes. *C. trachomatis* was detectable in most of the clinical specimens with large amounts of infectious particles. Also some isolation-negative specimens gave a positive signal in the test.

INTRODUCTION

Chlamydia trachomatis is currently the most common cause of sexually transmitted diseases in a majority of western countries. Definitive diagnosis of chlamydial infections is usually made by isolation of infectious organisms in cell culture (Darougar *et al.*, 1971) or by direct demonstration of specific antigens in inflammatory exudates using monoclonal antibody staining (Tam *et al.*, 1984). Both methods, although specific and sensitive, need microscopy for evaluation of test results and are therefore time consuming and require a trained specialist.

We recently described a spot hybridization test for detection of chlamydial DNA (Hyypiä *et al.*, 1984). In the present report we have further evaluated the sensitivity and specificity of the test, and used it to detect *C. trachomatis* DNA in clinical specimens.

METHODS

Specimens. McCoy cells, treated with 1 µg cycloheximide ml⁻¹, were infected with *Chlamydia trachomatis* LGV strains L1 (440-L), L2 (434-B) and L3 (404-L) at a m.o.i. of about 1·0 IFU (inclusion-forming unit) per cell. The cells were grown for 3 d, at which time 80–90% of them showed visible inclusions, harvested and stored at −70 °C until tested. Prior to the test, the cells were treated with 0·1 mg proteinase K ml⁻¹ (Merck) for 1 h at 37 °C, denatured by boiling in 0·3 M-NaOH, chilled on ice and neutralized with HCl.

Control microbial strains were obtained from routine isolations (Department of Medical Microbiology, University of Turku) and identified by standard procedures (Lennette *et al.*, 1980). The strains included isolates of *Escherichia coli*, *Klebsiella pneumoniae*, *Proteus mirabilis*, *Citrobacter freundii*, *Enterobacter aerogenes*, *Salmonella typhimurium*, *Yersinia enterocolitica*, *Pseudomonas aeruginosa*, *Gardnerella vaginalis*, *Neisseria gonorrhoeae*, *Staphylococcus aureus*, *Streptococcus agalactiae*, *Streptococcus faecalis*, *Streptococcus viridans*, *Corynebacterium* sp., *Lactobacillus* sp., *Bacteroides fragilis* and *Candida albicans*. The micro-organisms were grown on plates or in liquid medium. For the test, a suspension of each of the microbes was boiled at an alkaline pH, chilled on ice and neutralized, as described above.

Male urethral and female cervical specimens were taken with cotton-tipped swabs into transport medium [0·2 M-sucrose, 0·02 M-sodium phosphate buffer containing 3% (v/v) foetal calf serum, 2·5 IU nystatin ml⁻¹ and 50 µg gentamicin ml⁻¹] and sent to our laboratory for isolation of *C. trachomatis*. Specimens were stored at −20 °C

0001-2334 © 1985 SGM

976   *Short communication*

until tested. The isolation was carried out in irradiated McCoy cells as described by Terho (1978). For the hybridization assay, the specimens were treated with proteinase K and either extracted with phenol, ethanol precipitated and denatured as above or denatured directly on the filter (Totten *et al.*, 1983).

Hybridization probes. Purification of *C. trachomatis* DNA from elementary bodies and characterization of the cloned L2 plasmid have been described previously (Hyypiä *et al.*, 1984). For the hybridization experiments, the hybrid plasmid was digested with *Bam*HI (Boehringer-Mannheim) and the L2 plasmid insert was recovered from agarose gels by use of a DEAE-cellulose membrane (Dretzen *et al.*, 1981).

Cloned chromosomal probes were constructed as follows. DNA purified from elementary bodies of strain L1, and pBR322 plasmid DNA were digested with *Bam*HI and ligated using T4 DNA ligase (Boehringer-Mannheim). The recombinant DNA was used to transform *E. coli* HB101. The ampicillin resistant, tetracycline sensitive colonies were screened by colony hybridization using an L1 DNA probe. Recombinant plasmid DNA was purified from one of the positive colonies and used as a probe in the spot hybridization test.

For the hybridization assay, the probes were labelled with ^{32}P in a nick translation reaction (Rigby *et al.*, 1977) to a specific activity of $10-100 \times 10^6$ c.p.m. μg^{-1}.

Spot hybridization test. The specimens were either spotted onto nitrocellulose filters (Schleicher & Schüll, Dassel, FRG) after denaturation by boiling in alkali, or were denatured *in situ* after spotting onto the filters. The filters were baked for 1 h at 80 °C and hybridization was carried out as described by Wahl *et al.* (1979). The radioactivity bound onto the filters was revealed by autoradiography.

RESULTS AND DISCUSSION

Sensitivity and specificity of the hybridization test

The sensitivity and specificity of the spot hybridization assay were evaluated using purified homologous and control DNA, and *in vitro* infected and control McCoy cells (Fig. 1). When whole DNA from L1 elementary bodies was used as a probe the sensitivity was 10–100 pg L1 DNA and 10 infected cells. No binding of the probe to herpes simplex virus DNA or to pBR322 plasmid DNA was observed. Large amounts of uninfected cells caused a slight background signal. Similar sensitivity and specificity were observed with cloned chromosomal and plasmid DNA probes with the exception that pBR322 DNA on the filter was detected at a sensitivity of 10–100 pg.

The sensitivity obtained here (less than 100 pg, corresponding to about 10^5 molecules) is somewhat better than that obtained by a recently described sandwich hybridization assay for *C. trachomatis* (Palva *et al.*, 1984). This may be explained by the differing test principle or the different radiolabelling methods. The slight background reaction of the probes with large amounts of uninfected cells indicates that these should be included routinely as a control in the test.

The cross-reactivity of the probes with various micro-organisms was tested because urethral and cervical specimens are sometimes contaminated with bacteria and *Candida*. Whole L1 chromosomal DNA showed reactivity with large amounts of some enterobacterial and gonococcal DNA when the autoradiographs were overexposed (Fig. 2, A). When the purified L2 plasmid DNA insert was used as a probe no reaction occurred even after longer exposures (Fig. 2, B). Because the large amounts of microbes sufficient to cause detectable background reactions are not usually present in swab specimens, the whole chromosomal DNA probe was used in the further assays.

Detection of C. trachomatis in clinical specimens

In order to see whether the clinical specimens would contain enough chlamydial DNA to give a positive signal in the spot hybridization assay, we tested eight specimens with large amounts of infectious particles and eight isolation-negative specimens. The whole L1 chromosomal DNA probe recognized seven of the positive specimens, and one of the negative samples was also scored positive (Fig. 3).

In the second part of the study with clinical specimens, 30 isolation-positive and 30 isolation-negative specimens were tested directly after proteinase treatment and *in situ* denaturation. Ten of the isolation-positive samples gave a signal which exceeded the signal of any of the negatives (Fig. 4). Overall, 24 of the 30 positives and 7 of the negatives gave a signal in the test.

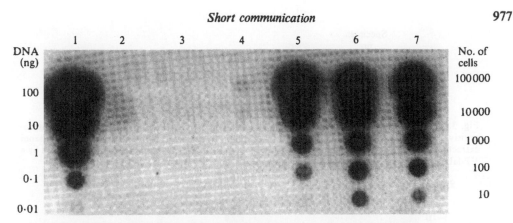

Fig. 1. Detection of chlamydial DNA by spot hybridization using whole L1 DNA probe. Various amounts of purified DNA or cells, treated with proteinase K, were denatured, applied to nitrocellulose filters and detected by a ^{32}P-labelled probe. The results were revealed by autoradiography. 1, Whole L1 DNA; 2, herpes simplex virus DNA; 3, pBR322 DNA; 4, McCoy cells; 5, L1 infected McCoy cells; 6, L2 infected McCoy cells; 7, L3 infected McCoy cells.

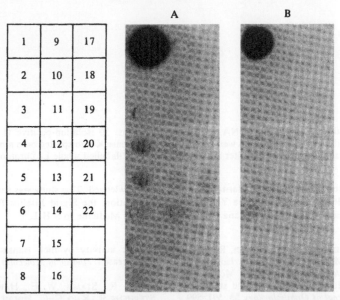

Fig. 2. Specificity of chlamydial probes. DNA or the cells were boiled in alkali, neutralized, applied to nitrocellulose filters and detected by ^{32}P-labelled probes. A, Whole L1 chromosomal DNA probe; B, L2 plasmid insert. 1, L1 infected McCoy cells; 2, McCoy cells; 3–5, *Escherichia coli*; 6, *Klebsiella pneumoniae*; 7, *Proteus mirabilis*; 8, *Citrobacter freundii*; 9, *Enterobacter aerogenes*; 10, *Salmonella typhimurium*; 11, *Yersinia enterocolitica*; 12, *Pseudomonas aeruginosa*; 13, *Gardnerella vaginalis*; 14, *Neisseria gonorrhoeae*; 15, *Staphylococcus aureus*; 16, *Streptococcus agalactiae*; 17, *Strep. faecalis*; 18, *Strep. viridans*; 19, *Corynebacterium* sp.; 20, *Lactobacillus* sp.; 21, *Bacteroides fragilis*; 22, *Candida albicans*.

Our findings confirm the earlier observations of Palva *et al.* (1984) that nucleic acid hybridization assays can be used to demonstrate the presence of *C. trachomatis* in clinical specimens. The sensitivity of the probe can be selected so that cross-reactivity with other microbial DNA usually present in the specimens does not occur. The problem, however, is the sensitivity. Some of the positive samples routinely taken for isolation do not contain sufficient amounts of specific DNA sequences to give a signal for specific diagnosis. Therefore either more sensitive probe systems or better sampling methods are required.

978   *Short communication*

Specimen
no.

1–8

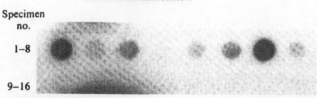

9–16

Fig. 3. Detection of chlamydial DNA in clinical specimens by spot hybridization. Eight isolation-positive (1–8) and eight isolation-negative (9–16) specimens were treated with proteinase K and phenol extracted. DNA was ethanol precipitated, denatured, applied to a nitrocellulose filter and detected by a ^{32}P-labelled whole L1 chromosomal DNA probe.

Specimen
no.

1–10

11–20

21–30

31–40

41–50

51–60

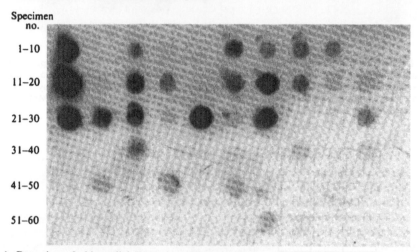

Fig. 4. Detection of chlamydial DNA in clinical specimens. 30 isolation-positive (1–30) and 30 isolation-negative (31–60) specimens were treated with proteinase K and spotted onto a nitrocellulose filter. DNA was denatured on the filter and detected by a ^{32}P-labelled whole L1 chromosomal DNA probe.

The excellent technical assistance of Marita Maaronen is acknowledged. This study was supported by grants from the Emil Aaltonen Foundation, the Yrjö Jahnsson Foundation, the Sigrid Juselius Foundation and a Program Development Award from Indiana University School of Medicine to S.H.L.

REFERENCES

DAROUGAR, S., TREHARNE, J. D., DWYER, R. ST C., KINNISON, J. R. & JONES, B. R. (1971). Isolation of TRIC agent (*Chlamydia*) in irradiated McCoy cell culture from endemic trachoma in field studies in Iran. *British Journal of Ophthalmology* **55**, 591–599.

DRETZEN, G., BELLARD, M., SASSONE-CORSI, P. & CHAMBON, P. (1981). A reliable method for the recovery of DNA fragments from agarose and acrylamide gels. *Analytical Biochemistry* **112**, 295–298.

HYYPIÄ, T., LARSEN, S. H., STÅHLBERG, T. & TERHO, P. (1984). Analysis and detection of chlamydial DNA. *Journal of General Microbiology* **131**, 3159–3164.

LENNETTE, E. H., BALOWS, A., HAUSLER, W. J., JR & TRUANT, J. P. (editors) (1980). *Manual of Clinical Microbiology*, 3rd edn. Washington, DC: American Society for Microbiology.

PALVA, A., JOUSIMIES-SOMER, H., SAIKKU, P., VÄÄNÄNEN, P., SÖDERLUND, H. & RANKI, M. (1984). Detection of *Chlamydia trachomatis* by nucleic acid sandwich hybridization. *FEMS Microbiology Letters* **23**, 83–89.

RIGBY, P. W. J., DIECMANN, M., RHODES, C. & BERG, P. (1977). Labelling deoxyribonucleic acid to high specific activity by nick translation with DNA-polymerase I. *Journal of Molecular Biology* **113**, 237–251.

TAM, M. R., STAMM, W. E., HANDSFIELD, H. H., STEPHENS, R., KUO, C.-C., HOLMES, K. K., DITZEN-BERGER, K., KRIEGER, M. & NOWINSKI, R. C. (1984). Culture-independent diagnosis of *Chlamydia trachomatis* using monoclonal antibodies. *New England Journal of Medicine* **310**, 1146–1150.

TERHO, P. (1978). Isolation techniques of *Chlamydia trachomatis* from patients with nonspecific urethritis. *Dermatologishe Monatsschrift* **164**, 515–520.

TOTTEN, P. A., HOLMES K. K., HANDSFIELD, H. H., KNAPP, J. S., PERINE, P. L. & FALKOW, S. (1983). DNA hybridization technique for the detection of *Neisseria gonorrhoeae* in men with urethritis. *Journal of Infectious Diseases* **148**, 462–471.

WAHL, G. M., STERN, M. & STARK, G. R. (1979). Efficient transfer of large DNA fragments from agarose gels to diazobenzyloxymethyl-paper and rapid hybridization by using dextran sulfate. *Proceedings of the National Academy of Sciences of the United States of America* **76**, 3683–3687.

COMMENTARY TO

page 489

39. Maskos, U. and Southern, E. M. (1992)
Oligonucleotide Hybridisations on Glass Supports: A Novel Linker for Oligonucleotide Synthesis and Hybridisation Properties of Oligonucleotides Synthesised In Situ. Nucleic Acids Research 20(7): 1679–1684.

Nucleic acid microarrays evolved from the technique of transfer or blotting of DNA onto membranes invented by Edwin Southern in 1975 (1). The transfer of DNA from electrophoresis gels to nitrocellulose membranes became known as Southern blotting after its inventor. From this technology various types of dot blots were developed. Kaftos and co-workers in 1979 spotted multiple targets on a membrane support in an array format for direct hybridization (2). Roger Ekins extended fluorescent immunoassays to microspot format on a solid support in 1989 (3). Laser-based confocal microscopy was used to detect fluorescent signals from 50-micron antibody spots after development. Large amounts of oligonucleotides or antibodies can be spotted onto membranes however they have limitations. One of the problems is that membranes change shape during the drying process. This makes registration of micron-sized spots difficult, especially as the spot density increases.

In the paper presented here Maskos and Southern describe a chemistry for the attachment of oligonucleotides to glass slides or beads. The desired oligonucleotides were synthesized directly on the slides. The authors used this technology to present the first application of a solid phase glass support for a DNA array. They studied the globin chain gene mutations in sickle cell hemoglobin in two latter papers (4,5). In 1995, Patrick O. Brown's laboratory at Stanford University reported on the development of a gene array on a solid glass slide (6).

In 2001 Southern wrote a chapter on the history of DNA microarrays in a book titled *DNA Arrays. Methods and Protocols* (7). His opening paragraph is an ideal end to this commentary on his paper presented here and in fact, could serve as an end, with slight modification, to this collection of landmark papers,

It may seem premature to be writing a history of DNA microarrays because this technology is relatively new and clearly has more of a future than a past. However readers could benefit from learning something about the technical basis of DNA microarrays, and younger readers may be curious to known something of the origins and antecedents of this new technology.

References

(1)   Southern, E.M. (1975) Detection of specific sequences among DNA fragments separated by gel electrophoresis. Journal of Molecular Biology. 98(3):503–517.

(2)   Kafatos, F.C., Jones, C.W., and Efstratiadis, A. (1979) Determination of nucleic acid sequence homologies and relative concentrations by a dot hybridization procedure. Nucleic Acids Research. 7(6):1541–1552.

(3)   Ekins, R., Chu, F., and Biggart, E. (1989) Development of microspot multi-analyte ratiometric immunoassay using dual fluorescent-labelled antibodies. Analytica Chimica Acta. 227:73–96.

(4)   Maskos, U. and Southern, E.M. (1993) A novel method for the analysis of multiple sequence variants by hybridisation to oligonucleotides. Nucleic Acids Research. 21(9):2267–2268.

(5)   Maskos, U. and Southern, E.M. (1993) A novel method for the parallel analysis of multiple mutations in multiple samples. Nucleic Acids Research. 21(9):2269–2270.

(6)   Schena, M., Shalon, D., Davis, R.W., and Brown, P.O. (1995) Quantitative monitoring of gene expression patterns with a complementary DNA microarray. Science. 270(5235):467–470.

(7)   Southern, E.M. (2001) DNA microarrays. History and overview, in *DNA Arrays. Methods and Protocols*. Rampal, J.B. (ed), Humana Press, Totowa, New Jersey, Chapter 1, pgs 1–15.

Nucleic Acids Res. 1992; 20(7); 1679–1684
Copyright 1992 Oxford University Press.
Reproduced with permission.

© 1992 Oxford University Press

Nucleic Acids Research, Vol. 20, No. 7 1679–1684

Oligonucleotide hybridisations on glass supports: a novel linker for oligonucleotide synthesis and hybridisation properties of oligonucleotides synthesised *in situ*

Uwe Maskos and Edwin M. Southern
Department of Biochemistry, Unversity of Oxford, South Parks Road, Oxford OX1 3QU, UK

Received December 12, 1991; Revised and Accepted March 9, 1992

ABSTRACT

A novel linker for the synthesis of oligonucleotides on a glass support is described. Oligonucleotides synthesised on the support remain tethered to the support after ammonia treatment and are shown to take part in sequence specific hybridisation reactions. These hybridisations were carried out with oligonucleotides synthesised on 'ballotini' solid sphere glass beads and microscope slides. The linker has a hexaethylene glycol spacer, bound to the glass via a glycidoxypropyl silane, terminating in a primary hydroxyl group that serves as starting point for automated or manual oligonucleotide synthesis.

INTRODUCTION

The rapid progress in molecular biology has depended on methods of detecting specific DNA or RNA sequences by molecular hybridisation, reviewed in (1). For the detection of mutations and polymorphisms for which the nucleotide sequence is known, Wallace and coworkers (2) introduced the differential hybridisation of oligodeoxyribonucleotide probes. The main advantage of oligonucleotide hybridisation lies in the possibility to detect single nucleotide changes, not feasible with longer probes that are less sensitive to mismatching.

For some applications DNA to be analysed is immobilised on nitrocellulose filters or nylon membranes and probed with labelled oligonucleotides (3). For other applications, e.g. isolation of mRNA (4) or DNA diagnostic purposes (5, 6), oligonucleotides are immobilised on solid supports and used to capture the macromolecule of interest. A number of methods have been reported for immobilising oligonucleotides on supports (5–11). Some of these use complicated chemistry (7, 8) or may lead to high levels of unspecific adsorption of oligonucleotide or probe material (9, 10). Furthermore, attachment through the bases as well as the ends (5) interferes with the hybridisation reaction. Only the method of Zhang *et al.* (6) seems straightforward although it requires the derivatisation of oligonucleotides with a chemical that is not yet commercially available. A method for immobilising oligonucleotides in a gel matrix attached to glass has recently been reported (11).

Solid phase methods for synthesising oligonucleotides offer the opportunity to make oligonucleotides directly on the support to be used for the hybridisation, obviating the need of detaching them from the matrix on which they were synthesised and re-attaching them to the hybridisation support.

The linker employed in standard oligonucleotide synthesis (12) is labile in the conc. ammonia used for the deprotection of the bases. The linker described here uses only the most stable bonds known in organic chemistry, viz. carbon–carbon, carbon–silicon, and ether bonds; it requires only readily available chemicals, is easy to synthesise and stable in the ammonia deprotection step.

MATERIALS AND METHODS

Linker synthesis

3-Glycidoxypropyltrimethoxysilane, hexaethylene glycol, pentaethylene glycol, ethylene glycol and HPLC grade acetonitrile were purchased from Aldrich and used without further purification. Underivatised CPG (controlled pore size glass, pore size 500Å, 37–74 μm particle size) was purchased from Pierce, ballotini glass beads (Grade 18, diameter range 40–75 μm, and Grade 13, diameter range 90–135 μm) were from Jencons. All DNA synthesis materials were purchased from Applied Biosystems.

In typical experiments, underivatised CPG (10 g) or ballotini glass beads (20 g) were suspended in a mixture of xylene (40 ml), 3-glycidoxypropyltrimethoxysilane (12 ml) (modified after [13]) and a trace of Hünig's base (after [14]) at 80°C overnight with stirring, then washed thoroughly with methanol, ether, air dried and dried *in vacuo*. Alternatively, beads were suspended in a solution of 5% of the silane in water, keeping the pH between 5.5 and 5.8 (after [15]). Reaction was for 30 minutes at 90°C. An alternative 'vapour deposition' strategy of derivatisation has also been reported (16).

In a second step these beads were heated with stirring in neat diol (hexaethylene glycol, pentaethylene glycol or ethylene glycol), containing a catalytic amount of conc. sulphuric acid, overnight at 80°C to yield alkyl hydroxyl derivatised support. After washing with methanol and ether the beads were air dried, then dried *in vacuo* and stored under Argon at −20°C. In addition to reaction with diol, a dilute solution of HCl in water was used to cleave the epoxide residue to yield a primary hydroxyl group (Figure 2). Alternatively, 40 ml of anhydrous dioxane containing 32 mmol of the alcohol and 1 ml boron trifluoride etherate were added and reaction carried out for 30 min at 90°C.

Microscope slides (BDH Super Premium, 76×26×1 mm) were derivatised in the same way. Reaction was carried out in a staining jar fitted with a drying tube. The slides were washed with methanol, ether, air dried and stored desiccated at −20°C until use.

Automated synthesis of oligonucleotides

Standard programmes were used for all oligonucleotide syntheses on an Applied Biosystems 381A automated synthesiser. Trityl effluents were collected, diluted to 10 ml with 0.1 M toluenesulfonic acid (Aldrich) in acetonitrile and absorbance measured in a 1 cm quartz cuvette on a Bausch & Lomb Spectronic 2000 spectrophotometer.

Manual synthesis of oligonucleotides

For the manual synthesis of oligonucleotides on a microscope slide, the synthesis cycle was performed as follows: The coupling solution was made up fresh for each step by mixing six volumes of 0.5 M tetrazole in anhydrous acetonitrile with five volumes of a 0.2 M solution of the required β-cyanoethylphosphoramidite (12). Coupling time was three minutes. Oxidation with a 0.1 M solution of iodine in tetrahydrofuran/pyridine/H_2O yielded a stable phosphotriester bond. Detritylation of the 5' end with 3% trichloroacetic acid in dichloromethane allowed further extension of the oligonucleotide chain. There was no capping step since the excess of phosphoramidites used over reactive sites on the slide was supposed to be large enough to drive the coupling to completion.

As the chemicals used in the coupling step are moisture-sensitive, this critical step must be performed under anhydrous conditions in a sealed container: The shape of the patch to be synthesised was cut out of a sheet of silicone rubber (76 ×26×0.5 mm) which was sandwiched between a derivatised microscope slide and a piece of teflon of the same size and thickness, to which was fitted a short piece of plastic tubing that allowed injection and withdrawal of the coupling solution by syringe, and flushing with argon. The assembly was held together by fold-back paper clips. After coupling the set-up was disassembled and the slide put through subsequent chemical reactions (oxidation with iodine, and detritylation by treatment with trichloroacetic acid) by dipping it into staining jars.

After the synthesis, the oligonucleotides were deprotected by putting the slide into a Schott bottle containing 30% NH_3 and incubating for 5 to 10 hours at 55°C in a water bath.

Set-up for column chromatography

A glass capillary (diameter 1.0 mm) was drawn out at one end to a narrow opening which was plugged with crushed glass particles and sintered in a flame. The inside of the capillary was silanised, approximately 40 mg of derivatised ballotini glass beads were layered on the glass frit and the top of the column connected to a syringe driven by an infusion pump. Hybridisation solution and washing solutions were applied to the column at rates in the range of 3−10 µl/min.

RESULTS AND DISCUSSION

Linker synthesis

The initial derivatisation of the solid support is carried out in two steps: The first reaction is a condensation of 3-glycidoxypropyltrimethoxysilane to the solid support bearing 'silanol' groups (Figure 1a). When 'wet' solvent is used or the

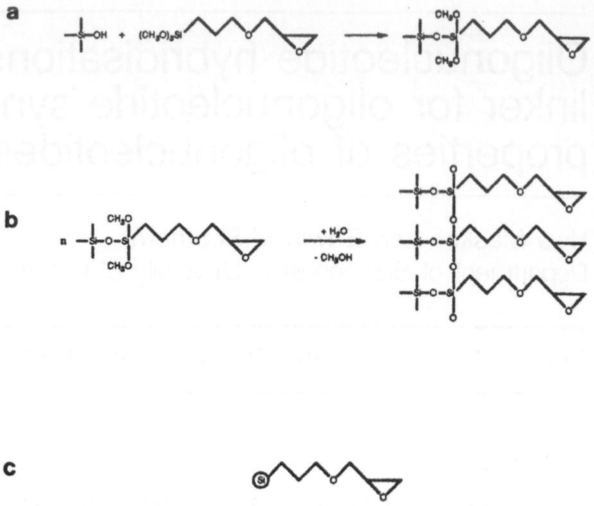

Figure 1. The first step in the derivatisation.

reaction is carried out in water, further crosslinking occurs (17), cf. Figure 1b, which can be abbreviated as in Figure 1c. In the second step the epoxide group is cleaved with a diol or water under acidic conditions. Several compounds were used to yield linker molecules with different chain lengths, cf. Figure 2.

Determination of loadings. The loadings achieved on CPG, ballotini beads and the slide were determined by carrying out standard oligonucleotide synthesis using both phosphoramidite and Hydrogen-phosphonate chemistry on an Applied Biosystems Model 381 A with derivatised CPG or ballotini beads in the column. Coupling yields were measured from the detritylation effluent using a standard procedure (18).

The first coupling step gave a high trityl yield, ca. 3−3.5 times that obtained with the commercially available LCAA-CPG support (Table I). Stepwise yields fell to between 85 and 95% for cycles 2 and 3, but in later steps coupling yields rose to close to 100%. The high first coupling could indicate that some diol is only loosely bound and removed during the synthesis cycles.

To test that the epoxide is still intact after attachment, epoxide-derivatised CPG was analysed in the same way. In this case, the coupling yields *increased* in the course of synthesis presumably because cleavage of the epoxide by the chemicals used in the reactions exposes hydroxy alkyl groups.

The properties of derivatised supports (Table I) show that the loadings per unit weight for ballotini beads (solid spheres) is only a fraction of that for porous glass, as expected from its smaller surface area. The yields for ethylene glycol, pentaethylene glycol and hexaethylene glycol derivatised CPG are very similar, whereas HCl/H_2O treated epoxide-support shows a lower loading consistent with the notion of increased steric interference with a short linker. The epoxide alone gives a yield only slightly lower than untreated CPG (which is presumably due to reaction with surface silanol groups). This small amount could be due to epoxide cleaved in the derivatisation or by the chemicals used in oligonucleotide synthesis or to incomplete derivatisation of the whole surface (some silanol groups remain unreacted). It is also an indication that most of the epoxide remains intact after the

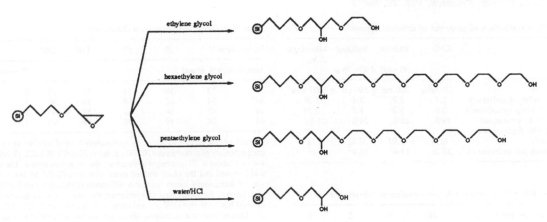

Figure 2. The second step in the derivatisation.

Table I. Loadings of linker molecules on glass supports

Molecules used for the derivatisation	Support	Loading nmol/mg	After ammonia treatment	Percentage remaining
CPG only	CPG	24.9	–	–
epoxide only	CPG	17.4	43.5	n/a
epoxide/HCl/H_2O	CPG	44.8	41.1	81
epoxide/ethylene glycol	CPG	48.9	47.5	94
epoxide/pentaethylene glycol	CPG	53.3	46.2	75
epoxide/hexaethylene glycol	CPG	50.8	45.4	81
commercial LCAA	CPG	14.2	–	–
epoxide/hexaethylene glycol	Ballotini 90–135 μm	0.07	–	–
epoxide/hexaethylene glycol	Ballotini 40–75 μm	0.05	–	–

Oligonucleotide syntheses were performed with approx. 4 mg CPG (or 120 mg ballotini) before and after ammonia treatment of the support and the detritylation effluent collected. Since cleavage is expected to produce silanol groups which will also take part in a coupling reaction coupling yields measured in this way are the sum of the reaction of these silanol and of linker hydroxy alkyl groups, given by the formula

$$OD_m = (1 - x)OD_a + xOD_s$$

with: x amount of cleavage; OD_m optical density measured at 498 nm per mg after deprotection; OD_a optical density of effluent from alkyl derivatised support before deprotection; OD_s optical density of effluent from silanol groups of underivatised CPG
The equation can be solved to give

$$x = \frac{OD_a - OD_m}{OD_a - OD_s}$$

The percentage remaining is then defined as $1 - x$, and results have been included in the table. The value for the third coupling steps was used.

initial condensation as proven by the much higher loading obtained after ammonia treatment.

Tests of linker stability. We have found that the linker is not completely stable to ammonia. Stability was determined in two different ways.

In a first series of experiments, the hexaethylene glycol derivatised beads were transferred into screw-capped microcentrifuge tubes after 'trityl-on' oligonucleotide synthesis

Table II. A time course of linker stability on glass supports.

Incubation time (hrs)	1	5	10	15
Stability on CPG (%)	45	19	11	7
Stability on ballotini (%)	37	28	23	25

Short oligonucleotides bearing a 5'-dimethoxytrityl group were synthesised on CPG (ca. 2 mg) or ballotini (ca. 90 mg) supports that were subsequently treated with conc. ammonia at 55°C for the times indicated. The supernatant was removed, the beads rinsed several times with conc. ammonia, the supernatants combined and evaporated. The residue of cleaved linker and the dried supports were treated with 80% acetic acid for 30 min to hydrolyse the dimethoxytrityl group and evaporated. 600 μl methanolic $HClO_4$ (60 ml 60% $HClO_4$ in 37 ml methanol) was added and the liberated dimethoxytrityl cation measured from the absorption at 498 nm. Stability is then defined as (trityl remaining on beads) / (trityl remaining on beads and in supernatant).

Control measurements were carried out on plain CPG (0.4% remaining after 5 hrs), standard ABI nucleoside derivatised support (0.8% after 1 hr), and LCAA−CPG (26% after 1 hr). 90% remain with the new linker after 30 min ammonia treatment at r.t.

(short runs of T or C) and treated with conc. ammonia at 55°C for different lengths of time. To test the linker stability, the amount of dimethoxytrityl groups in the ammonia supernatant and remaining on the support was determined as described in the Table II legend.

As an independent measure of linker stability, and to detect any diol adsorbed to the surface after derivatisation or extensive polymerisation (15, 17), an aliquot (~50 mg) of the linker derivatised CPG was put through a 16 hr ammonia deprotection to remove adsorbed diol and cleave any linker molecules labile to ammonia. Oligonucleotide synthesis on the treated supports showed slightly lower first coupling yields as compared with untreated support, with subsequent stepwise yields of 96−100%. Comparing the coupling values to those measured before ammonia treatment (the value for the third coupling was used) gives an indirect estimate of the linker stability as detailed in the caption to Table I. For the epoxide the measured loading rose 2.5 times after ammonia treatment because epoxides are hydrolysed under these conditions to yield glyceroloxypropyl-CPG (Figure 2), which serves as a substrate for oligonucleotide synthesis. The values for mono-, penta-, and hexaethylene glycol derivatised support are similar, which would indicate a high stability of the linkage.

Table III. A comparison of properties of different supports

Support	CPG	Ballotini 90−135 μ	Ballotini 40−75 μ	Microscope slide
Surface area	70 m²/g	200 cm²/g	350 cm²/g	1300 mm²
Highest loading (pmol/mm²)	1.1	5.2	2.4	1.08
Average loading (pmol/mm²)	0.6	3.3	1.5	0.39
Stability (5 hr incubation)	19%	28%	28%	(28%)
Loading after deprotection (pmol/mm²)	0.11	0.92	0.42	0.11
Surface area per molecule (Å²)	38.2²	13.4²	19.9²	39.1²

Table IV. Effect of oligonucleotide concentration on hybridisation rate

Relative concentration	50	20	10	2	1
% of input hybridised	23	14	5.5	0.8	0.6
Relative rate	38	23	9	1.5	1

Hybridisation for 2 hrs at 30°C in 0.1 M NaCl in 1.5 ml centrifuge tubes was followed by removal of the supernatant, three washes with 0.1 M NaCl at 0°C and determination of the amount of radioactivity associated with the beads by Čerenkov counting.

Table V. Effect of temperature on elution rate

Elution time 5′	20′	50′	110′	230′	
Temperature % eluted					% remaining
30°	23	38	54	71	85	0
36°	34	61	78	90	96	3
44°	42	54	67	87	97	2
65°	74	89	96	98	98	1.4

To each of five tubes was added an approximately equal number of beads and complementary oligonucleotide (30,000 c.p.m.) in 50 μl 0.1 M NaCl. Hybridisation was carried out at 30° overnight to maximise the amount of hybrid. The solution was removed and the beads washed twice with ice-cold 0.1 M NaCl. Elution was for increasingly longer times at a different temperature for each tube. After each interval the supernatant was removed, the beads were washed twice with ice-cold NaCl solution (100 μl), and eluted with prewarmed NaCl solution (100 μl). Elution times are cumulative times, i.e. the last time point was taken 230 min (~4 hrs) after the beginning, and 2 hrs after the previous one. The four temperatures indicated were kept constant for the four tubes. The values in the five columns denote the cumulative amount of material that had been eluted at the indicated times as determined by scintillation counting and listed as the fraction of total material bound. The last column gives the percentage of counts that remained after an elution in TE at 60°C (for the 30°C elution), or after 4 hrs of elution at the indicated temperatures, respectively.

Table VI. Properties of derivatised ballotini beads

Bead size (μm)	90−130	40−75
Surface are per bead (mm²)	0.038	0.011
Volume per bead (mm³)	6.97 × 10⁻⁴	1.13 × 10⁻⁴
Values calculated for radius (μm)	55	30
Density (g/cm³)	2.95	2.95
Mass per bead (ng)	2.06	0.33
Number of beads per mg	486	3000
Surface area per mg (mm²)	18.5	33.0
Oligonucleotide loading per mg (pmol)		
highest yields	~104	~81
average yields	~66	~51
yield after deprotection	~18	~14
Oligonucleotide loading per bead (fmol)	38	4.8

The values obtained in the two assays differ considerably (Tables I and II). However, the two assays are both indirect and experimental error could account for the differences. As will be seen, the amount of oligonucleotide remaining is sufficient to provide sensitive detection of hybridisation.

The stability on ballotini beads is higher than on CPG, possibly because the latter consists of almost pure silica (19), which might partially dissolve under these conditions, thus accounting for part of the instability. The other most likely point of instability is the attachment site of the silane to the glass, which is only a single silicon-oxygen-silicon bond if cross-linking is incomplete (see Figure 1a and b) or if polymerisation occurs (17).

Loading on a microscope slide. Oligonucleotide synthesis was carried out over the whole surface of a microscope slide and the detritylation effluent collected and quantified (Table III). The loadings per unit area on CPG and the microscope slide correspond well, whereas the ballotini loading is higher by a factor of approx. 4−8. This could be a sign of more successful derivatisation of ballotini beads since the value of one molecule per (13.4 Å)² is close to the (5 Å)² value for silanol groups in porous glass (14). (The value before ammonia treatment is (5.7 Å)².) The diffusion of reactants into the pores of CPG might be impeded by narrow pore size, and the derivatisation of microscope slides without stirring could lead to lower yields.

Hybridisation to tethered oligonucleotides

The linker was developed to provide a convenient way of making support bound oligonucleotides to use in nucleic acid hybridisation. The behaviour of the supports in hybridisation was explored using ballotini glass beads.

To permit the use of small numbers of beads, which would be difficult to handle in a column or centrifuge tube, beads were immobilised by fusing them to plastic cocktail sticks (2 cm long) which were dipped to ca. 1 mm in molten polypropylene, then brought into contact with a pile of derivatised glass beads and allowed to cool. Approximately 100−200 beads adhered to each stick. A stick with approx. 100 glass beads derivatised with the

sequence 3′-AGGTCGCCGCCC was dipped into 30 μl of a solution containing the complementary oligonucleotide 5′-TCC-AGCGGCGGG (30,000 c.p.m., 80 fmol in 30 μl of 0.1 M NaCl, 37°C). After hybridisation for various times and at different temperatures the stick was removed from the tube, rinsed, and the bound material eluted by dipping the stick into 0.1 M NaCl at 50°C. Typical hybridisation yields were: 4% at 37°C for 30 min; 5.3% at 30°C for 55 min; 13% at 30°C for 16 hrs. Nonspecific binding was between 0.05% and 0.2% of input counts. Under the same conditions (16 hrs at 30°C) the non-complementary oligonucleotide 5′-GGGCGGCGACCT showed only 0.2% binding of input counts.

Hybridisation and elution behaviour. For larger scale work hybridisations with beads—typically 1 mg—were carried out in 0.5 ml centrifuge tubes. The dependence of rates of hybridisation and elution on concentration and temperature were determined (Tables IV and V).

The concentration of input oligonucleotide over a 50-fold range showed an almost proportional change in yield of hybrid (i.e. rate of hybridisation, Table IV, compare 'relative concentration' to 'relative rate'), compatible with the expected pseudo first order reaction kinetics.

Nucleic Acids Research, Vol. 20, No. 7 **1683**

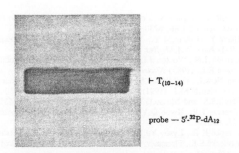

⊢ T$_{(10-14)}$

probe — 5'-^{32}P-dA$_{12}$

Figure 3. Hybridisation to a slide bearing sequences T$_{10}$-T$_{14}$. 10 pmol oligo-A$_{12}$, labelled to a total activity of 1.5 million c.p.m. was used in a hybridisation carried out in a perspex (Plexiglass) container made to fit a microscope slide filled wit 1.2 ml of 1 M NaCl in Tris—HCl pH 7.5, containing 0.1% SDS, for 5 minutes at 20°C. After a short rinse in the same solution without oligonucleotide more than 2,000 c.p.s. could be detected with a handheld monitor.

The highest concentration of oligonucleotide was 160 fmol (13,000 c.p.m.) in 20 μl. Only 0.03% bound non-specifically and could not be eluted. Under the same conditions only 0.02% non-complementary oligonucleotide bound to the beads (12 out of 65,000 c.p.m.) as compared to 23% for a similar concentration of complementary oligonucleotide, a further indication that this method of isolating DNA fragments is very specific and clean, as compared with other solid phase methods (9, 10). The elution rate increased with temperature, as expected (Table V). Three times as much material eluted at 65°C as at 30°C, within 5 min. Even at 30°C, which is below T$_m$, there is a non-negligible rate. Little material remains after 4 hrs under all conditions of elution.

Column chromatography. Another convenient way to isolate oligonucleotides and to test the hybridisation behaviour of the novel support is by column chromatography (see MATERIALS AND METHODS). In a typical experiment using the derivatised beads described above and the same complementary oligonucleotide, 0.2 pmol of labelled oligonucleotide (34,000 c.p.m.) in 1 ml of a solution of 0.1 M NaCl was applied to the column at a rate of 3 μl/min. The jacketed column was kept at 35°C. 90 μl fractions were collected in microcentrifuge tubes and the flow-through monitored by Čerenkov counting. The column was washed in 0.1 M NaCl at 35°C and then 48°C. 70% of the oligonucleotide bound to the beads, of which 99.9% was eluted at the higher temperature. The high specificity of hybridisation was demonstrated by carrying out a similar experiment with oligonucleotide 5'-GGGCGGTGACCT, and though this produces a single G-T mismatch at position 7, a mismatch that is the least destabilising (20), there was only 0.5% binding at 35° and 0.2% at 40°C.

Table VI contains a short summary of the physical parameters of ballotini glass beads. These experiments suggest that the derivatised beads will be useful in the chromatographic separation of nucleic acids.

Synthesis of oligonucleotides on a microscope slide and hybridisation behaviour

The experiments with beads demonstrate desirable properties of specific and fast hybridisations. For certain applications (21, 22) it would be desirable to synthesise a number of different oligonucleotides side by side on a flat surface, e.g. a slide. This format would permit rapid and simple analysis of many reactions

⊢ 3'CCC GCC GCT GGA
⊢ 3'CCC GCC t CT GGA

probe — 5'-^{32}P-GGG CGG CGA CCT

Figure 4. Hybridisation to a slide bearing *cosL* and a related sequence. The probe oligonucleotide *cosR* was kinase labelled with ^{32}P to 1.1 million c.p.m. Hybridisation was for 5 hours at 32°C in 0.1 M NaCl, Tris—HCl pH 7.5, 0.1% SDS.

simultaneously using techniques well-established for filter hybridisations. After the initial synthesis of oligonucleotides on the slide, it can essentially be treated like a membrane and analysed by autoradiography or phosphorimaging (23).

A slide bearing sequences T$_{10}$-T$_{14}$. Oligo-T$_{10}$ to oligo-T$_{14}$ were synthesised on a slide by gradually decreasing the level of the coupling solution in steps 10 to 14 (see MATERIALS AND METHODS). Thus the 10mer was synthesised on the upper part of the slide, the 14mer at the bottom and the 11-, 12- and 13mers were in between. The slide was probed with A$_{12}$. An autoradiograph (Figure 3) showed that the probe hybridised in high yield to the area where the oligonucleotide had been synthesised, with no detectable non-specific binding to the glass or to the region that had been derivatised with the linker alone.

By gradually heating the slide in the wash solution the T$_d$ was determined to be ca. 33°C (in 1 M NACl). No counts were detectable after incubation at 39°C. The cycle of hybridisation and melting was repeated eight times with no diminution of the signal. Around 5% of the input counts were taken up by the slide at each cycle.

Effect of a mismatch. The specificity of hybridisation seen on the columns was tested on the slide using a similar set of oligonucleotides. Two sequences, 3'-CCCGCCGCTGGA (*cosL*), complementary to the left cohesive end of bacteriophage λ, and 3'-CCCGCCtCTGGA were synthesised which differ by one base at position 7. All bases except the seventh were added in a rectangular patch as outlined in MATERIALS AND METHODS. At the seventh base, half of the rectangle was exposed in turn to add the two different bases, in two adjacent stripes. After hybridisation of *cosR* probe oligonucleotide (5'-GGGCGGCG-ACCT) 100 c.p.s. could be detected on the front of the slide after rinsing. Autoradiography showed that annealing occurred only to the part of the slide with the fully complementary oligonucleotide (Figure 4). No signal was detectable on the patch with the mismatched sequence.

CONCLUSIONS

The three requirements for a suitable linker, that it should be easy to synthesise, bearing a hydroxyl group, and be sufficiently stable to the harsh conditions used for the deprotection of the bases are satisfied by the two step synthesis described here. The chemicals required are readily available, the first step takes 30

494

minutes, the second can be conveniently carried out in an overnight reaction. The linker bears a hydroxyl group and can be used directly on CPG in automated DNA synthesisers. Some of the linker is removed by the ammonia deprotection step, but the oligonucleotide remaining after the treatment is sufficient for hybridisation detection. Base protecting groups are now commercially available (e.g. from Applied Biosystems) that can be removed by a very short treatment.

Initial yields of oligonucleotide synthesis are remarkably high, close to what is theoretically possible from the known chemical composition of the glass surface. The area at the glass surface occupied by each oligonucleotide is approx. $(6 \text{ Å})^2$ to $(12 \text{ Å})^2$. The length of the linker is ca. 45 Å, and each nucleotide residue adds ca. 20 Å to the length and considerable bulk. With such tight packing it is possible that they would interfere sterically with neighbouring oligonucleotides. Even with the losses due to ammonia treatment, the oligonucleotides are closely packed at one per $(40 \text{ Å})^2$ on CPG and the slide that give lower loadings (Table III). The probes used in the work described here were short, comparable in length with the tethered oligonucleotide. However, in work to be described elsewhere (22), we have shown that longer sequences, in excess of 100 bases, can form duplexes with oligonucleotides down to 12mers, tethered through a 20 atom linker, with little sign of steric interference from the solid support. In general, the stabilities of tethered duplexes are similar to those of oligonucleotides in solution. The kinetics of the hybridisation agree with the expected pseudo first order reaction. Hybridisation is highly specific and sensitive to a single base mismatch. In these senses, the system appears to be homogeneous, well behaved and predictable.

Two physical forms were developed, which will find quite different applications. Beads, either of porous or conventional glass, are readily produced in automated oligonucleotide synthesis machines and can be used in bulk procedures, such as columns, to isolate specific nucleic acid sequences. Support bound DNA is also used to isolate DNA binding proteins. Most DNA binding proteins bind to short sequences of double-stranded DNA. Such sequences could readily be made as 'fold-back' sequences using the method described here, though we have not yet tested this approach.

The linkage also allows the manual synthesis of oligonucleotides on slides as a pre-requisite for constructing large arrays of oligonucleotides to be used in DNA sequence analysis and the detection of mutations, to be reported elsewhere (21, 22).

ACKNOWLEDGEMENTS

We would like to thank Dr H.Blöcker for suggesting the 'trityl-on' assay to us and also for valuable comments on the manuscript. U.M. was supported by the Maximilianeum, Munich, Germany, the Studienstiftung des deutschen Volkes, the Bayerische Begabtenförderung, and the Deutscher Akademischer Austauschdienst.

REFERENCES

1. Matthews,J.A. and Kricka,L.J. (1988) Anal. Biochem. 169, 1–25.
2. Wallace,R.B., Shaffer,J., Murphy,R.F., Bonner,J., Hirose,T. and Itakura,K. (1979) Nucleic Acids Res. 6, 3543–3557.
3. Conner,B.J., Reyes,A.A., Morin,C., Itakura,K., Teplitz,R.L. and Wallace,R.B. (1983) Proc. Natl. Acad. Sci. USA 80, 278–282.
4. Aviv,H. and Leder,P. (1972) Proc. Natl. Acad. Sci. USA 69, 1408–1412.
5. Saiki,R.K., Walsh,P.S., Levenson,C.H. and Erlich,H.A. (1989) Proc. Natl. Acad. Sci. USA 86, 6230–6234.
6. Zhang,Y., Coyne,M.Y., Will,S.G., Levenson,C.H. and Kawasaki,E.S. (1991) Nucleic Acids Res. 19, 3929–3933.
7. Kremsky,J.N., Wooters,J.L., Dougherty,J.P., Meyers,R.E., Collins,M. and Brown,E.L. (1987) Nucleic Acids Res. 15, 2891–2909.
8. Van Ness,J., Kalbfleisch,S., Petrie,C.R., Reed,M.W., Tabone,J.C. and Vermeulen,N.M.J. (1991) Nucleic Acids Res. 19, 3345–3350.
9. Ghosh,S.S. and Musso,G.F. (1987) Nucleic Acids Res. 15, 5353–5372.
10. Gingeras,T.R., Kwoh,D.Y. and Davis,G.R. (1987) Nucleic Acids Res. 15, 5373–5390.
11. Khrapko,K.R., Lysov,Yu.P., Khorlin,A.A., Ivanov,I.B., Yershov,G.M., Vasilenko,S.K., Florentiev,V.L. and Mirzabekov,A.D. (1991) DNA Sequence 1, 375–388.
12. Sproat,B.S. and Gait,M.J. (1984) In Gait,M.J. (ed.), Oligonucleotide Synthesis–A Practical Approach, IRL Press, Oxford, pp. 83–115.
13. Engelhardt,H. and Mathes,D. (1977) J. Chromatogr. 142, 311–320.
14. Boksányi,L., Liardon,O. and Kováts,E. sz. (1976) Advances in Colloid and Interface Science 6, 95–137.
15. Chang,S.H., Gooding,K.M. and Regnier,F.E. (1976) J. Chromatogr. 120, 321–333.
16. Mandenius,C.F., Mosbach,K., Welin,S. and Lundström,I. (1986) Anal. Biochem. 157, 283–288.
17. Sherrington,D.C. (1980) In Hodge,P. and Sherrington,D.C. (eds), Polymer-supported reactions in Organic Synthesis. Wiley, Chichester, pp. 1–92.
18. Applied Biosystems (1987) User Bulletin, The Evaluation and Purification of Synthetic Oligonucleotides, Issue No. 13.
19. Pierce Chemical Company (1987) Handbook & General Catalog, Pierce & Warriner (UK) Ltd, pp. 66–69.
20. Aboul-ela,F., Koh,D., Tinoco,I.J. and Martin,F.H. (1985) Nucleic Acids Res. 13, 4811–4824.
21. Southern,E.M., Maskos,U. and Elder,J.K. (submitted).
22. Maskos,U. and Southern,E.M. (manuscript in preparation).
23. Johnston,R.F., Pickett,S.C. and Barker,D.L. (1990) Electrophoresis 11, 355–360.

Overview of the
Publication Source
of the Original Articles

Acta Endocrinologica (Copenhagen)   **(currently the European Journal of Endocrinology)**

20. Wide, L., Gemzell, C.A.
    An Immunological Pregnancy Test. 35:261–267 (1960) October

American Journal of Clinical Pathology

18. Natelson, S.
    Routine Use of Ultramicro Methods in the Clinical Laboratory. 21(12):1153–1172 (1951)

24. Skeggs, L.T.
    An Automatic Method for Colorimetric Analysis. 28(3):311–322 (1957)

32. Levey, S., Jennings, E.R.
    The Use of Control Charts in the Clinical Laboratory. 20(11):1059–1066 (1950)

33. Caraway, W.
    Chemical and Diagnostic Specificity of Laboratory Tests Effect of Hemolysis, Anticoagulants, Medications, Contaminants, and Other Variables. 37(5):445–464 (1962)

Analytical Biochemistry

27. Friedman, S.M., Nakashima, M.
    Single Sample Analysis with the Sodium Electrode. 2(6):568–575 (1961)

28. Anderson, N.G.
    Analytical techniques for Cell Fractions. XII. A Multiple-Cuvette Rotor for a New Microanalytical System. 28(1):545–562 (1969)

29. Glad, C., Grubb, A.O.
    Immunocapillary Migration- A New Method for Immunochemical Quantitation. 85(1):180–187 (1978)

Analytical Chemistry

31. Jorgenson, J.W., Lukacs, K. DeA.,
    Zone Electrophoresis in Open-Tubular Glass Capillaries. 53(8):1298–1302 (1981)

496

Biochemical and Biophysical Research Communications

2. Dandliker, W.B., Feigen, G.A.
   Quantification of the Antigen–Antibody Reaction by the Polarization of Fluorescence. 5(4):299–304 (1961) July 26

5. Rubenstein, K.E., Schneider, R.S., Ullman, E.F.
   "Homogeneous" Enzyme Immunoassay. A New Immunochemical Technique. 47(4):846–851 (1972)

Clinica Chimica Acta

21. Rahbar, S.
    An Abnormal Hemoglobin in Red Cells of Diabetics. 22(2):296–298

Clinical Chemistry

13. Mercer, D.W.
    Separation of Tissue and Serum Creatinine Kinase Isoenzymes by Ion-Exchange Column Chromatography. 20(1):36–40 (1974)

25. Free, A.H., Adams, E.C., Kercher, M.L., Free, H.M., Cook, M.H.
    Simple Specific Test for Urine Glucose. 3(3):163–168 (1957).

22. Friedewald, W.T., Levy, R.I., Fredrickson, D.S.
    Estimation of the concentration of Low-Density Lipoprotein Cholesterol in Plasma, Without Use of the Preparative Ultracentrifuge. 18(6):499–502 (1972)

35. Radin, N.
    What is a Standard? 13(1):55–76 (1967)

Journal of the American Chemical Society

23. Durrum, E.L.
    A Microelectrophoretic and Microionophoretic Technique. 72(7):2943–2948 (1950)

Journal of the American Medical Association

6. Wuth, O.
   Rational Bromide Treatment. 88(26):2013–2017 (1927)

Journal of Applied Physiology

26. Severinghaus, J., Bradley, A.F.
    Electrodes for Blood pO_2 and pCO_2 Determination. 13(3):515–520 (1958)

Journal of Biological Chemistry

11. Bessey, O.A., Lowry, O.H., Brock, M.J.
    A method for the Rapid Determination of Alkaline Phosphatase with Five Cubic Millimeters of Serum. 164(1):321–324 (1946)

14. Folin, O., Wu, H.
    A System of Blood Analysis. 38(1):81–110 (1919)

15. Van Slyke, D.D. Neill, J.M.
    The Determination of Gases in Blood and Other Solutions by Vacuum Extraction and Monometric Measurement. I. 61(2):523–573 (1924)

16. Malloy, H.T., Evelyn, K.A.
    The Determination of Bilirubin with the Photoelectric Colorimeter. 119(2):481–490 (1937)

17. Gornall, A.G., Bardawill, C.J., David, M.M.
    Determination of Serum Proteins by Means of the Biuret Reaction. 177(2):751–766 (1949)

7. Bratton, A.C., Marshall, E.K.
    A New Coupling Component for Sulfanilamide Determination. 128(2):537–550 (1939)

Journal of Clinical Investigation

1. Yalow, R.S., Berson, S.A.
    Immunoassay of Endogenous Plasma Insulin in Man. 39(7):1157–1175 (1960)

12. Karmen, A., Wroblewski, F., LaDue, J.S.
    Transaminase Activity in Human Blood. 34(1):126–133 (1955)

Journal of Clinical Endocrinology and Metabolism

3. Murphy, B.E.P., Pattee, C.J.
    Determination of Thyroxine Utilizing the Property of Protein-Binding. 24(2):187–196 (1964) February

Journal of General Microbiology   **(currently named Microbiology Reading)**

38. Hyypia, T., Jalava, A., Larsen, S.H., Terho, P., Hukkanen, V.
    Detection of Chlamydia Trachomatis in Clinical Specimens by Nucleic Acid Spot Hybridization. 131(Pt 4):975–978 (1985)

Journal of Immunology

4. Engvall, E., Perlmann, P.
    Enzyme-Linked Immunosorbent Assay, ELISA. III. Quantitation of Specific Antibodies by Enzyme-Labeled Anti-Immunoglobulin in Antigen-Coated Tubes. 109(1):129–135 (1972)

Journal of Laboratory and Clinical Medicine

10. Thompson, R.D., Nagasawa, H.T., Jenne, J.W.
    Determination of Theophylline and its Metabolites in Human Urine and Serum by High-Pressure Liquid Chromatography. 84(4):584–593 (1974)

498

Journal of Pharmacology and Experimental Therapeutics

8. Brodie, B.B., Udenfriend, S.
   The Estimation of Quinine in Human Plasma with a Note on the Estimation of Quinidine. 78(2):154–158 (1943)

The Lancet

19. Huggett, A.S., Nixon, D.A.
    Use of Glucose Oxidase, Peroxidase, and o-dianisidine in Determination of Blood and Urinary Glucose. 2(6991):368–370 (1957) 24 August 1957

New England Journal of Medicine

9. Smith, T.W., Butler, V.P., Haber, E.
   Determination of Therapeutic and Toxic Serum Digoxin Concentrations by Radioimmunoassay. 281(22): 1212–1216 (1969)

34. Vecchio, T.J.
    Predicitive Value of a Single Diagnostic Test in Unselected Populations. 274(21):1171–1173 (1966)

36. Barnett, R.N., Cash, A.D., Junghans, S.P.
    Performance of "kits" Used for Clinical Chemical Analysis of Cholesterol. (Special Article) 279(18):974–979 (1968)

Nucleic Acids Research

39. Maskos, U., Southern, E.M.
    Oligonucleotide Hybridizations on Glass Supports: A Novel Linker for Oligonucleotide Synthesis and Hybridization Properties of Oligonucleotide Synthesized in situ. 20(7):1679–1684 (1992)

Proceedings of the National Academy of Sciences of the United States of America

30. Towbin, H., Stachelin, T., Gordon, J.
    Electrophoretic Transfer of Proteins from Polyacrylamide Gels to Nitrocellulose Sheets: Procedure and some Applications. 76(9):4350–4354 (1979)

Science

37. Pauling, L., Itano, H.A., Singer, S.J., Wells, I.C.
    Sickle Cell Anemia, A Molecular Disease. 110(2865):543–548 (1949) November 25, 1949

Author Index

Subject Index

Printed and bound by CPI Group (UK) Ltd, Croydon, CR0 4YY

03/10/2024

01040330-0019